Caister Academic Press
Norfolk, UK

www.caister.com

British Library Cataloguing-in-Publication Data
A catalogue record for this book is available from the British Library

ISBN: 978-1-908230-14-0

Cover image courtesy of M. Madan Babu

Printed and bound in Great Britain

Bacterial Gene Regulation and Transcriptional Networks

Edited by

M. Madan Babu

MRC Laboratory of Molecular Biology
Cambridge
UK

Caister Academic Press

Contents

Kalai Mathee
Department of Molecular Microbiology and Infectious Diseases
Herbert Wertheim College of Medicine
Florida International University
Miami, FL
USA

kalai.mathee@fiu.edu

Stephen D. Minchin
School of Biosciences
University of Birmingham
Birmingham
UK

s.d.minchin@bham.ac.uk

Gabriel Moreno-Hagelsieb
Department of Biology
Wilfrid Laurier University
Walterloo, ON
Canada

gmoreno@wlu.ca

Senthil Kumar Murugapiran
Department of Molecular Microbiology and Infectious Diseases
Herbert Wertheim College of Medicine
Florida International University
Miami, FL
USA

senthil.murugapiran@unlv.edu

Georgi Muskhelishvili
School of Engineering and Sciences
Jacobs University
Bremen
Germany

g.muskhelishvili@jacobs-university.de

Kenta Nakai
Institute of Medical Science
University of Tokyo
Tokyo
Japan

knakai@ims.u-tokyo.ac.jp

Giri Narasimhan
School of Computing and Information Science
Florida International University
Miami, FL
USA

giri@cs.fiu.edu

Vincenzo Scarlato
Department of Biology
University of Bologna
Italy

vincenzo.scarlato@unibo.it

Lisa Schneper
Department of Molecular Microbiology and Infectious Diseases
Herbert Wertheim College of Medicine
Florida International University
Miami, FL
USA

lschneper@fiu.edu

Aswin Sai Narain Seshasayee
EMBL-European Bioinformatics Institute
Wellcome Trust Genome Campus
Hinxton
UK;
National Centre for Biological Sciences
Tata Institute of Fundamental Research
GKVK
Bengaluru
India

aswin@ncbs.res.in

Eugenia Silva-Herzog
Department of Molecular Microbiology and Infectious Diseases
Herbert Wertheim College of Medicine
Florida International University
Miami, FL
USA

esilvahe@fiu.edu

Zhengchang Su
Department of Bioinformatics and Genomics
College of Computing and Informatics
University of North Carolina
Charlotte, NC
USA

zcsu@uncc.edu

Gorakh Tatke
Department of Biological Sciences (College of Arts and Science)
Florida International University
Miami, FL
USA

gtatk001@fiu.edu

Andrew Travers
MRC Laboratory of Molecular Biology
Cambridge
UK

aat@mrc-lmb.cam.ac.uk

Ying Xu
Department of Biochemistry and Molecular Biology
Institute of Bioinformatics
University of Georgia
Athens, GA
USA;
College of Computer Science and Technology
Jilin University
Changchun
China

xyn@bmb.uga.edu

Xing Yang
School of Computing and Information Science
Florida International University
Miami, FL
USA

xyang006@fiu.edu

Preface

Our understanding of prokaryotic gene regulation has come a long way since the pioneering work by Jacob and Monod on the Lac operon in *E. coli*. In recent years, researchers have generated a wealth of data on bacterial gene regulation from biochemical investigations, structural studies, sequence analyses, comparative genomics and genome-scale experiments. This has provided us with unprecedented insights into bacterial gene regulation and transcriptional networks. This book is an effort to organize the knowledge on bacterial gene regulation through currently prevailing concepts, theories and methods.

The chapters in this book are organized around three major themes. The first six chapters describe the *trans*- and *cis*-acting components required for gene regulation. These components include the RNA polymerase, sequence-specific transcription factors, nucleoid-like structuring proteins, small molecule regulators, *cis*-regulatory elements and operon structures. Some of the chapters also discuss genome-scale theories of gene regulation and examples from the comparative investigation of a large number of completely sequenced bacterial genomes. They further highlight how properties of small gene circuits permit a bacterial population to achieve phenotypic diversity even though each member is genetically identical. Chapter 7 and Chapter 8 introduce experimental and computational methods. These can be used to study gene regulation on a genomic scale in the form of transcriptional regulatory networks. In such a network representation, transcription factors and their target genes are depicted as nodes while the regulatory interactions among them are denoted as links. Based on this network framework, Chapters 9–14 present insights into the principles of transcriptional regulation on a genomic scale, covering a wide range of model bacterial organisms.

Uncovering the molecular details and genomic principles of gene regulation in bacteria provides a unique opportunity to (i) develop novel clinical strategies for combating pathogenic bacteria, and (ii) engineer or design regulatory circuits for biotechnological applications and for other purposes such as diagnostic applications. Advances in experimental technologies such as 'next generation sequencing' will permit rapid sequencing of genomes and global quantification of transcript abundance, resulting in an explosion of information describing the process of gene regulation. The vast amount of data generated will describe bacterial gene regulation (i) at much higher temporal resolution, (ii) across multiple different stages during the bacterial life cycle and (iii) within different environmental niches. I hope that the concepts, theories and methods presented in this collection of chapters by various experts in the field will inspire the next generation of investigators to make the best use of this information and to pursue research on bacterial gene regulation in a multifaceted manner.

M. Madan Babu

The Bacterial Transcription Apparatus

1

L. Aravind and Lakshminarayan M. Iyer

Abstract

We provide a portrait of the bacterial transcription apparatus in light of recent structural studies, sequence analysis and comparative genomics to bring out several key underappreciated features. Comparisons of cellular RNA polymerase subunits with the RNA-dependent RNA polymerase involved in RNAi in eukaryotes and their homologues from newly identified bacterial selfish elements have helped in the identification of novel domains and the possible evolutionary stages leading to the RNA polymerases of extant life forms. We present the case for the ancient orthology of the basal transcription factors, the sigma factor and TFIIB, in the bacterial and the archaeo-eukaryotic lineages. Further, we furnish a synopsis of the structural and architectural taxonomy of specific transcription factors and their genome-scale demography. Although the proteome-wide trends in transcription factor distribution are generally invariant, there are certain notable deviations in firmicutes such as *Paenibacillus* and *Geobacillus* owing to unusual lineage-specifically expanded two-component signalling systems that might have a special biological significance. We then discuss the intersection between functional properties of transcription factors and the organization of transcriptional networks. Finally, we bring attention to some puzzling questions raised by our new understanding of the bacterial transcription apparatus and potential areas for future explorations.

Introduction

The flow of genetic information from a gene to its RNA or protein product is controlled at points, of which regulation at the transcriptional level is shared by all organisms. Transcription regulation is central to the conversion of signals from environmental stimuli and intracellular fluxes of metabolites to homeostatic responses (Watson, 2004). The general paradigms for transcription initiation and regulation first emerged from pioneering studies on gene expression in bacteria and phages (Jacob and Monod, 1961; Ptashne, 2004). Transcription in bacteria and several DNA viruses which infect them was found to be catalysed by a single multisubunit RNA polymerase. It is recruited to conserved DNA sequence elements upstream of genes, namely the promoter, by means of a DNA-binding protein, the σ factor, which specifically recognizes these elements. The σ factor and the RNA polymerase, together, constitute the 'basal transcription apparatus' required for baseline transcription of all genes (Fig. 1.1). In particular, the σ factor is identified as a 'general' or 'basal' transcription factor (TF) (Watson, 2004). Early studies, particularly in the *Bacillus subtilis* sporulation model, showed that there are several alternative sigma factors beyond the commonly used version, which might recruit the catalytic core of the RNA polymerase to specific sets of genes, resulting in temporally and spatially distinct alternative transcriptional programmes (Stragier and Losick, 1996; Ju *et al.*, 1999). Comparable control mechanisms are prevalent across bacteria and regulate broad changes in gene expression, which correlate with the different developmental or differentiation states of the organism. Starting with the classical studies of Jacob and Monod it became clear that functionally linked groups of genes are often simultaneously co-regulated by dedicated regulators. These functionally linked

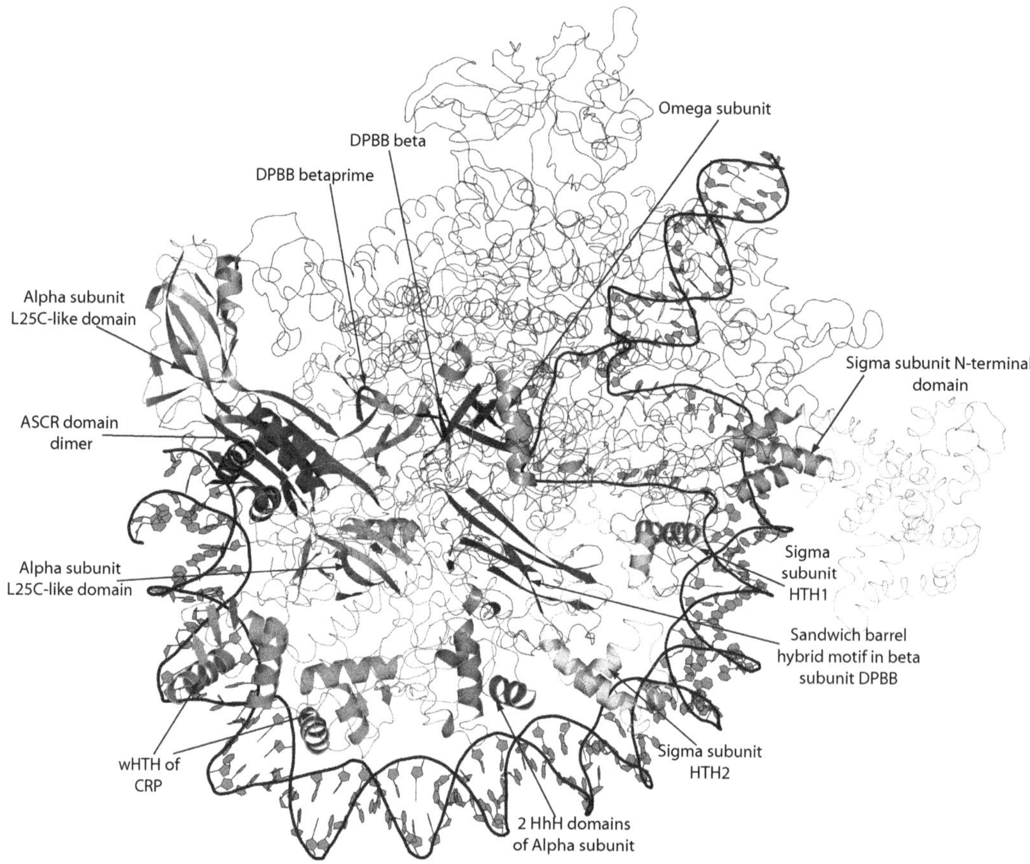

Figure 1.1 Structure of the bacterial transcription initiation complex. The cartoon representation was derived from an EM structure of the initiation complex (PDB: 3iyd) in association with DNA. The structure contains the α, β, β′, ω and σ^{70} of the bacterial DNA-dependent RNA polymerase and the wHTH domains of CRP (CAP) transcription factor. Only the key globular domains of these proteins are shown and labelled for better clarity. The remaining parts of the structure are shown as coils. A colour version of this figure is located in the plate section at the back of the book.

genes typically occur as collinear groups (operons) on the chromosome, and encode components of a pathway for the utilization of a particular metabolite (e.g. lactose), or constitute interacting components of a macromolecular complex or developmental pathway (e.g. lytic or lysogenic development of phages) (Jacob and Monod, 1961; Ptashne, 2004). In certain cases multiple operons might be co-regulated and such an assembly of genes might be terms a 'regulon'. Dedicated regulators of operons are usually DNA-binding proteins that bind specific DNA sequences associated with the operon, which are distinct from the promoter, and act as transcription regulatory switches. These proteins, termed the 'specific TFs' (as opposed to the general TFs mentioned above), might be repressors, which negatively regulate transcription, activators, which positively regulate transcription of their target genes or in some cases dual mode regulators which might be both repressors and activators. Affinities of specific TFs for their target sequences on DNA are often affected by their binding of low-molecular weight compounds (effectors) or phosphorylation and other post-transcriptional modifications. Thus, specific TFs are integral elements of the apparatus which 'converts' an intrinsic or extrinsic sensory input to a transcriptional response.

The rapid growth of structural studies in the past 20 years, primarily X-crystallography and site-direct mutagenesis, supplemented by NMR spectroscopy and electron microscopy, has elucidated these interactions at the atomic scale (Harrison, 1991; Latchman, 1997). Not only have the structures of exemplars of most of the DNA-binding and effector-binding domains of TFs and RNA polymerase subunits become available, but also structures of entire complexes, such as the transcription initiation complex have been published (Hudson *et al.*, 2009; Feklistov and Darst, 2011). These efforts allow the 'microscopic scrutiny' of the transcription apparatus and interpretation of various observations stemming from functional and evolutionary studies at the molecular level. In parallel, there have also been major advances in terms of our 'macroscopic' understanding of transcription regulation. At the 'systems' level, the total set of regulatory interactions mediated by the binding of general and specific TFs, either singly or in combination, to promoters and regulatory elements can be conceptualized as a network, termed the transcriptional regulatory network (Madan Babu *et al.*, 2007). The nodes of this network are target genes and TFs, and edges represent regulatory interactions between them. Advances in genomics over the past two decades have made the reconstruction and analysis of such networks a reality. At an abstract level they have architectures which can be approximated by scale-free networks that are also encountered in non-biological systems such as the internet (Barabasi and Bonabeau, 2003). They are characterized by the recursive patterns of interconnections called network motifs (Shen-Orr *et al.*, 2002; Madan Babu *et al.*, 2007). The study of the transcription network and its motifs are beginning to reveal genome-scale principles in associations between TFs, their response to external or internal changes and the mode of alteration of gene expression (i.e. activation or repression) (Babu *et al.*, 2004). Here, we mainly focus on the TF nodes of the transcription regulatory network, but interpret some of the observations on these nodes in light of our current knowledge of the architecture of the transcription network.

Our primary objective of this chapter is to provide a portrait of the transcription apparatus as from the vantage point of data coming from structural studies, sequence analysis and comparative genomics. Owing to constraints on space this portrait would necessarily be rendered in broad strokes, yet we attempt to bring out key features that are commonly overlooked by workers less familiar with evolutionary considerations. We hope that this distinct perspective would inspire a more natural vision of the transcription apparatus.

Basic anatomy of the RNA polymerase

The bacterial DNA-dependent RNA polymerase is a six-subunit complex, comprised of two identical α subunits and one subunit each of β, β′, σ and ω (Feklistov and Darst, 2011; Hudson *et al.*, 2009; Iyer *et al.*, 2004a; Watson, 2004). Most bacteria have a single gene for each of the RNA polymerase subunits. In some instances the genes for two subunits are fused, e.g. the endosymbiotic gamma-proteobacterium *Wolbachia* and several epsiloproteobacteria such as *Helicobacter* and *Wolinella*. Certain lineages of symbionts or parasites with degenerate genomes and the chloroflexi are exceptions in that the ω subunit is currently undetectable. Highly degenerate, cooperative intracellular symbionts like *Sulcia* (a bacteroidetes) and *Hodgkinia* (an alphaproteobacterium), which live in close association with each other have individually lost several components of essential functional systems, but complement each other by exchanging components such as tRNA synthetases and ribosomal subunits (McCutcheon *et al.*, 2009). Even these organisms encode their own α, β, β′ and σ subunits, though it appears that they share a common ω subunit (encoded by *Sulcia*). The active site for the nucleotidyltransferase activity of the RNA polymerase is constituted by residues from both the β and β′ subunits that together are termed the catalytic subunits (Cramer *et al.*, 2001; Vassylyev *et al.*, 2002; Iyer *et al.*, 2003; Opalka *et al.*, 2010). The α subunit does not directly contribute to the catalytic activity but is still absolutely required for the effective initiation and elongation by the polymerase. The σ factors are primarily needed for the initiation step to bind to the promoter. However, they have also been found to remain associated with the elongating polymerase and cause pausing at promoter proximal sites by rebinding DNA sequences resembling the −10 sites of the promoter (Mooney *et al.*, 2005). The ω subunit is the least understood of the subunits and is an entirely α-helical protein that is asymmetrically positioned in the complex. It primarily binds the catalytic domain of the β′ subunit, but makes more limited contacts with the two α subunits, the σ factor and specific activator TFs (Cramer *et al.*, 2001; Vassylyev *et al.*, 2002) (Fig. 1.1). The organizational logic of the bacterial RNA polymerase has become clear with the sequence-structure analysis of the crystal structures of the holoenzyme complexes and cryo-EM structure of the initiation complex (Fig. 1.1) (Cramer *et al.*, 2001; Vassylyev *et al.*, 2002; Iyer *et al.*, 2003; Hudson *et al.*, 2009; Opalka *et al.*, 2010). We describe this organization in terms of the constituent conserved domains and their functional properties, we consider below the major subunits with their key structural features.

The α-subunits

The bacterial α-subunit comprises three domains. The N-terminal unit has an α-subunit-core-related (ASCR) domain that is shared with the archaeo-eukaryotic RNA polymerase subunits RPB3 and RPB11 (Iyer *et al.*, 2003). Into the ASCR domain is inserted a distinct, previously unrecognized domain in the bacterial α-subunit and the archaeo-eukaryotic RPB3. Sequence and structure comparisons showed that this inserted domain shares a common fold with the C-terminal domain of the bacterial ribosomal subunit L25 (PDB: 1feu) and related proteins such as YbbR (Fig. 1.2). The C-terminal module of the α-subunit (CTD) comprises two helix–hairpin–helix (HhH) motifs (Mah *et al.*, 2000) (Fig. 1.2). In the transcriptional complex the two α-subunits dimerize via their ASCR domains, while the L25-like domains point in opposite directions (Fig. 1.1). The C-terminal HhH motifs contact the minor groove of DNA in a manner similar to HhH motifs found in several other DNA-binding proteins (Fromme *et al.*, 2004). The HhH motifs of the CTD of α also interact with the second helix–turn–helix (HTH) domain of the σ-factor, which binds the −35 promoter element in the major groove adjacent to the contact of the HhH motifs (Fig. 1.1). Similarly, the HhH motifs contact the specific activator TFs that bind their target elements upstream of the promoter (Fig. 1.1) (Hudson *et al.*, 2009). The α-dimer is asymmetrically positioned with respect to the homologous catalytic domains of the β and β′ subunits (see below). The ASCR domain from one of the α-subunits primarily contacts the catalytic domain of the β subunit, whereas that from the second α-subunit mainly interacts with the catalytic domain of the β′ subunit (Fig. 1.1). The newly identified L25-like domain from only one of the subunits makes a second major contact with the β catalytic domain, while the equivalent domain from the other α-subunit makes a distinct contact with the β′ subunit far away from its catalytic domain. The HhH motifs of the α-subunits do not conspicuously alter the curvature of the helical axis of DNA at the points of their individual DNA contacts. However, the layout of the α-dimer is such that it can accommodate specific TFs that bind target sequences to bend the DNA upstream of the promoter. Thus, the interaction of the α-dimer with both the specific and basal TFs is critical for engagement of the transcription initiation site by the RNA polymerase (Fig. 1.1).

The catalytic subunits β and β′

The β and β′ subunits share a homologous core comprised of a domain with the double-ψ-β-barrel fold (DPBB) (Castillo *et al.*, 1999; Iyer *et al.*, 2003; Hulko *et al.*, 2007) (Figs. 1.2 and 1.3). The DPBB domains from the two subunits are appressed against each other, with both of them providing key residues to the active site. The DPBB of the β′-subunit bears an absolutely conserved DxDxD signature (where x is any amino acid), which chelates a Mg^{2+} ion required for directing the phosphate of the incoming nucleotide to react with the 3′ hydroxyl of the initial nucleotide (Fig. 1.2). The DPBB of the β-subunit contains two absolutely conserved lysines that apparently stabilize the hypercharged reaction intermediate and interact with the negatively charged backbone of the elongating RNA-chain (Cramer *et al.*, 2001; Iyer *et al.*, 2003) (Fig. 1.2). Studies have suggested that homologues of the DPBB domains of the β and β′ subunits are also found in the eukaryotic RNA-dependent-RNA polymerases (RdRPs), which are involved in amplification of the siRNA pathway and related families proteins found in several bacteria and bacteriophages (Iyer *et al.*,

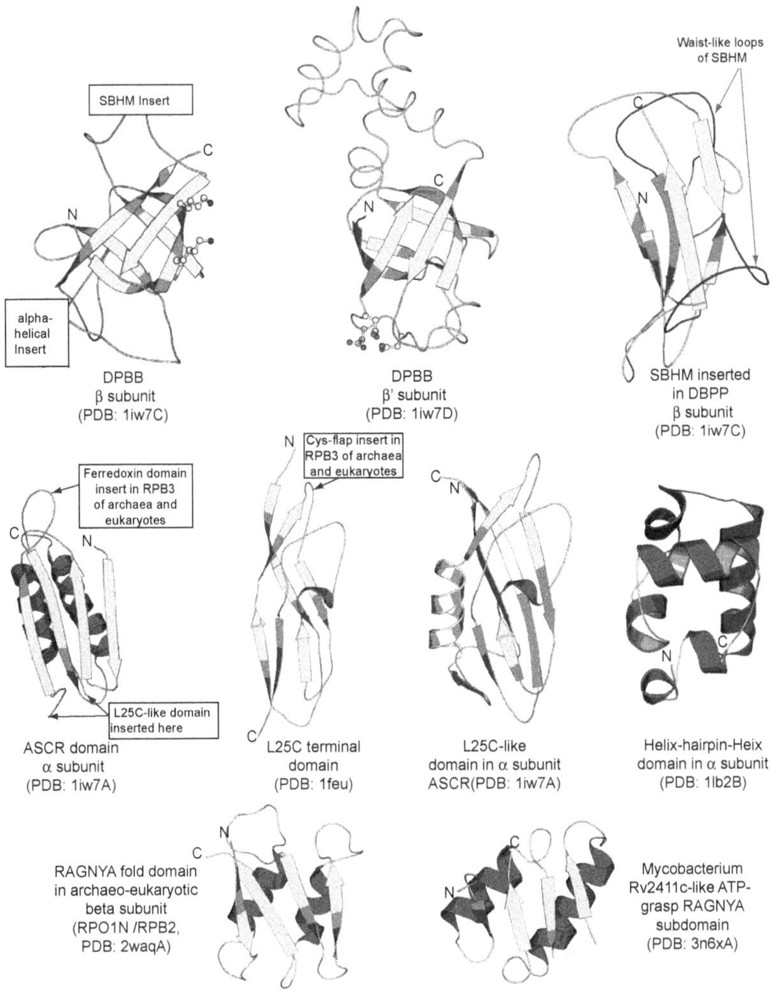

Figure 1.2 Structures of key conserved domains of the β, β′ and α subunits. Strands are coloured yellow, whereas helices are coloured red or blue. Only the core conserved regions of the domains are shown. Inserts in domains are mostly suppressed or excised as depicted. The C-terminal domain of the ribosomal L25 protein is depicted to illustrate its structural relationship with the conserved domain inserted into the ASCR domain of the α subunit (L25C-like domain). Also shown are the positions of the RPB3 (α) subunit-specific inserts in the archaeo-eukaryotes. Structural elements in the L25C-like domain of the α subunit that are not present in the ribosomal L25 protein are coloured grey. A colour version of this figure is located in the plate section at the back of the book.

2003; Salgado *et al.*, 2006; Ruprich-Robert and Thuriaux, 2010) (Figs. 1.2 and 1.3). In these proteins the DPBBs, which are equivalent to β and β′ are fused together in a single polypeptide, with the cognate of the β DPBB being the N-terminal domain and the one equivalent to the β′ DPBB being the C-terminal domain, connected by a long helical linker. Beyond the RdRP-like proteins, there are other single polypeptide RNA polymerases, such as those encoded by the fungal killer plasmids (e.g. the *Kluyveromyces* killer plasmid) and a group of bacterial proteins typified by *Corynebacterium glutamicum* NCgl1702, both of which are closer to the cellular DNA-dependent RNA polymerases (Iyer *et al.*, 2003). The domain architectures and gene-neighbourhoods suggest that most of these single polypeptide RNA polymerases are likely to be components of mobile

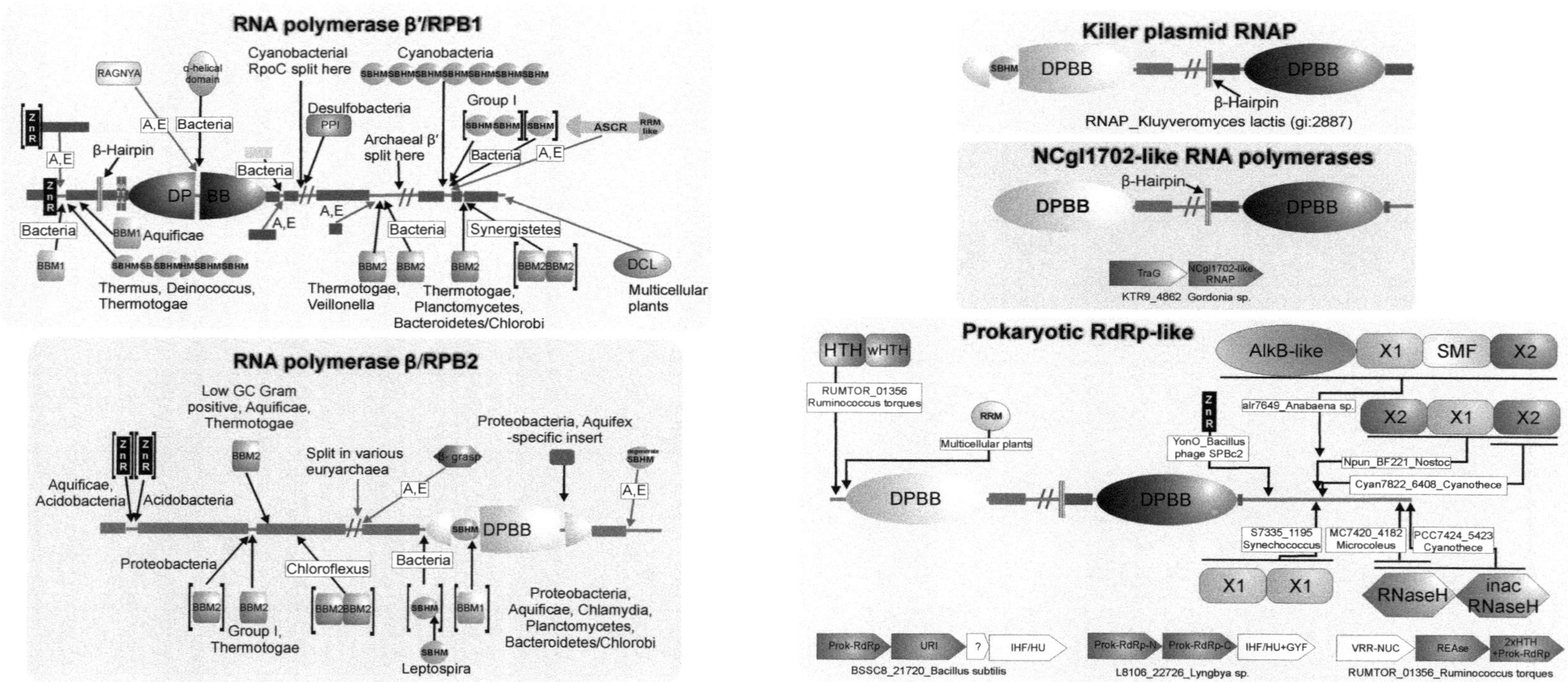

Figure 1.3 Domain architectures of the RNA polymerase β and β′ subunits, yeast killer plasmid RNA polymerase, NCgl1702-like RNA polymerases and the prokaryotic RdRP-like RNA polymerases. For the β and β′ subunits, the domain architecture reconstructed to the last universal common ancestor is shown in the centre and inserts in various lineages are shown around this core. Archaeo-eukaryotic domain inserts are indicated with a red arrow and bacterial inserts are marked with a black arrow. Lineages in which the inserts are observed are indicated near the arrows or architecture. Bacterial inserts, on occasions, differ within members of a closely related bacterial lineage. For a more detailed discussion of these variations, refer to (Lane and Darst, 2010a). A similar representation is used for the prokaryotic RdRP-like proteins, where lineage-specific inserts are marked with a representative gene and species name around a core conserved architecture. Genes in operons are shown as box-arrows with the arrow head pointing from the 5′ to the 3′ direction of the coding sequence. Operons are labelled with the gene name of the polymerase gene and species name. Refer to the supplement (ftp://ftp.ncbi.nih.gov/pub/aravind/PROKHTH/prok_trans.html) for more detailed domain architectures and gene neighbourhoods. Standard abbreviations are used for domain and lineage names. The DCL domain is an RNA binding domain which is also found in a stand-alone form in bacteria and in several eukaryotic rRNA biogenesis proteins. Other abbreviations: A, E, archaea and eukaryotes; ASCR, alpha subunit core related; ATL, AT-hook like motifs; PPI, peptidyl prolyl isomerase; ZnR, zinc ribbon. A colour version of this figure is located in the plate section at the back of the book.

selfish elements (Iyer and Aravind, 2011). Several prokaryotic RdRP-like proteins are encoded by bacteriophages (Iyer *et al.*, 2003), and might mediate transcription in these viruses. Of the remaining bacterial RdRP-like proteins, a subset typified by RUMTOR_01356 (gi: 153815131) are encoded by a predicted mobile element, which additionally code for at least three other proteins (Fig. 1.3) – two nucleases of the restriction endonuclease fold, one of which is related to the previously characterized VRR-Nuc family (Iyer *et al.*, 2006) and a third small α-helical protein (Iyer and Aravind, 2011). These RdRP-like proteins display fusions to two N-terminal transcription factor-related helix–turn–helix (HTH) domains that are predicted to bind DNA (Fig. 1.3). The cyanobacterial RdRP-like proteins may show fusions to a SMF/DprA-like Rossmann fold domain (Fig. 1.3) that is predicted to bind DNA (Aravind *et al.*, 2005; Smeets *et al.*, 2006). In several bacteria this domain participates in DNA uptake during transformation. Additionally, some of the cyanobacterial RdRP-like proteins display a fusion to one or more RNAseH domains (Iyer and Aravind, 2011). The genes for the RdRP-like proteins in certain Gram-positive bacteria are also present in a predicted mobile element which additionally encodes a nuclease with an UvrC-Intron homing endonuclease (URI) domain (Fig. 1.3) (Aravind *et al.*, 1999). The NCgl1702-like RNA polymerases are encoded by distinct mobile elements that also encode a DNA-pumping ATPase of the HerA-FtsK superfamily (Fig. 1.3) that is similar to those encoded by certain conjugative transposons and related mobile elements (Iyer *et al.*, 2004b). These domain architectures and gene neighbourhood contexts (e.g. RNaseH fusion, presence of DNA-binding HTH and SMF domains, endonucleases) suggest that the action of these single polypeptide RNA polymerases aids in the replication of the respective selfish elements by synthesizing a RNA primer. This priming reaction might be initiated by the nicking action nucleases encoded by some of these mobile elements or as these mobile elements are being taken up by a target cell.

We interpret the above single polypeptide RNA polymerases in selfish elements as late-surviving representatives of different stages leading to the ancestral RNA polymerase of cellular forms in course of the ancient diversification of RNA polymerases in early replicons. First, these enzymes suggest that the common ancestor of the DNA-dependent-RNA polymerases and the RdRP-like proteins emerged as a single protein. The adjacent dyad of DPBB domains corresponded to the catalytic domains of the β and β′ subunits seen in extant cellular enzymes. The evolution of both the RdRP-like proteins of the mobile elements and the cellular RNA polymerases of extant cellular organisms is dominated by the accretion of several accessory domains on either side of the two DPBBs, as well as even insertion within the DPBBs themselves (Iyer *et al.*, 2003, 2004a; Lane and Darst, 2010a; Opalka *et al.*, 2010). For example, the cyanobacterial RdRP-like proteins show an extraordinary diversity of architectures (Fig. 1.3), including accretion of an AlkB-like 2-oxoglutarate and iron dependent dioxygenases that might modify methylated DNA or RNA (Iyer *et al.*, 2010; Iyer and Aravind, 2011). The emergence of β and β′ subunits of cellular RNA polymerases were accompanied by an entirely different set of accretions. The single polypeptide RNA polymerase of the fungal killer plasmids contains several of these accretions and insertions (Fig. 1.3, see below), which suggest that the split of the ancestral protein into two distinct subunits happened only after these initial accretion events. Crystal structures of the bacterial RNA polymerase complexes have thrown considerable light on the significance of these inserts. One key insert, also called the 'flap domain', is that of the sandwich–barrel-hybrid motif (SBHM) domain in the DPBB of the β-subunit (Figs. 1.2 and 3). This insert is present in the polymerases of fungal killer plasmids, but is absent in the RdRP-like proteins and the NCgl1702-like RNA polymerases (Fig. 1.3). Thus, it was possibly acquired when the enzyme was still a single subunit polymerase with fused β and β′ cognates. In bacteria, it interacts specifically with the σ-factor (Fig. 1.1) (Kuznedelov *et al.*, 2002; Murakami *et al.*, 2002), while its cognates in archaea and eukaryotes interact with TFIIB (Kostrewa *et al.*, 2009). Hence, the emergence of this SBHM insert was probably the critical event that allowed the ancestral RNA polymerase of cellular life forms to be recruited to the basal TF that recognized the promoter. This insert forms a part of the RNA-exit channel (Toulokhonov *et al.*, 2001) and also makes conspicuous contacts with regulatory proteins such anti-σ factors (Pineda *et al.*, 2004), bacteriophage anti-termination proteins (Yuan *et al.*, 2009) and the elongation factor NusA (Toulokhonov *et al.*, 2001), suggesting that it is a nexus point for various transcription regulatory events.

The ancestral version of all RNA-polymerases [including the RdRP-like enzymes (Salgado *et al.*, 2006)] can be reconstructed as having a distinctive bihelical extension preceded by two extended segments forming a standalone β-hairpin at the N-terminus of the β′-DPBB domain (Iyer and Aravind, 2011). Specifically in DNA-dependent RNA polymerases of cellular life-forms (but not RdRP-like proteins, NCgl1702-like and killer plasmid RNA polymerases) the first long helix of this extension acquired a distinctive insert in the form two flap-like structures resembling the AT-hook DNA-binding motif (Iyer *et al.*, 2003). The above-mentioned β-hairpin and the AT-hook-like structures contact the template strand at the transcription start site and appear to be critical for melting dsDNA to allow the polymerase catalytic domains to access their template (Westover *et al.*, 2004; Vassylyev *et al.*, 2007). Thus the β-hairpin is likely to have been a template strand binding element that had already emerged in the common ancestor of all RNA polymerases (including RdRP-like proteins), while the innovation of the AT-hook-like flaps augmented this interaction in the common ancestor of the DNA-dependent RNA polymerases of cellular forms. Comparisons of the structures of the RdRP and the cellular RNA polymerases also show that the common ancestor of all RNA polymerases had a segment in the extended conformation at the C-terminus of the β DPBB that formed a brace to contact the β' DPBB. This feature might have been a key element that held the two DPBB domains in close proximity in the ancestral polymerase. C-terminal to the β′ DPBB there is a conserved extension that folds back and interacts with the β DPBB, which is shared by all cellular RNA polymerases and the versions encoded by the killer plasmids. We posit that this region might shield part of the active site and potentially exclude solvent from the active site to favour a more processive catalytic activity.

Both the β and the β′ subunits of the bacterial RNA polymerase have several insertions of additional domains that are not found in the archaeo-eukaryotic RNA polymerases and vice versa (Lane and Darst, 2010a,b). The β′ DPBB shows entirely distinct inserts in the bacterial and the archaeo-eukaryotic lineages: The bacteria acquired an all α-helical insert (Figs. 1.1 and 1.3). In contrast, the β' DPBB in archaeo-eukaryotic lineage acquired, in the equivalent position, an unrelated insert of a RAGNYA fold domain that is closely related in structure to the ATP-binding version found in the ATP-grasp module (Balaji and Aravind, 2007; Iyer and Aravind, 2011) (Fig. 1.2). In both cases the inserts are spatially directed in a manner similar to the SBHM of β DPBB and respectively recruit the ω-subunit in bacteria or its cognate RBP6 in archaea and eukaryotes by contacting them equivalently in the loop between their two conserved helices (Minakhin *et al.*, 2001). Given the nucleic acid-binding properties of certain representatives of the RAGNYA fold (Balaji and Aravind, 2007), it would be of interest to investigate if it might additionally interact with the emerging transcript in the archaeo-eukaryotic polymerases. The other major inserts, which differ between the bacterial and archaeo-eukaryotic lineages, include multiple SBHM domains and two small domains respectively known as the β-β-motif-1 (BBM1) and the β-β′-motif-2 (BBM2) (Iyer *et al.*, 2003, 2004a). The latter domains are comprised of long extended segments forming a highly curved hairpin, which is bounded on either side by helical segments. Several of the SBHM domain inserts show dramatic differences between various bacterial lineages in terms of their presence or absence or in the number of SBHM domain copies (Iyer *et al.*, 2003, 2004a; Lane and Darst, 2010a). Archaea, eukaryotes and the killer-plasmid β subunit have a C-terminal degenerate SBHM which appears to have been lost in the bacterial forms (Iyer and Aravind, 2011). The functions of the SBHM domains still remain incompletely understood. The conserved SBHMs found at the C-terminus of the bacterial β′ subunit have been shown to interact with the transcription elongation factors of the GreA/B family (Chlenov *et al.*, 2005; Lamour *et al.*, 2008). A set of lineage-specific SBHM inserts seen in the N-terminus of the β′ subunit of the *Thermus–Deinococcus* lineage and *Thermotoga* are known to contact the σ-factor (Vassylyev *et al.*, 2002; Chlenov *et al.*, 2005). Based on this, we suggest that the lineage-specific SBHM inserts might mediate interactions with transcription regulators that allow for control processes unique to specific groups of bacteria. Remarkably, the β′ subunit of the delta-proteobacterial lineage of Desulfobacterales shows an insertion, downstream of the catalytic DPPB domain, of a parvulin-like peptidyl prolyl isomerase domain. It would be of interest to investigate if this domain might provide an in-built prolyl isomerization chaperone function for the RNA polymerase in these organisms.

The ω subunit

The enigmatic α-helical ω subunit, which is a cognate of RPB6 in the archaeo-eukaryotic lineage, was for a long time even considered an impurity that associates with the purified RNA polymerase complex. However, number of studies have confirmed its role as a player in the assembly of the β′ subunit into the RNA polymerase complex by preventing its aggregation (Minakhin *et al.*, 2001; Mathew and Chatterji, 2006). Specifically in bacteria, the ω subunit is the focus of the stringent response, in which the metabolite (p)ppGpp produced by the SpoT/RelA-type enzymes (Potrykus and Cashel, 2008) causes a drastic global shift in the transcription profile from growth- and cell-division- related genes to amino acid synthesis genes. It appears that the ω subunit is the binding-site for (p)ppGpp and mediates the sensitivity of the polymerase to this metabolite (Mathew and Chatterji, 2006). While there is no comparable stringent response in archaea and most eukaryotes, the RBP6 subunit is likely to play a role comparable to the bacterial ω in assembly of the RNA polymerase by interacting with the insert domain in DPBB of the β′ subunit.

σ-factors

The σ-factor that is conserved in all bacterial genomes is σ^{70}, which initiates transcription of all or the majority of promoters in any given bacterium. Most bacteria, except symbionts and parasites with extremely reduced genomes, encode at least one alternative σ-factor. The majority of these alternative σ-factors are relatively close paralogues of σ^{70} and are collectively referred to as the σ^{70}-family (Gruber and Gross, 2003; Paget and Helmann, 2003). The remaining alternative σ-factors belong to the σ^{54}-family that bear multiple conserved HTH domains, but are only very distantly related to the σ^{70} family. Traditionally, the primary structure of the σ^{70}-family has been divided into four collinear regions, numbered 1–4, which were mapped on the basis of their functional properties and sequence conservation (Gruber and Gross, 2003; Paget and Helmann, 2003). While the structure-based dissection of the domains of the σ^{70}-family partly confirms this nomenclature, it provides a more natural way of visualizing these σ factors; hence, our discussion entirely follows the structural paradigm. The conserved core of σ^{70}-family proteins contains an N-terminal domain in the form of a 4-helical bundle. This domain maps to the only helix in region 1, which is conserved throughout the family, and the entire conserved region 2. The N-terminal domain of the primary σ-factor from several bacterial lineages usually contains a large helical insert of variable size (Iyer *et al.*, 2004a). The N-terminal 4-helical bundle inserts deeply into the DNA at the −10 element of the promoter and fosters melting of the double helix around the transcription start site (Feklistov and Darst, 2011) (Fig. 1.1). The primary σ-factor contains a further α-helical domain, N-terminal to the first core domain (mapping to the reminder of region 1), which functions as a negative regulator of its DNA-binding activity (Barne *et al.*, 1997). This additional N-terminal domain is entirely absent in the alternative σ-factors and also the primary σ-factor of the bacteroidetes–chlorobium–gemmatimonad lineage (Iyer *et al.*, 2004a). The first domain of the conserved core of the σ factor is immediately followed by the first HTH domain (domain 2 of the conserved core) that maps to the earlier defined region 3 (Aravind *et al.*, 2005). It binds the extended −10 element that is upstream of the −10 element (Barne *et al.*, 1997; Campbell *et al.*, 2002). Binding of this element by this HTH domain is particularly important in transcription initiation through promoters lacking the

−35 element. This HTH domain has completely degenerated in most members of the extracellular function (ECF; see below) clade of the σ^{70}-family (Gruber and Gross, 2003). Remarkably, in the *Dictyoglomus* lineage a further HTH domain is inserted between helix-2 and helix-3 of this HTH domain and is predicted to make a unique lineage-specific contact upstream of the extended −10 element (Iyer and Aravind, 2011). The C-terminal-most domain (domain 3) of the conserved σ core is the second HTH domain that interacts with the α-subunit and binds the −35 element (Gruber and Gross, 2003; Paget and Helmann, 2003).

Bacteriologists have classified the σ^{70}-family in groups 1–5 (Gruber and Gross, 2003; Paget and Helmann, 2003). It should be emphasized that this classification is partly inaccurate and misleading because groups 2 and 3 are not evolutionarily monophyletic assemblages within the σ^{70} family. Group 1 contains the classical σ^{70} and is typically present in a single copy in all bacterial genomes. Group 2 consists of σ factors closely related to σ^{70}; however, these function as alternative σ factors, for example in the initiation of the transcriptional programmes associated with stationary phase and stress response (e.g. σ^{S} of *Escherichia coli*). Examination of the phylogenetic trees of σ-factors (Gruber and Gross, 2003; Paget and Helmann, 2003) suggests that group 2 σ-factors have repeatedly arisen through lineage-specific duplications of the primary σ factor. The group 3 σ factors are a heterogeneous, non-monophyletic assemblage including several distinct families that are involved in initiating transcription of multigene batteries associated with major conditional and developmental programmes such as heat shock response (e.g. *E. coli* RpoH gene product), flagellar gene expression and motility (e.g. *E. coli* FliA product), sporulation in firmicutes (*Bacillus subtilis* SigE, SigF and Sig G) and stress response (e.g. *B. subtilis* SigB) (Gruber and Gross, 2003; Paget and Helmann, 2003). The group 4 or the ECF σ factors are a monophyletic clade of fast-evolving σ factors. They are typically associated with an anti-σ factor that might be a membrane protein with an extracellular domain (Helmann, 2002). The anti-sigma factor is dissociated from the cognate σ upon receiving a sensory stimulus, typically from the extracellular environment allowing the σ factor to initiate a transcriptional programme. Group 4 σ factors are major regulators of transcription in response to extrinsic sensory inputs such as iron availability, misfolded proteins in the periplasm, redox stress and host-derived signals in the case of pathogenic bacteria. However, a subset of these σ factors might also respond to intracellular sensory stimuli as seen in the case of the redox based regulation of σ^{R} of *Streptomyces coelicolor* (Paget *et al.*, 1998; Helmann, 2002) or downstream of two-component regulatory systems (see below) as seen in the case of σ^{E} from the same organism (Paget *et al.*, 1999; Helmann, 2002). Phylogenetic analysis shows that the group 5 sigma factors typified by TxeR of *Clostridium difficile* are merely a highly divergent group of ECF σ factors. Like them, they have been found to initiate the transcription of a small group of genes related to toxin and bacteriocin production (Mani and Dupuy, 2001). The ECF σ factors are greatly expanded in bacteria with complex metabolic and developmental features (see below for genomic scaling). Thus, the ECF σ-factors might be seen in functional terms as intermediates between specific TFs and conventional σ-factors.

The σ^{54}-family is typically present in a single copy per genome and is sporadically distributed across the bacterial tree – it is present in proteobacteria and their closest relatives (the group-I bacteria) and firmicutes among the group-II bacteria (Iyer *et al.*, 2004a). However, it is absent in most major group-II clades such as actinomycetes and cyanobacteria. The phyletic pattern of the σ^{54}-family is strictly correlated with a distinctive class of specific TFs, namely the NtrC family of ATPases (also called enhancer-binding proteins) (Aravind *et al.*, 2005; Ammelburg *et al.*, 2006; Hong *et al.*, 2009). A complete representative of the σ^{54}-family has not yet been structurally characterized. Analysis of the structurally characterized fragments along with sequence profile analysis suggests that σ^{54} is comprised of four distinct conserved regions. The N-terminal-most of these is a well-conserved α-helical segment, which binds the AAA+ domain of the NtrC-like protein and regulates its ATPase activity during the assembly of the σ^{54} initiation complex (Doucleff *et al.*, 2005). The second domain is a conserved HTH domain, which has been shown to interact with the RNA-polymerase core, though it could potentially make additional DNA contacts. The third conserved element is also a HTH domain that is likely to contact the −12 element of the σ^{54}-dependent promoters. The C-terminal-most domain is yet another HTH domain, which contacts the −24 element of these promoters (Doucleff *et al.*, 2005). Like in σ^{70} the two C-terminal HTHs respectively contact the 5′ and 3′ elements in an N- to C- terminal polarity (Hong *et al.*, 2009). Furthermore, σ^{54} also interacts with the SBHM domain inserted into the β subunit just as the σ70 family (Wigneshweraraj *et al.*, 2003). These observations suggest that there could be a potential common origin for the two families of σ-factors.

The Gram-positive RNA-polymerase delta subunit and related proteins

Gram-positive bacteria display a unique RNA polymerase subunit termed delta (RpoE), which has been shown to bind the RNA polymerase catalytic complex, reduce its affinity for nucleic acids and increase transcription specificity by promoting recycling (Lopez de Saro *et al.*, 1999; Motackova *et al.*, 2010). Specifically, this subunit inhibits the downstream propagation of the transcription bubble at the −10 region, with its acidic C-terminal tail mimicking RNA and interacting with the RNA polymerase catalytic complex. The delta subunit contains a novel winged HTH (wHTH) domain that is fused to a highly acidic C-terminal low-complexity tail (Motackova *et al.*, 2010). This wHTH domain is widely distributed in bacteria (also fused to restriction endonuclease domains) and eukaryotes (chromatin proteins like HB1 and ASXL1/2/3) and is termed the HB1, ASXL, Restriction Endonuclease (HARE)-HTH domain (Aravind and Iyer, 2012). Certain proteobacteria also contain a version of the HARE-HTH domain comparable to delta that instead has an acidic low-complexity tail at the N-terminus. A remarkable group of proteins found sporadically in actinobacteria, firmicutes and proteobacteria combine a C-terminal HARE-HTH to (1) an N-terminal region containing two or more repeats of the specialized helix–hairpin–helix (HhH) domain found in the CTD of the bacterial RNA polymerase α-subunit; (2) two additional HTH modules specifically related to those found in the regions 3 and 4 of the sigma factors (Aravind and Iyer, 2012). Thus, these proteins combine parts of the architecture of the RNA polymerase α and σ subunits with the HARE-HTH in a single polypeptide (Fig. 1.1). These bacterial proteins are striking because an examination of the RNA polymerase holoenzyme complex with the transcription start site (TSS) shows that the modules in this protein occupy successive sites on the DNA just upstream of the TSS (Fig. 1.1). Thus, these proteins are predicted to function as mimics of the α and σ subunits, with the C-terminal HARE-HTH, potentially occupying yet another site upstream of the TSS. Accordingly, these proteins could possibly function as a novel inhibitor of TSS-binding by the bacterial RNA polymerase, which might either function as a negative transcriptional regulator, or a negative regulator of improper transcription initiation.

Specific TFs and a structural portrait of their DNA-binding domains

Specific TFs are best classified based on their DNA-binding domains. The two prokaryotic superkingdoms are differentiated from eukaryotes in terms of the DNA-binding domains of their specific TFs. Most specific TFs of prokaryotes contain a version of the helix–turn–helix DNA-binding domain (Fig. 1.3) (Aravind *et al.*, 2005). In contrast, eukaryotic transcription factors show a great variety of DNA-binding domains belonging to many different folds not encountered in their bacterial counterparts (Iyer *et al.*, 2008). Even those HTH DNA-binding domains prevalent in specific TFs (e.g. Homeo or POU domains) in many eukaryotic lineages are distinct from the bacteria versions and show only a distant sequence relationship to them. Additionally, eukaryotes possess large numbers of Zn-chelating DNA-binding domains such as the C2H2 Zn-finger, the C6 fungal-type Zn-finger and the WRKY Zn finger, which are rare or entirely absent in the prokaryotic superkingdoms (Iyer *et al.*, 2008). The dominance of the HTH-containing specific TFs across bacteria considerably aids their computational detection as high-sensitivity sequence profiles have been developed for the HTH domain (Aravind and Koonin, 1999b; Babu *et al.*, 2004). Thus searches with such profiles allow rather accurate estimates of the specific TF complement of a given prokaryotic organism from its genome sequence. In this chapter we summarize the taxonomy and structural variations of the HTH domain that are observed among bacterial specific TFs.

Trihelical HTH domains

The simplest version of the HTH domain, the basic trihelical version, is comprised entirely of the three core helices with no additional elaborations (Fig. 1.4). This configuration is closest to the ancestral state of the HTH and is seen across the three super-kingdoms of life. The third helix of this unit, like in most other HTH domains plays a key role in contacting DNA via insertion into the major groove, and is called the recognition helix (Brennan and Matthews, 1989; Clark *et al.*, 1993). This version is seen in the Fis family of transcription factors (typified by the *E. coli* protein Fis), the first HTH domain of the σ^{70} family and the three HTH domains of the σ^{54} family (Fig. 1.5). The Fis HTH domains are typically found fused to the C-termini of the AAA+ domains of the NtrC-like proteins which bind 'enhancer elements' which are located at much greater distances from the promoter than conventional target sites bound by specific TFs (Morett and Bork, 1998; Rombel *et al.*, 1998). Also displaying this type of HTH domains are bacterial TFs of the Rok and YlxL/SwrB families. The Myb/SANT domain, which is very common in eukaryotic TFs and chromatin proteins is also a typical trihelical HTH domain (Aravind *et al.*, 2005). In bacteria the Myb/SANT domain is rarer than in eukaryotes and is found in TFs typified by the RsfA proteins, which are pre-spore transcription factors in firmicutes (Juan Wu and Errington, 2000) and the proteobacterial GcrA-like transcription factors (Holtzendorff *et al.*, 2004). More recently, using sequence profile searches we reported several proteins in bacteria with multiple Myb/SANT repeats, which are specifically related to those seen in eukaryotes (e.g. Fig. 1.5) (Iyer and Aravind, 2011). These versions are encoded in operons with integrases, endonucleases and DNA methylases in bacteriophages (e.g. gp65 of *Listeria* phage B054) and bacterial genomes (e.g. A33_2137; gi: 254286508 in *Vibrio cholerae*) or are fused to endonuclease domains of the HNH and the LAGLIDADG superfamilies (Iyer and Aravind, 2011). These observations suggest that they are DNA-binding domains of phages or novel mobile selfish elements, wherein they help recognize integration sites. The versions derived from such selfish elements appear to have given rise to the Myb/SANT domain of the eukaryotic transcription factors. The second HTH domain of σ^{70} family is a derived version of the trihelical HTH class, which shows an additional N-terminal helix also observed in the archaeo-eukaryotic TFIIB proteins (Fig. 1.4).

Tetrahelical HTH domains

The tetrahelical version of HTH domain is an elaboration of the simple trihelical version and is characterized by an additional C-terminal helix which packs against the shallow cleft formed owing to the open configuration of the trihelical core (Fig. 1.4). Several major families of bacterial transcription factors contain this version of HTH, which can be differentiated on the basis of sequence features. The cI-like family, prototyped by the phage lambda cI protein is one of the major families with this type of DNA-binding domain. Several distinct subfamilies can be recognized within this family. The largest of these is the repressor subfamily typified by the protein PbsX (Xre) from the *B. subtilis* prophage 168, which appears to represent the prototypical specific TFs that function as repressors in bacteria (Wood *et al.*, 1990). Another major assemblage within the tetrahelical class of HTHs contains the six major families of exclusively prokaryotic TFs. These are AraC, LuxR, LacI, DnaA, TrpR and TetR families (Aravind *et al.*, 2005). The first four of these families are nearly panbacterial in their distribution suggesting that they had probably diverged from each other even in the common ancestor of all bacteria (Fig. 1.4). The latter two lineages are more limited, being most prevalent in proteobacteria and firmicutes. DnaA is typically found in a single copy across bacteria, with a tetrahelical HTH occurring C-terminal to an AAA+ domain. The DnaA protein is primarily required in replication initiation, but it also functions as a transcription

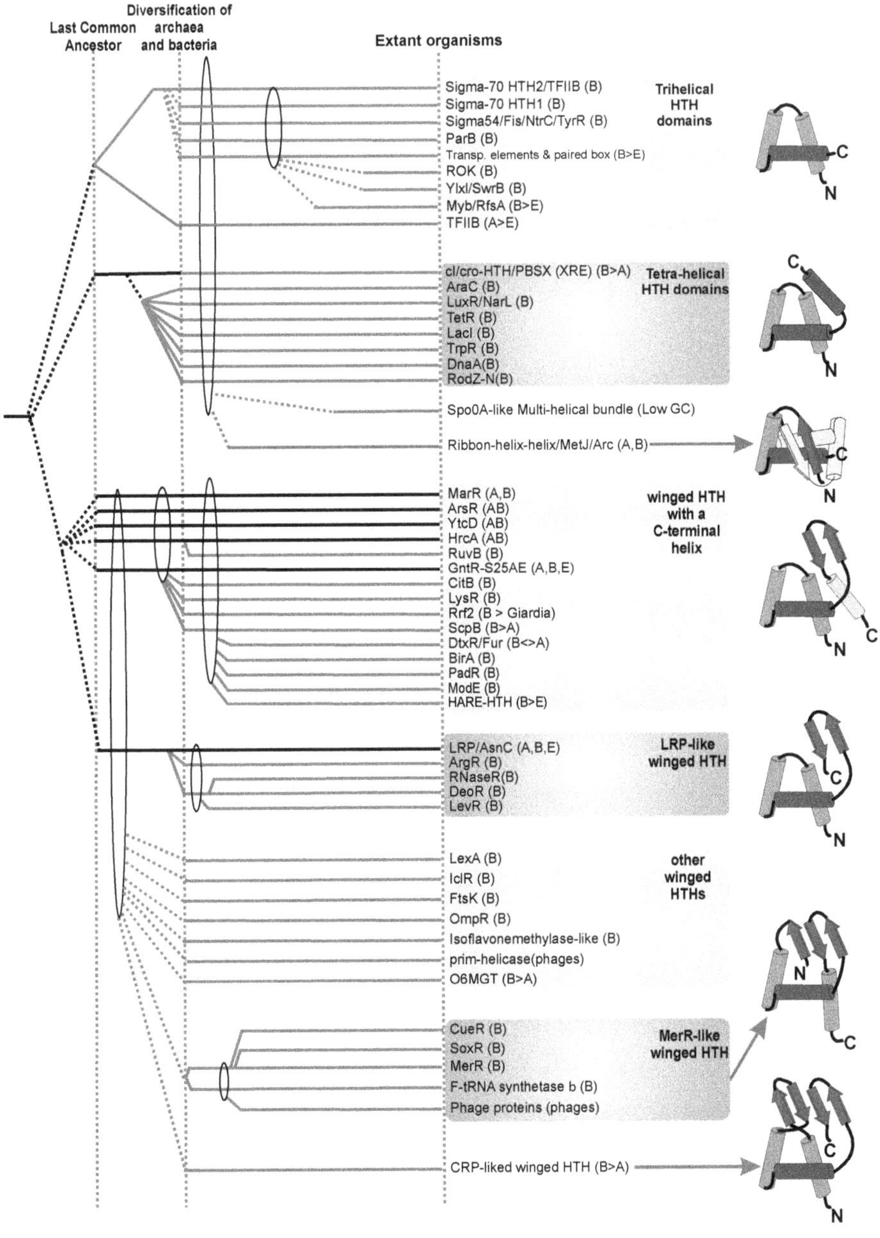

Figure 1.4 Higher order evolutionary relationships of bacterial specific transcription factors containing a HTH domain. The horizontal lines represent temporal epochs corresponding to major transitions in evolution of bacteria, namely the last universal common ancestor and the diversification of archaea and bacteria. Solid lines reflect the maximum depth of time to which a particular family can be traced. Broken lines indicate an uncertainty with respect to the exact point of origin of a lineage. The ellipses encompass groups of lineages from which a new lineage with relatively limited distribution could have potentially emerged. Lineages of archaeal origin are coloured blue, those of bacterial origin are coloured orange and those present in archaea and bacteria are coloured black. The phyletic distribution of the lineages are also shown in brackets. A, Archaea; B, bacteria; E, eukaryotes. The '>' reflects lateral transfer with the arrow head pointing to the potential direction of transfer. Also shown to the right are cartoon representations of the major structural types of HTH domains found in bacterial transcription factors. The TFIIB lineage of archaeo-eukaryotic HTHs is shown to illustrate its relationship with the sigma factor. A colour version of this figure is located in the plate section at the back of the book.

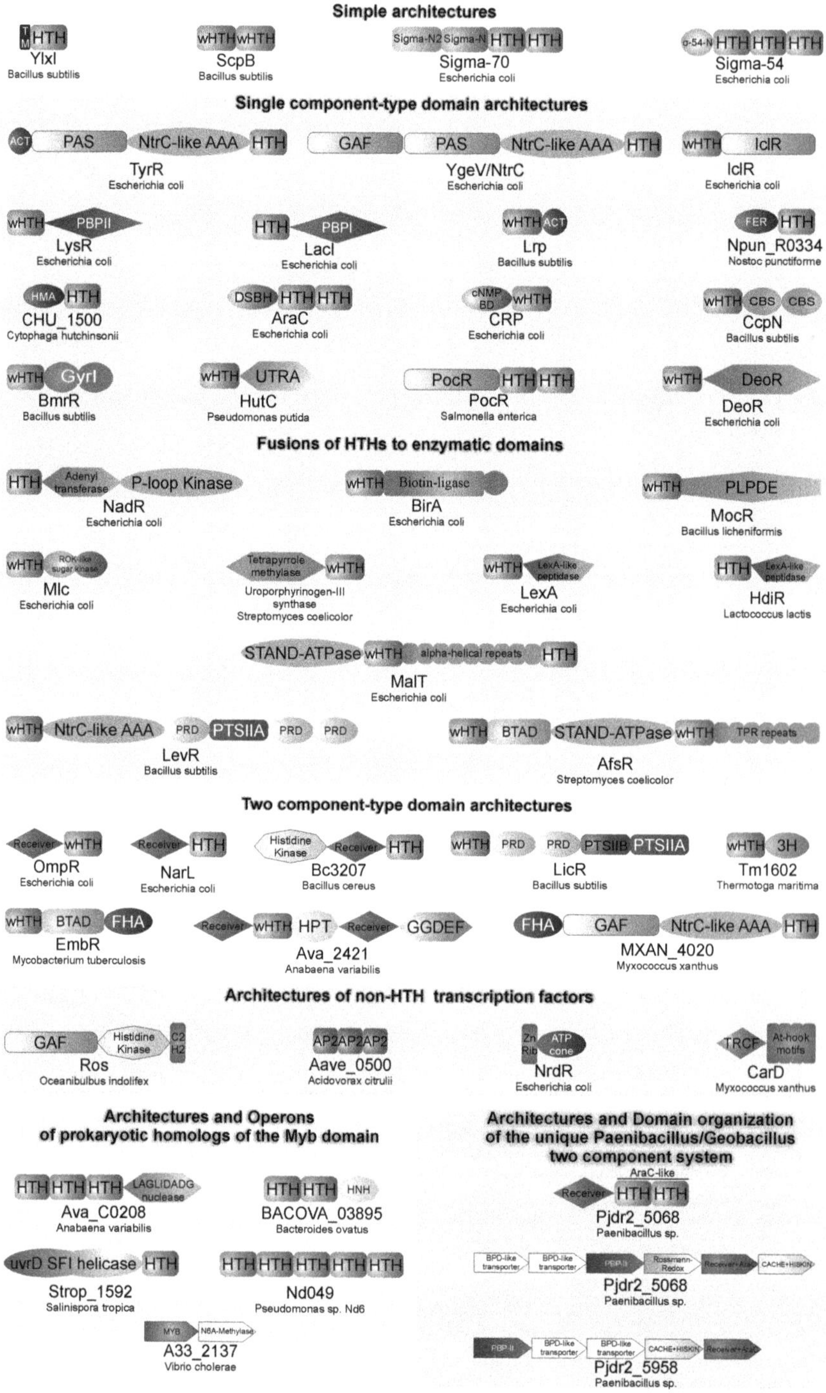

Figure 1.5 Examples of domain architectures of bacterial transcription factors described in the text. Proteins are labelled with their gene and species names. The domains are not drawn to scale. Standard nomenclatures were mostly used to depict the various domains. Additional abbreviations include: TM, transmembrane; σ-54N, globular domain found at the N-terminus of σ^{54}; Sigma-N2 and Sigma-N, conserved N-terminal domains found in σ^{70}; BTAD, conserved domain found in bacterial signalling proteins; ZnRib, zinc ribbon; FER, classical ferredoxin domain of the RRM fold. A colour version of this figure is located in the plate section at the back of the book.

factor (Fujikawa *et al.*, 2003; Messer and Weigel, 2003). Additionally, sporadic versions of the tetrahelical HTH are also seen in several phage transposases related to the Mu transposase, which in some cases also function as TFs (Wojciak *et al.*, 2001).

Winged HTH domains

The winged HTH (wHTH) domains are distinguished by the presence of a C-terminal β-strand hairpin unit (the wing) that packs against the shallow cleft of the partially open trihelical core (Brennan, 1993) (Fig. 1.4). The simplest versions of the wHTH domains contain a tight helical core similar to basic trihelical version followed by the two-strand hairpin. However, many wHTH domains display serial elaborations of the β-sheet (Fig. 1.4) (Aravind *et al.*, 2005). In the three-stranded version, the loop between helix-1 and helix-2 of the HTH assumes an extended configuration and is incorporated as the third strand in the sheet, via hydrogen-bonding with the basic C-terminal hairpin. In the 4-stranded version, the linker between helix-1 and helix-2 itself forms a hairpin with two β-strands, and along with the C-terminal wing forms an extended β-sheet (Fig. 1.4). The wing often provides an additional interface for substrate contact, typically by interacting with the minor groove of DNA through charged residues in the hairpin (Brennan, 1993; Clark *et al.*, 1993; Swindells, 1995). Majority of bacterial TFs contain the wHTH as their DNA-binding domain. Fourteen major families of prokaryotic TFs, namely the HARE-HTH (see above), BirA, ArsR, GntR, DtxR-FurR, CitB, LysR, ModE, MarR, PadR, YtcD, Rrf2, ScpB and HrcA-RuvB families, are unified by a helix after the C-terminal wing, and comprise the largest monophyletic assemblage within the wHTH superclass (Fig. 1.4). Of these the DtxR–Fur family appears to have specialized early in bacterial evolution in regulating metal-dependent transcription of genes (Hantke, 2001); here the wing is incorporated into a large sheet formed with additional C-terminal strands. Another major monophyletic assemblage within the wHTH superclass includes the DNA-binding domains of the DeoR, ArgR, LevR and Lrp-AsnC families of TFs. These families are unified by overall sequence similarity, and a conserved pattern with a conserved glutamine or arginine residue between helix-1 and helix-2 of the HTH domain (Aravind *et al.*, 2005). Other distinct families of wHTH TFs in bacteria, namely the LexA, OmpR, and IclR families, with two- or three-stranded wHTH domains, do not appear to belong to any of the aforementioned assemblages (Fig. 1.4). Of these the classical representatives of the LexA family appear to be involved in regulating responses to DNA damage in diverse lineages of bacteria (Peat *et al.*, 1996), whereas OmpR-like TFs are one of the largest group of specific TFs that function downstream of histidine kinases (Itou and Tanaka, 2001).

Distinct from all the above families is the Crp family that is typified by the presence of a four-stranded version of the wHTH domain (Fig. 1.4). This family has a pan-bacterial distribution and is typically fused to a C-terminal cNMP-binding domain (Korner *et al.*, 2003). These appear to have specialized early on as the primary cyclic nucleotide-dependent TFs in bacteria. Beyond these classical wHTH domains, several modified versions display highly derived versions of the wHTH (Fig. 1.3). These include the MerR-like family, which contains a truncated form of the three-stranded wHTH domain with a deletion of the first helix. Instead, these proteins show an additional helical element C-terminal to the wing that appears to have displaced the first helix. The MerR family has vastly proliferated into several distinct subfamilies, like the SoxR and CueR subfamilies (Brown *et al.*, 2003). A similar form of wHTH is also observed in the phage lambda excisionase and terminase proteins and the phage Mu-repressor family.

The ribbon–helix–helix or MetJ/Arc domain

The MetJ-Arc family (also known as ribbon-helix-helix/RHH family) of TFs is a uniquely prokaryotic family of TFs typified by the methionine operon repressor MetJ and the bacteriophage repressor Arc (Aravind and Koonin, 1999b; Aravind *et al.*, 2005). They function as obligate dimers, which pair through a single N-terminal strand, and possess a C-terminal helix–turn–helix unit (Fig. 1.4). The C-terminal helical unit is organized identically to corresponding unit in the HTH domain, and it shows the characteristic conserved sequence features of the HTH domain. The sheet formed by the N-terminal strands of the dimerized domain is inserted into the major groove of DNA (Gomis-Ruth *et al.*, 1998). Mutagenesis experiments have shown that even single mutations in the N-terminal strand convert the strand of the RHH domain to a helix, and result in a structural packing that is closer to the canonical HTH domain (Cordes *et al.*, 1999). This result, together with the notable structural and sequence similarities with HTH domains, suggest that the RHH domain was derived from the HTH domain through conversion of the N-terminal helix to a strand (Aravind *et al.*, 2005). Concomitant with this modification, the N-terminal strand, which came to lie atop the recognition helix, appears to have taken up the primary DNA-binding role in this domain. They are most frequently found as transcriptional regulators of the mobile toxin–antitoxin operons (Anantharaman and Aravind, 2003). Hence, it is possible that they were originally derived in such toxin–antitoxin systems, through rapid divergence from a conventional HTH. This appears to have happened early in the evolution of one of the prokaryotic lineages (Fig. 1.4), after which they were widely disseminated across the bacteria and archaea owing to the extensive horizontal mobility of toxin–antitoxin systems.

Other DNA-binding domains found in bacterial specific TFs

A small set of non-HTH DNA-binding domains are found in bacteria specific TFs. These include the YefM and AbrB-likes domain which were first identified in selfish toxin–antitoxin systems (Anantharaman and Aravind, 2003). Like the RHH DNA-binding domain described above, they appear to have diversified greatly in these systems and have been occasionally recruited as specific TFs with distinct cellular functions. While the C2H2 Zn-finger is probably the most prevalent DNA-binding domain of eukaryotic specific TFs, it is rare in prokaryotes. The Ros/MucR family of TFs is typified by the Ros protein of *Agrobacterium tumefaciens*, which regulates the expression of virulence genes on the Ti plasmid (Chou *et al.*, 1998), and MucR, which regulates the exopolysaccharide biosynthesis in various rhizobia (Keller *et al.*, 1995). These proteins contain a single copy of the C2H2 Zn-finger and, unlike their eukaryotic counterparts, have only 9 or 10 residues between the two pairs of metal-chelating ligands (Esposito *et al.*, 2006). These TFs are currently known only from proteobacteria. The Zn-ribbon is an ancient nucleic-acid-binding domain that is found in large number of nucleic acid metabolism proteins (Aravind and Koonin, 1999b; Krishna *et al.*, 2003). While it is found in the core transcriptional machinery, for example, as a domain of the β′ subunit and occasionally inserted into the β-subunit (in Aquificae and acidobacteria) of the RNA polymerase (Iyer *et al.*, 2004a; Lane and Darst, 2010a), it rarely used as the primary DNA-binding domain in specific TFs. Zn-ribbon TFs in bacteria are typified by the *E. coli* NrdR protein which is a regulator of the ribonucleotide reducatase operons (Grinberg *et al.*, 2006). Here it combined with a C-terminal ATP-cone domain which acts a nucleotide sensor (Fig. 1.5) (Aravind *et al.*, 2010). A few other specific TFs with the Zn-ribbon fused to other sensor domains (e.g. CBS domains) are also encountered in prokaryotes (Aravind and Koonin, 1999b). The AT-hook is a very common DNA-binding motif in eukaryotes that specifically contacts the minor groove (Aravind and Landsman, 1998). A small number of bacterial TFs with the AT-hook are currently known. The best example of this is the CarD protein from *Myxococcus xanthus* and other myxobacteria, which is known to function as a light-induced transcription factor (Penalver-Mellado *et al.*, 2006). Here, the AT-hooks, which bind the target sequences, are combined with a TRCF-like domain (Fig. 1.4) (Subramanian *et al.*, 2000). In the transcription repair-coupling helicase (TRCF) the same domain is fused to a superfamily-II helicase module and facilitates interaction with the RNA-polymerase holoenzyme (Westblade *et al.*, 2010). Outside of myxobacteria the CarD orthologues merely contain a TRCF-like domain but not AT-hooks (Subramanian *et al.*, 2000). In these organisms it is likely that these proteins associate with the RNA polymerase but do not bind DNA. Hence, these versions might not function as *bona fide* specific TFs. The AP2 domain is a DNA-binding domain which is found specific TFs of several eukaryotic lineages such as plants, stramenopiles and apicomplexans (Balaji *et al.*, 2005). In bacteria they are typically found associated with integrases and transposases of selfish elements such as phages and transposons. We have also identified versions in bacteria that resemble eukaryotic versions from plants, stramenopiles and apicomplexans in having multiple tandem copies of the AP2 domain and are independent of integrase or transposase catalytic domains (Fig. 1.4) (Iyer and Aravind, 2011). We predict that these versions are likely to function as novel specific TFs and might have been the progenitors of the TFs observed in the above-stated eukaryotic lineages.

RNA regulators of transcription that interact with the RNA polymerase

The *E. coli* 6S RNA, discovered over 40 years ago, remained mysterious in function until recently. It was shown to be the prototype of a class of widely conserved non-coding bacterial RNAs that directly interact with the RNA polymerase to regulate transcription (Willkomm and Hartmann, 2005; Wassarman, 2007). These RNAs are about 185 nucleotides in length and fold through complementary base-pairing to give rise to a structure, which contains a large central bulge which is believed to resemble the open promoter at the transcription start site. In *E. coli* the 6S RNA associates with the σ^{70}-containing holoenzyme and represses transcription from specific promoters in the stationary phase (Wassarman, 2007). While the 6S RNA homologues from other bacteria also associated with the RNA polymerase complex, their targets and the phase of the life cycle in which they function remain unclear. Some organisms, like *B. subtilis*, possess multiple 6S RNA homologues suggesting that there might be alternative regulation of transcription in different developmental phases by distinct 6S RNAs (Willkomm and Hartmann, 2005). The 6S RNA has been shown to potentially interact with the β, β' and σ subunits suggesting that it might interact in the region of the conserved SBHM in β (the so-called flap domain) (Wassarman, 2007). Its structural similarity to the open promoter has also been interpreted as a means of mimicking the former and thereby withholding the holoenzyme from the actual promoter. While most non-coding RNAs in bacteria work at the level of translation regulation (Gottesman, 2004), it is conceivable that there are other RNAs which operate similarly to the 6S RNA to regulate transcription.

An overview of the domain architectures of bacteria-specific TFs

The above DNA domains are often combined with other domains in the same protein giving rise to a remarkable array of domain architectures (Fig. 1.5). Despite their diversity, all architectures can be classified into a small number of generic archetypal classes, the members of each class being unified by certain general organizational and functional principles. Hence, these organizational principles serve as strong predictors of function for bacterial TFs (Aravind *et al.*, 2005). These convergent architectural classes illustrate how natural selection has engineered similar functional solutions using a relatively small repertoire of domains, with the most populated classes representing particularly successful functional solutions.

Specific TFs with simple domain architectures

The simplest architectures are the standalone copies of the DNA-binding domain as typified by proteins related to the cI repressors and Fis. These proteins are usually almost entirely comprised of just a standalone HTH, and might, at best, have some small extensions that play a role in dimerization or interactions with other components of the basal transcriptional machinery (Aravind *et al.*, 2005). A family of bacterial TFs prototyped by the *B. subtilis* sigma D regulator YlxL (SwrB) (Kearns and Losick, 2005) contains a HTH domain fused to a N-terminal transmembrane region (Fig. 1.5). These HTH proteins might regulate transcription under the influence of signalling events associated with the cell membrane. The next level of architectural diversification involves tandem duplications of HTH domains. Other than the σ-factors, duplicated HTH domains are encountered in a few bacterial DNA-binding proteins like ScpB that could potentially function as TF in addition to having a role as co-factors for the chromosome-condensing SMC proteins (Mascarenhas *et al.*, 2002; Soppa *et al.*, 2002).

TFs displaying single component-type domain architectures

The single-component systems are defined as signalling systems in which the transcription DNA-binding domain and the stimulus sensor module are combined into a single polypeptide. These architectures are by far the most prevalent class in bacteria. The simplest of these are no different from the above class in that they are simply comprised of DNA-binding domain that not only binds DNA but also directly interacts with small-molecule effectors. These minimal one-component regulators are prototyped by the MetJ-type RHH transcription factor, which, in addition to binding DNA, also senses S-adenosyl methionine directly (Augustus *et al.*, 2010). A more typical one component TF combines a HTH domain with a small molecule binding domain (SMBD, Fig. 1.5) (Aravind *et al.*, 2010). More complex architectures might involve multiple SMBDs or even additional domains such as the NtrC-like AAA+ ATPase domain. The most common SMBDs fused to HTHs in the single component systems are drawn from a relative small set of ancient protein folds (Fig. 1.5):

1 the PAS-like fold, with versions such as the PAS domain, the GAF domain, and the ligand-binding domains of the IclR-type transcription factors (Aravind *et al.*, 2010);
2 periplasmic-binding protein type I and type II domains, which include the ligand-binding domains of the LysR family (Vartak *et al.*, 1991; Tam and Saier, 1993; Tyrrell *et al.*, 1997);
3 the ferredoxin-like fold, which includes the ACT domain and related ligand-sensing domains of the Lrp-like transcription factors and the classic ferredoxins, which are fused to HTH domains in cyanobacterial proteins (Aravind and Koonin, 1999a; Brinkman *et al.*, 2003; Bull and Cox, 1994);
4 the double-stranded β-helix domain (cupin), which contains the AraC-type ligand-binding domains, as well as the cNMP-binding domains found in Crp/Cap/Fnr family TFs (Anantharaman *et al.*, 2001; Kannan *et al.*, 2007);
5 the CBS domain that occurs as an obligate dyad (Bateman, 1997);
6 the GyrI domain, contains two copies of the SHS2 structural module, appears to be one of the principal ligand-binding domains of the MerR family (Anantharaman *et al.*, 2001; Kannan *et al.*, 2007);
7 the UTRA domain, found in the HutC/FarR group of GntR family transcription factors possesses the same fold as chorismate lyase (Anantharaman and Aravind, 2003);
8 the DeoR ligand-binding domain shares a common α/β fold (the ISOCOT fold) with enzymes of the phosphosugar isomerase family such as ribose phosphate isomerase (Anantharaman and Aravind, 2006).

Several distinct clades of specific TFs, often defined by a specific architectural theme can be identified within this mélange of one-component systems. For example, the AraC family contains a duplication of the tetrahelical version of the HTH domain (Fig. 1.5) and typically occurs fused to the sugar-binding cupin domain suggesting that most representatives of the clade function as sugar-sensing transcription factors.

A variation on the single-component theme is the fusion of the DNA-binding domain to an enzymatic domain, which catalyses a reaction related to the biochemical pathway regulated by the specific TF (Fig. 1.4). Thus, such TFs are major players in the phenomenon of feedback regulation of metabolic pathways, in which the concentrations of the metabolites produced by the pathway regulate the activity of the TF. The archetypal representative of this architectural theme is the biotin operon repressor, BirA, which contains an N-terminal HTH domain fused to a C-terminal biotin ligase domain (Wilson *et al.*, 1992). In the presence of biotin the enzymatic domain synthesizes the co-repressor, and the HTH domain represses the transcription of the biotin biosynthesis genes (Wilson *et al.*, 1992). Comparative genomics suggests that architectures involving fusions to a range of enzymes from cofactor, nucleotide, amino acid and carbohydrate metabolism are fairly common in bacteria (Fig. 1.5) (Aravind and Koonin, 1999b; Aravind *et al.*, 2005). Some notable fusions include combination of the HTH with nicotinamide mononucleotide adenylyl transferase and a P-loop kinase in NadR, with the pyridoxal phosphate-dependent aminotransferase domain (TFs of the GntR family) and sugar kinases (Rok family) (Fig. 1.4) (Singh *et al.*, 2002). Some such architectures, like BirA, are widely distributed in prokaryotes and appear to be ancient, while others like the fusion of an OmpR family wHTH with the uroporphyrinogen-III synthase are found only in actinobacteria. These observations suggest that the combinations of HTHs with enzymatic domains have been repeatedly selected for throughout bacterial evolution. Yet another variation on the theme of enzyme-linked HTH domains is provided by the LexA protein, the repressor of several bacterial DNA repair genes (Fig. 1.4). It contains a protease domain of the signal peptidase fold fused to a wHTH domain. The protease domain catalyses an autocatalytic cleavage in response to a DNA-damage signal and triggers dissociation of its wHTH domain from target sequences, thereby allowing transcription of DNA repair genes (Peat *et al.*, 1996). Architectures analogous to LexA are also seen in the repressors typified by the heat-response transcription factor HdiR from the *Lactococcus lactis*, where a LexA-like protease domain is fused to a cI-like HTH instead of the wHTH seen in LexA (Savijoki *et al.*, 2003). This implies that the mechanism of transcription regulation with a proteolytic processing step was innovated at least twice independently.

TFs with specialized architectures involving ATPase domains

Two other specialized classes of domain architectures arise through fusions of the HTH domains with either of two types of P-loop NTPase domains, namely the NtrC-like AAA+ domains (Zhang *et al.*, 2002) and the related STAND (signal transduction ATPases with numerous domains) NTPase domain (Leipe *et al.*, 2004; Ammelburg *et al.*, 2006). These NtrC-like TFs typically sense various sensory inputs via their effector-binding domains and associate as a ring-shaped multimer with σ^{54} via their AAA+ ATPase domains (Wigneshweraraj *et al.*, 2008). The AAA+ ATPase domains of these proteins perform an ATP-dependent chaperone-like activity that converts the 'closed' σ^{54}-containing transcription complexes to an 'open' configuration, which is favourable for transcription initiation (Wigneshweraraj *et al.*, 2008). The NtrC-like AAA+ domains are fused to at least two different types of HTH domains. The versions like NtrC and TyrR are fused to C-terminal trihelical HTH domains of the Fis family (Wang *et al.*, 2001). The second version typified by the *Bacillus* levanase operon regulator, LevR, instead contains an N-terminal wHTH domain (Aravind *et al.*, 2005). Structural comparisons suggest that core NTPase module of the STAND superfamily has been derived from the Orc/Cdc6 family of AAA+ domains. These two share a unique configuration of the dyad of helices occurring after the core NTPase strand-2 and a distinctive winged HTH (wHTH) occurring C-terminal to AAA+ module [part of the HETHS module (Leipe *et al.*, 2004)]. Given that the Orc/CDC6 family of AAA+ NTPases is ancestrally present in the archaeo-eukaryotic lineage, it is likely that the STANDs emerged from them early in archaeal evolution. Indeed, several archaea show lineage-specific expansions of the basal versions of the STAND NTPases encoded by mobile elements (the MJ-, PH- and SSO- type ATPases) that still retain several features of the ancestral AAA+ ATPases (Leipe *et al.*, 2004), such as the position of the arginine finger. These archaeal versions are often linked in the same polypeptide with restriction endonuclease fold domains and are likely to catalyse the ATP-dependent assembly of complexes on DNA that allow the replication of the mobile elements that encode them. Hence, they are likely to retain the ancestral function of the Orc/Cdc6 family in assembling complexes on DNA.

However, from such precursors a distinct lineage of STAND NTPases with signalling functions arose in bacteria (Leipe *et al.*, 2004). As a rule they are large multidomain proteins that catalyse the ATP-dependent assembly of complexes in variety of signalling contexts. They typically contain superstructure-forming repeat domains, such as the WD and TPR domains, which may serve as surfaces for the assembly of multi-protein complexes (Leipe *et al.*, 2004). Archetypal members of the architectural class combining a DNA-binding HTH and STAND NTPases are *E. coli* MalT (Larquet *et al.*, 2004; Marquenet and Richet, 2010), *B. subtilis* GutR (Poon *et al.*, 2001) and *Streptomyces* AfsR proteins (Lee *et al.*, 2002). The DNA-binding HTH domains in these proteins are of several distinct types. The fusions involving the OmpR family of wHTH domains (e.g. in AfsR) usually link the HTH to the N-terminus of the STAND NTPase domain. In contrast, fusions involving the LuxR family of tetrahelical HTHs link it to the C-terminus of the STAND module, with a set of super-structure forming α-helical repeats occurring between these two modules (e.g. GutR and MalT) (Fig. 1.4). The STAND-domain-containing transcription regulators integrate signalling inputs sensed via their super-structure forming domains with an NTP-dependent switch provided by the STAND. The energetically demanding use of NTPs in STAND signalling suggests these switches are likely to control expression of metabolic states that might impose a high cost on the cell (Marquenet and Richet, 2010). The STAND regulators are particularly prevalent in developmentally or organizationally more complex bacteria like cyanobacteria and actinobacteria, suggesting that they might function as regulatory switches for key processes related to development or differentiation.

Specific TFs with architectures pertaining to two-component, phosphotransfer and serine/threonine kinase signalling systems

The core of the two component phospho-relay system comprises of a histidine kinase and the receiver domain, which is phosphorylated on a conserved aspartate. These represent one of the most prevalent signalling systems of the bacterial world (Pao and Saier, 1995; West and Stock, 2001; Ulrich and Zhulin, 2007). A large subset of the receiver components are specific TFs that convert the sensory input received from the histidine kinase into a transcriptional response (Ulrich and Zhulin, 2007). Such TFs are typified by fusions of the receiver domain to a HTH domain. Two of the most common architectures, seen in majority of bacteria, involve combinations of a single N-terminal receiver domain to either a LuxR-like tetrahelical HTH domain (e.g. UhpA and NarL) or a wHTH domain (e.g. OmpR and PhoB) (Fig. 1.5). Less frequent fusions involving HTH domains of the AraC and the CitB families are seen in certain bacteria. Other than these simple architectures, several more complicated architectures involving multiple receiver domains or even fusions to histidine kinase modules (e.g. *B. cereus* protein BC3207) and NtrC-like AAA+ ATPase (e.g. *E. coli* NtrC) domains are also observed (Fig. 1.5). The PTS sugar-transport systems use a phosphorelay cascade to transfer a phosphate from phosphoenol pyruvate to a histidine on the PTS regulatory domain (PRD), which often co-occurs in the same polypeptide with HTH domains (Stulke *et al.*, 1998; Barabote and Saier, 2005). The PRDs receive the phosphates from the HPr and EIIB proteins of the PTS system, and depending on their phosphorylation state regulate transcription. Architectures displaying the PRD domain are analogous to those involving the receiver domain of the two-component system (Barabote and Saier, 2005). The simplest versions contain an N-terminal wHTH domain fused to a C-terminal PRD domain (Aravind *et al.*, 2005). The more complex forms contain more than one PRD domains, or fusions to NtrC-like AAA+ domains and PTS system EIIB domains, which determine sugar specificity (Fig. 1.5). The *B. subtilis* LicR protein contains an N-terminal HTH fused to two PRDs and both EIIB and EIIA components of the PTS system, indicating that it is a multifunctional protein that directly regulates both sugar uptake and transcription of sugar-utilization genes (Tobisch *et al.*, 1999). The 3H domain, which is related to the HPr domain of the PTS system, is also found fused to a BirA-related wHTH domain in several bacterial proteins typified by Tm1602 from *Thermotoga maritima* (Fig. 1.5) (Anantharaman *et al.*, 2001; Weekes *et al.*, 2007). The 3H domain might also be regulated by phosphorylation on its conserved histidines, perhaps via a PTS-like system. The serine–threonine kinases are over-represented in certain organizationally complex bacteria, like the cyanobacteria, myxobacteria and the actinobacteria (Aravind *et al.*, 2010). In the latter group there is class of proteins, typified by the protein EmbR, containing a fusion of the HTH domain with the FHA domain (Hofmann and Bucher, 1995). The FHA domain in this protein binds phosphoserine peptides, and mediates its interaction with the upstream protein kinase in regulating the biogenesis of the mycobacterial cell wall (Molle *et al.*, 2003). The same SMBDs found in the single component systems may also occasionally be found fused to two-component and other phosphorylation-dependent regulators, where they might supply secondary allosteric inputs (Fig. 1.5).

The proteome-wide demographics and phyletic patterns of specific TFs

Availability of a large number and phyletic diversity of complete bacterial genome sequences allows robust estimation of the general trends in the proteome-wide distribution of TFs. Position-specific score matrices or sequence profiles for the various distinct families of DNA-binding domains found in TFs have proven to be an effective method to detect TFs in proteomes. These sequence profiles can be used to iteratively search the target proteomes with the PSI-BLAST program (Altschul *et al.*, 1997). Alternatively, the seed alignments for the different families can be used to generate hidden Markov models, which can be similarly used to search the proteomes with the HMMER package (Eddy, 2009). Over the years several independent studies on scaling of the number of transcription factors with proteome size in bacteria point to a very specific version of the power-law, $y = ax$ (where y is number of TFs per proteome, x is the proteome size, a is a constant and φ is the power, which is around 1.62) (van Nimwegen, 2003; Aravind *et al.*, 2005, 2010) (Fig. 1.6). Interestingly, examination of individual bacterial clades shows that this form of the power-law scaling of TFs is rather invariant across lineages (Fig. 1.6). Thus, irrespective of whether we are looking at proteobacteria, firmicutes, actinobacteria or cyanobacteria the exponent of this power-law remains similar, suggesting that this factor stems from a rather fundamental feature of the bacterial cell. This distribution function suggests that as gene number increases, a greater than linear number of TFs are required per operon/gene.

However, very distinct trends are observed when individual architectural classes of TFs are examined. In bacteria, two-component systems show a strong tendency for linear scaling with respect to proteome size (Fig. 1.6) (Aravind *et al.*, 2005, 2010). Thus, there is a strong tendency across bacterial lineages to show about one copy of a two component TF for every 175 genes. This scaling trend should be evaluated in light of the observation that the scaling of receiver domain proteins with respect to histidine kinases is generally linear in most bacteria (Aravind *et al.*, 2010). This suggests that each two-component system TF is strongly coupled with respect to its upstream signalling histidine kinase. This observation, together with the linear scaling of two-component system TFs with proteome size (Fig. 1.6), indicates that a similar constraint also operates with respect to the number of target genes downstream of the two-component TF. It implies that two-component TFs tend to regulate their own target operons to the exclusion of other two-component TFs. This exclusivity is likely to result in linear increase in the number of such TFs with increasing proteome size. Remarkably, the only notable exceptions to this situation is seen in certain sporulating firmicutes of the *Bacillus*-like clade *Paenibacillus* and *Geobacillus*, which display anomalously large number of two-component TFs for their proteome size (one every 47 and 50 genes respectively; Fig. 1.6). The excess in these organisms appear to stem from the lineage-specific expansion of a version of two-component TF that is relatively uncommon in other bacteria, namely the version combining the receiver domain to C-terminal AraC family HTH domains. Given this unusual violation of a strong trend, we propose that in these organisms these excess two component TFs do not function as distinct TFs in separate signalling processes, but more likely as alternative forms of the same TF in a single signalling process. This idea is supported by our observation that these TFs occur in a very stereotypic operon that also encodes a histidine kinase with an extracellular sensory CACHE domain, a multi-TM transporter and a

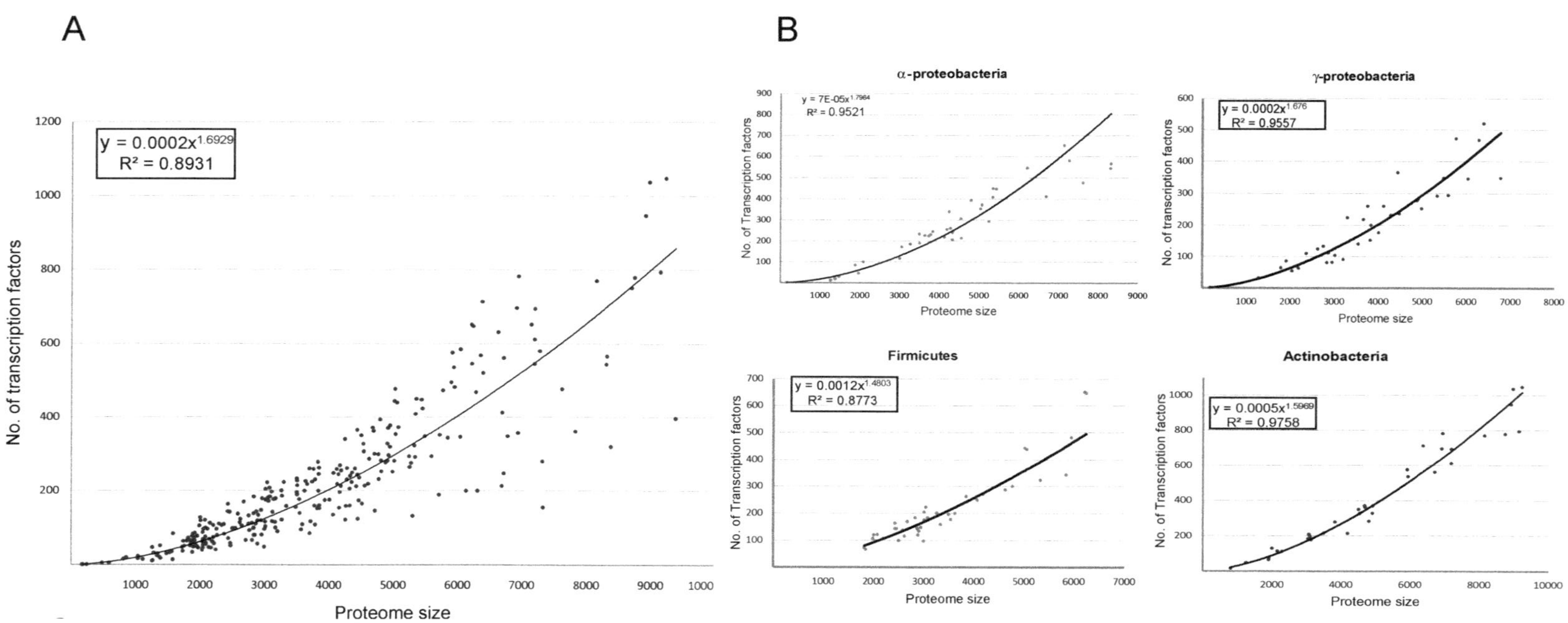

A
y = 0.0002x^1.6929
R² = 0.8931
No. of transcription factors
Proteome size
B
α-proteobacteria
y = 7E-05x^1.7964
R² = 0.9521
No. of Transcription factors
Proteome size
γ-proteobacteria
y = 0.0002x^1.676
R² = 0.9557
No. of transcription factors
Proteome size
Firmicutes
y = 0.0012x^1.4803
R² = 0.8773
No. of Transcription factors
Proteome size
Actinobacteria
y = 0.0005x^1.5969
R² = 0.9758
No. of Transcription factors
Proteome size

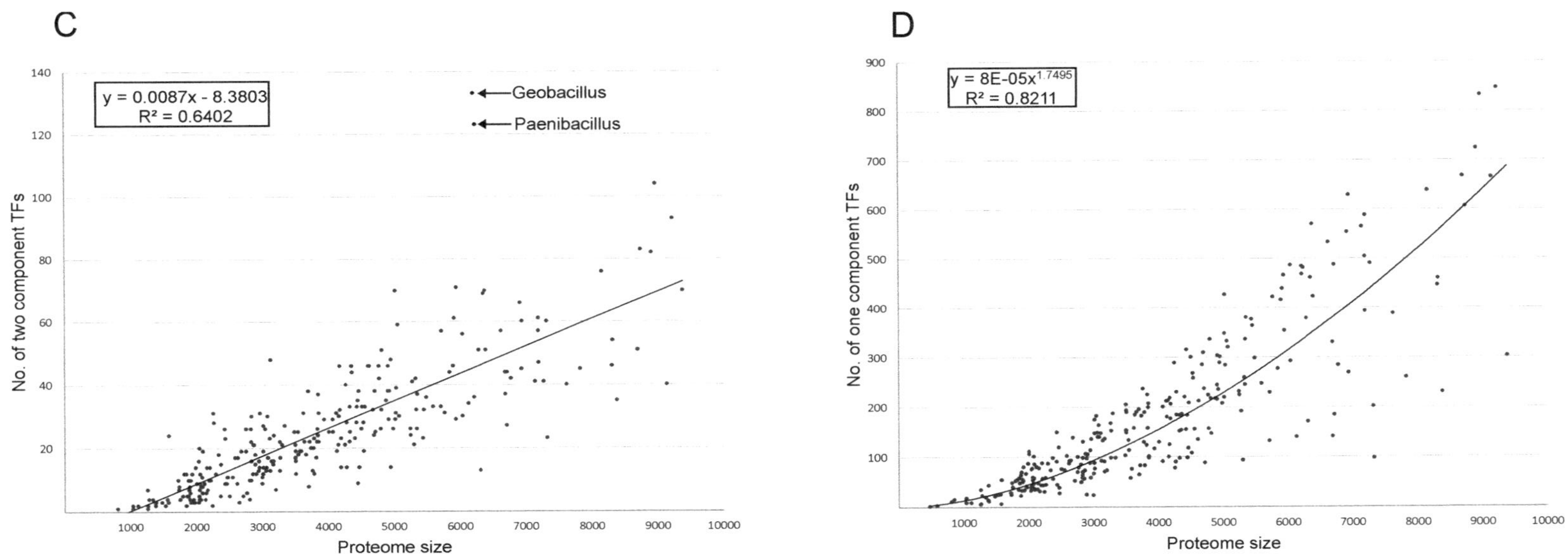

Figure 1.6 Scaling of bacterial transcription factors with proteome size. All graphs show a scatter plot of number of transcription factors in a given proteome (Y-axis) versus the number of protein-coding genes in that organism (X-axis). In (A) and (B), the Y-axis is the overall number of transcription factors across bacteria and in individual lineages respectively. In (C) The Y-axis is the number of predicted two-component system proteins. In particular, note the anomalous numbers in *Geobacillus* and *Paenibacillus*. (D) The Y-axis is the number of one-component system and other phospho-relay system proteins.

PBP-II-type solute-binding protein (Fig. 1.5; see acknowledgement for supplementary material). These operonic connections suggest that each isoform of this two-component system is a sensor that recognizes alternative versions of a variable soluble secreted signal. It is conceivable that the associated transporter and PBP-II domains are involved in the trafficking of the cognate version of the secreted signal. We propose that the diversification of this two component system might be related to the phenomenon of 'identity switching' (Ben-Jacob, 2003) and 'sibling rivalry' observed in *Paenibacillus*, in which under nutrient-poor conditions encroaching sibling colonies are killed by a secreted toxin (Be'er *et al.*, 2011). Such behaviour would be facilitated if the bacteria have a means of distinguishing self from non-self colonies. In light of this, it is conceivable that expression of different alternative versions of the above two component system operon from colony to colony might provide the necessary diversity for such discrimination. This remarkable system would benefit from further experimental exploration.

In contrast to the above picture, the one component TFs and σ factors, scaled non-linearly with proteome size and their distributions are best approximated by a power-law distribution comparable to that observed for the overall TF counts (Fig. 1.6) (Aravind *et al.*, 2005). This observation suggests that increase in genome size is accompanied by a greater than proportional increase in the numbers of one-component transcription factors. For instance, the GntR family has vastly proliferated in several bacteria giving rise to many of the major bacterial one-component transcription factors. This tendency might be related with the need to regulate specialized genes batteries by combining multiple, distinct inputs sensed by effector-binding domains of different sets of one-component TFs, especially in metabolically or organizationally complex bacteria with large genomes. This proposal is also consistent with other observations based on transcription networks, which suggest that one-component TFs are likely to be important for the fine tuning of gene expression in conjunction with more global changes mediated by two-component TFs (Balaji *et al.*, 2007). Non-linear scaling of the σ factors suggests that in the more complex genomes the additional genes are distributed amongst several functionally specialized gene batteries, which are under the regulation of devoted sigma factors responding to specific conditions. Interestingly, a few genomes display a significantly higher than expected number of σ factors. Strikingly, in *Phytoplasma asteris*, which, like other mycoplasmas, has a highly reduced genome with just over 700 genes (Aravind *et al.*, 2005), shows a significant excess of σ factors. Whereas the other mycoplasmas have only a basal σ-factor, *P. asteris* has a recent lineage-specific expansion of 11 sigma factors that are related to the *Bacillus* σ^F. Likewise, *Bacteroides thetaiotaomicron* and *Nitrosomonas* show recent lineage-specific expansions of ECF-type sigma factors that have given rise to at least ten closely related paralogous members in their proteomes (Aravind *et al.*, 2005). In the case of *P. asteris* there is evidence that the sigma factors may constitute a novel transposon (Lee *et al.*, 2005). While these supernumerary σ factors from the other bacteria with comparable expansions might indeed represent transposons, it is also possible that they might be conventional transcriptional regulators recruited for a distinctive sensory signalling pathway.

The logic of the overall organization of the transcriptional regulatory interactions in bacteria

Until recently it was thought that the transcription regulatory network (TRN) of both eukaryotes and bacteria are essentially similarly organized as approximations of scale-free networks (Thieffry *et al.*, 1998; Guelzim *et al.*, 2002; Balazsi *et al.*, 2005). However, new studies exploring their fine structure revealed that despite their superficial similarities, the organizational principles of the TRN of the model bacterium *E. coli* notably differs from that of the model eukaryote *Saccharomyces cerevisiae* (Balaji *et al.*, 2007). From such studies several principles pertaining to TRN organization might be discerned. In eukaryotes, highly connected TFs or hubs of the TRN, i.e. those that regulate a large number of genes are not typically those that integrate disparate transcriptional responses (Balaji *et al.*, 2006a,b). In contrast, in the bacterial TRN hubs function both as global regulators and as integrators of diverse transcriptional responses (Balaji *et al.*, 2007). By linking multiple TFs that regulate the same genes in the TRN one can reconstruct the underlying 'co-regulatory network' (CRN), which defines how TFs intersect in their regulatory actions. In the *E. coli* network, the degree distribution of TFs in this CRN (i.e. the number of regulatory intersections they make with other TFs) approximately follows a power law (Balaji *et al.*, 2007). On the other hand, the yeast CRN displays a discernible central tendency in the degree distribution (Balaji *et al.*, 2006a). These organizational differences appear to be related to the fact the bacterial genes are primarily organized as operons or regulons with their own dedicated specific TFs (Collado-Vides *et al.*, 2009). Though *S. cerevisiae* and *E. coli* have a comparable number of predicted TFs, the organization of the bacterial genome into operons, with several genes sharing a common set of regulatory elements, effectively reduces the set of targets available for TFs. Hence, in the bacterial TRN the global TFs would also have a propensity for being required for across-operon integration of gene regulation. In eukaryotes, absence of operonic organization, with co-expressed genes scattered around the chromosome, might have selected for a preferred number of co-regulatory associations between different TFs to allow co-regulation of a group of genes in different sets of conditions (Balaji *et al.*, 2007).

Further, the hubs in the bacterial TRN are enriched in specific TFs that have a dual function as both activators and repressors and are significantly underrepresented in TFs that are either dedicated activators or repressors. Likewise, the CRN hubs are significantly enriched in TFs that can function as both repressors and activators (Balaji *et al.*, 2007). The enrichment of the dual mode regulators in TRN hubs suggests that TFs mediating large-scale physiological state changes primarily do so by causing large-scale bi-directional changes in gene expression. Further, their prevalence in the CRN implies that these changes are likely to involve cooperative action with other specific TFs, wherein the dedicated activator and repressor TFs might provide further fine-tuning and amplification of the original effects. Interestingly, two-component systems tend not be pure repressors and are evenly distributed amongst activators or dual regulators. In contrast, one-component TFs depending on import of external effector metabolites by transporters are rarely dual mode regulators and are evenly distributed amongst dedicated repressors and activators (Balaji *et al.*, 2007). Thus, the distinct modes of signal sensing, namely via two-component systems or via one-component systems are distinguished by their mode of action. Two-component TFs are also enriched in hubs when compared to one-component TFs that depend on the import of external metabolites into the cell by transporters (Balaji *et al.*, 2006a). Hence, the former TFs appear to have been optimized for signalling larger scale changes. The latter category, in contrast, tend to regulate small group of genes specifically required for processing a given metabolite, and appear to do so by merely turning them 'on' or 'off'. Hubs in the TRN appear to be preferentially retained across genomes at small phylogenetic distances (e.g. within a well-defined lineage such as gamma-proteobacteria) (Balaji *et al.*, 2006b). In contrast, at larger phylogenetic distances there is no evidence for preferential retention of hubs amongst bacteria. At large phylogenetic distances hubs are only about as well conserved as any other TF suggesting that there are major differences in the global regulators between major clades of bacteria (Madan Babu *et al.*, 2006). Given the strong correlation between TFs and proteome size across all bacterial lineages (i.e. the linear scaling for two-component TFs and a gentle power-law increase for one-component TFs), it is quite likely that the general features of the transcriptional network inferred from *E. coli* are generally relevant for bacteria. In specific terms, at smaller phylogenetic distances the large-scale, bi-directional transcriptional responses appear to be preserved through vertical inheritance, but at larger phylogenetic distances there appear to be regular 'regime changes' that result in the emergence of new TFs for such large-scale transcriptional changes. The extensively studied model organisms do not capture the true extent of the diversity in bacterial signalling mechanisms. Particularly, certain lineages like cyanobacteria, myxobacteria and filamentous actinomycetes display complex signalling cascades involving STAND superfamily NTPase, eukaryote-type serine/threonine kinases, and caspase-like proteases, which are rare or entirely absent in *E. coli* (Aravind *et al.*, 2010). Hence, it is conceivable that certain optimizations of the TRNs in these bacteria are notably different.

Comparative and evolutionary perspectives

Early studies on the bacterial transcription apparatus saw it as model for all of life, indeed keeping with the adage of Monod: 'anything that is true of *E. coli* must be true of elephants, except more so'. As subsequent studies indicated the archaeal and eukaryotic systems are noticeably more complex than bacterial systems, they came to be seen as simplified models from which several basic mechanistic conclusions could be extrapolated to the other systems (Ptashne, 2004; Watson, 2004). This belief turned out to be partly true at least in the case of the core RNA polymerase complex (Cramer *et al.*, 2001; Cramer, 2002; Vassylyev *et al.*, 2002). With respect to the RNA polymerase complex, while the archaea and eukaryotes share orthologues of the α, β, β′ and ω subunits with the bacteria; thus, in the last universal common ancestor (LUCA) the RNA polymerase might be reconstructed as having at least four distinct subunits (however, see below for discussion on the α-subunit). Comparisons with the RdRPs and the RNA polymerases of selfish elements help reconstruct the potential pre-LUCA stages in the evolution of these enzymes. The earliest precursor was probably a DPBB domain that bound nucleic acids as a dimer and probably facilitated replication or transcription as a protein cofactor (Iyer *et al.*, 2003). Subsequently, this DPBB domain duplicated, and the two copies diverged, with each acquiring a distinct set of residues to respectively constitute the Mg^{2+}-chelating and negative-charge-stabilizing parts of the polymerase active site. These forms probably functioned in priming replication of DNA replicons that, unlike RNAs, cannot initiate unprimed daughter strand synthesis (Iyer *et al.*, 2005). This activity is predicted to be still retained by the RNA polymerases found in the bacterial selfish elements described above. The transcripts produced by the RNA polymerases of the selfish elements were probably also used by RNA-dependent restriction systems [e.g. such as the CRISPR system (Makarova *et al.*, 2011)] to control their activity. This type of an activity appears to have found a niche in the RNAi system of eukaryotes, where the polymerases were recruited as RNA-replicating enzymes that catalyse the primed or unprimed synthesis of dsRNA from diverse templates ranging from small siRNAs to mRNA.

Finally, the polymerase increased in complexity via domain accretion and became the primary catalyst of transcription. By the time of the LUCA it split up into two separate catalytic subunits and two additional subunits in the form of α and ω were added to the catalytic core. Architectures of the α-subunit cognates suggest that this process might not have been necessarily straightforward. The archaea and eukaryotes contain two distinct α-subunit cognates, prototyped by RPB3 and RPB11. Of these, the RPB11 cognates are comprised of just the ASCR domain. RPB3 cognates are instead closer to the bacterial α-subunits in possessing an insert of the L25-like domain. Additionally, RPB3 acquired two inserts in the form of a classical ferredoxin domain (which has degenerated or is entirely lost independently in several archaea and eukaryotes) and a novel C4 Zn-cluster early in the evolution of the archaeo-eukaryotic lineage. Thus, of all the α-subunits cognates the eukaryotic RPB11 is architecturally closest to the common ancestor of all α-subunits. This, together with the structurally asymmetric contacts of the bacterial α-subunits, raises that possibility that the archaeo-eukaryotic lineage more recapitulates the ancestral condition of the α-like subunits in the LUCA – one subunit was a minimal α with just the ASCR (like RPB11) and the second acquired an insert of the L25-like domain (like α and RPB3). This would imply that in bacteria the minimal version was lost, with both the role of both the α-subunits being taken up by the version with the L25 insert. Nevertheless, on the whole, the bacterial enzyme remained more or less close to the state in LUCA (probably simpler in light of the above), whereas the archaea and eukaryotes added several additional subunits to this core, which are highly conserved in those two superkingdoms (Cramer, 2002; Werner and Grohmann, 2011).

With regards to TFs, both general and specific, and the organization of the transcriptional network profound differences emerged in each of the three superkingdoms, whose full magnitude has only recently become clear with availability of genomic data from a wide phylogenetic spectrum. In terms of the actual protein components there are four major areas of difference between the bacterial and archaeo-eukaryotic systems: (1) the subunit complexity of the RNA polymerase, (2) the nature of basal TFs, (3) the specific TFs and (4) the role of chromatin-associated proteins (Iyer *et al.*, 2008). With respect to basal TFs, the archaeo-eukaryotic system possesses two distinct TFs, namely TFIIB and the TATA-binding protein (TBP) that apparently have no orthologues in the bacteria (Burley, 1996; Aravind and Koonin, 1999b). However, reanalysis of the structures of the respective RNA polymerases complexes with the basal TFs suggests that the picture might be more complex. Firstly, the TFIIB protein contains two HTH domains, by means of which it makes a direct contact with the promoter elements on either side of the TBP-binding site (Nikolov *et al.*, 1995). This contact of the promoter region by means of the two HTH domains of the archaeo-eukaryotic TFIIB is reminiscent of the situation in bacteria, wherein the two HTH domains of the σ-factor mediate two major DNA contacts associated with the two separated promoter elements (Hudson *et al.*, 2009). Furthermore, both TFIIB and σ-factor also make comparable contacts with the conserved SBHM domain (the so called 'flap' region) of the β (or its orthologues) catalytic region. This observation suggests that the SBHM insert of the RNA polymerase in the LUCA was already recruiting the primary basal TF that was bound to the promoter. Further, in light of the above, it is likely that the basal TF in the LUCA was potentially comprised of two HTH domains contacting DNA; hence, TFIIB and the σ-factor are likely to be ancient orthologues. Thus, the RNA polymerase complex in the LUCA can be reconstructed as having not just the core conserved subunits but also a two-HTH-domain basal TF that enabled it to become the primary catalyst of transcription. In bacteria, the basal TF evolved into the σ factor by accretion of an additional N-terminal helical domain, which performed the function of −10 element recognition and initiation of promoter melting. On the other hand in the archaeo-eukaryotic lineage the RNA polymerase complex appears to have recruited a new promoter-binding protein in the form of TBP (Cramer *et al.*, 2001; Cramer, 2002; Vassylyev *et al.*, 2002). TBP belongs to the larger helix-grip fold that includes proteins with various binding capabilities (Iyer *et al.*, 2001). Of these, the CCTBP domain found in association with sulfur transfer systems and the RNAseHIII nucleic acid-binding domains are closest to TBP (Aravind and Koonin, 2001; Burroughs *et al.*, 2009). The RNAseHIII TBP-like domain might interact with DNA–RNA hybrid molecules. Hence, it is possible that the TBP domain arose from a nucleic acid binding protein that recognized unusual nucleic acid structures such as DNA–RNA hybrids.

Specific TFs appear to have followed a different evolutionary course – eukaryotes possess very different specific TFs, but bacteria and archaea share several families of specific TFs, especially those with HTH domains (Aravind and Koonin, 1999b; Aravind *et al.*, 2005; Iyer *et al.*, 2008). Though several of the specific TF families shared by bacteria and archaea can be easily explained as arising from relatively recent lateral transfer between the prokaryotic super-kingdoms, some others like the MarR, ArsR, YctD, Lrp, HrcA and GntR families appear to show distinct pan-archaeal and pan-bacterial groups. This suggests that they were present from very early in the evolution of each of the prokaryotic super-kingdoms (Aravind *et al.*, 2005). As a corollary we are presented with an apparent evolutionary conundrum because the evolutionary picture of these specific TFs is not congruent with that of the basal TFs and the RNA polymerase complex, which point to a greater and hence possibly much earlier divergence. This paradox is further accentuated by the fact that the specific TFs of bacteria and the archaea interact with the RNA polymerase core in very distinct ways – for example, the archaeo-eukaryotic orthologues of the α-subunit lack the HhH motifs (CTD) found in the C-terminus of the bacterial α subunit that interact with the specific TFs. While number of scenarios can be conceived to account for this situation, the one that resorts to least number of unusual events depends on two considerations: (1) The core RNA polymerase and basal TF form a tightly interacting system (both in terms of interactions between the polymerase subunits and between the polymerase and the basal TF) that does not tolerate much xenologous displacement following lateral transfer. (2) The specific TFs interact less tightly and do not require conserved interfaces for these interactions. Thus, they are liable to lateral transfer. Hence, the families of specific TFs, which are shared widely by the two prokaryotic superkingdoms, might be interpreted as very early lateral transfers that happened between the two prokaryotic superkingdoms. The spread of these TFs through lateral transfer could be related to their adaptive value given that they (usually one-component TFs) often confer ability to alter gene expression in response to specific environmental compounds (Madan Babu *et al.*, 2006). The origin of eukaryotes through the symbiosis of an archaeal and bacterial progenitor generated a compartmentalized cell. This appears to have rendered most prokaryote-type one-component systems superfluous (Aravind *et al.*, 2005). Furthermore, emergence of histone- modification-mediated chromatin-based gene repression (see below) in the eukaryotes appears to have made the prokaryote-type repressors otiose. As consequence, early in eukaryotic evolution there appears to have been massive loss of the specific TFs inherited from the two prokaryotic progenitors, clearing the way for the recruitment and innovation of new types of eukaryote-specific TFs (Iyer *et al.*, 2008). Our studies suggest that some of these eukaryotic TFs might have been recruited from DNA-binding domains that were already present in bacterial TFs (e.g. AP2 and Myb) but with a marginal phyletic spread. However, in certain eukaryotic lineages they expanded to give rise to some of the largest families of paralogous specific TFs encoded by those genomes (Iyer *et al.*, 2008).

Finally, although bacteria possess chromatin proteins that package genomic DNA in a functionally analogous manner to the eukaryotes, with some exceptions, they do not possess the remarkable array of chromatin- remodelling and modifying enzyme complexes that are conserved throughout eukaryotes (Aravind *et al.*, 2011; Iyer *et al.*, 2011). These eukaryotic complexes include Swi2/Snf2 ATPases (a specific version of the superfamily-II helicases), acetylases, methylases, ubiquitin-conjugating enzymes, deacetylases, demethylases and deubiquitinating enzymes, which remodel chromatin proteins in an ATP-dependent manner or modify histone-side chains covalently or remove such covalent modifications. Bacteria possess two major types of Swi2/Snf2 ATPases, namely RapA/HepA and those associated with restriction-modification systems. The RapA/HepA protein is highly conserved in bacteria and associates with dsDNA and the RNA polymerase. In bacteria, the RNA polymerase after performing a single or a limited set of transcription cycles become incapable of further activity unless it is taken off the template and allowed to re-associate with σ and this recycling is catalysed by the RapA/HepA Swi2/Snf ATPase (Nechaev and Severinov, 2008; Shaw *et al.*, 2008). Thus, this bacterial Swi2/Snf2 ATPase is mechanistically similar to the eukaryotic Swi2/Snf2 ATPases in reorganizing protein-DNA contacts in an ATP-dependent manner which might involve their helicase activity. However, the bacterial version appears to play no such role with respect to the bacterial chromatin. The Swi2/Snf2 ATPases associated with restriction-modification systems appear to be required for remodelling the protein–DNA complexes in facilitating restriction enzymes that cut sites distant to their recognition site (Iyer *et al.*, 2006). Thus, while these systems again mechanistically resemble their eukaryotic counterparts they do not appear to have any dedicated transcription related function. Likewise, while some bacteria possess chromatin-modifying SET domain methylases (e.g. in *Chlamydia*) (Koonin *et al.*, 2001), which might function in conjunction with a SWIB domain protein (also found in eukaryotic chromatin remodelling complexes) and a topoisomerase (Aravind *et al.*, 2011; Iyer *et al.*, 2011). However, this does not appear to be a widely used regulatory mechanism. Similarly, other covalent modification of chromatin proteins, like that seen in eukaryotes, is not prevalent in bacteria (Aravind *et al.*, 2011; Iyer *et al.*, 2011).

Future directions

With the recent advances we have come a long way in our understanding of the bacterial transcription apparatus since the proposal of the operon theory of bacterial gene regulation and the discovery of the RNA polymerase. Yet, the obsessive focus of biomedical research on eukaryotic transcription systems has resulted in the more interesting problems in bacterial transcription regulation being considerably neglected. In particular, the discoveries of a rather invariant scaling of TFs in bacterial genomes and differences in the underlying architecture of bacterial and eukaryotic TRNs emphasize the need for more studies on bacterial TRNs. These need to be directed at questions such as:

1. Why exactly do these scaling laws hold across widely different bacteria?
2. Do bacteria with more complex signalling systems (e.g. actinobacteria, cyanobacteria and myxobacteria) and architecturally complex specific TFs (i.e. the STAND domain TFs) possess differences in the organization of the TRNs?
3. Are there any discernible patterns in terms of the TRN hubs that emerge in different bacterial lineages?
4. Can the binding sites of TFs be identified on a genome scale?
5. Can a comprehensive catalogue of the effectors bound by bacterial single-component systems be developed?
6. What do the archaeal TRNs look like and do they differ in any way from the bacterial versions?

These and other questions firstly require a dedicated experimental programme that is ready to explore systems beyond model bacteria such as *E. coli*, *Pseudomonas aeruginosa* and *B. subtilis*. The existence of genome sequences and reverse-genetics approaches for a wide range of bacteria make these studies at least technically feasible. The computational analysis of the data emerging from such studies is likely to open unexpected vistas and offer some of the most fundamental insights into the functions and evolution of prokaryotes.

Chapter highlights

- The bacterial RNA polymerase is a six subunit complex, comprised of two identical α subunits and one subunit each of β, β′, σ and ω.
- The conserved cores of the α, β, β′ and σ subunits are each composed of several domains, which have distinct roles in catalysis, and in contacting DNA elements and other transcriptional regulators. The ω subunit is required for RNA polymerase complex assembly and is the target of the stringent response alarmone (p)ppGpp.
- Majority of σ factors belong to the σ^{70}-like family, which additionally includes various types of alternative σs and the ECF σ factors. The σ^{54}-like family shows a distinct domain architecture and binds its obligate functional partner, the AAA+ domain of the NtrC-like or enhancer-binding proteins.
- The Gram-positive RNA polymerase δ subunit is a winged HTH that is widely distributed in bacteria and eukaryotes. Versions of this domain are fused to domains of the α and σ subunits and might function as novel inhibitors of the RNA polymerase in certain bacteria.
- The HTH is the dominant TF fold in prokaryotes and includes four major structural classes of HTH domains. Other DNA-binding domains of bacterial TFs include the C2H2, Zn-ribbon, AT-hook and AP2 domains. The 6S RNA is a RNA regulator of transcription.
- There are four major architectural categories among specific TFs namely solo domains, single component TFs, two component TFs and TFs fused to ATPase modules of the AAA+ or STAND superfamilies.

- The number of transcription factors scale with proteome size in bacteria as a specific version of the power-law: $y=a \cdot x^{\psi}$; ψ is usually around 1.62 irrespective of the bacterial lineage. Two-component systems scale linearly with respect to proteome size showing about one copy of a two component TF for every 175 genes.
- In the *E. coli* network, the degree distribution of TFs in the co-regulatory network follows a power law, whereas in the yeast CRN it shows a detectable central tendency.
- In bacteria at smaller phylogenetic distances hubs tend to be conserved but at larger phylogenetic distance there is no evidence for preferential retention of hubs.
- TFIIB and the σ-factors are likely to be ancient orthologues that have diverged extensively. In contrast the specific TFs shared by archaea and bacteria show lower levels of divergence.

Acknowledgement

Work by the authors is supported by the intramural funds of the National Library of Medicine, National Institutes of Health, USA. Supplementary material can be accessed at ftp://ftp.ncbi.nih.gov/pub/aravind/PROKHTH/prok_trans.html

References

Altschul, S.F., Madden, T.L., Schaffer, A.A., Zhang, J., Zhang, Z., Miller, W., and Lipman, D.J. (1997). Gapped BLAST and PSI-BLAST: a new generation of protein database search programs. Nucleic Acids Res. *25*, 3389–3402.

Ammelburg, M., Frickey, T., and Lupas, A.N. (2006). Classification of AAA+ proteins. J. Struct. Biol. *156*, 2–11.

Anantharaman, V., and Aravind, L. (2003). New connections in the prokaryotic toxin–antitoxin network: relationship with the eukaryotic nonsense-mediated RNA decay system. Genome Biol. *4*, R81.

Anantharaman, V., and Aravind, L. (2006). Diversification of catalytic activities and ligand interactions in the protein fold shared by the sugar isomerases, eIF2B, DeoR transcription factors, acyl-CoA transferases and methenyltetrahydrofolate synthetase. J. Mol. Biol. *356*, 823–842.

Anantharaman, V., Koonin, E.V., and Aravind, L. (2001). Regulatory potential, phyletic distribution and evolution of ancient, intracellular small-molecule-binding domains. J. Mol. Biol. *307*, 1271–1292.

Aravind, L., and Landsman, D. (1998). AT-hook motifs identified in a wide variety of DNA-binding proteins. Nucleic Acids Res. *26*, 4413–4421.

Aravind, L., and Koonin, E.V. (1999a). Gleaning non-trivial structural, functional and evolutionary information about proteins by iterative database searches. J. Mol. Biol. *287*, 1023–1040.

Aravind, L., and Koonin, E.V. (1999b). DNA-binding proteins and evolution of transcription regulation in the archaea. Nucleic Acids Res. *27*, 4658–4670.

Aravind, L., and Koonin, E.V. (2001). A natural classification of ribonucleases. Methods Enzymol. *341*, 3–28.

Aravind, L., and Iyer, L.M. (2012). The HARE-HTH and associated domains: Novel modules in the coordination of epigenetic DNA and protein modifications. Cell Cycle *11*, 1–13.

Aravind, L., Walker, D.R., and Koonin, E.V. (1999). Conserved domains in DNA repair proteins and evolution of repair systems. Nucleic Acids Res. *27*, 1223–1242.

Aravind, L., Anantharaman, V., Balaji, S., Babu, M.M., and Iyer, L.M. (2005). The many faces of the helix–turn–helix domain: transcription regulation and beyond. FEMS Microbiol. Rev. *29*, 231–262.

Aravind, L., Iyer, L.M., and Anantharaman, V. (2010). Natural history of sensor domains in bacterial signaling systems. In Sensory Mechanisms in Bacteria: Molecular Aspects of Signal Recognition, Spiro, S., and Dixon, R., eds. (Caister Academic Press, London), pp. 1–38.

Aravind, L., Abhiman, S., and Iyer, L.M. (2011). Natural history of the eukaryotic chromatin protein methylation system. Prog. Mol. Biol. Transl. Sci. *101*, 105–176.

Augustus, A.M., Sage, H., and Spicer, L.D. (2010). Binding of MetJ repressor to specific and nonspecific DNA and effect of S-adenosylmethionine on these interactions. Biochemistry *49*, 3289–3295.

Babu, M.M., Luscombe, N.M., Aravind, L., Gerstein, M., and Teichmann, S.A. (2004). Structure and evolution of transcriptional regulatory networks. Curr. Opin. Struct. Biol. *14*, 283–291.

Balaji, S., and Aravind, L. (2007). The RAGNYA fold: a novel fold with multiple topological variants found in functionally diverse nucleic acid, nucleotide and peptide-binding proteins. Nucleic Acids Res. *35*, 5658–5671.

Balaji, S., Babu, M.M., Iyer, L.M., and Aravind, L. (2005). Discovery of the principal specific transcription factors of Apicomplexa and their implication for the evolution of the AP2-integrase DNA binding domains. Nucleic Acids Res. *33*, 3994–4006.

Balaji, S., Iyer, L.M., Aravind, L., and Babu, M.M. (2006a). Uncovering a hidden distributed architecture behind scale-free transcriptional regulatory networks. J. Mol. Biol. *360*, 204–212.

Balaji, S., Babu, M.M., Iyer, L.M., Luscombe, N.M., and Aravind, L. (2006b). Comprehensive analysis of combinatorial regulation using the transcriptional regulatory network of yeast. J. Mol. Biol. *360*, 213–227.

Balaji, S., Babu, M.M., and Aravind, L. (2007). Interplay between network structures, regulatory modes and sensing mechanisms of transcription factors in the transcriptional regulatory network of *E. coli*. J. Mol. Biol. 372, 1108–1122.

Balazsi, G., Barabasi, A.L., and Oltvai, Z.N. (2005). Topological units of environmental signal processing in the transcriptional regulatory network of *Escherichia coli*. Proc. Natl. Acad. Sci. U.S.A. *102*, 7841–7846.

Barabasi, A.L., and Bonabeau, E. (2003). Scale-free networks. Sci. Am. *288*, 60–69.

Barabote, R.D., and Saier, M.H., Jr. (2005). Comparative genomic analyses of the bacterial phosphotransferase system. Microbiol. Mol. Biol. Rev. *69*, 608–634.

Barne, K.A., Bown, J.A., Busby, S.J., and Minchin, S.D. (1997). Region 2.5 of the *Escherichia coli* RNA polymerase sigma70 subunit is responsible for the recognition of the 'extended-10' motif at promoters. EMBO J. *16*, 4034–4040.

Bateman, A. (1997). The structure of a domain common to archaebacteria and the homocystinuria disease protein. Trends Biochem. Sci. *22*, 12–13.

Be'er, A., Florin, E.L., Fisher, C.R., Swinney, H.L., and Payne, S.M. (2011). Surviving bacterial sibling rivalry: inducible and reversible phenotypic switching in *Paenibacillus dendritiformis*. MBio *2*, e00069–11.

Ben-Jacob, E. (2003). Bacterial self-organization: co-enhancement of complexification and adaptability in a dynamic environment. Phil. Trans. A Math. Phys. Eng. Sci. *361*, 1283–1312.

Brennan, R.G. (1993). The winged-helix DNA-binding motif: another helix–turn–helix takeoff. Cell *74*, 773–776.

Brennan, R.G., and Matthews, B.W. (1989). The helix–turn–helix DNA binding motif. J. Biol. Chem. *264*, 1903–1906.

Brinkman, A.B., Ettema, T.J. de Vos, W.M. van der Oost, J. (2003). The Lrp family of transcriptional regulators. Mol. Microbiol. *48*, 287–294.

Brown, N.L., Stoyanov, J.V., Kidd, S.P., and Hobman, J.L. (2003). The MerR family of transcriptional regulators. FEMS Microbiol. Rev. *27*, 145–163.

Bull, P.C., and Cox, D.W. (1994). Wilson disease and Menkes disease: new handles on heavy-metal transport. Trends Genet. *10*, 246–252.

Burley, S.K. (1996). The TATA box binding protein. Curr. Opin. Struct. Biol. *6*, 69–75.

Burroughs, A.M., Iyer, L.M., and Aravind, L. (2009). Natural history of the E1-like superfamily: implication for adenylation, sulfur transfer, and ubiquitin conjugation. Proteins 75, 895–910.

Campbell, E.A., Muzzin, O., Chlenov, M., Sun, J.L., Olson, C.A., Weinman, O., Trester-Zedlitz, M.L., and Darst, S.A. (2002). Structure of the bacterial RNA polymerase promoter specificity sigma subunit. Mol. Cell *9*, 527–539.

Castillo, R.M., Mizuguchi, K., Dhanaraj, V., Albert, A., Blundell, T.L., and Murzin, A.G. (1999). A six-stranded double-psi beta barrel is shared by several protein superfamilies. Structure 7, 227–236.

Chlenov, M., Masuda, S., Murakami, K.S., Nikiforov, V., Darst, S.A., and Mustaev, A. (2005). Structure and function of lineage-specific sequence insertions in the bacterial RNA polymerase beta′ subunit. J. Mol. Biol. *353*, 138–154.

Chou, A.Y., Archdeacon, J., and Kado, C.I. (1998). Agrobacterium transcriptional regulator Ros is a prokaryotic zinc finger protein that regulates the plant oncogene ipt. Proc. Natl. Acad. Sci. U.S.A. *95*, 5293–5298.

Clark, K.L., Halay, E.D., Lai, E., and Burley, S.K. (1993). Co-crystal structure of the HNF-3/fork head DNA-recognition motif resembles histone H5. Nature *364*, 412–420.

Collado-Vides, J., Salgado, H., Morett, E., Gama-Castro, S., Jimenez-Jacinto, V., Martinez-Flores, I., Medina-Rivera, A., Muniz-Rascado, L., Peralta-Gil, M., and Santos-Zavaleta, A. (2009). Bioinformatics resources for the study of gene regulation in bacteria. J. Bacteriol. *191*, 23–31.

Cordes, M.H., Walsh, N.P., McKnight, C.J., and Sauer, R.T. (1999). Evolution of a protein fold *in vitro*. Science *284*, 325–328.

Cramer, P. (2002). Multisubunit RNA polymerases. Curr. Opin. Struct. Biol. *12*, 89–97.

Cramer, P., Bushnell, D.A., and Kornberg, R.D. (2001). Structural basis of transcription: RNA polymerase II at 2.8 angstrom resolution. Science 292, 1863–1876.

Doucleff, M., Malak, L.T., Pelton, J.G., and Wemmer, D.E. (2005). The C-terminal RpoN domain of sigma54 forms an unpredicted helix–turn–helix motif similar to domains of sigma70. J. Biol. Chem. *280*, 41530–41536.

Eddy, S.R. (2009). A new generation of homology search tools based on probabilistic inference. Genome Inform. 23, 205–211.

Esposito, S., Baglivo, I., Malgieri, G., Russo, L., Zaccaro, L., D'Andrea, L.D., Mammucari, M., Di Blasio, B., Isernia, C., Fattorusso, R., and Pedone, P.V. (2006). A novel type of zinc finger DNA binding domain in the *Agrobacterium tumefaciens* transcriptional regulator Ros. Biochemistry 45, 10394–10405.

Feklistov, A., and Darst, S.A. (2011). Structural basis for promoter-10 element recognition by the bacterial RNA polymerase sigma subunit. Cell *147*, 1257–1269.

Fromme, J.C., Banerjee, A., Huang, S.J., and Verdine, G.L. (2004). Structural basis for removal of adenine mispaired with 8-oxoguanine by MutY adenine DNA glycosylase. Nature 427, 652–656.

Fujikawa, N., Kurumizaka, H., Nureki, O., Terada, T., Shirouzu, M., Katayama, T., and Yokoyama, S. (2003). Structural basis of replication origin recognition by the DnaA protein. Nucleic Acids Res. *31*, 2077–2086.

Gomis-Ruth, F.X., Sola, M., Acebo, P., Parraga, A., Guasch, A., Eritja, R., Gonzalez, A., Espinosa, M., del Solar, G., and Coll, M. (1998). The structure of plasmid-encoded transcriptional repressor CopG unliganded and bound to its operator. EMBO J. *17*, 7404–7415.

Gottesman, S. (2004). The small RNA regulators of *Escherichia coli*: roles and mechanisms*. Annu. Rev. Microbiol. 58, 303–328.

Grinberg, I., Shteinberg, T., Gorovitz, B., Aharonowitz, Y., Cohen, G., and Borovok, I. (2006). The *Streptomyces*

NrdR transcriptional regulator is a Zn ribbon/ATP cone protein that binds to the promoter regions of class Ia and class II ribonucleotide reductase operons. J. Bacteriol. *188*, 7635–7644.

Gruber, T.M., and Gross, C.A. (2003). Multiple sigma subunits and the partitioning of bacterial transcription space. Annu. Rev. Microbiol. *57*, 441–466.

Guelzim, N., Bottani, S., Bourgine, P., and Kepes, F. (2002). Topological and causal structure of the yeast transcriptional regulatory network. Nat. Genet. *31*, 60–63.

Hantke, K. (2001). Iron and metal regulation in bacteria. Curr. Opin. Microbiol. *4*, 172–177.

Harrison, S.C. (1991). A structural taxonomy of DNA-binding domains. Nature *353*, 715–719.

Helmann, J.D. (2002). The extracytoplasmic function (ECF). sigma factors. Adv. Microb. Physiol. *46*, 47–110.

Hofmann, K., and Bucher, P. (1995). The FHA domain: a putative nuclear signalling domain found in protein kinases and transcription factors. Trends Biochem. Sci. *20*, 347–349.

Holtzendorff, J., Hung, D., Brende, P., Reisenauer, A., Viollier, P.H., McAdams, H.H., and Shapiro, L. (2004). Oscillating global regulators control the genetic circuit driving a bacterial cell cycle. Science *304*, 983–987.

Hong, E., Doucleff, M., and Wemmer, D.E. (2009). Structure of the RNA polymerase core-binding domain of sigma(54). reveals a likely conformational fracture point. J. Mol. Biol. *390*, 70–82.

Hudson, B.P., Quispe, J., Lara-Gonzalez, S., Kim, Y., Berman, H.M., Arnold, E., Ebright, R.H., and Lawson, C.L. (2009). Three-dimensional EM structure of an intact activator-dependent transcription initiation complex. Proc. Natl. Acad. Sci. U.S.A. *106*, 19830–19835.

Hulko, M., Lupas, A.N., and Martin, J. (2007). Inherent chaperone-like activity of aspartic proteases reveals a distant evolutionary relation to double-psi barrel domains of AAA-ATPases. Protein Sci. *16*, 644–653.

Itou, H., and Tanaka, I. (2001). The OmpR-family of proteins: insight into the tertiary structure and functions of two-component regulator proteins. J. Biochem. *129*, 343–350.

Iyer, L.M., and Aravind, L. (2011). Insights from the architecture of the bacterial transcription apparatus. J. Struct. Biol.

Iyer, L.M., Koonin, E.V., and Aravind, L. (2001). Adaptations of the helix-grip fold for ligand binding and catalysis in the START domain superfamily. Proteins 43, 134–144.

Iyer, L.M., Koonin, E.V., and Aravind, L. (2003). Evolutionary connection between the catalytic subunits of DNA-dependent RNA polymerases and eukaryotic RNA-dependent RNA polymerases and the origin of RNA polymerases. BMC Struct. Biol. *3*, 1.

Iyer, L.M., Koonin, E.V., and Aravind, L. (2004a). Evolution of bacterial RNA polymerase: implications for large-scale bacterial phylogeny, domain accretion, and horizontal gene transfer. Gene *335*, 73–88.

Iyer, L.M., Makarova, K.S., Koonin, E.V., and Aravind, L. (2004b). Comparative genomics of the FtsK-HerA superfamily of pumping ATPases: implications for the origins of chromosome segregation, cell division and viral capsid packaging. Nucleic Acids Res. *32*, 5260–5279.

Iyer, L.M., Koonin, E.V., Leipe, D.D., and Aravind, L. (2005). Origin and evolution of the archaeo-eukaryotic primase superfamily and related palm-domain proteins: structural insights and new members. Nucleic Acids Res. *33*, 3875–3896.

Iyer, L.M., Babu, M.M., and Aravind, L. (2006). The HIRAN domain and recruitment of chromatin remodeling and repair activities to damaged DNA. Cell Cycle *5*, 775–782.

Iyer, L.M., Anantharaman, V., Wolf, M.Y., and Aravind, L. (2008). Comparative genomics of transcription factors and chromatin proteins in parasitic protists and other eukaryotes. Int. J. Parasitol. *38*, 1–31.

Iyer, L.M., Abhiman, S. de Souza, R.F., and Aravind, L. (2010). Origin and evolution of peptide-modifying dioxygenases and identification of the wybutosine hydroxylase/hydroperoxidase. Nucleic Acids Res. *38*, 5261–5279.

Iyer, L.M., Abhiman, S., and Aravind, L. (2011). Natural history of eukaryotic DNA methylation systems. Prog. Mol. Biol. Transl. Sci. *101*, 25–104.

Jacob, F., and Monod, J. (1961). Genetic regulatory mechanisms in the synthesis of proteins. J. Mol. Biol. *3*, 318–356.

Ju, J., Mitchell, T., Peters, H., 3rd, and Haldenwang, W.G. (1999). Sigma factor displacement from RNA polymerase during *Bacillus subtilis* sporulation. J. Bacteriol. *181*, 4969–4977.

Juan Wu, L., and Errington, J. (2000). Identification and characterization of a new prespore-specific regulatory gene, rsfA, of *Bacillus subtilis*. J. Bacteriol. *182*, 418–424.

Kannan, N., Wu, J., Anand, G.S., Yooseph, S., Neuwald, A.F., Venter, J.C., and Taylor, S.S. (2007). Evolution of allostery in the cyclic nucleotide binding module. Genome Biol. *8*, R264.

Kearns, D.B., and Losick, R. (2005). Cell population heterogeneity during growth of *Bacillus subtilis*. Genes Dev. *19*, 3083–3094.

Keller, M., Roxlau, A., Weng, W.M., Schmidt, M., Quandt, J., Niehaus, K., Jording, D., Arnold, W., and Puhler, A. (1995). Molecular analysis of the Rhizobium meliloti mucR gene regulating the biosynthesis of the exopolysaccharides succinoglycan and galactoglucan. Mol. Plant Microbe Interact. *8*, 267–277.

Koonin, E.V., Makarova, K.S., and Aravind, L. (2001). Horizontal gene transfer in prokaryotes: quantification and classification. Annu. Rev. Microbiol. *55*, 709–742.

Korner, H., Sofia, H.J., and Zumft, W.G. (2003). Phylogeny of the bacterial superfamily of Crp-Fnr transcription regulators: exploiting the metabolic spectrum by controlling alternative gene programs. FEMS Microbiol. Rev. *27*, 559–592.

Kostrewa, D., Zeller, M.E., Armache, K.J., Seizl, M., Leike, K., Thomm, M., and Cramer, P. (2009). RNA polymerase II-TFIIB structure and mechanism of transcription initiation. Nature *462*, 323–330.

Krishna, S.S., Majumdar, I., and Grishin, N.V. (2003). Structural classification of zinc fingers: survey and summary. Nucleic Acids Res. *31*, 532–550.

Kuznedelov, K., Minakhin, L., Niedziela-Majka, A., Dove, S.L., Rogulja, D., Nickels, B.E., Hochschild, A., Heyduk, T., and Severinov, K. (2002). A role for interaction of the RNA polymerase flap domain with the sigma subunit in promoter recognition. Science *295*, 855–857.

Lamour, V., Rutherford, S.T., Kuznedelov, K., Ramagopal, U.A., Gourse, R.L., Severinov, K., and Darst, S.A. (2008). Crystal structure of *Escherichia coli* Rnk, a new RNA polymerase-interacting protein. J. Mol. Biol. *383*, 367–379.

Lane, W.J., and Darst, S.A. (2010a). Molecular evolution of multisubunit RNA polymerases: sequence analysis. J. Mol. Biol. *395*, 671–685.

Lane, W.J., and Darst, S.A. (2010b). Molecular evolution of multisubunit RNA polymerases: structural analysis. J. Mol. Biol. *395*, 686–704.

Larquet, E., Schreiber, V., Boisset, N., and Richet, E. (2004). Oligomeric assemblies of the *Escherichia coli* MalT transcriptional activator revealed by cryo-electron microscopy and image processing. J. Mol. Biol. *343*, 1159–1169.

Latchman, D.S. (1997). Transcription factors: an overview. Int. J. Biochem. Cell Biol. *29*, 1305–1312.

Lee, I.M., Zhao, Y., and Bottner, K.D. (2005). Novel insertion sequence-like elements in phytoplasma strains of the aster yellows group are putative new members of the IS3 family. FEMS Microbiol. Lett. *242*, 353–360.

Lee, P.C., Umeyama, T., and Horinouchi, S. (2002). afsS is a target of AfsR, a transcriptional factor with ATPase activity that globally controls secondary metabolism in *Streptomyces coelicolor* A3(2). Mol. Microbiol. *43*, 1413–1430.

Leipe, D.D., Koonin, E.V., and Aravind, L. (2004). STAND, a class of P-loop NTPases including animal and plant regulators of programmed cell death: multiple, complex domain architectures, unusual phyletic patterns, and evolution by horizontal gene transfer. J. Mol. Biol. *343*, 1–28.

Lopez de Saro, F.J., Yoshikawa, N., and Helmann, J.D. (1999). Expression, abundance, and RNA polymerase binding properties of the delta factor of *Bacillus subtilis*. J. Biol. Chem. *274*, 15953–15958.

McCutcheon, J.P., McDonald, B.R., and Moran, N.A. (2009). Convergent evolution of metabolic roles in bacterial co-symbionts of insects. Proc. Natl. Acad. Sci. U.S.A. *106*, 15394–15399.

Madan Babu, M., Teichmann, S.A., and Aravind, L. (2006). Evolutionary dynamics of prokaryotic transcriptional regulatory networks. J. Mol. Biol. *358*, 614–633.

Madan Babu, M., Balaji, S., and Aravind, L. (2007). General trends in the evolution of prokaryotic transcriptional regulatory networks. Genome Dyn. *3*, 66–80.

Mah, T.F., Kuznedelov, K., Mushegian, A., Severinov, K., and Greenblatt, J. (2000). The alpha subunit of *E. coli* RNA polymerase activates RNA binding by NusA. Genes Dev. *14*, 2664–2675.

Mani, N., and Dupuy, B. (2001). Regulation of toxin synthesis in *Clostridium difficile* by an alternative RNA polymerase sigma factor. Proc. Natl. Acad. Sci. U.S.A. *98*, 5844–5849.

Marquenet, E., and Richet, E. (2010). Conserved motifs involved in ATP hydrolysis by MalT, a signal transduction ATPase with numerous domains from *Escherichia coli*. J. Bacteriol. *192*, 5181–5191.

Mascarenhas, J., Soppa, J., Strunnikov, A.V., and Graumann, P.L. (2002). Cell cycle-dependent localization of two novel prokaryotic chromosome segregation and condensation proteins in *Bacillus subtilis* that interact with SMC protein. EMBO J. *21*, 3108–3118.

Mathew, R., and Chatterji, D. (2006). The evolving story of the omega subunit of bacterial RNA polymerase. Trends Microbiol. *14*, 450–455.

Messer, W., and Weigel, C. (2003). DnaA as a transcription regulator. Methods Enzymol. *370*, 338–349.

Minakhin, L., Bhagat, S., Brunning, A., Campbell, E.A., Darst, S.A., Ebright, R.H., and Severinov, K. (2001). Bacterial RNA polymerase subunit omega and eukaryotic RNA polymerase subunit RPB6 are sequence, structural, and functional homologs and promote RNA polymerase assembly. Proc. Natl. Acad. Sci. U.S.A. *98*, 892–897.

Molle, V., Kremer, L., Girard-Blanc, C., Besra, G.S., Cozzone, A.J., and Prost, J.F. (2003). An FHA phosphoprotein recognition domain mediates protein EmbR phosphorylation by PknH, a Ser/Thr protein kinase from *Mycobacterium tuberculosis*. Biochemistry *42*, 15300–15309.

Mooney, R.A., Darst, S.A., and Landick, R. (2005). Sigma and RNA polymerase: an on-again, off-again relationship? Mol. Cell *20*, 335–345.

Morett, E., and Bork, P. (1998). Evolution of new protein function: recombinational enhancer Fis originated by horizontal gene transfer from the transcriptional regulator NtrC. FEBS Lett. *433*, 108–112.

Motackova, V., Sanderova, H., Zidek, L., Novacek, J., Padrta, P., Svenkova, A., Korelusova, J., Jonak, J., Krasny, L., and Sklenar, V. (2010). Solution structure of the N-terminal domain of *Bacillus subtilis* delta subunit of RNA polymerase and its classification based on structural homologs. Proteins *78*, 1807–1810.

Murakami, K.S., Masuda, S., Campbell, E.A., Muzzin, O., and Darst, S.A. (2002). Structural basis of transcription initiation: an RNA polymerase holoenzyme–DNA complex. Science *296*, 1285–1290.

Nechaev, S., and Severinov, K. (2008). RapA: completing the transcription cycle? Structure *16*, 1294–1295.

Nikolov, D.B., Chen, H., Halay, E.D., Usheva, A.A., Hisatake, K., Lee, D.K., Roeder, R.G., and Burley, S.K. (1995). Crystal structure of a TFIIB-TBP-TATA-element ternary complex. Nature *377*, 119–128.

van Nimwegen, E. (2003). Scaling laws in the functional content of genomes. Trends Genet. *19*, 479–484.

Opalka, N., Brown, J., Lane, W.J., Twist, K.A., Landick, R., Asturias, F.J., and Darst, S.A. (2010). Complete structural model of *Escherichia coli* RNA polymerase from a hybrid approach. PLoS Biol. *8*, e1000483.

Paget, M.S., and Helmann, J.D. (2003). The sigma70 family of sigma factors. Genome Biol. *4*, 203.

Paget, M.S., Kang, J.G., Roe, J.H., and Buttner, M.J. (1998). sigmaR, an RNA polymerase sigma factor that modulates expression of the thioredoxin system in response to oxidative stress in *Streptomyces coelicolor* A3(2). EMBO J. *17*, 5776–5782.

Paget, M.S., Leibovitz, E., and Buttner, M.J. (1999). A putative two-component signal transduction system regulates sigmaE, a sigma factor required for normal cell wall integrity in *Streptomyces coelicolor* A3(2). Mol. Microbiol. *33*, 97–107.

Pao, G.M., Saier, M.H., Jr. (1995). Response regulators of bacterial signal transduction systems: selective domain shuffling during evolution. J. Mol. Evol. *40*, 136–154.

Peat, T.S., Frank, E.G., McDonald, J.P., Levine, A.S., Woodgate, R., and Hendrickson, W.A. (1996). Structure of the UmuD′ protein and its regulation in response to DNA damage. Nature *380*, 727–730.

Penalver-Mellado, M., Garcia-Heras, F., Padmanabhan, S., Garcia-Moreno, D., Murillo, F.J., and Elias-Arnanz, M. (2006). Recruitment of a novel zinc-bound transcriptional factor by a bacterial HMGA-type protein is required for regulating multiple processes in *Myxococcus xanthus*. Mol. Microbiol. *61*, 910–926.

Pineda, M., Gregory, B.D., Szczypinski, B., Baxter, K.R., Hochschild, A., Miller, E.S., and Hinton, D.M. (2004). A family of anti-sigma70 proteins in T4-type phages and bacteria that are similar to AsiA, a Transcription inhibitor and co-activator of bacteriophage T4. J. Mol. Biol. *344*, 1183–1197.

Poon, K.K., Chu, J.C., and Wong, S.L. (2001). Roles of glucitol in the GutR-mediated transcription activation process in *Bacillus subtilis*: glucitol induces GutR to change its conformation and to bind ATP. J. Biol. Chem. *276*, 29819–29825.

Potrykus, K., and Cashel, M. (2008). (p)ppGpp: still magical? Annu. Rev. Microbiol. *62*, 35–51.

Ptashne, M. (2004). A Genetic Switch: Phage Lambda Revisited, 3rd edn (Cold Spring Harbor Laboratory Press, Cold Spring Harbor, NY).

Rombel, I., North, A., Hwang, I., Wyman, C., and Kustu, S. (1998). The bacterial enhancer-binding protein NtrC as a molecular machine. Cold Spring Harb. Symp. Quant. Biol. *63*, 157–166.

Ruprich-Robert, G., and Thuriaux, P. (2010). Non-canonical DNA transcription enzymes and the conservation of two-barrel RNA polymerases. Nucleic Acids Res. *38*, 4559–4569.

Salgado, P.S., Koivunen, M.R., Makeyev, E.V., Bamford, D.H., Stuart, D.I., and Grimes, J.M. (2006). The structure of an RNAi polymerase links RNA silencing and transcription. PLoS Biol. *4*, e434.

Savijoki, K., Ingmer, H., Frees, D., Vogensen, F.K., Palva, A., and Varmanen, P. (2003). Heat and DNA damage induction of the LexA-like regulator HdiR from *Lactococcus lactis* is mediated by RecA and ClpP. Mol. Microbiol. *50*, 609–621.

Shaw, G., Gan, J., Zhou, Y.N., Zhi, H., Subburaman, P., Zhang, R., Joachimiak, A., Jin, D.J., and Ji, X. (2008). Structure of RapA, a Swi2/Snf2 protein that recycles RNA polymerase during transcription. Structure *16*, 1417–1427.

Shen-Orr, S.S., Milo, R., Mangan, S., and Alon, U. (2002). Network motifs in the transcriptional regulation network of *Escherichia coli*. Nat. Genet. *31*, 64–68.

Singh, S.K., Kurnasov, O.V., Chen, B., Robinson, H., Grishin, N.V., Osterman, A.L., and Zhang, H. (2002). Crystal structure of Haemophilus influenzae NadR protein. A bifunctional enzyme endowed with NMN adenyltransferase and ribosylnicotinimide kinase activities. J. Biol. Chem. *277*, 33291–33299.

Smeets, L.C., Becker, S.C., Barcak, G.J., Vandenbroucke-Grauls, C.M., Bitter, W., and Goosen, N. (2006). Functional characterization of the competence protein DprA/Smf in *Escherichia coli*. FEMS Microbiol. Lett. *263*, 223–228.

Soppa, J., Kobayashi, K., Noirot-Gros, M.F., Oesterhelt, D., Ehrlich, S.D., Dervyn, E., Ogasawara, N., and Moriya, S. (2002). Discovery of two novel families of proteins that are proposed to interact with prokaryotic SMC proteins, and characterization of the *Bacillus subtilis* family members ScpA and ScpB. Mol. Microbiol. *45*, 59–71.

Stragier, P., and Losick, R. (1996). Molecular genetics of sporulation in *Bacillus subtilis*. Annu. Rev. Genet. *30*, 297–241.

Stulke, J., Arnaud, M., Rapoport, G., and Martin-Verstraete, I. (1998). PRD – a protein domain involved in PTS-dependent induction and carbon catabolite repression of catabolic operons in bacteria. Mol. Microbiol. *28*, 865–874.

Subramanian, G., Koonin, E.V., and Aravind, L. (2000). Comparative genome analysis of the pathogenic spirochetes *Borrelia burgdorferi* and *Treponema pallidum*. Infect. Immun. *68*, 1633–1648.

Swindells, M.B. (1995). Identification of a common fold in the replication terminator protein suggests a possible mode for DNA binding. Trends Biochem. Sci. *20*, 300–302.

Tam, R., and Saier, M.H., Jr. (1993). Structural, functional, and evolutionary relationships among extracellular solute-binding receptors of bacteria. Microbiol. Rev. *57*, 320–346.

Thieffry, D., Huerta, A.M., Perez-Rueda, E., and Collado-Vides, J. (1998). From specific gene regulation to genomic networks: a global analysis of transcriptional regulation in *Escherichia coli*. Bioessays *20*, 433–440.

Tobisch, S., Stulke, J., and Hecker, M. (1999). Regulation of the lic operon of *Bacillus subtilis* and characterization of potential phosphorylation sites of the LicR regulator protein by site-directed mutagenesis. J. Bacteriol. *181*, 4995–5003.

Toulokhonov, I., Artsimovitch, I., and Landick, R. (2001). Allosteric control of RNA polymerase by a site that contacts nascent RNA hairpins. Science *292*, 730–733.

Tyrrell, R., Verschueren, K.H., Dodson, E.J., Murshudov, G.N., Addy, C., and Wilkinson, A.J. (1997). The structure of the cofactor-binding fragment of the LysR family member, CysB: a familiar fold with a surprising subunit arrangement. Structure 5, 1017–1032.

Ulrich, L.E., and Zhulin, I.B. (2007). MiST: a microbial signal transduction database. Nucleic Acids Res. *35*, D386–D390.

Vartak, N.B., Reizer, J., Reizer, A., Gripp, J.T., Groisman, E.A., Wu, L.F., Tomich, J.M., and Saier, M.H., Jr. (1991). Sequence and evolution of the FruR protein of *Salmonella typhimurium*: a pleiotropic transcriptional regulatory protein possessing both activator and repressor functions which is homologous to the periplasmic ribose-binding protein. Res. Microbiol. *142*, 951–963.

Vassylyev, D.G., Sekine, S., Laptenko, O., Lee, J., Vassylyeva, M.N., Borukhov, S., and Yokoyama, S. (2002). Crystal structure of a bacterial RNA polymerase holoenzyme at 2.6 A resolution. Nature *417*, 712–719.

Vassylyev, D.G., Vassylyeva, M.N., Perederina, A., Tahirov, T.H., and Artsimovitch, I. (2007). Structural basis for transcription elongation by bacterial RNA polymerase. Nature *448*, 157–162.

Wang, Y., Zhao, S., Somerville, R.L., and Jardetzky, O. (2001). Solution structure of the DNA-binding domain of the TyrR protein of *Haemophilus influenzae*. Protein Sci. *10*, 592–598.

Wassarman, K.M. (2007). 6S RNA: a regulator of transcription. Mol. Microbiol. 65, 1425–1431.

Watson, J.D. (2004). Molecular Biology of the Gene, 5th edn (Pearson/Benjamin Cummings, San Francisco, CA, and Cold Spring Harbour Laboratory Press, Woodbury, NY).

Weekes, D., Miller, M.D., Krishna, S.S., McMullan, D., McPhillips, T.M., Acosta, C., Canaves, J.M., Elsliger, M.A., Floyd, R., and Grzechnik, S.K. (2007). Crystal structure of a transcription regulator (TM1602). from Thermotoga maritima at 2.3 A resolution. Proteins *67*, 247–252.

Werner, F., and Grohmann, D. (2011). Evolution of multisubunit RNA polymerases in the three domains of life. Nat. Rev. Microbiol. 9, 85–98.

West, A.H., and Stock, A.M. (2001). Histidine kinases and response regulator proteins in two-component signaling systems. Trends Biochem. Sci. 26, 369–376.

Westblade, L.F., Campbell, E.A., Pukhrambam, C., Padovan, J.C., Nickels, B.E., Lamour, V., and Darst, S.A. (2010). Structural basis for the bacterial transcription-repair coupling factor/RNA polymerase interaction. Nucleic Acids Res. *38*, 8357–8369.

Westover, K.D., Bushnell, D.A., and Kornberg, R.D. (2004). Structural basis of transcription: separation of RNA from DNA by RNA polymerase II. Science *303*, 1014–1016.

Wigneshweraraj, S., Bose, D., Burrows, P.C., Joly, N., Schumacher, J., Rappas, M., Pape, T., Zhang, X., Stockley, P., Severinov, K., and Buck, M. (2008). Modus operandi of the bacterial RNA polymerase containing the sigma54 promoter-specificity factor. Mol. Microbiol. *68*, 538–546.

Wigneshweraraj, S.R., Kuznedelov, K., Severinov, K., and Buck, M. (2003). Multiple roles of the RNA polymerase beta subunit flap domain in sigma 54-dependent transcription. J. Biol. Chem. *278*, 3455–3465.

Willkomm, D.K., and Hartmann, R.K. (2005). 6S RNA – an ancient regulator of bacterial RNA polymerase rediscovered. Biol. Chem. *386*, 1273–1277.

Wilson, K.P., Shewchuk, L.M., Brennan, R.G., Otsuka, A.J., and Matthews, B.W. (1992). *Escherichia coli* biotin holoenzyme synthetase/bio repressor crystal structure delineates the biotin- and DNA-binding domains. Proc. Natl. Acad. Sci. U.S.A. *89*, 9257–9261.

Wojciak, J.M., Iwahara, J., and Clubb, R.T. (2001). The Mu repressor–DNA complex contains an immobilized 'wing' within the minor groove. Nat. Struct. Biol. *8*, 84–90.

Wood, H.E., Devine, K.M., and McConnell, D.J. (1990). Characterisation of a repressor gene (xre). and a temperature-sensitive allele from the *Bacillus subtilis* prophage, PBSX. Gene *96*, 83–88.

Yuan, A.H., Nickels, B.E., and Hochschild, A. (2009). The bacteriophage T4 AsiA protein contacts the beta-flap domain of RNA polymerase. Proc. Natl. Acad. Sci. U.S.A. *106*, 6597–6602.

Zhang, X., Chaney, M., Wigneshweraraj, S.R., Schumacher, J., Bordes, P., Cannon, W., and Buck, M. (2002). Mechanochemical ATPases and transcriptional activation. Mol. Microbiol. *45*, 895–903.

DNA Structure and Bacterial Nucleoid-associated Proteins

2

Georgi Muskhelishvili and Andrew Travers

Abstract

In the bacterial nucleoid different configurations of negatively supercoiled DNA are constrained by different NAPs. Thus while H-NS can constrain, by bridging, the interwindings of plectonemic structure, HU induces a left-handed coiled configuration while FIS can bind within DNA loops. The topological and dynamic interconvertibility of these structures contributes substantially to the regulation of gene expression in bacteria. We review here some of the mechanisms involved and argue that they form the basis for a coordination of gene transcription which results in the establishment of a single interconnected heterarchical control system that is responsible both for maintaining, when appropriate, a homoeostatic control of growth and also for mediating transitions between different cellular physiological states.

Introduction

The genome of *E. coli* consists of a single covalently closed circular DNA molecule, which has to be compacted by about one thousand-fold to fit inside the cell (Travers and Muskhelishvili, 2005a). Despite this dramatic packaging of the molecule, regulatory sites have to be kept in a state remaining accessible to the transcription machinery in order to govern the expression of genes involved in cellular growth and self-reproduction. This double need is satisfied by utilizing the superhelical energy of DNA as a major force driving the compaction of genome in three-dimensional space. The level of DNA supercoiling is homeostatically regulated by DNA topoisomerases (Menzel and Gellert, 1983), but the overall DNA superhelicity varies with the physiological state of the cell (Balke and Gralla, 1987; van Workum *et al.*, 1996; Snoep *et al.*, 2002). The highly dynamic free DNA supercoils are constrained in *E. coli* and other bacteria by nucleoid-associated proteins (NAPs), counterparts of eukaryotic histones. These abundant proteins are expressed in a growth phase-dependent manner (Azam *et al.*, 1999) potentially covering most of the genomic DNA and acting both as modulators of topoisomerase activity and also as global regulators of transcription (McLeod and Johnson 2001; Dorman and Deighan 2003; Travers and Muskhelishvili, 2005b). Importantly, some of the major NAPs bind DNA ubiquitously, while others exhibit a less random distribution in the genome with preferred sequence motifs (normally AT-rich) for binding (Betermier *et al.*, 1994; Azam and Ishihama, 1999; Azam *et al.*, 2000; Ussery *et al.*, 2001; Lang *et al.*, 2007). Nevertheless almost all the NAPs – although to different extents – modulate the supercoil dynamics of DNA and stabilize different three-dimensional structures on cooperative binding at multiple DNA sites (Broyles and Pettijohn, 1986; Schneider *et al.*, 2001; Dame, 2005; Maurer *et al.*, 2009).

NAPs

In both the bacterial nucleoid and the eukaryotic nucleus DNA is compacted by packaging by abundant DNA-binding proteins. In the nucleus the histones, in the form of the histone octamer and the linker histones, are by far the dominant species. In contrast, in the bacterial nucleoid a wide variety of such proteins are present. The most widely distributed are members of the HU, LRP and H-NS families while others, for example FIS, are more restricted in their occurrence (Luijsterburg *et al.*, 2008; Browning *et al.*, 2010; Rimsky and Travers 2011). Most, although not all, of these proteins bend DNA substantially and consequentially bind with higher affinity to pre-bent DNA. They thus act as architectural proteins, in the sense that they stabilize a particular configuration of the trajectory of the DNA double helix. A corollary of this function is that although certain proteins, such as H-NS and FIS, can bind to particular DNA sequences with high affinity *in vitro* (Lazarus *et al.*, 1993; Bouffartigues *et al.*, 2007; Lang *et al.*, 2007), and because they predominantly recognize local DNA conformations rather than the DNA bases per se, NAPs generally exhibit a wide and quasi-continuous range of DNA sequence-dependent affinity. Another property, recognized for eukaryotic histones (Travers *et al.*, 2010), is that the interaction between the DNA and protein is potentially 'tunable' in the sense that precise binding preferences are not fixed but instead respond differently to changes in the immediate environment of the DNA–protein interaction.

The ordering of DNA structure by nucleoid-associated proteins can be considered to act at several levels: short range determination of local conformations (twisting and untwisting or DNA bending over short distances), medium range determination of more extensive configurations (stabilization of particular supercoil-dependent structures) and finally long range organization of large domains. It is the synthesis of this ordering that is a crucial determinant of the overall pattern of gene regulation in bacteria.

In the local context of a promoter certain NAPs, including FIS and CRP may also interact directly with RNA polymerase (RNAP) (Bokal *et al.*, 1995; Muskhelishvili *et al.*, 1995; Busby and Ebright, 1999) and thus, in this situation, act formally as transcription factors. In other situations, for example, at the origin of replication *OriC*, NAPs may act simply as local DNA bending proteins (Skarstad *et al.*, 1990; Roth *et al.*, 1994; Ryan *et al.*, 2004). In their primary role the cellular abundances of different NAPs correlate closely with the overall level of negative superhelicity (Travers and Muskhelishvili, 2005b; Muskhelishvili and Travers, 2009). In *Escherichia coli* FIS and the HUα dimer are most abundant when the overall superhelical density is high, while others including Dps, IHF and the HUβ dimer are favoured by lower superhelical density. The abundances of yet others, including H-NS, StpA and CRP are less sensitive to superhelicity. Not only do the abundances of NAPs correlate with in vivo supercoiling levels but also NAPs, with the exceptions of Dps and IHF, constrain negatively supercoiled DNA. The properties of the *E. coli* NAPs are summarized in Table 2.1.

DNA supercoiling

DNA supercoiling profoundly influences the structure of the polymer. With the exception of some hyperthermophilic microorganisms, nuclear and nucleoid DNA is negatively supercoiled – that is, it is unwound relative to its unconstrained state. This unwinding is manifest in two ways – as an enhanced rate of local transient untwisting culminating in strand separation, especially at TpA steps (Drew and Travers, 1985) and, a coiling of the trajectory of the DNA double helix. Within a closed topological domain these two properties are linked such that the sum of the untwisting and coiling, termed the linking number deficit, is, in the absence of topoisomerization, a constant. In more formal terms:

$$\Delta L = \Delta Tw + Wr \qquad (2.1)$$

where T*w* is the twist (approximately the number of double helical turns per unit length) and *Wr* the writhe (approximately the coiling) of the DNA within the domain. Otherwise expressed, an untwisting of the DNA is compensated by less negative writhe and vice versa.

The repartitioning of the superhelicity enables, in principle, the probability of local melting at particular sequences to be finely tuned. For promoters this is particularly important since the −10 hexamer sequences at which DNA untwisting is initiated (Auner *et al.*, 2003) often have the lowest melting energy of any sequence within the promoter region (Fig. 2.1). Melting at these sites

Table 2.1 *E. coli* NAPs

NAP	Max. abundance	DNA bending	Constraint of negative superhelicity[1]	Probable function	Eukaryotic counterpart[2]
FIS	Early exp	60–90°/dimer	+	Bend and loop stabilization	Histone octamer
CRP	?	60–90°/dimer	+	Bend and loop stabilization	
LRP	Exp/stat transition	~55°/dimer	+	Bend (?) and loop stabilization	
$HU\alpha_2$	Early exp	~60°/dimer[3]	++	Untwisting, left-hand toroidal bending	HMGB proteins
$HU\alpha\beta$	Mid exp → stat		++		
$HU\beta_2$	Stat		None observed		
IHF	Mid exp → stat	~160°/dimer	None observed		
StpA	Mid → late exp	Bends, angle ND	++	Bridging	
H-NS (aka H1)	Stat (marginally)	Bends, angle ND	++	Bridging, stabilizes plectonemes	Linker histone
Dps	Stat	Not observed	Not observed		?

[1]+and ++ indicates constraint of low and high superhelical densities respectively. Direct experimental comparisons are not available

[2]Eukaryotic counterparts indicate conservation of function and may apply to more than one *E. coli* NAP. For example the histone octamer stabilizes a toroidal loop, like FIS, while the linker histone bridges two DNA duplexes, like H-NS.

[3]The bend angle for HU is taken from a determination for an HU homodimer isolated from *Anabaena* (Swinger *et al*. 2003).

ND, not determined.

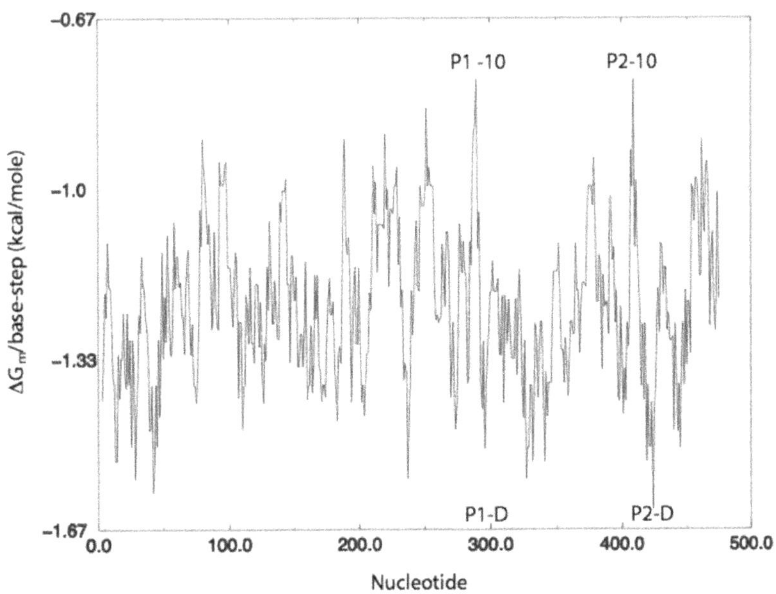

Figure 2.1 Melting energy for the DNA sequence at the *rrnA* P1 and P2 promoters. The DNA sequences was scanned with moving six base-step window using the calculated energies for the melting of individual base-steps taken from Protozanova *et al*. (2004). The −10 regions of the *rrnA* P1 and P2 promoters are indicated (P1–10 and P2 −10) as also are the corresponding discriminator regions (p1-D and p2-D). Nucleotide 1 corresponds to *E. coli* coordinate 4032488. The melting energy is averaged across six base steps.

is thought be coupled to the superhelical configuration of the flanking DNA regions such that removal of a negative node from these regions may compensate for the formation of a transcription bubble of 10–12 bp (Travers and Muskhelishvili, 2007).

Modulation of DNA supercoiling by NAPs

NAPs can play a crucial role in regulating promoter opening. Negatively supercoiled DNA adopts two principal configurations, a plectonemic or interwound form and a toroidal form. At a constant superhelical density these two forms are interconvertible (Fig. 2.2A) and this transition itself will, in most cases, repartition writhe and twist. *In vitro* at DNA concentrations substantially lower than occur in vivo, the toroidal form is metastable to the plectonemic form (Boles *et al.*, 1990). However *in vitro* HU can both untwist and confer a toroidal configuration on DNA (Swinger *et al.*, 2003; Guo and Adhya, 2007) while H-NS, by virtue of its ability to bridge between two DNA duplexes (Dame *et al.*, 2000; Schneider *et al.*, 2001) stabilizes a plectonemic configuration (Lang *et al.*, 2007; Maurer *et al.*, 2009). H-NS generally acts as a transcriptional repressor (Dorman, 2004; Dame, 2005; Fang and Rimsky, 2008) while HU function has been correlated with the activation of expression from some promoters, e.g. proU (Oberto *et al.*, 2009; Berger *et al.*, 2010) and also the initiation of DNA replication (Hwang and Kornberg, 1992). Both of these processes require strand separation. Interconversion of plectonemic and toroidal supercoils by competition between the binding NAPs (Fig. 2.2B) can thus both, sustain compaction and modulate the accessibility of DNA supercoils to transcription machinery (Broyles and Pettijohn, 1986; Travers and Muskhelishvili, 2007; Maurer *et al.*, 2009; Muskhelishvili and Travers, 2009).

At many of the most active promoters in *Escherichia coli* the DNA sequences immediately upstream are either intrinsically bent (Ussery *et al.*, 2001) or can be tightly bent by NAPs such as CRP or FIS acting as transcription factors. In a supercoiled DNA the loop formed by this bending could be located either at the base of a plectoneme or within a toroidal coil. However, by linking the two DNA duplexes in a plectoneme H-NS has the potential to interfere with the coupling of promoter opening to the loss of a negative superhelical node. Indeed at the proU promoter H-NS antagonizes promoter opening by RNA polymerase (Tupper *et al.*, 1994). Similarly LRP,

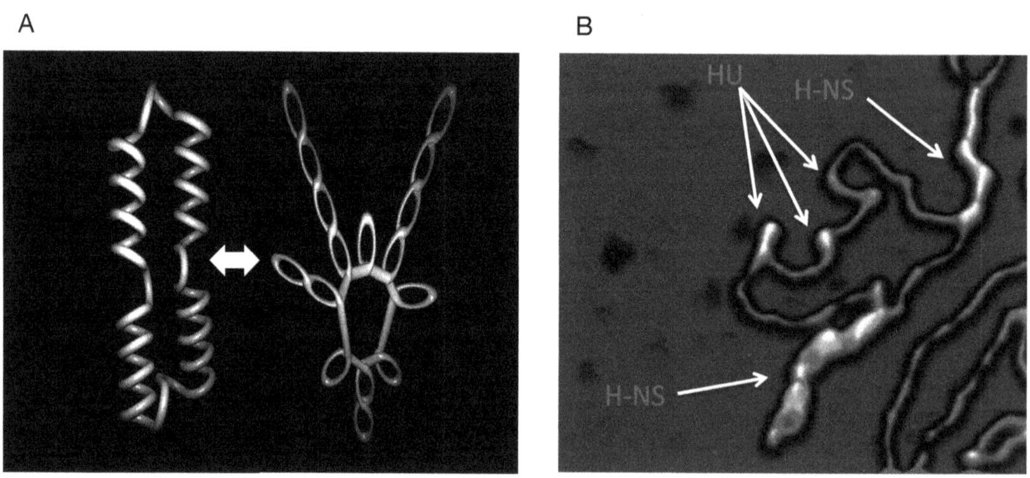

Figure 2.2 Regulation of supercoil forms by competition between NAPs. (A) Model of the interconversion of toroidal and plectonemic supercoils. For clarity the NAPs are omitted from the drawing. (B) Atomic force microscopy image of adjacent alternative structures stabilized by H-NS and HU on binding linear lambda DNA. The structures stabilized by H-NS and HU are thought to correspond to plectonemic and toroidal coiling of the DNA, respectively (Maurer *et al.*, 2009).

which also acts as a repressor, wraps negatively supercoiled DNA as a loop and could antagonise access of RNA polymerase (Pul *et al.*, 2008). Were HU to displace H-NS a more open structure would be created which would not be subject to these constraints.

Although topological transitions between plectonemic and toroidal forms of negatively superhelical DNA are possible another important variable *in vivo* is the superhelical density itself. In *E. coli* this varies with growth rate such that the highest densities occur during early exponential phase (Balke and Gralla, 1987; Bordes *et al.*, 2003). Changes in superhelical density have structural consequences for DNA. At low superhelical densities the configuration of the DNA would be less coiled than at high superhelical densities and the differential energy required for localized strand separation, e.g. at gene promoters, would be greater. This implies that under these conditions DNA loops, in the absence of constraining proteins, would be larger and consequently less likely to loop around RNA polymerase (Maurer *et al.*, 2006). However, when a bend can be induced by a NAP such as FIS posing as a transcription factor, promoter activity is then buffered against abrupt changes in superhelical density (Rochman *et al.*, 2002, 2004; Auner *et al.*, 2003). These changes in the differential energy for promoter opening might be compensated for by changes in the transcription apparatus itself – for example, by substitution of σ^S for σ^{70}.

The topological organization of the bacterial nucleoid

The DNA of the bacterial nucleoid is highly condensed. To achieve this, the DNA must not only be organized at the local level but also must adopt a higher-level ordering to enable tight packaging. The recent identification of a small number of macrodomains, extensive discrete regions of the *E. coli* chromosome up to ~1.5 Mb long is consistent with this notion (Valens *et al.*, 2004). In these macrodomains genes that are separated by substantial lengths of chromosomal DNA are in spatial proximity. Thus in the rrn 'macrodomain' which incorporates three of the macrodomains characterized by Valens *et al.* (2004) *OriC* and the *rrn* transcription units are located in very close reach to each other within the overall structure of the chromosome even though the domain itself extends for > 2 Mb (Berger *et al.*, 2010). The organization of the macrodomain associated with the replication terminus requires a sequence-specific DNA binding protein MatP (Mercier *et al.*, 2008). While other such proteins may be involved in the specification of other macrodomains, they may also require the participation of NAPs (Dame *et al.*, 2011).

The low-resolution structure of bacterial chromosome is now becoming apparent. Two independent studies conclude that the highest order of folding of the chromosome is best described as a simple plectoneme. This configuration was deduced both from 5C analysis of the *Caulobacter crescentus* chromosome (Umbarger *et al.*, 2011) and from functional interconnections within the *Escherichia coli* chromosome (Sobetzho *et al.*, 2012). Owing to symmetry considerations the sense of coiling of the plectoneme could not be established in either study but because overall the chromosome is negatively supercoiled it seems likely that the coiling is right-handed. The organization of the individual interwound arms of the chromosome is less well established. Junier *et al.* (2012) show that the periodic occurrence of functionally related genes in the *E. coli* chromosome can be related to a coiled structure, either toroidal or plectonemic, while Wiggins *et al.* (2010) favour short plectonemic loops.

Integration of chromatin structure and metabolic regulation

Since in growing bacterial cells both the NAP composition and the overall superhelicity of DNA depend on growth conditions it is thought that dynamic constraint of DNA supercoils by the NAPs and the resultant topological differentiation of the chromosome serve as a means of coordinating genomic expression with physiological state (Travers and Muskhelishvili, 2005a). Indeed, previous observations of variation of the superhelical density of the genomic DNA with physiological state (Balke and Gralla, 1987; Hsieh *et al.*, 1991; van Workum *et al.*, 1996; Snoep *et al.*, 2002), were supported by later data suggesting that in *E. coli* the changes of metabolism and nucleoid structure are tightly coupled (Blot *et al.*, 2006; Sonnenschein *et al.*, 2011). More specifically, the biosynthetic genes are found preferentially transcribed under conditions of high negative superhelicity, whereas transcription of catabolic genes involved in production of energy equivalents is activated under conditions of DNA relaxation (Blot *et al.*, 2006). Similarly, the dependence of circadian clocks in Cyanobacteria on varying levels of negative superhelicity during the diurnal cycle is again consistent with varying energy inputs, with levels of superhelicity maximal at the subjective dusk when biosynthetic genes, for example ribosomal proteins, are activated and lowest at the subjective dawn when the genes for the photosynthetic machinery are activated (Vijayan *et al.*, 2009). Importantly, a direct connection between the central metabolism and chromatin structure appears to be a general phenomenon implicated also in regulating mammalian gene expression (Ladburner, 2009).

Cooperation between RNAP and the NAPs

During cellular growth the composition of RNAP holoenzyme changes coordinately with relative abundance of the NAPs (Ishihama, 2000; Muskhelishvili and Travers, 2009) and overall superhelicity of the DNA (Balke and Gralla, 1987). Unsurprisingly, different holoenzymes have distinct supercoiling preferences for transcription (Kusano *et al.*, 1996; Bordes *et al.*, 2003), such that $E\sigma^{70}$ is the major form during fast growth, whereas $E\sigma^{S}$ is required under conditions of stationary growth and environmental stress. Nevertheless, recent studies suggest that these holoenzymes have overlapping functions such that different effective compositions of the holoenzyme can be adapted to sustain optimal growth (Weber *et al.*, 2005; Geertz *et al.*, 2011). Furthermore, mutations of NAPs and alterations of the RNAP holoenzyme composition were both found to affect the expression of genes encoding dedicated transcription factors (Blot *et al.*, 2006), thus fine-tuning the composition of regulatory nucleoprotein complexes depending on the favoured superhelical density of the DNA (Muskhelishvili and Travers, 2009; Muskhelishvili *et al.*, 2010; Geertz *et al.*, 2011). This notion is consistent with selective spatial binding of distinct chromatin proteins and organization of topological domains and transcription units of variable size in the genome (Browning *et al.*, 2004; Grainger *et al.*, 2006; Marr *et al.*, 2008; Cho *et al.*, 2009; Janga *et al.*, 2009; Vora *et al.*, 2009).

The homeostatic network regulating DNA topology

The topological alterations of DNA affect the expression of NAPs, RNAP subunits and topoisomerases (Menzel and Gellert, 1983; Schneider *et al.*, 2000; Travers *et al.*, 2001; Blot *et al.*, 2006). It is assumed therefore that genes encoding these factors are connected in a homeostatic network regulating the superhelical density of the DNA, such that coordination of genomic transcription is achieved by their interdependent expression (Travers and Muskhelishvili, 2005b; Muskhelishvili *et al.*, 2010). Available data indicate that while mutations of NAPs can affect RNAP composition and DNA topology, compositionally altered holoenzymes can optimize their transcriptional environments by selecting appropriate NAPs and DNA topology (Arnold and Tessmann, 1988; Tupper *et al.*, 1994; Barth *et al.*, 1995; Bensaid *et al.*, 1996; Malik *et al.*, 1996; Schneider *et al.*, 1997; Bouvier *et al.*, 1998; Balandina *et al.*, 2001; Blot *et al.*, 2006; Muskhelishvili and Travers, 2009; Geertz *et al.*, 2011). The corollary to this interdependent expression of the NAPs and transcription machinery components is that at each instant the RNAP holoenzyme composition on the one hand and the NAPs and DNA topology (chromatin architecture) on the other determine each other reciprocally (Fig. 2.3), explaining the coordinated changes of gene expression (Muskhelishvili *et al.*, 2010; Geertz *et al.*, 2011).

It is noteworthy that the electronically compiled hierarchical transcriptional regulatory network (TRN) of *E. coli* (Salgado *et al.*, 2006) describes directional interactions between the genes and thus employs lineal causality (Fig. 2.4A), whereas the homeostatic (and so by definition, self-referential) network regulating

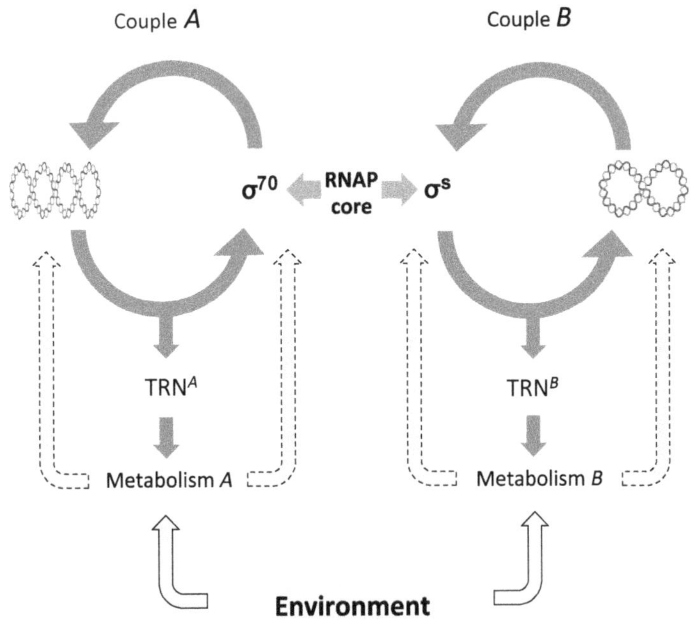

Figure 2.3 Relationship of reciprocal determination between the composition of RNAP holoenzyme and DNA topology in *E. coli*. The switch in the sigma factor composition of the RNAP holoenzyme induces different DNA topologies and vice versa (the NAPs corresponding to different DNA topologies are omitted for clarity). The structurally coupled combinations (A and B) produce corresponding gene expression programmes governing cellular metabolism. A switch in the growth environment can produce changes of metabolism affecting both the RNAP composition and DNA topology. Facile adaptation is enabled by relationship of reciprocal determination mediated by structural coupling between the RNAP and the DNA topology. Note that this thermodynamically open system is operationally closed.

DNA superhelicity is operationally closed onto itself and so employs circular logic (Fig. 2.4B–D) adopted for the description of self-reproducing systems (Varela *et al.*, 1974). Furthermore, in a hierarchical network the directional transmission of information involves sequential steps, whereas the heterarchical organization enables the network to respond to perturbations instantly, as a whole. This is because the sequential interactions implicating time are substituted by logic of structural coupling and reciprocal determination between the interacting components of the network implicating their interdependent alterations (albeit to different degrees) at any instant of information processing. Put another way, the heterarchical network provides a global response to any local perturbation. Indeed, the bacterial chromosome is thought to respond to environmental challenge as a whole, either by overall relaxation or by hypernegative supercoiling of chromosomal DNA depending on the type of the stress (McClellan *et al.*, 1990; Dorman, 1996; Tse-Dinh *et al.*, 1997). Another example consistent with global response of the nucleoid is formation of transcription foci as distinct structures in the fast growing, but not starving cells, and association of overall DNA relaxation and global reorganization of genomic transcription with impaired foci forming capability (Berger *et al.*, 2010).

Transitions between states

Chromosomes may be viewed essentially as topological machines reflecting the chirality of the DNA double helix. Conformational transitions in chromatin structure accompanying DNA replication and changes in gene expression in both bacteria and eukaryotes are powered by ATP-dependent molecular motors including both DNA and RNA polymerases, chromatin

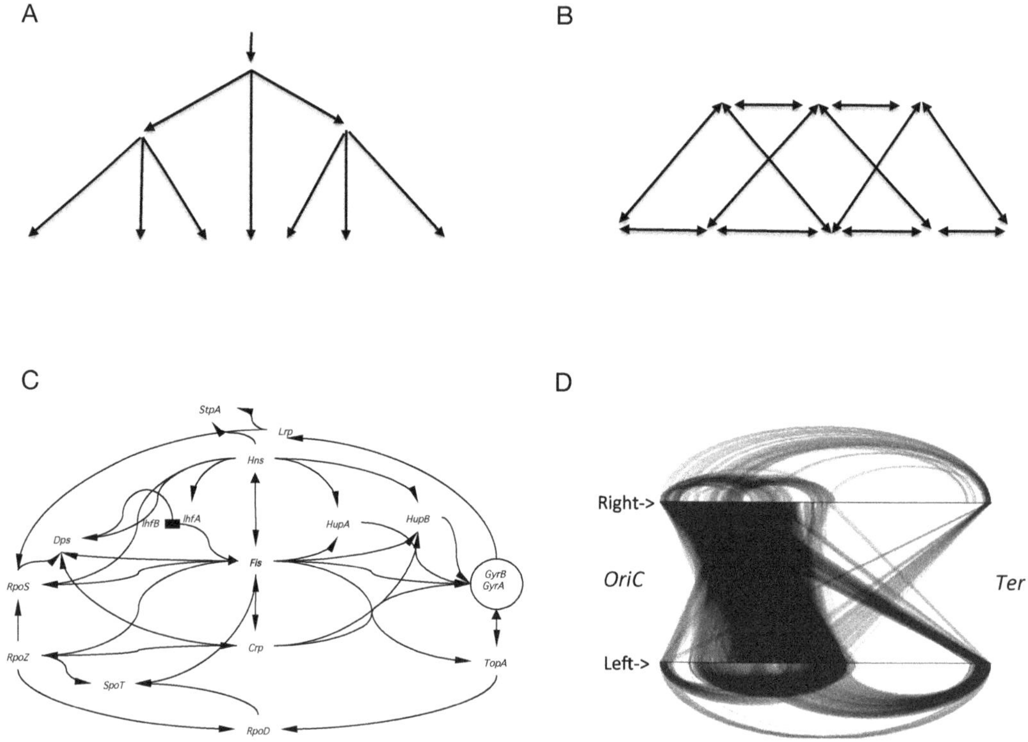

Figure 2.4 Distinction between the hierarchical and heterarchical networks. (A) A typical hierarchical network comprising directional connections assembled in subsequent steps. (B) In a typical heterarchical network the connections are bidirectional enabling the network to process new information as a unity. (C) Heterarchical network of DNA architectural proteins, DNA topoisomerases and transcription machinery components in *E. coli*. (D) Heterarchical network representing all significant functional communications between the loci in the *E. coli* chromosome (Sobetzko *et al.*, 2012). The right and left replichores are drawn separately to illustrate the *trans* (across the replichores) and *cis* (along the replichores) communications. The OriC and Ter ends of the chromosome are indicated.

remodelling assemblies (in eukaryotes) and DNA gyrase (in bacteria). For example, in *E. coli* both DNA replication and transcription may reorganize local domains of superhelicity (Deng *et al.*, 1995).

A heterarchical organization of communications in the chromosome (Fig. 2.4D) is well suited to maintain homeostasis but the network must also be capable of responding to changes in external inputs – for example, the availability of nutrients. These act to alter the balance between the different components of the network. Examples of such transitions during the bacterial growth cycle would include the relatively abrupt entry into exponential growth on nutrient shift-up and a more protracted change from exponential to stationary phases. These transitions are accompanied by substantial changes in nucleoid dynamics and overall DNA superhelicity. Put another way, the transitions in the state of the heterarchical network would be reflected in distinct changes of the chromosomal nucleoprotein complex and ultimately the shape of the chromosome, implying that these dynamical changes are ordered.

An obvious variable affecting nucleoid dynamics is the rate of initiation of DNA replication. In this context it is striking that the ordering, relative to the replication origin, of the genes encoding the NAPs approximately reflects their relative abundance during the growth cycle (Sobetzko *et al.*, 2012). Thus *hupA*, encoding HUα is closer to the origin than *hupB*, encoding HUβ (Fig. 2.5).

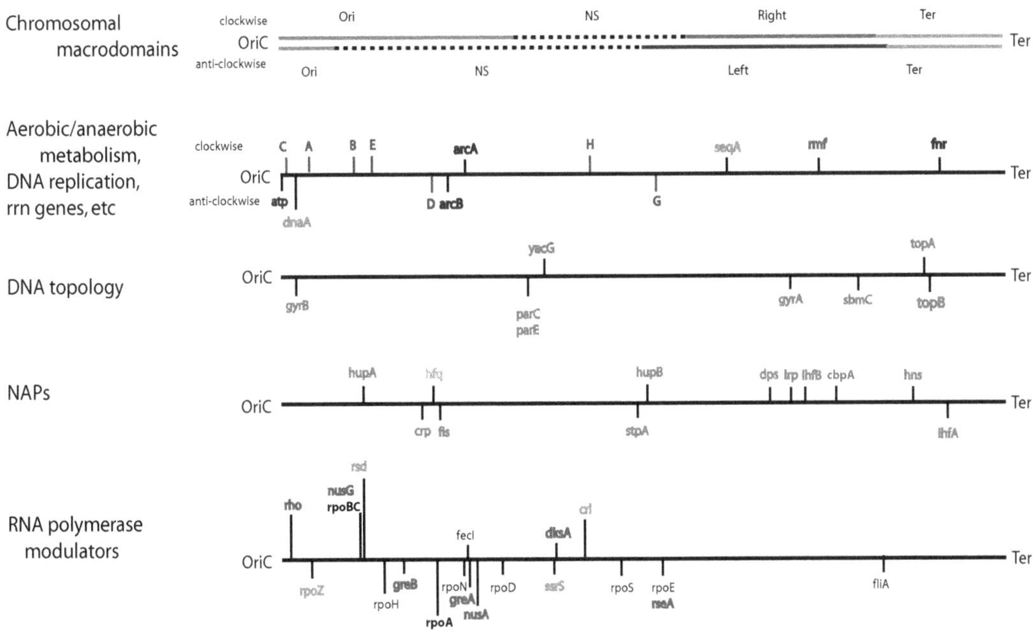

Figure 2.5 Spatial ordering of regulatory genes on the *E. coli* chromosome along the OriC–Ter axis. For full details see Sobetzko *et al*., (2012). Top line: Correspondence of macrodomains defined by Valens *et al*., (2004) to linear map. First bar: Selected genes involved in aerobic/anaerobic metabolism (dark blue), DNA replication (orange), *rrn* genes (red) and transition phase (brown). Genes on clockwise (right) replichore are above the bar and genes on anti-clockwise (left) replichore below the bar. Second bar: Selected genes involved in control of DNA topology. *gyrB*, a component of DNA gyrase, responsible for increasing negative superhelicity, maps close to the origin while gyrase inhibitor *sbmC*, and *topA* and *topB*, both responsible for relaxing DNA map either close to or within the Ter macrodomain. Third bar: Selected genes encoding NAPs. The NAP-encoding gene closest to OriC is *hupA*, encoding HUα. Its early expression relative to *hupB*, encoding HUβ , could buffer high negative superhelicity generated by DNA gyrase. Fourth bar: Selected genes involved in modulating RNA polymerase activity, including sigma factor utilization regulators, secondary channel binding proteins, termination/elongation factors and RNA polymerase subunits. Sigma factor utilization regulators (light green): *rpoZ*, mapping close to the origin, encodes the ω subunit of RNA polymerase, which confers a preference for utilization of σ^{70}. A colour version of this figure is located in the plate section at the back of the book.

Similarly *stpA*, whose product is more abundant during mid-exponential growth is closer to the origin than *hns*, encoding a paralogue whose function is correlated with the transition to stationary phase (Barth *et al.*, 1995). More generally, those NAPs associated with the higher overall superhelical density of exponential growth are closer to the origin and those associated with the lower superhelical density of stationary phase are closer to the replication terminus (Fig. 2.5). The ordering of the genes for the transcriptional machinery exhibits a similar pattern to those of the NAPs with *rpoD*, encoding the sigma factor for vegetative growth, being located closer to the replication origin that *rpoS*, encoding the stationary phase sigma factor. Notably, *rpoZ*, whose product facilitates utilization of the vegetative sigma factor in preference to the stationary phase sigma factor (Geertz *et al.*, 2011) and which is strongly expressed during exponential but not stationary phase (Geertz, 2004), is located close to the origin.

Furthermore, *gyrB* (but not *gyrA*) encoding subunit B of DNA gyrase is located in close proximity of OriC. Gyrase increases negative superhelicity especially with the higher ATP/ADP ratios prevailing on nutritional shift-up (van Workum *et al.*, 1996). In contrast both *topA* and *topB*, encoding the DNA relaxing topoisomerases, are closer to Ter. Spatial organization of genes modifying the transcription machinery (*rho*, the *nus* factors, the *gre* factors, *dksA*, *crl*), and genes sustaining catabolism and energy production under aerobic (*atp* operon), microaerobic (*arcA/B*) and anaerobic (*fnr*) conditions, as well as those involved in activation and negative modulation of replication (*dnaA*, *seqA*) and translation (*rrn* operons, *rmf*), exhibit a similar chromosomal ordering pattern. Importantly this relative ordering of different classes of regulatory genes is largely conserved throughout Gram-negative and Gram-positive bacteria (Sobetzko *et al.*, 2012).

Not only are the genes coordinating the major regulatory pathways ordered but so also are their targets. Analyses of the distribution density of binding sites for $E\sigma^{70}$ and $E\sigma^{S}$ holoenzymes compiled in RegulonDB (Gama-Castro *et al.*, 2008) show opposite spatial biases. For the vegetative σ^{70} factor the highest percentage of targets is found around the origin, whereas for the stationary phase σ^{S} factor the highest target density is close to the terminus, consistent with both the closer location of *rpoD* to OriC and the temporal division of labour between σ^{70} and σ^{S} during the bacterial growth cycle (Ishihama, 2000). Similarly, the average density of binding sites for DNA gyrase diminishes by 5- to 10-fold from OriC to Ter (Sobetzko *et al.*, 2012). This organization could generate a gradient of superhelical density, correlating with that of $E\sigma^{70}$ targets and anticorrelating with $E\sigma^{S}$ targets, as expected from the opposite supercoiling preferences of these holoenzymes (Bordes *et al.*, 2003; Geertz *et al.*, 2011) and in keeping with the requirement of high negative superhelicity for initiation of OriC replication.

For the major NAPs, despite distinct chromosomal location and abundances during the growth cycle, a high percentage of binding sites compiled in RegulonDB occur around the origin. Among the NAPs encoded in the Ter-proximal region only IHF targets activating binding sites in the vicinity of Ter, whereas the stationary phase regulator LRP (Tani *et al.*, 2002) and the global repressor H-NS (Dorman 2004) both preferentially target the Ori-proximal region. Additionally HU, the major supercoil-constraining NAP for which no binding site information is available, has distinct and opposite functional effects at the Ori and Ter ends of the chromosome, respectively reducing and increasing transcription.

To what extent, if any, does the linear ordering of genes in this way have a functional significance? The passage of the replication forks through a bacterial chromosome has two major consequences. First the copy number of replicated genes will be higher than that of unreplicated genes and second the movement of the replication forks will in principle generate positive supercoils ahead of the replisome (and so act to reduce negative superhelicity) and negative supercoils behind (Liu and Wang, 1987). This latter effect is significant, since not only is topoisomerase IV necessary for the decatenation the positively supercoiled braids generated by converging replication forks (Zechiedrich and Cozzarelli, 1995) but also it possesses a strong chiral selectivity for positive, rather than negative, braids (Stone *et al.*, 2003). Thus DNA replication may redistribute the average negative superhelicity such that regions near the origin will be more negatively supercoiled, on average, than those close to the terminus. The extent of such an effect would depend on the rate of replication initiation and the consequent number of forks in a chromosome. This would be a global effect and distinct from the intrinsic local response of transcription units to negative superhelicity.

Nutrient shift up is accompanied by an increase in average negative superhelicity (Balke and Gralla, 1987) and a transition to a state that supports multiple replication initiation events. The latter events are themselves dependent on negative superhelicity and require HU (Skarstad *et al.*, 1990; Hwang and Kornberg, 1992; von Freiesleben and Rasmussen, 1992). The initial generation of such superhelicity must thus be independent of DNA replication. A possible scenario is that the high adenylate energy charge induced on shift up activates DNA gyrase (van Workum *et al.*, 1996). Since the local concentration of gyrase binding sites is highest in the region flanking *OriC* (Berger *et al.*, 2010; Sobetzko *et al.*, 2012), gyrase action could again create a gradient of negative superhelicity along the chromosome and thus facilitate replication initiation. The promoters for some transcription units close to the replication origin – for example, the *atp* operon and *dnaA* – have an organization that is characteristic for promoters that require high levels of negative superhelicity for transcription initiation (Travers and Muskhelishvili, 2005b) and would thus likely be activated at this step. However, this does not exclude that the steady-state levels of expression might, owing to more complex regulation, have a different response to changes in superhelicity levels.

Similar considerations would apply to the transition from exponential to stationary phase. In this case, however, the rate of DNA replication initiation would decrease more gradually as the nutrient availability diminished. This would act to restore the copy number balance between the origin and terminal regions and also to reduce the redistribution of superhelicity generated by the movement of replisomes.

Organization of information in the transcriptional regulation system

The data on the directional (one-to-one) interactions between unique genes compiled in the electronic Regulon database belongs to the discontinuous, or 'digital' type of information (von Neumann, 1958), whereas integration of the TRN necessitates information of continuous, or 'analogue' type (Marr *et al.*, 2008; Muskhelishvili *et al.*, 2010). Such continuous information is provided by genome-wide fluctuations of supercoil energy reflecting the crosstalk between NAPs, RNAP and topoisomerases acting as interdependent components of the heterarchical network (Travers and Muskhelishvili, 2005b; Blot *et al.*, 2006; Muskhelishvili and Travers, 2009; Muskhelishvili *et al.*, 2010). From these considerations it is apparent that any holistic description of the transcriptional regulation system has to provide means for the conversion of the digital information into analogue and vice versa. In practice, the analogue regulation of gene expression is the predominant mode during exponential growth (Sonnenschein *et al.*, 2011) that can be approximated from determinations of specific combinations of NAPs and RNAP holoenzymes impacting the effective transcription profiles. This is achieved by representing the *E. coli* RegulonDB dataset (Salgado *et al.*, 2006) as a system coordinated by two axes, one of which corresponds to NAPs and global transcriptional regulators, and the other to seven alternative σ factors of the *E. coli* RNAP holoenzyme (Fig. 2.6A). Both the NAPs and RNAP holoenzymes are expressed over a range of concentrations and so represent analogue components of the system (von Neumann, 1958). Intersections of these axes determine subsets of genes under the control of specific combinations of two analogue components of the transcription system (one for initiation of DNA untwisting and one DNA architectural). Determination of these subsets of unique genes coupled by a common control mechanism and dubbed 'couplons', thus enables conversion of the analogue into digital information (Muskhelishvili *et al.*, 2010). Couplon matrix represents a heterarchical network, the topological closure of which becomes conspicuous when the square matrix is wrapped around both axes generating a torus (Fig. 2.6B). This torus generated from the orthogonally interdigitated regulon circuits is thus formally analogous to the topological closure of the physical *E. coli* chromosome.

Importantly, analyses of couplons demonstrating significant changes in effective transcript profiles (compared to the null model containing same number of randomly picked genes) reveal patterns of metabolic function coordinated by particular combinations of regulators and thus have predictive power. Couplons are entities independent of the parent regulons (Muskhelishvili *et al.*, 2010) and so their analysis reveals a new flexibility in organizing a genetic programme (Geertz *et al.*, 2011). Furthermore, using the couplons for mapping functionally related domains on the physical chromosome appears instrumental for revealing links between the gene positions, nucleoid dynamics and the physiological state transitions (Sobetzko *et al.*, 2012).

Conclusions

We have argued that a primary role of NAPs is to constrain the dynamics of DNA to produce differentiated DNA structures within the bacterial nucleoid. These enable the structural transitions that facilitate regulatory responses to changes in environmental conditions. Notably, this organization allows the nucleoid and its associated expression machinery to act as a single

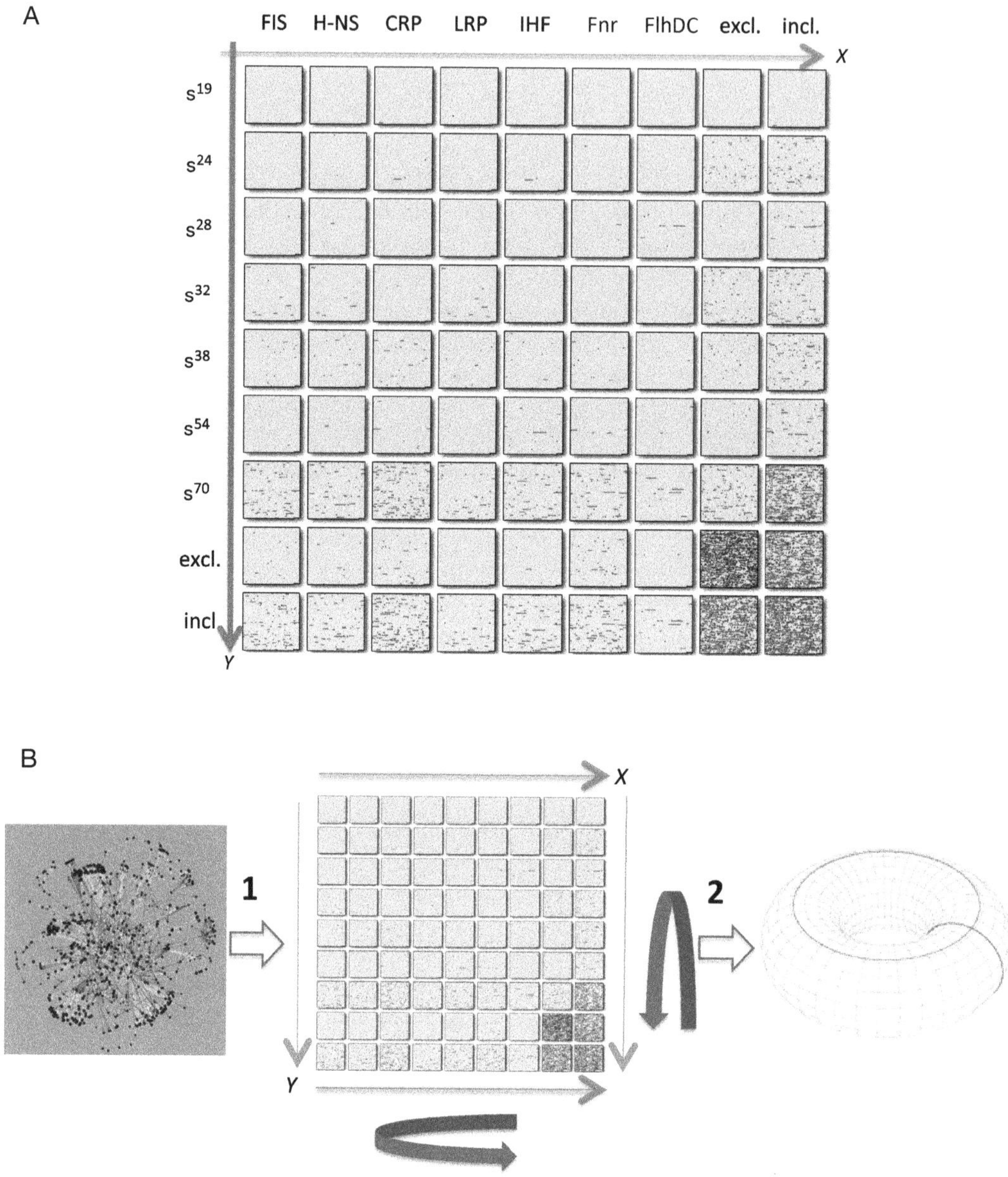

Figure 2.6 (A) Couplon matrix of σ factors and abundant NAPs. FlhDC is not a NAP but is incorporated in the matrix as a well-known regulator of a specific function – biosynthesis of flagella. Fnr is a global transcriptional regulator closely related to CRP. Each square or couplon holds all genes regulated by the two intersecting factors indicated on each axis. Genes organized in operons are shown in red colour, independent genes are in blue. The genes in each couplon are topologically ordered and arranged from left to right in rows for each couplon square. The 'excl' columns or rows represent the genes regulated by the column or row factor but by no listed factor intersecting with the column or row. The 'incl' columns or rows hold the genes that are regulated by the column or row factor without any additional restrictions. 'incl/excl' and 'excl/incl' couplons hold the sum of respective 'incl' row or column. 'incl/incl' holds all genes involved in creating this matrix e.g. all genes regulated by a sigma factor or a NAP. (B) The holistic model of topologically closed transcriptional regulation system. The RegulonDB dataset (Salgado *et al.*, 2006) is organized in the couplon matrix (1). Wrapping of the square scheme of couplon matrix (2) around both the vertical and horizontal axes (arrows) generates a torus of orthogonally operating regulon circuits (Muskhelishvili *et al.*, 2010). Note that this torus represents a logical construct, rather than mathematical object. A colour version of this figure is located in the plate section at the back of the book.

interconnected heterarchical control system rather than as a set of disconnected or loosely connected hierarchical systems. Transitions between the cellular physiological states are thus explained by state transitions in this single heterarchical control system.

Chapter highlights

- Holistic view of prokaryotic gene regulation requires integration of the chromosome structure and metabolic function reflected in the DNA superhelical density.
- The effective DNA superhelical density is determined by the crosstalk between the nucleoid-associated proteins acting as global transcriptional regulators, DNA topoisomerases, and transcription machinery components.
- Holistic approach assumes that the DNA superhelical density and composition of RNA polymerase holoenzyme are interdependent (relationship of reciprocal determination).
- Genetic regulation necessitates to be represented as a heterarchical, rather than hierarchical network.
- The couplon matrix is a means for assessing the changing states of the heterarchical network.
- Holistic analysis of gene expression in *E. coli* suggest that transcriptional regulation is largely determined by spatial ordering of genes and a temporal gradient of DNA superhelicity along the OriC–Ter axis.

References

Arnold, G.F., and Tessman, I. (1988). Regulation of DNA superhelicity by rpoB mutations that suppress defective Rho-mediated transcription termination in *Escherichia coli*. J. Bacteriol. *170*, 4266–4271.

Auner, H., Buckle, M., Deufel, A., Kutateladze, T., Lazarus, L., Mavathur, R., Muskhelishvili, G., Pemberton, I., Schneider, R., and Travers, A. (2003). Mechanism of transcriptional activation by FIS: role of core promoter structure and DNA topology. J. Mol. Biol. *331*, 331–344.

Azam, T.A., and Ishihama, A. (1999). Twelve species of the nucleoid-associated protein from *Escherichia coli*. Sequence recognition specificity and DNA binding affinity. J. Biol. Chem. *274*, 33105–33113.

Azam, T.A., Iwata, A.Q., Nishimura, A., Ueada, S., and Ishihama, A. (1999). Growth phase dependent variation in protein composition of the *Escherichia coli* nucleoid. J. Bacteriol. *181*, 6361–6370.

Azam, T.A., Hiraga, S., and Ishihama, A. (2000). Two types of localization of the DNA-binding proteins within the *Escherichia coli* nucleoid. Genes Cells *5*, 613–626.

Balandina, A., Claret, L., Hengge-Aronis, R., and Rouvière-Yaniv, J. (2001). The *Escherichia coli* histone-like protein HU regulates rpoS translation. Mol. Microbiol. *39*, 1069–1079.

Balke, V.L., and Gralla, J.D. (1987). Changes in the linking number of supercoiled DNA accompany growth transitions in *Escherichia coli*. J. Bacteriol. *169*, 4499–4506.

Barth, M., Marschall, C., Muffler, A., Fischer, D., and Hengge-Aronis, R. (1995). Role for the histone-like protein H-NS in growth phase-dependent and osmotic regulation of sigma S and many sigma S-dependent genes in *Escherichia coli*. J. Bacteriol. *177*, 3455–3464.

Bensaid, A., Almeida, A., Drlica, K., and Rouvière-Yaniv, J. (1996). Cross-talk between topoisomerase I and HU in *Escherichia coli*. J. Mol. Biol. *256*, 292–300.

Berger, M., Farcas, A., Geertz, M., Zhelyazkova, P., Brix, K., Travers, A., and Muskhelishvili, G. (2010). Coordination of genomic structure and transcription by the main bacterial nucleoid-associated protein HU. EMBO Rep. *11*, 59–64.

Bétermier, M., Galas, D.J., and Chandler, M. (1994). Interaction of Fis protein with DNA: bending and specificity of binding. Biochimie *76*, 958–967.

Blot, N., Mavathur, R., Geertz, M., Travers, A., and Muskhelishvili, G. (2006). Homeostatic regulation of supercoiling sensitivity coordinates transcription of the bacterial genome. EMBO Rep. *7*, 710–715.

Bokal, A.J., 4th, Ross, W., and Gourse, R.L. (1995). The transcriptional activator protein FIS: DNA interactions and cooperative interactions with RNA polymerase at the *Escherichia coli rrnB* P1 promoter. J. Mol. Biol. *245*, 197–207.

Boles, T.C., White, J.H., and Cozzarelli, N.R. (1990). The structure of plectonemically supercoiled DNA. J. Mol. Biol. *213*, 931–951.

Bordes, P., Conter, A., Morales, V., Bouvier, J., Kolb, A., and Gutierrez, C. (2003). DNA supercoiling contributes to disconnect sigmaS accumulation from sigmaS-dependent transcription in *Escherichia coli*. Mol. Microbiol. *48*, 561–571.

Bouffartigues, E., Buckle, M., Badaut, C., Travers, A., and Rimsky, S. (2007). H-NS cooperative binding to high-affinity sites in a regulatory element results in transcriptional silencing. Nat. Struct. Mol. Biol. *14*, 441–448.

Bouvier, J., Gordia, S., Kampmann, G., Lange, R., Hengge-Aronis, R., and Gutierrez, C. (1998). Interplay between global regulators of *Escherichia coli*: effect of RpoS, Lrp and H-NS on transcription of the gene osmC. Mol. Microbiol. *28*, 971–980.

Browning, D.F., Cole, J.A., and Busby, S.J.W. (2004). Transcription activation by remodelling of a nucleoprotein assembly: the role of NarL at the FNR-dependent *Escherichia coli nir* promoter. Mol. Microbiol. *53*, 203–215.

Browning, D.F., Grainger, D.C., and Busby, S.J.W. (2010). Effects of nucleoid-associated proteins on bacterial chromosome structure and gene expression. Curr. Opin. Microbiol. *13*, 773–780.

Broyles, S.S., and Pettijohn, D.E. (1986). Interaction of the *Escherichia coli* HU protein with DNA. Evidence for formation of nucleosome-like structures with altered DNA helical pitch. J. Mol. Biol. *187*, 47–60.

Busby, S., and Ebright, R.H. (1999). Transcription activation by catabolite activator protein (CAP). J. Mol. Biol. *293*, 199–213.

Cho, B.K., Zengler, K., Qiu, Y., Park, Y.S., Knight, E.M., Barrett, C.L., Gao, Y., and Palsson, B.Ø. (2009). The transcription unit architecture of the *Escherichia coli* genome. Nat. Biotechnol. *27*, 1043–1049.

Dame, R.T. (2005). The role of nucleoid-associated proteins in the organization and compaction of bacterial chromatin. Mol. Microbiol. *56*, 858–870.

Dame, R.T., Wyman, C., and Goosen, N. (2000). H-NS mediated compaction of DNA visualised by atomic force microscopy. Nucleic Acids Res. *28*, 3504–3510.

Dame, R.T., Kalmykoya, O.J., and Grainger, D.C. (2011). Chromosomal macrodomains and associated proteins: implications for gene organisation and replication in Gram-negative bacteria. PLoS Genet. *7*, 1002123.

Deng, S., Stein, R.A., and Higgins, N.P. (2005). Organization of supercoil domains and their reorganization by transcription. Mol. Microbiol. *57*, 1511–1521.

Dorman, C.J. (1996). Flexible response: DNA supercoiling, transcription and bacterial adaptation to environmental stress. Trends Microbiol. *4*, 214–216.

Dorman, C.J. (2004). H-NS: a universal regulator for a dynamic genome. Nat. Rev. Microbiol. *2*, 391–400.

Dorman, C.J., and Deighan, P. (2003). Regulation of gene expression by histone-like proteins in bacteria. Curr. Opin. Gen. Dev. *13*, 179–184.

Drew, H.R., and Travers, A.A. (1985). DNA bending and its relation to nucleosome positioning. J. Mol. Biol. *186*, 773–790.

Fang, C.F., and Rimsky, S. (2008). New insights into transcriptional regulation by H-NS. Curr. Opin. Microbiol. *11*, 113–120.

von Freiesleben, U., and Rasmussen, K.V. (1992). The level of supercoiling affects the regulation of DNA replication in *Escherichia coli*. Res. Microbiol. *143*, 655–663.

Gama-Castro, S., Jiménez-Jacinto, V., Peralta-Gil, M., Santos-Zavaleta, A., Peñaloza-Spinola, M.I., Contreras-Moreira, B., Segura-Salazar, J., Muñiz-Rascado, L., Martínez-Flores, I., Salgado, H., *et al.* (2008). RegulonDB (version 6.0): gene regulation model of *Escherichia coli* K-12 beyond transcription, active (experimental) annotated promoters and Textpresso navigation. Nucleic Acids Res. *36*, D120–D124.

Geertz, M. (2004). Elucidation of the functional role of the ω subunit of *Escherichia coli* RNA polymerase. Diploma Thesis. Philipps University Marburg, Marburg, Germany.

Geertz, M., Travers, A., Janga, S.C., Mehandziska, S., Lungu, C., Sobetzko, P., Shimamoto, N., and Muskhelishvili, G. (2011). Structural coupling between RNA polymerase composition and DNA supercoiling coordinates bacterial transcription. mBio *2*, e00034–11.

Grainger, D.C., Hurd, D., Goldberg, M.D., and Busby, S. (2006). Association of nucleoid proteins with coding and non-coding segments of the *Escherichia coli* genome. Nucleic Acids Res. *34*, 4642–4652.

Guo, F., and Adhya, S. (2007). Spiral structure of *Escherichia coli* HUαβ provides foundation for DNA supercoiling. Proc. Natl. Acad. Sci. U.S.A. *104*, 4309–4314.

Hsieh, L.S., Burger, R.M., and Drlica, K. (1991). Bacterial DNA supercoiling and [ATP]/[ADP]. Changes associated with a transition to anaerobic growth. J. Mol. Biol. *219*, 443–450.

Hwang, D.S., and Kornberg, A. (1992). Opening of the replication origin of *Escherichia coli* by DnaA protein with protein HU or IHF. J. Biol. Chem. *267*, 23083–23086.

Ishihama, A. (2000). Functional modulation of *Escherichia coli* RNA polymerase. Annu. Rev. Microbiol. *54*, 499–518.

Janga, S.C., Salgado, H., and Martínez-Antonio, A. (2009). Transcriptional regulation shapes the organization of genes on bacterial chromosomes. Nucleic Acids Res. *37*, 3680–3688.

Junier, I., Hérisson, J., and Képès, F. (2012). Genomic organization of evolutionrily correlated genes in bacteria: limits and strategies. J. Mol. Biol. doi: 10.1016/j.jmb.2012.03.009.

Kusano, S., Ding, Q., Fujita, N., and Ishihama, A. (1996). Promoter selectivity of *Escherichia coli* RNA polymerase $E\sigma^{70}$ and $E\sigma^{38}$ holoenzymes. Effect of DNA supercoiling. J. Biol. Chem. *271*, 1998–2004.

Ladburner, A.G. (2009). Chromatin places metabolism center stage. Cell *138*, 18–20.

Lang, B., Blot, N., Bouffartigues, E., Buckle, M., Geertz, M., Gualerzi, C.O., Mavathur, R., Muskhelishvili, G., Pon, C.L., Rimsky, S., *et al.* (2007). High-affinity DNA binding sites for H-NS provide a molecular basis for selective silencing within proteobacterial genomes. Nucleic Acids Res. *35*, 6330–6337.

Lazarus, L.R., and Travers, A.A. (1993). The *Escherichia coli* FIS protein is not required for the activation of *tyrT* transcription on entry into exponential growth. EMBO J. *12*, 2483–2494.

Liu, L.F., and Wang, J.C. (1987). Supercoiling of the DNA template during transcription. Proc. Natl. Acad. Sci. U.S.A. *84*, 7024–7027.

Luijsterburg, M.S., White, M.F., van Driel, R., and Dame, R.T. (2008). The major architects of chromatin: architectural proteins in bacteria, archaea and eukaryotes. Crit. Rev. Biochem. Mol. Biol. *43*, 393–418.

McClellan, J.A., Boublíková, P., Palecek, E., and Lilley, D.M.J. (1990). Superhelical torsion in cellular DNA responds directly to environmental and genetic factors. Proc. Natl. Acad. Sci. U.S.A. *87*, 8373–8377.

McLeod, S.M., and Johnson, R.C. (2001). Control of transcription by nucleoid proteins. Curr. Opin. Microbiol. *4*, 152–159.

Malik, M., Bensaid, A., Rouvière-Yaniv, J., and Drlica, K. (1996). Histone-like protein HU and bacterial DNA topology: suppression of an HU deficiency by gyrase mutations. J. Mol. Biol. *256*, 66–76.

Marr, C., Geertz, M., Hütt, M.T., and Muskhelishvili, G., (2008). Dissecting the logical types of network control in gene expression profiles. BMC Syst. Biol. *2*, 18.

Maurer, S., Fritz, J., Muskhelishvili, G., and Travers, A. (2006). RNA polymerase and an activator form discrete subcomplexes in a transcription initiation complex. EMBO J. *25*, 3784–3790.

Maurer, S., Fritz, J., and Muskhelishvili, G. (2009). A systematic *in vitro* study of nucleoprotein complexes formed by bacterial nucleoid associated proteins revealing novel types of DNA organization. J. Mol. Biol. *387*, 1261–1276.

Menzel, R., and Gellert, M. (1983). Regulation of the genes for *E. coli* DNA gyrase: homeostatic control of DNA supercoiling. Cell *34*, 105–113.

Mercier, R., Petit, M.A., Schbath, S., Robin, S., El Karoui, M., Boccard, F., and Espéli, O. (2008). The MatP/*matS* site-specific system organizes the terminus region of the *E. coli* chromosome into a macrodomain. Cell *135*, 475–485.

Muskhelishvili, G., and Travers, A. (2009). Intrinsic *in vivo* modulators: negative supercoiling and the constituents of the bacterial nucleoid. In RNA Polymerases as Molecular Motors, Buc, H., and Strick, T., eds. (RSC publishing, Cambridge, UK). pp. 69–95.

Muskhelishvili, G., Travers, A.A., Heumann, H., and Kahmann, R. (1995). FIS and RNA polymerase form a specific nucleoprotein complex at a stable RNA promoter. EMBO J. *14*, 1446–1452.

Muskhelishvili, G., Sobetzko, P., Geertz, M., and Berger, M., (2010). General organisational principles of the transcriptional regulation system: a tree or a circle? Mol. Biosyst. *6*, 662–676.

von Neumann, J. (1958). The Computer and the Brain (Yale University Press, New Haven, CT).

Oberto, J., Nabti, S., Jooste, V., Mignot, H., and Rouvière-Yaniv, J. (2009). The HU regulon is composed of genes responding to anaerobiosis, acid stress, high osmolarity and SOS induction. PLoS One *4*, e4367.

Protozanova, E., Yakovchuk, P., and Frank-Kamenetskii, M.D. (2004). Stacked- unstacked equilibrium at the nick site of DNA. J. Mol. Biol. *342*, 775–785.

Pul, U., Lux, B., Wurm, R., and Wagner, R. (2008). Effect of upstream curvature and transcription factors H-NS and LRP on the efficiency of *Escherichia coli* rRNA promoters P1 and P2 – a phasing analysis. Microbiology *154*, 2546–2558.

Rimsky, S., and Travers, A. (2011). Pervasive regulation of nucleoid structure and function by nucleoid-associated proteins. Curr. Opin. Micro. *14*,136–141.

Rochman, M., Aviv, M., Glaser, G., and Muskhelishvili, G. (2002). Promoter protection by a transcription factor acting as a local topological homeostat. EMBO Rep. *3*, 355–360.

Rochman, M., Blot, N., Dyachenko, M., Glaser, G., Travers, A., and Muskhelishvili, G. (2004). Buffering of stable RNA promoter activity against DNA relaxation requires a far upstream sequence. Mol. Microbiol. *53*, 143–152.

Roth, A., Urmoneit, B., and Messer, W. (1994). Functions of histone-like proteins in the initiation of DNA replication at *oriC* of *Escherichia coli*. Biochimie *76*, 917–923.

Ryan, V.T., Grimwade, J.E., Camara, J.E., Crooke, E., and Leonard, A.C. (2004). *Escherichia coli* prereplication complex assembly is regulated by dynamic interplay among Fis, IHF and DnaA. Mol. Microbiol. *51*, 1347–1359.

Salgado, H., Gama-Castro, S., Peralta-Gil, M., Díaz-Peredo, E., Sánchez-Solano, F., Santos-Zavaleta, A., Martínez-Flores, I., Jiménez-Jacinto, V., Bonavides-Martínez, C., Segura-Salazar, J., *et al.* (2006). RegulonDB (version 5.0): *Escherichia coli* K-12 transcriptional regulatory network, operon organization, and growth conditions. Nucleic Acids Res. *34*, D394–D397.

Schneider, R., Travers, A.A., and Muskhelishvili, G. (1997). FIS regulates the bacterial growth phase-dependent topological transitions in *Escherichia coli*. Mol. Microbiol. *26*, 519–530.

Schneider, R., Travers, A., and Muskhelishvili, G. (2000). The expression of the *Escherichia coli fis* gene is strongly dependent on the superhelical density of DNA. Mol. Microbiol. *38*, 167–175.

Schneider, R., Lurz, R., Lüder, G., Tolksdorf, C., Travers, A., and Muskhelishvili, G. (2001). An architectural role of the *Escherichia coli* protein FIS in organising DNA. Nucleic Acids Res. *29*, 5107–5114.

Skarstad, K., Baker, T.A., and Kornberg, A. (1990). Strand separation required for initiation of replication at the chromosomal origin of *E. coli* is facilitated by a distant RNA–DNA hybrid. EMBO J. *9*, 2341–2348.

Snoep, J.L., van der Weijden, C.C., Andersen, H.W., Westerhoff, H.V., and Jensen, P.R. (2002). DNA supercoiling in *Escherichia coli* is under tight and subtle homeostatic control, involving gene-expression and metabolic regulation of both topoisomerase I and DNA gyrase. Eur. J. Biochem. *269*, 1662–1669.

Sobetzko, P., Travers, A., and Muskhelishvili, G. (2012). Gene order and chromosome dynamics coordinate spatiotemporal gene expression during the bacterial growth cycle. Proc. Natl. Acad. Sci. U.S.A. *109*, E42–E50.

Stone, M.D., Bryant, Z., Crisona, N.J., Smith, S.B., Vologodskii, A., Bustamante, C., and Cozzarelli, N.R. (2003). Chirality sensing by *Escherichia coli* topoisomerase IV and the mechanism of type II topoisomerases. Proc. Natl. Acad. Sci. U.S.A. *100*, 8654–8659.

Swinger, K.K., Lemberg, K.M., Zhang, Y., and Rice, P.A. (2003). Flexible DNA bending in HU-DNA cocrystal structures. EMBO J. *22*, 3749–3760.

Tani, T.H., Khodursky, A., Blumenthal, R.M., Brown, P.O., and Matthews, R.G. (2002). Adaptation to famine: a family of stationary-phase genes revealed by microarray analysis. Proc. Natl. Acad. Sci. U.S.A. *99*, 13471–13476.

Travers, A., and Muskhelishvili, G. (2005a). Bacterial chromatin. Curr. Opin. Genet. Dev. *15*, 507–514.

Travers, A., and Muskhelishvili, G. (2005b). DNA supercoiling – a global transcriptional regulator for enterobacterial growth? Nat. Rev. Microbiol. 3, 157–169.

Travers, A., and Muskhelishvili, G. (2007). A common topology for bacterial and eukaryotic transcription initiation? EMBO Rep. *8*, 147–151.

Travers, A., Schneider, R., and Muskhelishvili, G. (2001). DNA supercoiling and transcription in *Escherichia coli* – the FIS connection. Biochimie *83*, 213–217.

Travers, A., Hiriart, E., Churcher, M., Caserta, M., and Di Mauro, E. (2010). The DNA sequence-dependence of. nucleosome positioning *in vivo* and *in vitro*. J. Biomol. Struct. Dyn. 27, 713–724.

Tse-Dinh, Y.C., Qi, H., and Menzel, R. (1997). DNA supercoiling and bacterial adaptation: thermotolerance and thermoresistance. Trends Microbiol. *5*, 323–326.

Tupper, A.E., Owen-Hughes, T.A., Ussery, D.W., Santos, D.S., Ferguson, D.J., Sidebotham, J.M., Hinton, J.C., and Higgins, C.F. (1994). The chromatin-associated protein H-NS alters DNA topology *in vitro*. EMBO J. *13*, 258–268.

Umbarger, M.A., Toro, E., Wright, M.A., Porreca, G.J., Baù, D., Hong, S.H., Fero, M.J., Zhu, L.J., Marti-Renom, M.A., McAdams, H.H., *et al.* (2011). The three-dimensional architecture of a bacterial genome and its alteration by genetic perturbation. Mol. Cell *44*, 252–264.

Ussery, D., Larsen, T.S., Wilkes, K.T., Friis, C., Worning, P., Krogh, A., and Brunak, S. (2001). Genome organisation and chromatin structure in *Escherichia coli*. Biochimie *83*, 201–212.

Valens, M., Penaud, S., Rossignol, M., Cornet, F., and Boccard, F. (2004). Macrodomain organization of the *Escherichia coli* chromosome. EMBO J. *23*, 4330–4341.

Varela, F.G., Maturana, H.R., and Uribe, R. (1974). Autopoiesis: the organization of living systems, its characterization and a model. Curr. Mod. Biol. *5*, 187–196.

Vijayan, V., Zuzow, R., and O'Shea, E.K. (2009). Oscillations in supercoiling drive circadian gene expression in cyanobacteria. Proc. Natl. Acad. Sci. U.S.A. *106*, 22564–22568.

Vora, T., Hottes, A.K., and Tavazoie, S. (2009). Protein occupancy landscape of a bacterial genome. Mol. Cell. *35*, 247–253.

Weber, H., Polen, T., Heuveling, J., Wendisch, V.F., and Hengge, R. (2005). Genome-wide analysis of the general stress response network in *Escherichia coli*: sigmaS-dependent genes, promoters, and sigma factor selectivity. J. Bacteriol. *187*, 1591–1603.

Wiggins, P.A., Cheveralls, K.C., Martin, K.S., Lintner, R., and Kondev, J. (2010). Strong intranucleoid interactions organize the *Escherichia coli* chromosome into a nucleoid filament. Proc. Natl. Acad. Sci. U.S.A. *107*, 4991–4995.

van Workum, M., van Dooren, S.J.M., Oldenburg, N., Molenaar, D., Jensen, P.R., Snoep, J.L., and Westerhoff, H.V. (1996). DNA supercoiling depends on the phosphorylation potential in *Escherichia coli*. Mol. Microbiol. *20*, 351–360.

Zechiedrich, E.L., and Cozzarelli, N.R. (1995). Roles of topoisomerase IV and DNA gyrase in DNA unlinking during replication in *Escherichia coli*. Genes Dev. *9*, 2859–2869.

Structure and Evolution of Prokaryotic Transcription Factor Binding Sites

3

Rekin's Janky

Abstract

With the ever-increasing number of available sequenced bacterial genomes and the availability of high-throughput experimental approaches such as chromatin immunoprecipitation sequencing (ChIP-seq), it has become possible to extend our knowledge about the transcriptional regulation from model organisms to other bacteria. Recent research has focused especially on the comparison of closely related species, allowing us to get insight into the different regulatory evolutionary events creating phenotypic diversity and involved in the evolution of bacterial gene regulatory networks.

Introduction

An estimated one billion bacterial species occupy virtually all ecological niches on our planet (Toussaint *et al.*, 2003). Bacteria show in particular a high capacity of adaptation to environmental changes and extreme conditions such as high temperature, high atmospheric or osmotic pressure, and pH. This adaptive behaviour suggests that various responses have been evolutionary selected needing different levels of regulation from the signal reception to the gene expression. Gene expression is controlled at multiple levels: (i) transcriptional level which can be regulated at the initiation and the termination steps, (ii) translational level where codon usage bias can optimize the translation, (iii) and post-translational modifications such as phosphorylation and methylation of the protein. In this chapter, we will focus on regulatory signals which can be detected from the genomic sequences.

At the level of the transcription initiation, a protein called transcription factor (TF) binds to the promoter region of the gene, activating or repressing its expression. The DNA signal where the TF binds is called a transcription factor binding site (TFBS) and is characterized by its sequence and its position. TFBS are also known as cis-acting regulatory elements where *cis* refers to the fact that the transcription factor binds to the same DNA molecule as the one on which the regulated gene resides. On the contrary, TF are *trans*-acting elements as they are coming from a different molecule than the target gene. Sigma factors and RNA polymerase (RNAP) are other *trans* actors that play an important role in the transcriptional regulation as described in the first chapter.

Core promoter

The core RNA polymerase has a strong affinity to the DNA but no signal specificity (Browning and Busby, 2004). To initiate the transcription, the polymerase needs to be properly positioned in the promoter region of the gene. To do this, the RNA polymerase makes a complex with sigma factors which bind upstream of the transcription start site (TSS). The most common sigma factor in *E. coli* is σ^{70}, but six other factors are known to be involved in response to diverse environmental stimuli (Rhodius *et al.*, 2006). They can be regulated by anti-sigma factors. When the polymerase is in limited concentration in the cell, the availability of certain sigma factors can put in competition their specific promoter regions.

In the common promoter region, two signals are necessary as they correspond to the binding of two domains of the σ^{70} factor family. The first one, bound by the σ^{2} domain, is located at −10 bp to the TSS, and is called TATA box as the binding motif is like TATAAT. The second one is also a hexamer TTGACA. It is located at −35 bp and bound by the σ^4 domain (Browning and Busby, 2004). The binding consensus of these two motifs is rather degenerated, one supplying the degeneracy of the other one. Many housekeeping genes present strong promoters. However, the optimal spacing between these two boxes is about 17 bp (deHaseth *et al.*, 1998) and has been used to design vectors for overexpression. Two additional elements can reinforce the strength of these promoters: an AT-rich 'UP' element located upstream of the −35 box and the extended −10 box which consensus is TRTG (R standing for A or G[1]). Thus, the structure of the core promoter region can make the transcription easier or not. To detect strong promoters, many prediction tools have recently emerged using motif consensus and their spacing (Dekhtyar *et al.*, 2008), spaced over-represented words (Touzain *et al.*, 2008), Hidden Markov Model and combinatorial methods (Eng *et al.*, 2009).

RNA binding signals

While the initiation of the transcription is a key step for the regulation, other steps of the gene expression can be under regulatory control. Transcription and translation are coupled in bacteria, i.e. ribosomes start the translation while the mRNA is synthesized (Burmann *et al.*,; Proshkin *et al.*, 2010). Thus, translational regulatory signals can be detected on the intergenic sequences such as the riboswitch (Abreu-Goodger and Merino, 2005; Abreu-Goodger *et al.*, 2004). The termination of the transcription can be regulated by a Rho factor. This is the Rho-dependent termination. The Rho factor will bind to the end of the newly transcribed RNA and will induce the dissociation of the ternary complex (mRNA/DNA/RNAP), preventing the transcription of downstream genes. Studies on the RNA binding sites of Rho factors show that they are enriched in cytosine (Alifano *et al.*, 1991; Ciampi, 2006). Rho-independent termination involved a stem–loop enriched in cytosine and guanine followed by a poly-T. Detection of such termination signals has been proven to be useful in order to predict the transcriptional units in Bacteria (Ermolaeva *et al.*, 2000).

Other type of regulation involving small RNAs will be discussed in the next chapter. In the next section, we will focus on transcription factor regulatory signals: their structure, their identification and their evolution.

Experimental approaches

How to identify TFBS? For this purpose, we will require the purified regulatory protein of interest and DNA fragments to be tested for binding. We will briefly present different experimental methods widely used for identification of TFBS.

Local analysis

EMSA (electrophoretic mobility shift assay) is the simplest method. It relies on the fact that the protein–DNA complex will migrate slower than the DNA alone. If there is a gel shift in the migration of a DNA fragment, it means that the protein binds to this fragment. The specificity of this interaction can be checked by incubating the complex with the antibody specific to the protein. The migration of such ternary complex is slower than the protein–DNA complex and is called a *supershift* (Brown, 2002). To localize more precisely the binding site on the DNA, we use the *DNAse footprinting*. Such method uses DNAse, or other chemical reagents, to digest free DNA while the residues involved in the binding site will be protected by the protein. After comparison of the migration profiles of labelled DNA fragments, the protected region is characterized by the absence of bands; this is the 'footprint' (Brown, 2002). The drawback of using such methods is that the binding affinity *in vitro* is not expected to be the same in the cell. Indeed, we may consider the difference of concentrations, the total number of accessible sites, the impact of the surrounding region and the genome organization.

Global analysis

To detect *in vivo* DNA–protein interactions at the genomic level, ChIP experiments have been widely used in the last decades. This method proceeds in several steps: (1) to cross-link using

1 http://www.chem.qmul.ac.uk/iupac/misc/naabb.html#p321

UV or formaldehyde the interaction between proteins and the target DNA; then (2) to make DNA fragments (~500 bp) by sonication; and finally (3) to use chromatin immunoprecipitation (ChIP) to select the bound fragments. In the last step, antibodies against the DNA-binding protein of interest are coupled to beads, retaining the bound DNA fragments after multiple washes. These DNA sequences can be amplified by PCR. Two strategies are then possible to identify the enriched sequences: ChIP-on-chip (Horak and Snyder, 2002) and ChIP-Seq. In the first one, the amplified sequences are hybridized to a microarray (chip) spotted with thousands of short genomic sequences. The enrichment is measured by comparison with a control of non-precipitated DNA. Tiling arrays can also be used as a substitute of the microarray in ChIP-on-chip. It contains range from 10,000 to greater than 6,000,000 probes covering the whole genome without any feature assumption and can give more precise information about the mapping of TFBS (Mockler *et al.*, 2005). ChIP-Seq method gained popularity by taking advantage of the efficiency of Next Generation Sequencing technology to sequence the immunoprecipitated fragments (Johnson *et al.*, 2007). This second method has a high sensitivity and specificity (Robertson *et al.*, 2007) and allows detection of binding sites with a good resolution (±50 bp). Limitations of ChIP-Seq methods are the variability of fragment sizes obtained by ultrasonication and the need of a ChIP antibody good enough to select relevant interactions among transient interactions.

Representation of a binding motif

The sequence of the TFBS is usually represented as a stretch of adjacent letters corresponding to the DNA alphabet (A, C, G and T). Depending on the experimental evidence, annotated binding sites from different databases are more or less reliable (Table 3.1). They used to be annotated with their flanking regions in order to take into account their genomic context in sequence alignments. The multiple sequence alignment of a collection of annotated binding sites allows representing at an abstract level the proportion of the nucleotides at each position as described in Table 3.2.

Table 3.2 illustrates the different representations of a regulatory motif with the example of LexA TF. This regulator represses several genes involved in the SOS response to DNA damage. Among nine regulated genes, 10 binding sites have been annotated in *Escherichia coli K12*, including the gene coding for LexA (autoregulated). The resulting alignment shows that the motif is composed of two cores of conserved nucleotides spaced by ten less conserved nucleotides (Table 3.2b). In bacteria, more than 80% of TFs belong to the helix–turn–helix (HTH) family and bind as dimers on spaced motifs (Huffman and Brennan, 2002; Minezaki *et al.*, 2005; Perez-Rueda *et al.*, 2004). Such motif can be represented as a consensus sequence, as a position specific matrix or graphically as a sequence logo.

The consensus sequence displays the most

Table 3.1 TFBS databases

Name	Organisms	Access	URL
RegulonDB	*E. coli* K12 and gamma-proteobacteria	Public	http://regulondb.ccg.unam.mx/
–	*E. coli* K12	Public	http://bayesweb.wadsworth.org/binding_sites/index.html
DBTBS	*B. subtilis* and Gram-positives	Public	http://dbtbs.hgc.jp
PRODORIC	Prokaryotes (*Pseudomonas aeruginosa*, *B. subtilis*, *E. coli*)	Private	http://prodoric.tu-bs.de/
RegPrecise	Prokaryotes	Public	http://regprecise.lbl.gov/RegPrecise/
MycoperonDB	Mycobacteria	Public	http://cdfd.org.in/mycoperondb
MtbRegList	*M. tuberculosis*	Public	http://www.USherbrooke.ca/vers/MtbRegList
CoryneRegNet	Corynebacteria, *E. coli*	Public	http://www.coryneregnet.de

Table 3.2 Representations of the LexA binding motif in *E. coli*: Alignment of annotated binding sites (A), consensus (B) and sequence logo built using *RSAT convert-matrix* with default parameters (C)

Gene	Aligned sequence
(a) Transcription factor binding sites	
phr	**GCCTGGCTTTCAGGGCAG**
recA	**TACTGTATGCTCATACAG**
rpsU	**AGCTGGCGTTGATGCCAG**
ssb	**ACCTGAATGAATATACAG**
sulA	**TACTGTATGGATGTACAG**
uvrA	**ACCTGAATGAATATACAG**
uvrB	**AACTGTTTTTTTATCCAG**
uvrD	**ATCTGTATATATACCCAG**
lexA	**TGCTGTATATACTCACAG**
lexA	**AACTGTATATACACCCAG**
(b) Consensus	
Strict	**NNCTGNNNNNNNNNNCAG**
IUPAC	**DNCTGDHKDNNHDBVCAG**
(c) Logo	

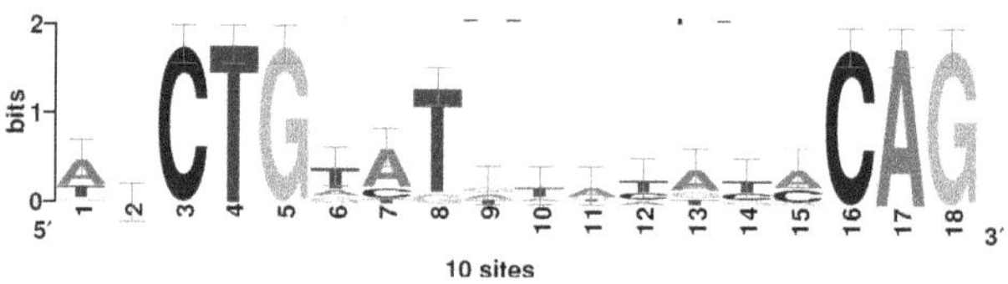

represented nucleotides at each position and uses the IUPAC-IUB[2] nomenclature to encode the degenerated positions. Because of its simplicity the consensus is popularly used by the biologists, but it is important to stress that it does not show the frequency differences in nucleotides, i.e. the letter R used for a purine position can be used if the ratio A/G is either 10% or 90%. Thus, several consensus sequences can be calculated from a permissive threshold on the nucleotide proportion (IUPAC consensus) to the most stringent (strict consensus) (Table 3.2b). One way to represent those differences is to use matrices. A position frequency matrix (PFM) shows the frequencies of the four nucleotides (row) for each position (column) (Table 3.3b). Berg and von Hippel (1987) have shown that binding activity is often proportional to the logarithm of the observed frequency of the nucleotides. We call position-specific scoring matrix (PSSM), or position weight matrix (PWM), a matrix of score values which represents the log-odds ratio of the observed frequency and prior probability from a background model (Table 3.3c). A pseudo-weight is generally applied to this calculation to take into account binding sites which are not known yet. Those matrices can be used to compute the informational content of a binding motif. This measures the nucleotide conservation at a given position using information theory. The sequence logo is a graphic representation which displays the product of the information and the occurrence for each nucleotide (in bits) as shown in Table 3.2c. This visual representation gives a good indication of the motif specificity.

Binding specificity

How to determine the binding specificity of a TF? Site-directed mutagenesis or SELEX (systematic evolution of ligands by exponential enrichment)

2 http://www.chem.qmul.ac.uk/iupac/misc/naabb.html#p321

Table 3.3 Matrix representations of the LexA binding motif in *E. coli*

(a) Position count matrix

Position	1	2	3	4	5	6	7	8	9	10	11	12	13	14	15	16	17	18
A	6	4	0	0	0	2	7	0	3	2	6	2	6	0	5	0	10	0
C	0	3	10	0	0	0	2	0	0	1	1	3	0	3	4	10	0	0
G	1	2	0	0	10	2	0	1	4	1	1	0	2	2	1	0	0	10
T	3	1	0	10	0	6	1	9	3	6	2	5	2	5	0	0	0	0
Sum	10	10	10	10	10	10	10	10	10	10	10	10	10	10	10	10	10	10
consensus	D	N	C	T	G	D	H	K	D	N	N	H	D	B	V	C	A	G

(b) Position frequency matrix using for background frequencies F(A) = F(T) = 0.28, F(C) = F(G) = 0.22 and a pseudo-weight of 1

Position	1	2	3	4	5	6	7	8	9	10	11	12	13	14	15	16	17	18
A	0.57	0.39	0.03	0.03	0.03	0.21	0.66	0.03	0.30	0.21	0.57	0.21	0.57	0.03	0.48	0.03	0.93	0.03
C	0.02	0.29	0.93	0.02	0.02	0.02	0.20	0.02	0.02	0.11	0.11	0.29	0.02	0.29	0.38	0.93	0.02	0.02
G	0.11	0.20	0.02	0.02	0.93	0.20	0.02	0.11	0.38	0.11	0.11	0.02	0.20	0.20	0.11	0.02	0.02	0.93
T	0.30	0.12	0.03	0.93	0.03	0.57	0.12	0.84	0.30	0.57	0.21	0.48	0.21	0.48	0.03	0.03	0.03	0.03
Sum	1	1	1	1	1	1	1	1	1	1	1	1	1	1	1	1	1	1

(c) Position weight matrix calculated using log(observed frequency/background frequency)

Position	1	2	3	4	5	6	7	8	9	10	11	12	13	14	15	16	17	18
A	0.71	0.33	−2.40	−2.40	−2.40	−0.30	0.86	−2.40	0.06	−0.30	0.71	−0.30	0.71	−2.40	0.54	−2.40	1.21	−2.40
C	−2.40	0.29	1.44	−2.40	−2.40	−2.40	−0.09	−2.40	−2.40	−0.68	−0.68	0.29	−2.40	0.29	0.56	1.44	−2.40	−2.40
G	−0.68	−0.09	−2.40	−2.40	1.44	−0.09	−2.40	−0.68	0.56	−0.68	−0.68	−2.40	−0.09	−0.09	−0.68	−2.40	−2.40	1.44
T	0.06	−0.88	−2.40	1.21	−2.40	0.71	−0.88	1.10	0.06	0.71	−0.30	0.54	−0.30	0.54	−2.40	−2.40	−2.40	−2.40
Sum	−2	0	−6	−6	−6	−2	−3	−4	−2	−1	−1	−2	−2	−2	−2	−6	−6	−6

has been widely used to test *in vitro* affinity of a given TF on various sequences. Starting from a known binding site, site-directed mutagenesis can be used to introduce specific nucleotide substitutions in DNA sequences (Hutchison *et al.*, 1978). The binding affinity can be measured by testing the impact of the mutated promoter on the expression of a reporter gene in a plasmid. For example, the effect of site mutations on the *recA* promoter for LexA regulator has been tested in Mycobacteria (Davis *et al.*, 2002). Bulyk *et al.* (2001) developed a DNA microarray-based *in vitro* technology, called protein-binding microarrays (PBMs), that allows rapid characterization of the sequence specificities of TFs (Bulyk *et al.*, 2001; Mukherjee *et al.*, 2004). Recent application of this method with 8-mer oligonucleotides revealed distinct binding affinity profiles for half of tested mouse TFs (Badis *et al.*, 2009). SELEX is a large-scale method which was developed in order to identify the DNA binding specificity of a TF (Ellington and Szostak, 1990; Tuerk and Gold, 1990). Starting from a pool of random synthetic oligonucleotides, the DNA sequences which are bound by the TF are enriched after multiple cycles of three steps: binding with the protein, selection by EMSA and DNA amplification by PCR. The most enriched fragments will be the most specific ones. This method can easily be used to find the optimal consensus, but this consensus may not be biologically relevant. Recent adaptation of this method allows the determination of much more accurate models (Liu and Stormo, 2005). Finally, ChIP-Seq experiments have been recently suggested to be used for such purpose as tag densities at the binding sites can be a good indicator of protein–DNA binding affinity (Jothi *et al.*, 2008). Such experiments when done *in vitro* often characterize for a given purified TF the sequence that will give the best affinity. However, this optimal sequence may not be represented in the biological sites. It is important to stress the difference between the optimal sequence and the consensus. Theoretically, the optimal sequence of a PWM will be the sequence which gives the best score, while the consensus includes the information from all the sites that can be bound by the TF. Moreover, even if well conserved, binding sites can be more or less close to the optimal sequence and show different binding affinities.

The binding probability depends on the binding energy on a sigmoid way, thus generating a threshold between weak and strong-affinity TFBS. It is hypothesized that TFBS with low binding affinity require the TF to be present at sufficient concentrations to affect the transcription. On the opposite, strong sites will be very sensitive to low concentrations of the regulator. The last may be more represented in repressor binding sites.

Location of TF binding sites

In Bacteria, the regulatory region or promoter region, where TF binds to the DNA, is usually on the upstream region of the gene under regulation, *aka* target gene (TG). However, a small fraction (15%) of the known binding sites is located in the coding regions but is not necessarily functional (Shimada *et al.*, 2008; Terai *et al.*, 2001). The size of this promoter region is in average 400 bp. We previously also show that a threshold of 55 bp on the intergenic distance is necessary to predict the transcriptional units in *B. subtilis* and *E. coli* with 80% of accuracy (Janky and van Helden, 2008).

Previous studies on *E. coli* (Collado-Vides *et al.*, 1991; Madan Babu and Teichmann, 2003), gamma-proteobacteria (Espinosa *et al.*, 2005) and *B. subtilis* (Moreno-Campuzano *et al.*, 2006) have shown that binding sites are mainly distributed between −300 and +300 bp relative to the transcription start site (TSS). Binding sites corresponding to activators are generally located upstream of the TSS as they make the binding of the RNA polymerase and then the initiation of the transcription easier. On the other hand, TFs repressing the gene expression bind to different sites close to the TSS, and mostly downstream the TSS, preventing by steric hindrance the RNA polymerase to bind or to progress along the DNA sequence.

The main bioinformatics approach used to locate the potential binding sites in a genome from a known motif is called pattern matching. This approach consists in scanning the genome sequence with a matrix built from the known binding sites assigning a score to each potential binding site. This score reflects the probability of being a binding site instead of a background motif. For example, one can use the matrix of LexA described above to scan the promoter regions of *E. coli* to detect the putative sites which will have the highest scores, and then the target genes of the LexA regulon. The strategy uses here a PWM from a known motif, but it can be used also for the *de novo* prediction of TFBS as described in the next paragraph.

Discovering *de novo* TF binding sites

Various computational approaches have been developed to predict *de novo* TFBS in completely sequenced genomes. Most of the available motif discovery algorithms have been initially developed to predict regulatory motifs from co-expression data (chip). The first approach relies on the assumption that co-expressed genes are regulated by at least one common TF. The second approach relies on the hypothesis that, under selective pressure, TFBS will evolve at a slower rate than the surrounding non-coding region. It consists in applying the motif discovery algorithms on promoter sequences of orthologous genes to predict conserved regulatory motifs. Owing to the ever-increasing number of sequenced genomes, this last strategy also called phylogenetic footprinting (Tagle *et al.*, 1988) is particularly promising.

The motif discovery problem can be stated as follow: How to detect a biologically relevant DNA motif which can be as short as a hexamer and as large as 25 bp, which can be a spaced motif (with a space as large as 20 bp) and often degenerated (between five to ten informative nucleotides)? Two computational approaches have been developed to address this question: string based approach and matrix based approach.

Matrix based approach uses PWM to detect potential binding sites (Stormo, 2000). The PWM can be from a known motif as described in the section above (pattern-matching). On the other hand, a matrix can be artificially built from the set of promoter sequences and optimized to get the PWM which can have the most optimal score when scanning the promoter sequences. This optimization step can use probabilistic or deterministic methods. The methods using the Gibbs sampling can be classified as probabilistic methods [*Gibbs* (Lawrence *et al.*, 1993; Neuwald *et al.*, 1995), *Motif Sampler* (Thijs *et al.*, 2002)]. Another probabilistic method is using the greedy algorithm as implemented in the program *consensus* (Hertz *et al.*, 1990; Hertz and Stormo, 1999). One example of a deterministic method is the program MEME (Bailey and Elkan, 1995) using the principle of expectation maximization.

The matrix-based approaches are really sensitive for the detection of highly degenerated binding sites. However, these approaches are poorly specific as they return a lot of false positives. The optimal score threshold which discriminate the signal to noise depends on the quality of the matrix and should be assessed for each matrix separately. Additionally, it is difficult to estimate prior the analysis parameters such as the number of expected binding sites per set of sequences and the size of the motif. Last but not least, such analyses are highly demanding in computational time, being a substantial drawback for large scale analysis.

The string-based approaches used to detect statistically over-represented k-words (of size k) in the set of input sequences. In comparison to the matrix-based approaches, they are fast and exhaustive, i.e. all potential k-words are tested which allows the detection of distinct regulatory motifs without need to estimate the expected number of sites (van Helden *et al.*, 1998). A significance score assesses the over-representation of the k-word indicating the biological relevance of motif. Variant of this approach uses a template of spaced pair of tri-nucleotides (*dyads*) (van Helden *et al.*, 2000) which is particularly powerful for detection of bacterial spaced motifs (Janky and van Helden). This last tool is able to take into account the palindromic nature of TFBS. Another one permits the detection of under-represented motifs such as restriction sites (Vandenbogaert and Makeev, 2003) which is not possible with PWM matrices.

The main difficulty with this approach is to extract from a collection of exact words the real biological motif including its variability. There are different ways to circumvent this problem. A first possibility developed in RSAT web server (*pattern-assembly;* Thomas-Chollier *et al.*, 2008) is to start from the assembly of the overlapping words to extract and align the putative sites in the input sequences, and then extract a count-matrix which can be converted into a sequence logo. Another possibility is to allow the degeneracy since the description of the motif model (Sinha and Tompa, 2003); or one can use a suffix tree from a consensus sequence (Pavesi *et al.*, 2004).

One current trend is to combine several motif discovery tools in order to run them on the same input sequences at the same time and to compare the results: SCOPE runs three string-based methods allowing the detection of oligonucleotides (BEAM), allowing degeneracy (PRISM) or spaced motifs (SPACER) (Chakravarty *et al.*, 2007); Webmotifs was developed for on line queries (Romer *et al.*, 2007) and is running four methods (Weeder + Phylocon + AlignACE + MDscan); MotifVoter with 11 programs (Wijaya *et al.*, 2008); and Tmod which runs 12 motif discovery programs (Sun *et al.*, 2010). Such an approach sounds powerful but it tends to lose the control of the input parameters. In addition, tools which require specific parameters such as those dedicated to phylogenetic footprinting (e.g. phylocon) may not be used optimally. String based methods are poorly represented in such approach and may be included in the future.

All motif representations described above relied on the assumption of the independence of individual base-pair contributions while this is not always the case (Badis *et al.*, 2009; Berg and von Hippel, 1987). The interdependencies between nucleotides can be characterized by Markov models (Bulyk *et al.*, 2002; Wang and Hannenhalli, 2005). This computes the probability of each nucleotide at a given position and given the *m* previous nucleotides (*m* corresponds to the order of the Markov chain). These representations also assume that all primary sequences will have the same specific affinity while the difference between spurious and functional TFBSs largely depends on their degeneracy, their location relative to the transcription start site (Madan Babu and Teichmann, 2003), the pleiotropic role and the concentration of the corresponding regulator (Janga *et al.*, 2009; van Hijum *et al.*, 2009). Indeed, global regulators such as CRP tend to have low-affinity binding sites and must be expressed at high concentration to regulate their multiple target genes. On the other hand, specific regulators have high-affinity binding sites (Lozada-Chavez *et al.*, 2008; Martınez-Antonio *et al.*,2008).

Evolution of TF binding sites

Changes in *cis*-regulatory sequences constitute an important part of the genetic basis for adaptation.

Bacterial regulatory networks evolve by gene duplication

Gene duplication often occurs in prokaryotes (at least 50% of the genes) (Perez and Groisman, 2009a). Genes that have diverged after a duplication event are called paralogues. The question whether they may inherit the regulatory interaction was evaluated considering different scenarios (Teichmann and Babu): gain or loss of regulatory interaction of either a TF or a target gene, or both of them. As a result, more than one third of the known regulatory interactions in *E. coli* were inherited by gene duplication. This study also showed that half of the interactions were gained during divergence after duplication. Recently, an update of this evolutionary model has been suggested to integrate other regulatory actors such as sigma factors, small RNAs, RNA-binding proteins, and other reactions such as DNA supercoiling and protein–protein interactions (Martinez-Nunez *et al.*, 2010). In this review, all these regulatory elements were shown to potentially shape the regulatory networks of paralogous genes as illustrated in *B. subtilis* and *E. coli* (see also their integrative example of OmpF-OmpC).

The previous evolutionary models assumed that proximal TFBS will be duplicated with the target gene leading to the conservation of the regulatory interaction. However, conservation of the regulatory interaction does not necessarily imply conservation of the binding sites that are subject to less selective pressure than coding region. The fact that regulatory binding sites of biotin, arginine and ribonucleotide reductase are highly conserved may remain exceptional (Balleza *et al.*, 2009). In fact, even if the general properties of gene regulatory networks are conserved, the underlying cis-regulatory network is believed to diverge extensively (Zinzen and Furlong, 2008) and may play an important early role in the quick adaptation to new environments. To investigate the evolution of TFBS, it is thus necessary to perform comparative analyses within closely related species while early studies on bacterial regulatory network evolution have focused on a relatively small number of phylogenetically distant model organisms. This is now possible with the effort to sequence genomes of closely related species, but also genomes corresponding to various strains of a same species.

Several evolutionary models for regulatory changes have been proposed (Perez and Groisman, 2009a; Tanay *et al.*, 2005). Four categories of regulatory changes contributing to phenotypic diversity within closely related species have been recently reviewed (Perez and Groisman, 2009a): the presence or loss of TFBS for a conserved TF–TG interaction, embedding via Horizontal Gene Transfer (HGT), promoter architecture restructuring and coupled TF-TFBS modifications. Additionally, binding site divergence can be considered as a distinct class of regulatory change. See the following paragraphs and Fig. 3.1 for their description.

Rewiring by gain and loss

A regulon is the set of genes under regulation of the same TF. The loss or the acquisition of TFBS corresponding to a conserved TF can change its regulon even between close species having similar gene content (Lozada-Chavez *et al.*, 2006; Madan Babu *et al.*, 2006; Price *et al.*, 2007). For example, only 30% of the target genes are shared between the PhoP regulons of *S. enterica* and *Y. pestis* (Perez *et al.*, 2009). It has been suggested that the gain and loss of TFBS is the major mechanism contributing to the evolution of gene regulation in higher

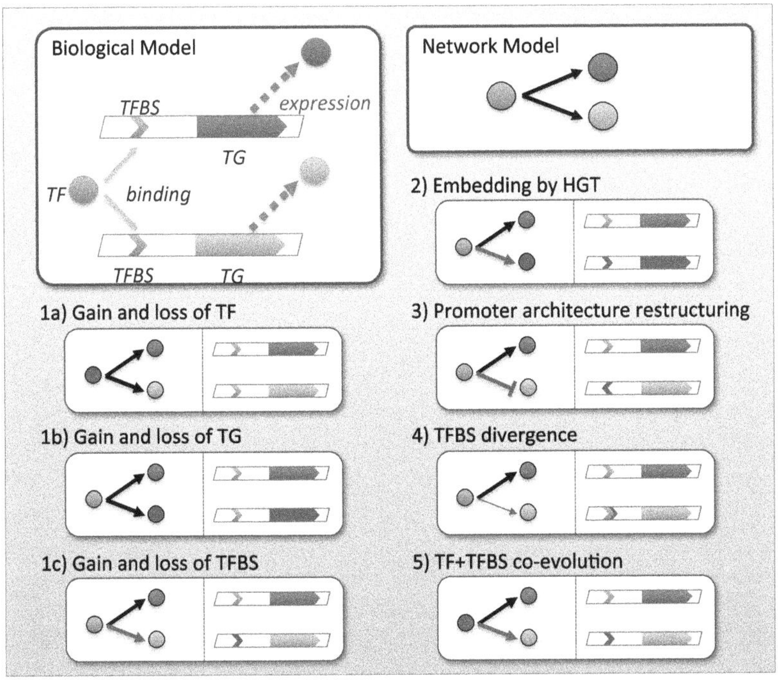

Figure 3.1 Evolutionary regulatory models in Bacteria. The biological model on the top left describes the regulation of two target genes (in blue and yellow) by the binding of a given TF (green sphere) on a specific TFBS (green chevron). Transcriptional regulatory networks use to represent a regulatory interaction by an edge directed from the TF to the TG (top right). We labelled from 1 to 5 the different models using red colour to highlight the regulatory change affecting one of the two genes being recently acquired in a bacterial species. We also represented how this corresponding novelty can affect the network model. Class 1: gain and loss of one of the regulatory actors such as (a) TF, (b) TG and (c) TFBS. Class 2: embedding of the new regulatory targets via HGT, i.e. the recently acquired TFBS and its new gene will evolve quickly to get regulated by a pre-existing TF. This situation is different to the class 1b where a new TG can be included within an operon with pre-existing TFBS. Class 3: Restructuring of the promoter architecture which implies changes in the orientation of the TFBS (as illustrated), but also the positioning of the TFBS, and to some extend their occurrence in the promoter region. The change in the orientation can have a consequence on the regulatory effect of a TF which can be seen on a transcriptional regulatory circuit showing activation and repression. Class 4: TFBS divergence such as mutation can lead to changes in the affinity of the TF or to gradually changes to an innovative TFBS. This can be observed in a weighted regulatory network where the weight on the edge indicates the strength of the regulation. Class 5: The TF and the TFBS can co-evolve together in general across different taxonomical groups. A colour version of this figure is located in the plate section at the back of the book.

eukaryotes (Carroll, 2008; Jeong *et al.*, 2008; Wray, 2007) as it can lead to a global rewiring of the transcriptional network (Tanay *et al.*, 2005). On the other side, a minimal regulatory change such as the loss of a key TF can be enough to affect the host specificity of different *Vibrio fisheri* strains (Mandel *et al.*, 2009).

Embedding by horizontal gene transfer

Horizontal gene transfer (HGT) is a common process among bacteria to transfer genetic material between different species. This process allows bacterial species to evolve rapidly by acquiring new functional genes and new phenotype that will bring species specificity. As the newly acquired genes by HGT are often controlled by ancestral transcription factors (Dorman, 2009), Perez and Groisman (2009a) suggested an embedding of the new regulatory targets into the ancestral regulatory networks. This is in particular due to the fast evolution of TFBSs of the transferred genes increasing the complexity of cis-regulatory interactions and improving co-regulation of physically interacting proteins (Lercher and Pal, 2008). As it allows genes to join or leave regulons independently depending on environmental circumstances, such evolutionary mechanism is suggested to play a major role in the evolution of TRN (Balleza *et al.*, 2009; Lercher and Pal, 2008; Madan Babu *et al.*, 2006).

Promoter architecture restructuring

The promoter architecture, i.e. position and orientation of the TFBSs, is crucial for gene transcription (Browning and Busby, 2004) and modifications in the promoter architecture can be a source of phenotypic variation. Perez and Groisman performed the comparison of the PhoP regulons in two enteric pathogens, *Salmonella enterica* and *Yersinia pestis*. This regulon is involved in the virulence and the adaptation to low-Mg^{2+} environments of both bacteria. However, functional changes in the architecture of the promoter regions as well as in the PhoP regulatory protein were observed (Perez and Groisman, 2009a). The orthologous TFs can bind to the same regulatory motif (PhoP box) preserving the control of the expression of ancestral common target genes, i.e. the core regulon. On the other hand, the location and the orientation of the PhoP Box differ in the promoter region of recently acquired target genes such as *Salmonella ugtL* and *Yersinia mgtC*. It is also possible to conceive that the regulatory effect of a given TF will be affected by the repositioning of TFBS. Examples of such regulatory divergence on the expression of paralogous genes has been highlighted by Martinez-Nunez *et al.* (2010). This regulatory effect may have also an impact on the evolution of the TFs as activators may be lost easily while repressors with many targets will tend to be conserved (Hershberg and Margalit, 2006).

Binding site divergence

The transcriptional regulation can be seen as a complex and dynamic equilibrium between the TF, the RNA polymerase and the cis-regulatory binding sites which output is the expression of the gene. It has been hypothesized that while paralogues may retain similar functions, their gene expression may diverge rapidly (Li *et al.*, 2005). This implies that single point mutations in the cis-regulatory region leading to changes of the expression will precede functional change of the coding region. Gradual divergence has been predicted for the regulatory motif of IFHL in upstream region of ribosomal proteins in Yeast species (Tanay *et al.*, 2005) and for the LexA auto-regulation in Gram-positives (Janky and van Helden, 2008). On the other side, a single cis-regulatory mutation is sufficient to change the phenotype of an organism as recently identified in the *Salmonella* genus (Osborne *et al.*, 2009). The authors show that a mutation in the promoter region of the gene *srfN* confers patho-adaptative fitness during enteric and typhoid disease in animals. Owing to the variability of the binding sites for a given TF and to the computational comparison of such motifs, it is still a challenge to identify such functional regulatory divergences.

Co-evolution of TF and TFBS

Divergence of TFBS may not necessarily lead to functional change as they may maintain the regulation of an orthologous target gene by a modified orthologous TF. Such co-evolution model is observed in general by comparing different taxonomic phyla while the regulatory motif will tend to be conserved across close related species. For example, distinct DNA binding motifs specific to various bacterial taxa have been characterized for the LexA regulator whose peptidic sequence is highly conserved across bacteria (Mazon *et al.*, 2004). Such co-evolution can be accurately predicted by *de novo* taxon-specific prediction of binding sites (Janky and van Helden, 2008). Finally, this model may apply to most conserved repressors which have been shown to co-evolve tightly with their target genes within close species, while activators tend to be lost easily (Hershberg and Margalit, 2006).

Conclusion

Many characteristics of TFBS such as the genomic position, the nucleotide content, the shape, the specificity, the regulatory effect and the evolution have been widely studied allowing us to get insight of the complexity of the transcriptional regulation in Prokaryotes. We focused in this chapter on a simple regulatory model involving only one TF having one TFBS at the promoter region of a given target gene. However, other levels of complexity can be added for a complete understanding of the evolution of regulatory networks and the phenotypic diversity. For example, genes are often under regulation of several TFs and the combinatorial impact of multiple TFBS in a given promoter region on the output, also called control logic, can be studied using Boolean logic gates (van Hijum *et al.*, 2009). As mentioned in the introduction, transcriptional initiation is a major step for regulation among others for the regulation of the gene expression. The importance of small molecules such as small RNA has emerged in the control of stress adaptation and pathogenesis (Liu and Camilli, 2010; Martinez-Nunez *et al.*, 2010).

Chapter highlights

- DNA and RNA binding sites can be detected in intergenic regions from the genomic sequence.
- Next generation sequencing (NGS) allows the *in vivo* detection of binding sites with a good resolution and a high sensitivity and specificity.
- Most bacterial binding sites are spaced motifs, also called dyads.
- Binding sites corresponding to activators are generally located upstream of the TSS while those corresponding to repressors are close to the TSS or downstream the TSS.
- Despite the large number of programs for *de novo* motif discovery, it is still a challenge to distinguish functional from spurious binding sites due essentially to the presence of low-affinity binding sites.
- Most regulatory changes contributing to phenotypic diversity within and across prokaryotic species can be put in different classes: (1) gain and loss of TFBS, (2) embedding via HGT, (3) promoter architecture restructuring, (4) binding site divergence and (5) co-evolution of TF and TFBS.

Acknowledgements

The author would like to thank Aiko Gryspeirt and A.J. Venkatakrishnan for critically reading this manuscript. This work was supported by the Medical Research Council Career Development Fellowship.

References

Abreu-Goodger, C., and Merino, E. (2005). RibEx: a web server for locating riboswitches and other conserved bacterial regulatory elements. Nucleic Acids Res. *33*, W690–692.

Abreu-Goodger, C., Ontiveros-Palacios, N., Ciria, R., and Merino, E. (2004). Conserved regulatory motifs in bacteria: riboswitches and beyond. Trends Genet. *20*, 475–479.

Alifano, P., Rivellini, F., Limauro, D., Bruni, C.B., and Carlomagno, M.S. (1991). A consensus motif common to all Rho-dependent prokaryotic transcription terminators. Cell *64*, 553–563.

Badis, G., Berger, M.F., Philippakis, A.A., Talukder, S., Gehrke, A.R., Jaeger, S.A., Chan, E.T., Metzler, G., Vedenko, A., Chen, X., *et al.* (2009). Diversity and complexity in DNA recognition by transcription factors. Science *324*, 1720–1723.

Bailey, T.L., and Elkan, C. (1995). The value of prior knowledge in discovering motifs with MEME. Proc Int Conf Intell Syst Mol Biol *3*, 21–29.

Balleza, E., Lopez-Bojorquez, L.N., Martinez-Antonio, A., Resendis-Antonio, O., Lozada-Chavez, I., Balderas-Martinez, Y.I., Encarnacion, S., and Collado-Vides, J. (2009). Regulation by transcription factors in bacteria: beyond description. FEMS Microbiol. Rev. *33*, 133–151.

Berg, O.G., and von Hippel, P.H. (1987). Selection of DNA binding sites by regulatory proteins. Statistical-mechanical theory and application to operators and promoters. J. Mol. Biol. *193*, 723–750.

Brown, T.A. (2002). Genomes, 2nd edn (Oxford, BIOS,).

Browning, D.F., and Busby, S.J. (2004). The regulation of bacterial transcription initiation. Nat. Rev. Microbiol. *2*, 57–65.

Bulyk, M.L., Huang, X., Choo, Y., and Church, G.M. (2001). Exploring the DNA-binding specificities of zinc fingers with DNA microarrays. Proc. Natl. Acad. Sci. U.S.A. *98*, 7158–7163.

Bulyk, M.L., Johnson, P.L., and Church, G.M. (2002). Nucleotides of transcription factor binding sites exert interdependent effects on the binding affinities of transcription factors. Nucleic Acids Res. *30*, 1255–1261.

Burmann, B.M., Schweimer, K., Luo, X., Wahl, M.C., Stitt, B.L., Gottesman, M.E., and Rosch, P. (2010).

A NusE:NusG complex links transcription and translation. Science *328*, 501–504.

Carroll, S.B. (2008). Evo-devo and an expanding evolutionary synthesis: a genetic theory of morphological evolution. Cell *134*, 25–36.

Chakravarty, A., Carlson, J.M., Khetani, R.S., DeZiel, C.E., and Gross, R.H. (2007). SPACER: identification of cis-regulatory elements with non-contiguous critical residues. Bioinformatics *23*, 1029–1031.

Ciampi, M.S. (2006). Rho-dependent terminators and transcription termination. Microbiology *152*, 2515–2528.

Collado-Vides, J., Magasanik, B., and Gralla, J.D. (1991). Control site location and transcriptional regulation in *Escherichia coli*. Microbiol. Rev. *55*, 371–394.

Davis, E.O., Dullaghan, E.M., and Rand, L. (2002). Definition of the mycobacterial SOS box and use to identify LexA-regulated genes in *Mycobacterium tuberculosis*. J. Bacteriol. *184*, 3287–3295.

deHaseth, P.L., Zupancic, M.L., and Record, M.T., Jr. (1998). RNA polymerase–promoter interactions: the comings and goings of RNA polymerase. J. Bacteriol. *180*, 3019–3025.

Dekhtyar, M., Morin, A., and Sakanyan, V. (2008). Triad pattern algorithm for predicting strong promoter candidates in bacterial genomes. BMC Bioinformatics *9*, 233.

Dorman, C.J. (2009). Regulatory integration of horizontally-transferred genes in bacteria. Front Biosci *14*, 4103–4112.

Ellington, A.D., and Szostak, J.W. (1990). *In vitro* selection of RNA molecules that bind specific ligands. Nature *346*, 818–822.

Eng, C., Asthana, C., Aigle, B., Hergalant, S., Mari, J.F., and Leblond, P. (2009). A new data mining approach for the detection of bacterial promoters combining stochastic and combinatorial methods. J Comput Biol *16*, 1211–1225.

Ermolaeva, M.D., Khalak, H.G., White, O., Smith, H.O., and Salzberg, S.L. (2000). Prediction of transcription terminators in bacterial genomes. J. Mol. Biol. *301*, 27–33.

Espinosa, V., Gonzalez, A.D., Vasconcelos, A.T., Huerta, A.M., and Collado-Vides, J. (2005). Comparative studies of transcriptional regulation mechanisms in a group of eight gamma-proteobacterial genomes. J. Mol. Biol. *354*, 184–199.

Hershberg, R., and Margalit, H. (2006). Co-evolution of transcription factors and their targets depends on mode of regulation. Genome Biol. *7*, R62.

Hertz, G.Z., Hartzell, G.W.d., and Stormo, G.D. (1990). Identification of consensus patterns in unaligned DNA sequences known to be functionally related. Comput Appl Biosci *6*, 81–92.

Hertz, G.Z., and Stormo, G.D. (1999). Identifying DNA and protein patterns with statistically significant alignments of multiple sequences. Bioinformatics *15*, 563–577.

Horak, C.E., and Snyder, M. (2002). ChIP-chip: a genomic approach for identifying transcription factor binding sites. Methods Enzymol. *350*, 469–483.

Huffman, J.L., and Brennan, R.G. (2002). Prokaryotic transcription regulators: more than just the helix–turn–helix motif. Curr. Opin. Struct. Biol. *12*, 98–106.

Hutchison, C.A., 3rd, Phillips, S., Edgell, M.H., Gillam, S., Jahnke, P., and Smith, M. (1978). Mutagenesis at a specific position in a DNA sequence. J. Biol. Chem. *253*, 6551–6560.

Janga, S.C., Salgado, H., and Martinez-Antonio, A. (2009). Transcriptional regulation shapes the organization of genes on bacterial chromosomes. Nucleic Acids Res. *37*, 3680–3688.

Janky, R., and van Helden, J. (2008). Evaluation of phylogenetic footprint discovery for predicting bacterial cis-regulatory elements and revealing their evolution. BMC Bioinformatics *9*, 37.

Jeong, S., Rebeiz, M., Andolfatto, P., Werner, T., True, J., and Carroll, S.B. (2008). The evolution of gene regulation underlies a morphological difference between two Drosophila sister species. Cell *132*, 783–793.

Johnson, D.S., Mortazavi, A., Myers, R.M., and Wold, B. (2007). Genome-wide mapping of in vivo protein–DNA interactions. Science *316*, 1497–1502.

Jothi, R., Cuddapah, S., Barski, A., Cui, K., and Zhao, K. (2008). Genome-wide identification of in vivo protein-DNA binding sites from ChIP-Seq data. Nucleic Acids Res. *36*, 5221–5231.

Lawrence, C.E., Altschul, S.F., Boguski, M.S., Liu, J.S., Neuwald, A.F., and Wootton, J.C. (1993). Detecting subtle sequence signals: a Gibbs sampling strategy for multiple alignment. Science *262*, 208–214.

Lercher, M.J., and Pal, C. (2008). Integration of horizontally transferred genes into regulatory interaction networks takes many million years. Mol Biol Evol *25*, 559–567.

Li, W.H., Yang, J., and Gu, X. (2005). Expression divergence between duplicate genes. Trends Genet. *21*, 602–607.

Liu, J., and Stormo, G.D. (2005). Combining SELEX with quantitative assays to rapidly obtain accurate models of protein–DNA interactions. Nucleic Acids Res. *33*, e141.

Liu, J.M., and Camilli, A. (2010). A broadening world of bacterial small RNAs. Curr. Opin. Microbiol. *13*, 18–23.

Lozada-Chavez, I., Janga, S.C., and Collado-Vides, J. (2006). Bacterial regulatory networks are extremely flexible in evolution. Nucleic Acids Res. *34*, 3434–3445.

Madan Babu, M., and Teichmann, S.A. (2003). Functional determinants of transcription factors in *Escherichia coli*: protein families and binding sites. Trends Genet. *19*, 75–79.

Madan Babu, M., Teichmann, S.A., and Aravind, L. (2006). Evolutionary dynamics of prokaryotic transcriptional regulatory networks. J. Mol. Biol. *358*, 614–633.

Mandel, M.J., Wollenberg, M.S., Stabb, E.V., Visick, K.L., and Ruby, E.G. (2009). A single regulatory gene is sufficient to alter bacterial host range. Nature *458*, 215–218.

Martinez-Nunez, M.A., Perez-Rueda, E., Gutierrez-Rios, R.M., and Merino, E. (2010). New insights into the regulatory networks of paralogous genes in bacteria. Microbiology *156*, 14–22.

Mazon, G., Erill, I., Campoy, S., Cortes, P., Forano, E., and Barbe, J. (2004). Reconstruction of the evolutionary history of the LexA-binding sequence. Microbiology *150*, 3783–3795.

Minezaki, Y., Homma, K., and Nishikawa, K. (2005). Genome-wide survey of transcription factors in prokaryotes reveals many bacteria-specific families not found in archaea. DNA Res *12*, 269–280.

Mockler, T.C., Chan, S., Sundaresan, A., Chen, H., Jacobsen, S.E., and Ecker, J.R. (2005). Applications of DNA tiling arrays for whole-genome analysis. Genomics *85*, 1–15.

Moreno-Campuzano, S., Janga, S.C., and Perez-Rueda, E. (2006). Identification and analysis of DNA-binding transcription factors in *Bacillus subtilis* and other Firmicutes--a genomic approach. BMC Genomics *7*, 147.

Mukherjee, S., Berger, M.F., Jona, G., Wang, X.S., Muzzey, D., Snyder, M., Young, R.A., and Bulyk, M.L. (2004). Rapid analysis of the DNA-binding specificities of transcription factors with DNA microarrays. Nat. Genet. *36*, 1331–1339.

Neuwald, A.F., Liu, J.S., and Lawrence, C.E. (1995). Gibbs motif sampling: detection of bacterial outer membrane protein repeats. Protein Sci. *4*, 1618–1632.

Osborne, S.E., Walthers, D., Tomljenovic, A.M., Mulder, D.T., Silphaduang, U., Duong, N., Lowden, M.J., Wickham, M.E., Waller, R.F., Kenney, L.J., *et al.* (2009). Pathogenic adaptation of intracellular bacteria by rewiring a cis-regulatory input function. Proc. Natl. Acad. Sci. U.S.A. *106*, 3982–3987.

Pavesi, G., Mereghetti, P., Mauri, G., and Pesole, G. (2004). Weeder Web: discovery of transcription factor binding sites in a set of sequences from co-regulated genes. Nucleic Acids Res. *32*, W199–203.

Perez-Rueda, E., Collado-Vides, J., and Segovia, L. (2004). Phylogenetic distribution of DNA-binding transcription factors in bacteria and archaea. Comput Biol. Chem. *28*, 341–350.

Perez, J.C., and Groisman, E.A. (2009a). Evolution of transcriptional regulatory circuits in bacteria. Cell *138*, 233–244.

Perez, J.C., and Groisman, E.A. (2009b). Transcription factor function and promoter architecture govern the evolution of bacterial regulons. Proc. Natl. Acad. Sci. U.S.A. *106*, 4319–4324.

Perez, J.C., Shin, D., Zwir, I., Latifi, T., Hadley, T.J., and Groisman, E.A. (2009). Evolution of a bacterial regulon controlling virulence and Mg(2+). homeostasis. PLoS Genet 5, e1000428.

Price, M.N., Dehal, P.S., and Arkin, A.P. (2007). Orthologous transcription factors in bacteria have different functions and regulate different genes. PLoS Comput Biol 3, 1739–1750.

Proshkin, S., Rahmouni, A.R., Mironov, A., and Nudler, E. (2010). Cooperation between translating ribosomes and RNA polymerase in transcription elongation. Science *328*, 504–508.

Rhodius, V.A., Suh, W.C., Nonaka, G., West, J., and Gross, C.A. (2006). Conserved and variable functions of the sigmaE stress response in related genomes. PLoS Biol. *4*, e2.

Robertson, G., Hirst, M., Bainbridge, M., Bilenky, M., Zhao, Y., Zeng, T., Euskirchen, G., Bernier, B., Varhol, R., Delaney, A., *et al.* (2007). Genome-wide profiles of STAT1 DNA association using chromatin immunoprecipitation and massively parallel sequencing. Nat Methods *4*, 651–657.

Romer, K.A., Kayombya, G.R., and Fraenkel, E. (2007). WebMOTIFS: automated discovery, filtering and scoring of DNA sequence motifs using multiple programs and Bayesian approaches. Nucleic Acids Res. *35*, W217–220.

Shimada, T., Ishihama, A., Busby, S.J., and Grainger, D.C. (2008). The *Escherichia coli* RutR transcription factor binds at targets within genes as well as intergenic regions. Nucleic Acids Res. *36*, 3950–3955.

Sinha, S., and Tompa, M. (2003). YMF: A program for discovery of novel transcription factor binding sites by statistical overrepresentation. Nucleic Acids Res. *31*, 3586–3588.

Stormo, G.D. (2000). DNA binding sites: representation and discovery. Bioinformatics *16*, 16–23.

Sun, H., Yuan, Y., Wu, Y., Liu, H., Liu, J.S., and Xie, H. (2010). Tmod: toolbox of motif discovery. Bioinformatics *26*, 405–407.

Tagle, D.A., Koop, B.F., Goodman, M., Slightom, J.L., Hess, D.L., and Jones, R.T. (1988). Embryonic epsilon and gamma globin genes of a prosimian primate (Galago crassicaudatus). Nucleotide and amino acid sequences, developmental regulation and phylogenetic footprints. J. Mol. Biol. *203*, 439–455.

Tanay, A., Regev, A., and Shamir, R. (2005). Conservation and evolvability in regulatory networks: the evolution of ribosomal regulation in yeast. Proc. Natl. Acad. Sci. U.S.A. *102*, 7203–7208.

Teichmann, S.A., and Babu, M.M. (2004). Gene regulatory network growth by duplication. Nat. Genet. *36*, 492–496.

Terai, G., Takagi, T., and Nakai, K. (2001). Prediction of co-regulated genes in *Bacillus subtilis* on the basis of upstream elements conserved across three closely related species. Genome Biol. *2*, RESEARCH0048.

Thijs, G., Marchal, K., Lescot, M., Rombauts, S., De Moor, B., Rouze, P., and Moreau, Y. (2002). A Gibbs sampling method to detect overrepresented motifs in the upstream regions of coexpressed genes. J Comput Biol 9, 447–464.

Thomas-Chollier, M., Sand, O., Turatsinze, J.V., Janky, R., Defrance, M., Vervisch, E., Brohee, S., and van Helden, J. (2008). RSAT: regulatory sequence analysis tools. Nucleic Acids Res. *36*, W119–127.

Toussaint, A., Ghigo, J.M., and Salmond, G.P. (2003). A new evaluation of our life-support system. EMBO Rep *4*, 820–824.

Touzain, F., Schbath, S., Debled-Rennesson, I., Aigle, B., Kucherov, G., and Leblond, P. (2008). SIGffRid: a tool to search for sigma factor binding sites in bacterial genomes using comparative approach and biologically driven statistics. BMC Bioinformatics *9*, 73.

Tuerk, C., and Gold, L. (1990). Systematic evolution of ligands by exponential enrichment: RNA ligands to bacteriophage T4 DNA polymerase. Science *249*, 505–510.

van Helden, J., Andre, B., and Collado-Vides, J. (1998). Extracting regulatory sites from the upstream region of yeast genes by computational analysis of oligonucleotide frequencies. J. Mol. Biol. *281*, 827–842.

van Helden, J., Rios, A.F., and Collado-Vides, J. (2000). Discovering regulatory elements in non-coding sequences by analysis of spaced dyads. Nucleic Acids Res. *28*, 1808–1818.

van Hijum, S.A., Medema, M.H., and Kuipers, O.P. (2009). Mechanisms and evolution of control logic in prokaryotic transcriptional regulation. Microbiol. Mol. Biol. Rev. *73*, 481–509, Table of Contents.

Vandenbogaert, M., and Makeev, V. (2003). Analysis of bacterial RM-systems through genome-scale analysis and related taxonomy issues. In Silico Biol *3*, 127–143.

Wang, J., and Hannenhalli, S. (2005). Generalizations of Markov model to characterize biological sequences. BMC Bioinformatics *6*, 219.

Wijaya, E., Yiu, S.M., Son, N.T., Kanagasabai, R., and Sung, W.K. (2008). MotifVoter: a novel ensemble method for fine-grained integration of generic motif finders. Bioinformatics *24*, 2288–2295.

Wray, G.A. (2007). The evolutionary significance of cis-regulatory mutations. Nat. Rev. Genet. *8*, 206–216.

Zinzen, R.P., and Furlong, E.E. (2008). Divergence in cis–regulatory networks: taking the 'species' out of cross-species analysis. Genome Biol. *9*, 240.

Operons and Prokaryotic Genome Organization

4

Sarath Chandra Janga and Gabriel Moreno-Hagelsieb

Abstract

An average of 60% of prokaryotic genes are organized into operons-polycistronic transcription units, making them a very important feature of their genomic organization. Operons most commonly contain genes whose products have functional associations and are abundant because they constitute an easy means for coregulation and the associated genes can act as a functional unit with a higher success rate in horizontal gene transfer events than single genes. Operons are transcribed from a single promoter, thus rarely needing genomic features between their constituting genes, naturally resulting in shorter distances between genes in operons than between adjacent genes in different transcription units. Thus, operons can be predicted based on distances between adjacent genes in the same DNA strand. This feature, intergenic distance, is the most informative criterion for predicting operons. However, predictions based on conservation of gene order followed by phylogenetic profiles, provide cleaner predictions, albeit with much lower coverage. Transcriptional terminators and other sequence features might add quality to operon predictions, but the gain is minimal for most prokaryotes. Operon organization is not well conserved with evolutionary divergence. However, operons rearrange in a functionally coherent manner. Thus, the combination of operon predictions with operon rearrangements constitutes the most powerful source for the prediction of functional associations by genomic context in prokaryotes.

Introduction

The first article describing operons defined them as 'a group of genes with expression coordinated by an operator' (Jacob *et al.*, 1960; Jacob *et al.*, 2005). This definition implies that only transcription units (TUs) containing more than one gene would be called operons. The definition has changed over time, with some authors arguing that the original definition included the regulatory protein. However, such a notion is not described in the original article. What might probably still be in dispute is whether an operator should be considered part of the operon or not. Today, most authors use operon to mean a TU producing a polycistronic mRNA, with other authors defining operons as the longest of a set of overlapping TUs, regardless of such TUs producing mono- or polycistronic mRNAs (Salgado *et al.*, 2006) (see Fig. 4.1a for definitions). These definitions would be more useful than one pretending to add regulatory proteins, and/or operators to the definition, since they allow for the discussion of TUs as operons whether the regulatory regions (promoters and operators) have been mapped or not. Leaving those arguments aside, in this chapter we will refer to operons as TUs producing polycistronic mRNAs unless stated otherwise.

The first operon to be described was the *lac* operon, whose three genes are involved in the utilization of lactose (Jacob and Monod, 1961). Over the years several examples of operons have been characterized both experimentally and computationally and it is now known that most genes in prokaryotes are organized into operons (Cherry, 2003; Moreno-Hagelsieb, 2006; Salgado *et al.*, 2000). Perhaps the articles listing most of

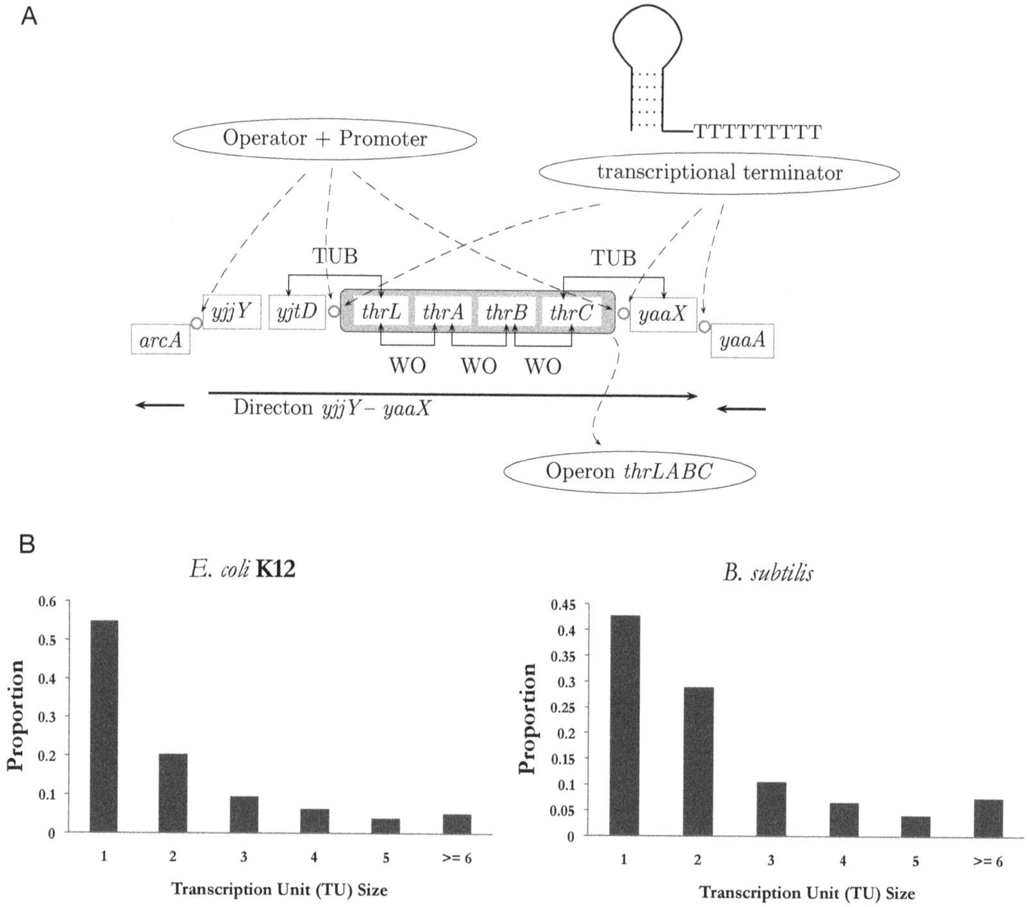

Figure 4.1 (a) Schematic figure showing the different features that have been used for predicting transcription units. The figure shows the idealized region between genes *arcA* and *yaaA*, which flanks the directon (a stretch of genes transcribed in the same direction with no intervening gene transcribed in the opposite direction) stretching from gene *yjjY* to gene *yaaX*. Solid horizontal arrows indicate direction of transcription. Transcription units are flanked by a promoter at the 5′ terminus, and a terminator signal at the 3′ end (terminator here exemplified by a Rho-independent terminator). The promoter can be accompanied by an operator sequence, where transcriptional regulators bind to DNA thus regulating transcription. The figure also shows within operon (WO) pairs of genes, and transcription unit boundary (TUB) pairs, commonly used for training methods for predicting operons. (b) Distribution of the sizes of experimental characterized transcriptional units in *E. coli* and *B. subtilis* genomes as documented in RegulonDB and DBTBS databases, respectively (Gama-Castro *et al.*, 2008; Sierro *et al.*, 2008).

the explanations for the existence of operons are those proposing the idea of the selfish operon (Lawrence, 1999; Lawrence, 1997; Lawrence and Roth, 1996). These explanations mix two things: [a] Mechanisms responsible for finding the genes together, and [b] the selective advantage of the operon structure. The prevailing notion about the advantage of operon organization is that it provides a means to coordinated expression of genes involved in a given function (Jacob and Monod, 1961; Jacob *et al.*, 1960, 2005; Price *et al.*, 2005b, 2006a). The selfish operon theory shifts the advantage of operon organization from the organism carrying the operon to the genes in the operon themselves, which would be more successful than separate genes in horizontal gene transfer (HGT) events by bringing a complete function into a naïve genome (Lawrence, 1997, 1999; Lawrence and Roth, 1996). These notions will be further discussed below as appropriate.

Operon structure in prokaryotic genomes

Operons might be a consequence of a higher order tendency of genes with related functions to be located somewhat close in prokaryotic genomes. For instance, it has been shown before that genes catalysing nearby biochemical steps in small molecule metabolism tend to be located nearby in the chromosome, apparently, operons explained part, but not all, of this tendency (Rison *et al.*, 2002). Other researchers have noted that genes whose products work in the same pathways annotated in EcoCyc (Keseler *et al.*, 2009) also tend to be located nearby in the chromosome (Yellaboina *et al.*, 2007). In the same way, TUs regulated by low-concentration transcription factors (TFs), tend to be closer together than those regulated by global high-concentration TFs (Janga *et al.*, 2009). Thus, operons might be part of such higher order genome organization, but not the whole story.

It is possible to estimate the total number of genes to be found in operons (Cherry, 2003; Moreno-Hagelsieb, 2006), as well as the total number of TUs (Cherry, 2003; Ermolaeva *et al.*, 2001; Moreno-Hagelsieb, 2006), by looking for simple clues before even attempting to predict the operons themselves. These estimates consist on counting the total number of directons, stretches of genes transcribed in the same direction with no intervening gene in the opposite strand (Fig. 4.1a) (Salgado *et al.*, 2000), containing a single gene, which would most probably correspond to monocistronic TUs. Since the probability for any TU to be isolated into a directon is 0.25, multiplying these monogenetic directons by 4 should give us the total number of monocistronic TUs. The remaining number of genes would then correspond to genes in operons. By this method, the estimated average proportion of genes in operons in Prokaryotes is 0.59, 0.60 in Bacteria, and 0.48 in Archaea (see supplementary website). In *E. coli* and *B. subtilis* these estimated proportions stand at 0.52 and 0.56, respectively. In agreement, the observed proportions for currently documented operons in *E. coli* (Gama-Castro *et al.*, 2008) and *B. subtilis* (Sierro *et al.*, 2008) stand at 0.50 and 0.57 (Fig. 4.1b).

The calculations above might correctly predict the total number of operons if they are defined as the longest of all overlapping TUs. However, it seems like a large proportion of operons contain internal TUs encompassing fewer genes. According to the annotations in RegulonDB (Gama-Castro *et al.*, 2008), ~10% of the operons of *E. coli* K12 contain internal TUs, while according to the DBTBS (Sierro *et al.*, 2008), ~24% of the operons of *B. subtilis* contain internal TUs. Recent tiling array experiments confirm the abundance of alternative TUs at 43% in the Archaea *Halobacterium salinarum* NRC-1 (Koide *et al.*, 2009). These data mean that the transcriptome organization might not be as simple as previously thought, and that it is possible that the transcription of subsets of genes within an operon will happen under different conditions, thus further complicating the picture of gene expression and its regulation.

At a more detailed level of organization, operons tend to contain, not only genes with related functions, but also genes used in nearby biochemical steps (Kovacs *et al.*, 2009; Rison *et al.*, 2002). Thus, a following question would be that of collinearity: whether the genes in the operon tend to occur in the same order as that in which the encoded proteins would be used to accomplish an overall function. A recent report has shown that genes in lowly expressed operons tend to show more significant collinearity than genes in highly expressed genes, as predicted with stochastic models of enzyme kinetics which predicted the need for collinearity in such operons to avoid stochastic stalling (Kovacs *et al.*, 2009). Table 4.1 summarizes some of the properties that distinguish genes in operons from adjacent genes occurring in different TUs.

Computational methods for identifying operons

Predicted operons can be useful for two main reasons: (i) for complementing the regulatory network and patterns of gene expression of a Prokaryote of interest; and (ii) for finding functional relationships among gene products. However, it should be clear that, while most operons contain genes with related functions, contradictory examples of genes with no evident related functions exist, and that adjacent, independently transcribed, TUs can still contain genes

Table 4.1 Some common features of genes and their intergenic regions in operons as understood from genome-scale studies in various bacterial genomes. Note that most of these properties are derived using the currently available documented regulatory and operon maps for *E. coli* (Gama-Castro *et al.*, 2008) and *B. subtilis* (Sierro et *al.*, 2008) genomes

Property	Observations
Average size	2.2 genes per operon
Longest operon	15 genes in *E. coli* and 22 genes in *B. subtilis*
Oligonucleotide composition	Upstream regions of genes within operons show different signatures compared to those which are in different transcription units, as the later contain promoter-like and other regulatory elements in their intergenic regions (Huerta *et al.*, 2006; Janga *et al.*, 2006)
Intergenic distance	50bp threshold is typically used as a rule of thumb for predicting adjacent genes to be operons
Average number of regulatory signals	Bigger operons tend to have more complex and higher number of regulatory signals than smaller ones (Price *et al.*, 2005b, 2006a)
Terminator type	No known association between the operon structure and the type of terminator. Firmicutes were found to be abundant in rho-independent terminators (de Hoon *et al.*, 2005)
Expression level	Expression of the genes in operon decreases from 5′ to 3′
Half-lives of transcripts	Stability of the transcripts encoded by genes in operons decrease from 5′ to 3′ (Selinger *et al.*, 2003)
Functional relationship	Genes are generally associated with the same functional category but there are many examples of unrelated genes clustered into operons suggested to be due to the formation of new operons (Price *et al.*, 2006a; Salgado *et al.*, 2000)

with functions related to each other. Thus, the correct prediction of genes transcribed into a single mRNA might somewhat conflict with the correct prediction of functional associations. The distinction is also important because methods using functional classifications, such as annotations from GenProtEc (Serres *et al.*, 2004), EcoCyc (Keseler *et al.*, 2009), clusters of orthologous groups (COGs) (Tatusov *et al.*, 2003; Tatusov *et al.*, 1997, 2001), and GO terms (Gene-Ontology-Consortium, 2010; Harris *et al.*, 2004), would be inadequate for predicting operons, given that common annotations already imply a functional association. Thus, using these features to predict operons seems somewhat circular. Not surprisingly, methods incorporating such annotations into their framework tend to have the highest accuracies at predicting operons (Dam *et al.*, 2007; Salgado *et al.*, 2000). However, the use of these features might be justified when the aim is to model transcriptomes later on. Methods based on structural features, such as intergenic distances or probable signals, plus evolutionary features, such as conservation of gene order and similarity of presence/absence vectors (phylogenetic profiles), would instead be more adequate for predicting functional associations and in understanding uncharacterized genomes.

The first article centred at predicting TUs in general used Hidden Markov models (HMMs) mixing features such as ribosomal binding sites, promoters, terminators, and coding sequences into a single predicting package (Yada *et al.*, 1999). The authors ended with a low proportion of correctly predicted TUs (0.6) and the proposal that such predictions were an open problem (Yada *et al.*, 1999). However, a previous work, embedded within the publication of the *Escherichia coli* K12 genome (Blattner *et al.*, 1997), had attained somewhat better results using an intergenic distance of 50 bp for predicting TUs. The idea that genes inside operons would be separated by short distances came from these regions not being expected to contain any signals, while the distances between genes at transcription unit boundaries would be longer because they would contain at least a promoter region, often also an operator (a binding site for a transcription factor), and perhaps transcription termination signals. Thus, the next study centred on predicting operons used intergenic distances and functional annotations (Salgado *et al.*, 2000) to predict operons in *E. coli* K12. The method consisted on calculating a log-likelihood (*llh*), given a distance (*d*), for two adjacent genes in the same strand to be in the same operon using the formula:

$$(llh|d) = \log(Pop/Ptub) \qquad (4.1)$$

where *Pop* and *Ptub* are the proportions of pairs of genes in the same operon (within-operon or WO pairs) and pairs of adjacent genes in the same strand, but different TU (TU boundaries or TUB pairs), respectively (Fig. 4.1a), found at such distance, in the training datasets. The attained accuracies (proportion of correctly predicted WO pairs and TUB pairs), using distance alone, were above 0.8. The expected distances for genes in the same operon were later confirmed for pairs of genes showing conservation of adjacency at long evolutionary distances (see next paragraph) (Moreno-Hagelsieb and Collado-Vides, 2002a,b; Moreno-Hagelsieb *et al.*, 2001), thus allowing for the prediction of TUs using intergenic distances in most of the available prokaryotic genomes (Moreno-Hagelsieb and Collado-Vides, 2002b). Later methods incorporated, again, several features aimed at improving the prediction of operons. All of these works confirmed intergenic distances as the most informative feature in the prediction of operons (Bockhorst *et al.*, 2003a,b; Dam *et al.*, 2007; Price *et al.*, 2005a), with other features improving accuracy.

The second most successful feature for predicting operons has been conservation of adjacency (Ermolaeva *et al.*, 2001; Huynen *et al.*, 2000; Janga and Moreno-Hagelsieb, 2004; Marcotte *et al.*, 1999b; Overbeek *et al.*, 1999; Pertea *et al.*, 2009). In this regard, one of the most elegant ideas was published by Ermolaeva *et al.* (Ermolaeva *et al.*, 2001). Basically, the method consisted on comparing the conservation of pairs of adjacent genes in the same strand, against the conservation of adjacent genes at different strands, with the latter representing random conservation. These predictions attained a small coverage (proportion of known WO pairs predicted), but high positive predictive value (true positives divided by the sum of true and false positives). Despite the low coverage, these predictions have been useful at confirming the tendencies for genes in operons to follow an intergenic distance distribution similar to those of *E. coli* K12 (Moreno-Hagelsieb, 2006; Moreno-Hagelsieb and Collado-Vides, 2002b). The idea also allowed for the development of methods for predicting operons using genome-specific features estimated using these two types of data sets as guidance and logistic regression for the proper weighting of each feature (Price *et al.*, 2005a). Further extension of the idea allowed for the evaluation of predictions using phylogenetic profiles (Moreno-Hagelsieb and Janga, 2008). This involved the assumption that most similar phylogenetic profiles would tend to increase the number of predicted associations between adjacent gene pairs from the same strand compared to those between opposite strand pairs (Moreno-Hagelsieb and Janga, 2008). It should be noted that operon predictions based on conservation of gene order and on phylogenetic profiles, as guided by these data sets, produced excellent results for difficult genomes with unusual intergenic distances (Ermolaeva *et al.*, 2001; Janga and Moreno-Hagelsieb, 2004; Moreno-Hagelsieb and Janga, 2008), albeit with low coverage.

As mentioned, other features are related to detecting signals such as promoters or transcription terminators (Bockhorst *et al.*, 2003a,b; Craven *et al.*, 2000; De Hoon *et al.*, 2004; de Hoon *et al.*, 2005; Yada *et al.*, 1999), or just assuming that such signals might render a different DNA signature (frequencies of oligonucleotides) for WO pairs as compared to those of TUBs (Janga *et al.*, 2006). It is worth mentioning that while most of these signals add slightly to the accuracies attained with intergenic distances alone, a method based only on predicting Rho-independent transcription terminators, using a decision rule based on the Gibbs energy of formation of the terminator stem–loop and the length of the T-stretch following the stem–loop, achieved accuracies of about 0.94 on average among the 57 Firmicutes, such as *B. subtilis*, available at the time of the study (de Hoon *et al.*, 2005).

The post-genomic era has seen the development of a number of resources and databases for understanding operon structure across prokaryotes. Table 4.2 summarizes some of the publicly available resources for the analysis of a large number of sequenced genomes, along with

Table 4.2 Databases and resources for understanding operon organization across prokaryotes. Last column briefly details the method of prediction and the number of genomes for which predictions are available in the resource

Database	URL	Description
DOOR (Mao *et al.*, 2009)	http://csbl1.bmb.uga.edu/OperonDB	This prediction program employs two classifiers, one for genomes with substantial number of experimentally validated operons and the other for genomes with only limited or no experimental data. For the first case, the program trains on a subset of the known operons, using a non-linear (decision tree-based) classifier utilizing both general features of genomes and genome-specific features, while for the second case, it is trained only on general features of genomes using a linear (logistic function-based) classifier. Predictions are available for 675 genomes
OperonDB (Pertea *et al.*, 2009)	http://operondb.cbcb.umd.edu/	This operon prediction method uses a purely computational and statistical approach, relying on conservation of gene order and orientation in two or more species to infer operon structure. Although highly specific (with fewer than 2% false positives), the sensitivity of the algorithm is estimated at 30–50% for the *Escherichia coli* genome. The current version of OperonDB claims to have greater sensitivity while maintaining a high level of specificity. Predictions are available for 550 genomes
ODB (Okuda *et al.*, 2006)	http://www.genome.sk.ritsumei.ac.jp/odb/	This system integrates four types of associations: genomic context, gene co-expression obtained from microarray data, functional links in biological pathways and the conservation of gene order across genomes. Combination of these indicators is employed for the reliable prediction of operons. Predictions are validated against known operon information obtained from the literature. Predicted and experimentally validated operons, wherever documented in the literature, are available for 203 genomes
Microbesonline (Dehal *et al.*, 2010)	http://www.microbesonline.org/	This system uses genome-specific features like the intergenic distance distribution specific to the organism of interest, estimated using a set of high-confidence operons identified by conservation of gene order and then applies logistic regression for the proper weighting of different features. Predictions are currently available for more than 1000 genomes
OperomeDB (Moreno-Hagelsieb and Collado-Vides, 2002b; Salgado *et al.*, 2000)	http://www.mrc-lmb.cam.ac.uk/genomes/sarath/genomeorg/or http://microbiome.wlu.ca/research/predicting-transcription-units/	This method is the first powerful approach for predicting operons across genomes and employs the idea that the intergenic distance distribution of genes in operons is different from that of genes which occur in different transcription units. It scores the predictions using a log-likelihood approach and attains a minimum accuracy of 85% when benchmarked against characterized genomes. Operon predictions are currently available for 1032 genomes

the approaches implemented in them to predict operons.

Evolution of operon organization

Clustering of functionally related genes plays an important role in the organization of the bacterial chromosome and several mechanisms have been proposed to explain it (Janga and Moreno-Hagelsieb, 2004; Lawrence and Roth, 1996; Ochman *et al.*, 2000; Omelchenko *et al.*, 2003; Price *et al.*, 2005a, 2006a; Rocha, 2008). There have been controversies about the sufficiency of each of these mechanisms to explain both the appearance of such gene clusters and their co-regulation via a single operator upstream of the first gene. In particular, as mentioned above, researchers have proposed that the main selective advantage of operons can be either of two: the convenience of co-regulation, or success in Horizontal Gene Transfer (HGT) and/or in genetic duplication.

Studies by Lawrence and co-workers claimed that there was no published explanation that could account for both gene clustering and an advantage for co-regulation from a single operator in operons. They suggested that organization into operons was selfish, meaning that it was advantageous for the genes in the operon rather than to the host genome, as such an organization resulting in co-transcription would increase the probability of success in HGT because an operon would more easily add a function to a naïve genome, compared to isolated, and/or independently transcribed, genes. Indeed, operons were shown to undergo HGT to a considerable extent, in a variety of genomic studies (Ochman *et al.*, 2000; Omelchenko *et al.*, 2003). However, one of the premises of the selfish operon hypothesis was that the clustering of genes and their subsequent organization into operons was particularly beneficial for weakly selected but functionally coupled genes. Yet, an analysis of the operon organization of the informational genes in *E. coli* has shown that such strongly selected genes can be more clustered or at least as likely to be clustered as are other less-essential genes, suggesting that selfishness does not provide a general explanation for operons (Pál and Hurst, 2004). In addition, the selfish operon also failed to explain the many experimentally known operons that contained functionally unrelated genes grouped together (Price *et al.*, 2006a). A second hypothesis about how operons might be convenient extends from the rationale that operons can be duplicated just as genes do thus easily providing groups of reactions that can work under different conditions. Group duplications have been observed both at the level of particular pathways as well as at the genomic level (Janga and Moreno-Hagelsieb, 2004; Xie *et al.*, 2003). Further support for this hypothesis is that ~20–25% of the operons have formed by duplication in most bacterial genomes (Janga and Moreno-Hagelsieb, 2004).

Operons require not only the coming together of genes but also the establishment and maintenance of coordinated regulation. Therefore, a likely explanation for the establishment of operons must be concerned with co-regulation. Indeed, the regulatory model provides several potential explanations for the existence of operons (Rocha, 2008). A favourable argument in support of this model is that a set of genes with expression coordinated from a single regulatory region would be easier to keep working together despite addition of more complex regulatory signals in comparison to the required concerted evolution necessary for maintaining the co-regulation of independent TUs for a particular environmental condition. Therefore, the dependence of several genes on a single regulatory sequence is both easier to maintain and allow for the emergence of more complex strategies over time. The finding that TUs with more complex regulatory regions tend to also be those TUs with a higher number of genes further supports this model (Price *et al.*, 2005b, 2006a). One reason for Lawrence and collaborators to reject co-regulation as a main advantage of operons was the supposition that optimal co-regulation was easier to attain using separate regulatory signals (Lawrence, 1997, 1999; Lawrence and Roth, 1996). The results just described support the idea that optimality in regulation might come with a cost for evolutionary stability and evolvability, which suggested that operons might be suboptimal in the sense of sacrificing optimal regulation for ease of co-regulation (Price *et al.*, 2006a; Price *et al.*, 2005b). However, it has been shown that highly expressed operons tend to have longer spacing between genes than other operons, suggesting the existence of internal signals for regulatory fine-tuning (Price *et al.*, 2006a). In addition laboratory studies have shown that the expression level of the *lac* operon evolves to optimality in just a few hundred generations suggesting that the optimal spacing observed in operons could reflect fine tuning of expression levels (Dekel and Alon, 2005). These observations are in line with more recent results from high-quality transcriptomic studies showing the existence of extensive internal regulatory elements in operons which are employed in a condition specific manner, suggesting a previously under-appreciated level of complexity in operon regulation in bacterial genomes (Guell *et al.*, 2009; Koide *et al.*, 2009; McGrath *et al.*, 2007; Sorek and Cossart, 2010).

The regulatory principles governing the organization of operons such as rapid and reliable expression control may also be required for the transcription factor gene to be in close proximity to the operators within the genome to which it binds (Janga *et al.*, 2009; Kolesov *et al.*, 2007; Korbel *et al.*, 2004). In prokaryotes, transcription and translation are coupled spatially and temporally, which could enable an efficient synthesis of transcription factors (TFs) near the genes that they regulate, enabling rapid binding of co-localized sites and tight coordinated regulation. This becomes an important constraint for lowly expressed TFs regulating few genes, in contrast to global regulators which are expressed at much higher levels (Janga *et al.*, 2009).

An additional argument in support of the regulatory principles governing the evolution of operons is that transcription of genes into a single mRNA is expected to diminish gene expression noise and ensure more precise stoichiometry (Rocha, 2008). Indeed, highly conserved operons tend to code for components of protein complexes or are part of metabolic operons where stoichiometry plays an important role (Dandekar *et al.*, 1998; Pál and Hurst, 2004; Zheng *et al.*, 2002). Nevertheless, only a very small fraction of the high-throughput determined protein–protein interactions involve genes encoded by the same operon (Butland *et al.*, 2005; Hu *et al.*, 2009) suggesting that more generally protein–protein interactions cannot be predicted using operonic or genomic context unlike previously thought (Enright *et al.*, 1999; Marcotte *et al.*, 1999a).

The birth and death of operons is a continuous process and the availability of genome sequences for hundreds of Prokaryotes provided a wealth of information to address these questions on a genome-wide scale by comparing the genome organization of organisms to infer the evolutionary life cycle of operons (Price *et al.*, 2006a). According to the selfish operon theory, HGT represents the prime force for the formation of operons and if this is the case the genes within operons having homology with genes from other sequenced organisms should be readily identified as the products of HGT. However, the finding of a number of new operons containing functionally unrelated genes as well as ORFans – defined as those which do not have identifiable homologues outside a group of related bacteria – supported the idea that selfish operons can not alone explain the formation of operons (Price *et al.*, 2005b, 2006a). Following these arguments, Price *et al.* (2006b) proposed that three main mechanisms are responsible for the formation of operons (Fig. 4.2). These include insertion of ORFan genes, which are frequently acquired with phages, onto the 3′ end of the native promoter of an already existing transcript (Fig. 4.2a). This arrangement may be selected for because the ORFan gene is transcribed from a

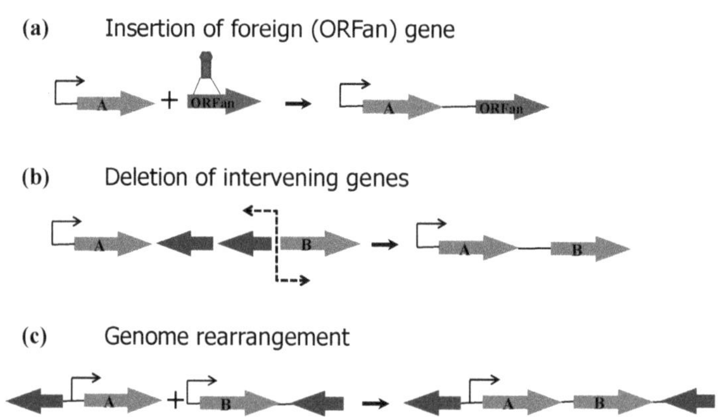

Figure 4.2 Different mechanisms responsible for the formation of operons in prokaryotes. Operons can be formed (a) by the insertion of a novel ORFan gene, obtained from phages, downstream of an already existing native gene so that it is transcribed under the influence of its native promoter. In this scenario, it is also possible that two or more ORFans which have already formed an operon in another organism could be transferred horizontally into the vicinity of a native gene so that the cluster is under the influence of the native promoter after purifying selection; (b) by the deletion of intervening genes (shown as unlabelled genes) on the chromosome which can then lead to the formation of an operon; (c) by a series of genomic re-arrangements that can bring together genes under the influence of different promoters.

native promoter without perturbing the expression of the native gene. A second mechanism could be the deletion of intervening genes or regions between two different genes, which are otherwise chromosomally far apart (Fig. 4.2b). Finally, operons can also be formed owing to a series of genomic rearrangements that can bring relevant genes together (Fig. 4.2c).

Given that very few operons are conserved across several genomes, it is clear that gene order is not conserved and hence some operons die after they are formed (Mushegian and Koonin, 1996; Price *et al.*, 2006a). Just like operons can be formed, they can be lost by the deletion of one or more genes or, alternatively, by splitting the operon apart. Since operon formation often brings functionally related genes together, it seems unlikely for rearrangements to be neutral processes. If operon formation is driven by gene expression, then it should be associated with changes in the expression patterns of the constituent genes. Preliminary studies using experimental data on the expression patterns of genes in *E. coli* operons and comparing their expression levels to orthologous genes in not-yet operons in *Shewanella oneidensis* MR-1 suggested that operon formation has a major effect on gene expression patterns (Price *et al.*, 2006a). However, it is not clear from these studies if these principles are more generic. It is neither apparent whether and how do factors such as phylogenetic distance and similarity in lifestyle play a role in the evolution of operon structures.

Based on these studies, it is possible to imagine that operon evolution could be accompanied by switching between constitutive and inducible expression. Constitutive expression appears a fairly expensive process thus questioning its possibility especially for newly formed operons. Nevertheless, favourable gene combinations cannot be selected for unless the genes are co-expressed in the first place. Therefore, constitutive gene expression, even at very low levels, would logically be expected to be necessary to enable the selection-mediated establishment of beneficial combinations of genes with subsequent fine-tuning of expression. In this context it is worth mentioning that genes within the same operon are under much stronger selection to remain together than genes that are not in operons but are otherwise adjacently located on the same strand of the chromosome or are in different operons (de Daruvar *et al.*, 2002; Ermolaeva *et al.*, 2001; Moreno-Hagelsieb *et al.*, 2001). Likewise, there is extensive evidence for a higher-level association between operons (Che *et al.*, 2006; Janga *et al.*, 2005; Lathe *et al.*, 2000) suggesting that although operon structure is unstable it provides an excellent platform for generating relevant information to assign functions to uncharacterized genes in prokaryotic genomes (Gonzalez *et al.*, 2006; Hu *et al.*, 2009; Janga *et al.*, 2005) (see Fig. 4.3 for an example of such higher level organization and section below).

Functional associations from operon rearrangements

One of the early discoveries of comparative genomics was the little conservation of gene order across bacterial genomes (Mushegian and Koonin, 1996). A later study concluded that this lack of conservation extended to the operon structure (Itoh *et al.*, 1999). Other studies seemed to contradict this lack of conservation. For instance, the comparison of adjacent genes in the same operon against genes at TUBs shows that WO pairs tend to be better conserved than genes at TUBs; and we have discussed that pairs of genes adjacently conserved in evolutionarily distant genomes indicate association into operons. However, there is no contradiction: WO pairs are better conserved than TUB pairs, yet, their conservation is still low (Janga and Moreno-Hagelsieb, 2004; Moreno-Hagelsieb and Collado-Vides, 2002a; Moreno-Hagelsieb *et al.*, 2001); this is why predictions based on conservation of gene order, while very clean, lack in coverage (Ermolaeva *et al.*, 2001; Janga and Moreno-Hagelsieb, 2004). The little conservation was soon noted as an advantage for functional inference (Galperin and Koonin, 2000; Lathe *et al.*, 2000). If operons are not conserved, but they reorganize into new operons, their reorganization might reveal other members of a coordinated, or functionally associated group of genes that might be dispersed among different TUs in one genome, but belong to

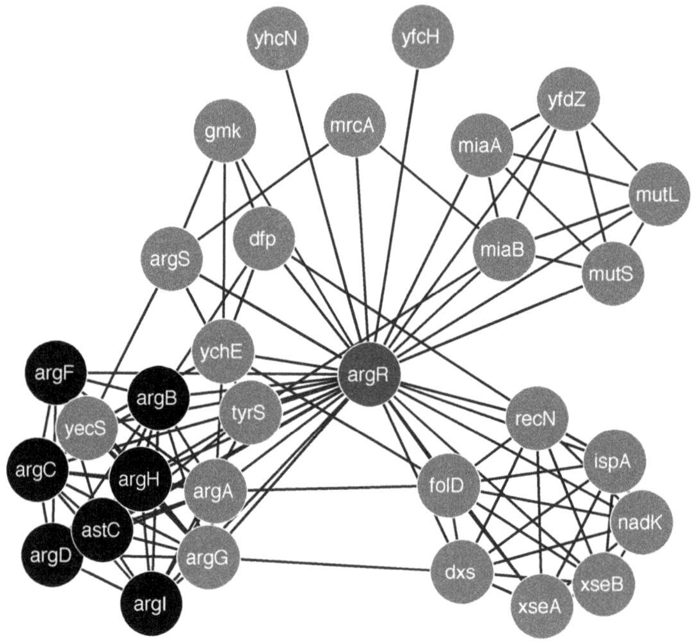

Figure 4.3 Example showing predictions produced using the higher-level reorganization of the genomic context of the *argR* gene across different prokaryotic genomes. This gene encoding for a transcription factor is not found in the same operon as the genes regulated by it in *Escherichia coli*. However, its re-association into different operons across other available genomes allows for the prediction of its association with genes known to be regulated by ArgR in *E. coli* (shown in black are those experimentally confirmed to be transcriptionally controlled by ArgR), and other groups of genes, mostly related to DNA recombination and repair, which have been shown to also be regulated by ArgR in other organisms.

a single operon in another (Fig. 4.3). Galperin and Koonin (2000) noted that the lack of a method for predicting operons by the time they suggested this idea was still an obstacle for finding these operon rearrangements. The concept of a group of genes linked by their associations into different operons in different organisms was dubbed a überoperon (Lathe *et al.*, 2000).

The two groups suggesting the idea later developed methods for extending functional associations by looking for rearranged operons (Rogozin *et al.*, 2002; Snel *et al.*, 2002), their methods were based on operon predictions based on conservation of gene order. The full power of the rearrangement of operons was realized later with the addition of operons predicted by intergenic distances in the Nebulon system (Janga *et al.*, 2005). These predictions have been instrumental, along with functional associations identified by the experimental determination of physical protein–protein interactions and other computational genomic context techniques, to the assignment of probable functions to more than half of the functional orphans of *E. coli* K12 (Hu *et al.*, 2009). The power of this genomic-context method for functional inference is evident from a comparison of the predictions shared between different genomic-context approaches. Operon reassociations clearly dominate in number of unique and hence likely novel predictions (Fig. 4.4a). As expected, the number of predictions from this method also increases with genome size, with *Burkholderia xenovorans* LB400 exhibiting as many as 136,573 predicted functional associations at a confidence threshold of 0.85. Bacteria have an average of ~66,000 predictions per genome, while Archaea average ~25,000, probably reflecting differences in genome sample phylogenetic coverage for each domain (Fig. 4.4b).

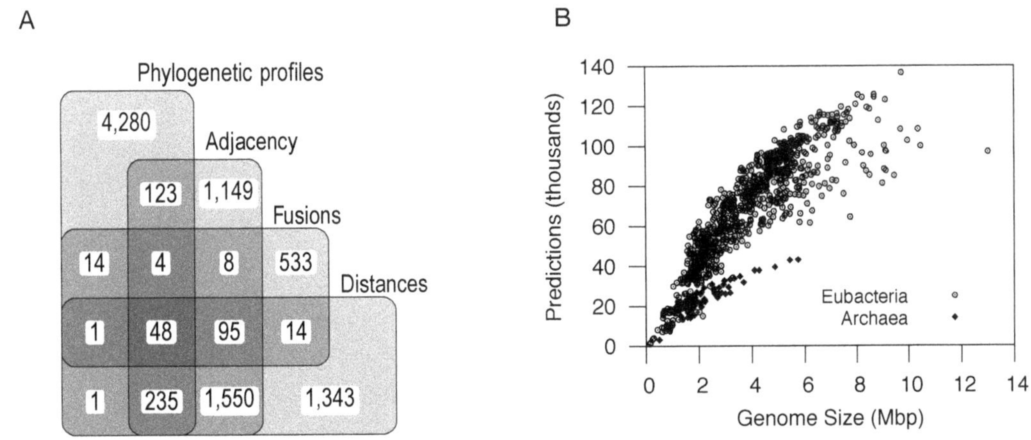

Figure 4.4 (a) Venn diagram showing the overlap between the numbers of predicted functional associations by different genomic-context methods. All predictions have been filtered at the same confidence threshold (0.85) when compared against a common set of gold standard associations identified in *E. coli* as described in a recent study (Hu *et al*., 2009). Rearrangement of operons or operon re-associations, predicted by intergenic distances, produces the highest number of predictions, followed by operons predicted by conservation of adjacency. (b) Number of functional interactions identified by operon rearrangements as a function of genome size. Note that the basis of this method is to exploit the evolutionary instability of the operon structure, unlike other approaches which strongly depend on conserved neighbourhood (Janga *et al*., 2005).

Conclusion and discussion

While the post-genomic era has introduced the genomic complement of hundreds of microbes, it has also left us with several unanswered questions regarding the functional relevance of the genes an organism encodes or principles that govern the organization of the genes on them. It is noteworthy to mention that even in a model organism like *E. coli*, which has been the workhorse for molecular genetics for more than 100 years, nearly one-third of the genes have no experimental evidence supporting their biological role (Hu *et al.*, 2009; Riley *et al.*, 2006). Understanding the organizational principles of genomes should help us devise approaches for predicting the functional associations of uncharacterized genes either to already characterized processes, or as new groups associated with yet to know processes. These approaches can be a big leap forward in uncovering the functional complements of a genome in years to come, such as has been demonstrated by the assignment of process membership to half of the functional orphans of *E. coli* by integrating a number of approaches, including operonic context which only depends on the availability of sequence information (Hu *et al.*, 2009). The number of uncharacterized genes across the prokaryotic lineage is considerably high, thus demanding the development of novel strategies. A better understanding of the principles governing operon evolution and their relevance to functional association between genes, together with advances in metagenomics, can increase our ability to improve the functional annotations of genes.

Although the field of microbial genome organization has come a long way with the advent of genomics, our understanding on the dynamic nature of transcriptomes is far from limited. However, with the developments of sequencing technologies extended to the RNA pools, there is a rapid increase in the number of high-quality genome-wide transcriptome maps available for prokaryotic genomes (Sorek and Cossart). These maps are not only going to revolutionize our current understanding of bacterial genome organization in coming years, but are also going to deliver improvements in the annotation of protein-coding genes – which have long been done primarily by computational approaches, identification of novel genes encoding for small regulatory RNAs, elucidation of condition-specific transcription start sites and potential regulatory elements which have been previously overlooked even in well-characterized model systems (Sorek and Cossart). Therefore, it is clear that new technologies such as RNA-seq (Wang *et al.*, 2009) are going to become an invaluable means for further understanding the principles governing operon structure and genome organization, and uncovering RNA-based regulatory mechanisms in a genome-wide manner. Given the unprecedented detail at which these high-throughput technologies can reveal the regulatory elements and expression levels specific to environmental conditions, it is possible to use these approaches to interrogate the prevalence of these phenomena in different states and thereby study their relevance to bacterial regulation, physiology and pathogenicity (Toledo-Arana *et al.*, 2009; Yoder-Himes *et al.*, 2009).

Chapter highlights

- Operons might be abundant because of their advantages for both coregulation and successful horizontal gene transfer.
- The most informative feature for predicting operons is the distance between genes.
- The features producing the cleanest operon predictions, albeit with low coverage, are predictions based on conservation of gene order followed by phylogenetic profiles.
- Transcriptional terminators and other sequence features might add quality to operon predictions, but the gain is minimal for most Prokaryotes.
- Operon rearrangements constitute the most useful source for the prediction of functional associations by genomic context in Prokaryotes.

Acknowledgements

SCJ acknowledges support from the School of Informatics at IUPUI in the form of start-up funds. GM-H acknowledges research support from the Natural Sciences and Engineering Research Council of Canada (NSERC), the Canadian Institutes of Health Research (CIHR), and computational facilities from the Shared Hierarchical Academic Research Computing Network (SHARCNET). We would also like to thank AJ Venkatakrishnan, Ritesh Kumar and Guilham Chalancon for critically reading the manuscript and providing helpful comments.

Supplementary material

Additional figures, tables and supporting material related to this work can be obtained from the URLs:

http://www.mrc-lmb.cam.ac.uk/genomes/sarath/genomeorg/

http://microbiome.wlu.ca/

References

Blattner, F.R., Plunkett, G., 3rd, Bloch, C.A., Perna, N.T., Burland, V., Riley, M., Collado-Vides, J., Glasner, J.D., Rode, C.K., Mayhew, G.F., *et al.* (1997). The complete genome sequence of *Escherichia coli* K-12. Science *277*, 1453–1474.

Bockhorst, J., Craven, M., Page, D., Shavlik, J., and Glasner, J. (2003a). A Bayesian network approach to operon prediction. Bioinformatics *19*, 1227–1235.

Bockhorst, J., Qiu, Y., Glasner, J., Liu, M., Blattner, F., and Craven, M. (2003b). Predicting bacterial transcription units using sequence and expression data. Bioinformatics *19 Suppl 1*, i34–43.

Butland, G., Peregrin-Alvarez, J.M., Li, J., Yang, W., Yang, X., Canadien, V., Starostine, A., Richards, D., Beattie, B., Krogan, N., *et al.* (2005). Interaction network containing conserved and essential protein complexes in *Escherichia coli*. Nature *433*, 531–537.

Che, D., Li, G., Mao, F., Wu, H., and Xu, Y. (2006). Detecting uber-operons in prokaryotic genomes. Nucleic Acids Res. *34*, 2418–2427.

Cherry, J.L. (2003). Genome size and operon content. J Theor Biol *221*, 401–410.

Craven, M., Page, D., Shavlik, J., Bockhorst, J., and Glasner, J. (2000). A probabilistic learning approach to whole-genome operon prediction. Proc Int Conf Intell Syst Mol Biol *8*, 116–127.

Dam, P., Olman, V., Harris, K., Su, Z., and Xu, Y. (2007). Operon prediction using both genome-specific and general genomic information. Nucleic Acids Res. *35*, 288–298.

Dandekar, T., Snel, B., Huynen, M., and Bork, P. (1998). Conservation of gene order: a fingerprint of proteins that physically interact. Trends Biochem. Sci. *23*, 324–328.

de Daruvar, A., Collado-Vides, J., and Valencia, A. (2002). Analysis of the cellular functions of *Escherichia coli* operons and their conservation in *Bacillus subtilis*. J. Mol. Evol. *55*, 211–221.

De Hoon, M.J., Makita, Y., Imoto, S., Kobayashi, K., Ogasawara, N., Nakai, K., and Miyano, S. (2004). Predicting gene regulation by sigma factors in *Bacillus subtilis* from genome-wide data. Bioinformatics (Oxford, England). *20 Suppl 1*, I101-I108.

de Hoon, M.J., Makita, Y., Nakai, K., and Miyano, S. (2005). Prediction of transcriptional terminators in *Bacillus subtilis* and related species. PLoS Comput Biol *1*, e25.

Dehal, P.S., Joachimiak, M.P., Price, M.N., Bates, J.T., Baumohl, J.K., Chivian, D., Friedland, G.D., Huang, K.H., Keller, K., Novichkov, P.S., *et al.* (2010). MicrobesOnline: an integrated portal for comparative and functional genomics. Nucleic Acids Res. *38*, D396–400.

Dekel, E., and Alon, U. (2005). Optimality and evolutionary tuning of the expression level of a protein. Nature *436*, 588–592.

Enright, A.J., Iliopoulos, I., Kyrpides, N.C., and Ouzounis, C.A. (1999). Protein interaction maps for complete genomes based on gene fusion events. Nature *402*, 86–90.

Ermolaeva, M.D., White, O., and Salzberg, S.L. (2001). Prediction of operons in microbial genomes. Nucleic Acids Res. *29*, 1216–1221.

Galperin, M.Y., and Koonin, E.V. (2000). Who's your neighbor? New computational approaches for functional genomics. Nat Biotechnol *18*, 609–613.

Gama-Castro, S., Jimenez-Jacinto, V., Peralta-Gil, M., Santos-Zavaleta, A., Penaloza-Spinola, M.I., Contreras-Moreira, B., Segura-Salazar, J., Muniz-Rascado, L., Martinez-Flores, I., Salgado, H., *et al.* (2008). RegulonDB (version 6.0): gene regulation model of *Escherichia coli* K-12 beyond transcription, active (experimental). annotated promoters and Textpresso navigation. Nucleic Acids Res. *36*, D120–124.

Gene-Ontology-Consortium (2010). The Gene Ontology in 2010: extensions and refinements. Nucleic Acids Res. *38*, D331–335.

Gonzalez, V., Santamaria, R.I., Bustos, P., Hernandez-Gonzalez, I., Medrano-Soto, A., Moreno-Hagelsieb, G., Janga, S.C., Ramirez, M.A., Jimenez-Jacinto, V., Collado-Vides, J., *et al.* (2006). The partitioned Rhizobium etli genome: genetic and metabolic redundancy in seven interacting replicons. Proc. Natl. Acad. Sci. U.S.A. *103*, 3834–3839.

Guell, M., van Noort, V., Yus, E., Chen, W.H., Leigh-Bell, J., Michalodimitrakis, K., Yamada, T., Arumugam, M., Doerks, T., Kuhner, S., *et al.* (2009). Transcriptome complexity in a genome-reduced bacterium. Science *326*, 1268–1271.

Harris, M.A., Clark, J., Ireland, A., Lomax, J., Ashburner, M., Foulger, R., Eilbeck, K., Lewis, S., Marshall, B., Mungall, C., *et al.* (2004). The Gene Ontology (GO). database and informatics resource. Nucleic Acids Res. *32 Database issue*, D258–261.

Hu, P., Janga, S.C., Babu, M., Diaz-Mejia, J.J., Butland, G., Yang, W., Pogoutse, O., Guo, X., Phanse, S., Wong, P., *et al.* (2009). Global functional atlas of *Escherichia coli* encompassing previously uncharacterized proteins. PLoS Biol. *7*, e96.

Huerta, A.M., Francino, M.P., Morett, E., and Collado-Vides, J. (2006). Selection for unequal densities of sigma70 promoter-like signals in different regions of large bacterial genomes. PLoS Genet *2*, e185.

Huynen, M., Snel, B., Lathe, W., 3rd, and Bork, P. (2000). Predicting protein function by genomic context: quantitative evaluation and qualitative inferences. Genome Res *10*, 1204–1210.

Itoh, T., Takemoto, K., Mori, H., and Gojobori, T. (1999). Evolutionary instability of operon structures disclosed by sequence comparisons of complete microbial genomes. Mol Biol Evol *16*, 332–346.

Jacob, F., and Monod, J. (1961). Genetic regulatory mechanisms in the synthesis of proteins. J. Mol. Biol. *3*, 318–356.

Jacob, F., Perrin, D., Sanchez, C., and Monod, J. (1960). [Operon: a group of genes with the expression coordinated by an operator.]. C R Hebd Seances Acad Sci *250*, 1727–1729.

Jacob, F., Perrin, D., Sanchez, C., Monod, J., and Edelstein, S. (2005). [The operon: a group of genes with expression coordinated by an operator. C.R.Acad. Sci. Paris 250 (1960). 1727–1729]. C R Biol *328*, 514–520.

Janga, S.C., Collado-Vides, J., and Moreno-Hagelsieb, G. (2005). Nebulon: a system for the inference of functional relationships of gene products from the rearrangement of predicted operons. Nucleic Acids Res. 33, 2521–2530.

Janga, S.C., Lamboy, W.F., Huerta, A.M., and Moreno-Hagelsieb, G. (2006). The distinctive signatures of promoter regions and operon junctions across prokaryotes. Nucleic Acids Res. *34*, 3980–3987.

Janga, S.C., and Moreno-Hagelsieb, G. (2004). Conservation of adjacency as evidence of paralogous operons. Nucleic Acids Res. *32*, 5392–5397.

Janga, S.C., Salgado, H., and Martinez-Antonio, A. (2009). Transcriptional regulation shapes the organization of genes on bacterial chromosomes. Nucleic Acids Res. *37*, 3680–3688.

Keseler, I.M., Bonavides-Martinez, C., Collado-Vides, J., Gama-Castro, S., Gunsalus, R.P., Johnson, D.A., Krummenacker, M., Nolan, L.M., Paley, S., Paulsen, I.T., *et al.* (2009). EcoCyc: a comprehensive view of *Escherichia coli* biology. Nucleic Acids Res. *37*, D464–470.

Koide, T., Reiss, D.J., Bare, J.C., Pang, W.L., Facciotti, M.T., Schmid, A.K., Pan, M., Marzolf, B., Van, P.T., Lo, F.Y., *et al.* (2009). Prevalence of transcription promoters within archaeal operons and coding sequences. Mol Syst Biol *5*, 285.

Kolesov, G., Wunderlich, Z., Laikova, O.N., Gelfand, M.S., and Mirny, L.A. (2007). How gene order is influenced by the biophysics of transcription regulation. Proc. Natl. Acad. Sci. U.S.A. *104*, 13948–13953.

Korbel, J.O., Jensen, L.J., von Mering, C., and Bork, P. (2004). Analysis of genomic context: prediction of functional associations from conserved bidirectionally transcribed gene pairs. Nat Biotechnol 22, 911–917.

Kovacs, K., Hurst, L.D., and Papp, B. (2009). Stochasticity in protein levels drives colinearity of gene order in metabolic operons of *Escherichia coli*. PLoS Biol. *7*, e1000115.

Lathe, W.C., 3rd, Snel, B., and Bork, P. (2000). Gene context conservation of a higher order than operons. Trends Biochem. Sci. *25*, 474–479.

Lawrence, J. (1999). Selfish operons: the evolutionary impact of gene clustering in prokaryotes and eukaryotes. Curr Opin Genet Dev 9, 642–648.

Lawrence, J.G. (1997). Selfish operons and speciation by gene transfer. Trends Microbiol. *5*, 355–359.

Lawrence, J.G., and Roth, J.R. (1996). Selfish operons: horizontal transfer may drive the evolution of gene clusters. Genetics *143*, 1843–1860.

Mao, F., Dam, P., Chou, J., Olman, V., and Xu, Y. (2009). DOOR: a database for prokaryotic operons. Nucleic Acids Res. *37*, D459–463.

Marcotte, E.M., Pellegrini, M., Ng, H.L., Rice, D.W., Yeates, T.O., and Eisenberg, D. (1999a). Detecting protein function and protein–protein interactions from genome sequences. Science *285*, 751–753.

Marcotte, E.M., Pellegrini, M., Thompson, M.J., Yeates, T.O., and Eisenberg, D. (1999b). A combined algorithm for genome-wide prediction of protein function. Nature *402*, 83–86.

McGrath, P.T., Lee, H., Zhang, L., Iniesta, A.A., Hottes, A.K., Tan, M.H., Hillson, N.J., Hu, P., Shapiro, L., and McAdams, H.H. (2007). High-throughput identification of transcription start sites, conserved promoter motifs and predicted regulons. Nat. Biotechnol 25, 584–592.

Moreno-Hagelsieb, G. (2006). Operons across prokaryotes: genomic analyses and predictions 300+ genomes later. Current Genomics 7, 163–170.

Moreno-Hagelsieb, G., and Janga, S.C. (2008). Operons and the effect of genome redundancy in deciphering functional relationships using phylogenetic profiles. Proteins 70, 344–352.

Moreno-Hagelsieb, G., and Collado-Vides, J. (2002a). Operon conservation from the point of view of *Escherichia coli*, and inference of functional interdependence of gene products from genome context. In Silico Biol. *2*, 87–95.

Moreno-Hagelsieb, G., and Collado-Vides, J. (2002b). A powerful non-homology method for the prediction of operons in prokaryotes. Bioinformatics *18 Suppl 1*, S329–336.

Moreno-Hagelsieb, G., Trevino, V., Perez-Rueda, E., Smith, T.F., and Collado-Vides, J. (2001). Transcription unit conservation in the three domains of life: a perspective from *Escherichia coli*. Trends Genet. *17*, 175–177.

Mushegian, A.R., and Koonin, E.V. (1996). Gene order is not conserved in bacterial evolution. Trends Genet. *12*, 289–290.

Ochman, H., Lawrence, J.G., and Groisman, E.A. (2000). Lateral gene transfer and the nature of bacterial innovation. Nature *405*, 299–304.

Okuda, S., Katayama, T., Kawashima, S., Goto, S., and Kanehisa, M. (2006). ODB: a database of operons accumulating known operons across multiple genomes. Nucleic Acids Res. *34*, D358–362.

Omelchenko, M.V., Makarova, K.S., Wolf, Y.I., Rogozin, I.B., and Koonin, E.V. (2003). Evolution of mosaic operons by horizontal gene transfer and gene displacement in situ. Genome Biol. *4*, R55.

Overbeek, R., Fonstein, M., D'Souza, M., Pusch, G.D., and Maltsev, N. (1999). The use of gene clusters to infer functional coupling. Proc. Natl. Acad. Sci. U.S.A. *96*, 2896–2901.

Pál, C., and Hurst, L.D. (2004). Evidence against the selfish operon theory. Trends Genet. *20*, 232–234.

Pertea, M., Ayanbule, K., Smedinghoff, M., and Salzberg, S.L. (2009). OperonDB: a comprehensive database of predicted operons in microbial genomes. Nucleic Acids Res. *37*, D479–482.

Price, M.N., Arkin, A.P., and Alm, E.J. (2006a). The life cycle of operons. PLoS Genet 2, e96.

Price, M.N., Arkin, A.P., and Alm, E.J. (2006b). OpWise: operons aid the identification of differentially expressed genes in bacterial microarray experiments. BMC Bioinformatics 7, 19.

Price, M.N., Huang, K.H., Alm, E.J., and Arkin, A.P. (2005a). A novel method for accurate operon predictions in all sequenced prokaryotes. Nucleic Acids Res. *33*, 880–892.

Price, M.N., Huang, K.H., Arkin, A.P., and Alm, E.J. (2005b). Operon formation is driven by co-regulation and not by horizontal gene transfer. Genome Res *15*, 809–819.

Riley, M., Abe, T., Arnaud, M.B., Berlyn, M.K., Blattner, F.R., Chaudhuri, R.R., Glasner, J.D., Horiuchi, T., Keseler, I.M., Kosuge, T., *et al.* (2006). *Escherichia coli* K-12: a cooperatively developed annotation snapshot--2005. Nucleic Acids Res. *34*, 1–9.

Rison, S.C., Teichmann, S.A., and Thornton, J.M. (2002). Homology, pathway distance and chromosomal localization of the small molecule metabolism enzymes in *Escherichia coli*. J. Mol. Biol. *318*, 911–932.

Rocha, E.P. (2008). The organization of the bacterial genome. Annu. Rev. Genet. *42*, 211–233.

Rogozin, I.B., Makarova, K.S., Murvai, J., Czabarka, E., Wolf, Y.I., Tatusov, R.L., Szekely, L.A., and Koonin, E.V. (2002). Connected gene neighborhoods in prokaryotic genomes. Nucleic Acids Res. *30*, 2212–2223.

Salgado, H., Gama-Castro, S., Peralta-Gil, M., Diaz-Peredo, E., Sanchez-Solano, F., Santos-Zavaleta, A., Martinez-Flores, I., Jimenez-Jacinto, V., Bonavides-Martinez, C., Segura-Salazar, J., *et al.* (2006). RegulonDB (version 5.0): *Escherichia coli* K-12 transcriptional regulatory network, operon organization, and growth conditions. Nucleic Acids Res. *34*, D394–397.

Salgado, H., Moreno-Hagelsieb, G., Smith, T.F., and Collado-Vides, J. (2000). Operons in *Escherichia coli*: genomic analyses and predictions. Proc. Natl. Acad. Sci. U.S.A. *97*, 6652–6657.

Selinger, D.W., Saxena, R.M., Cheung, K.J., Church, G.M., and Rosenow, C. (2003). Global RNA half-life analysis in *Escherichia coli* reveals positional patterns of transcript degradation. Genome Res *13*, 216–223.

Serres, M.H., Goswami, S., and Riley, M. (2004). GenProtEC: an updated and improved analysis of functions of *Escherichia coli* K-12 proteins. Nucleic Acids Res. *32 Database issue*, D300–302.

Sierro, N., Makita, Y., de Hoon, M., and Nakai, K. (2008). DBTBS: a database of transcriptional regulation in *Bacillus subtilis* containing upstream intergenic conservation information. Nucleic Acids Res. *36*, D93–96.

Snel, B., Bork, P., and Huynen, M.A. (2002). The identification of functional modules from the genomic association of genes. Proc. Natl. Acad. Sci. U.S.A. *99*, 5890–5895.

Sorek, R., and Cossart, P. (2010). Prokaryotic transcriptomics: a new view on regulation, physiology and pathogenicity. Nat. Rev. Genet. *11*, 9–16.

Tatusov, R.L., Fedorova, N.D., Jackson, J.D., Jacobs, A.R., Kiryutin, B., Koonin, E.V., Krylov, D.M., Mazumder, R., Mekhedov, S.L., Nikolskaya, A.N., *et al.* (2003). The COG database: an updated version includes eukaryotes. BMC Bioinformatics *4*, 41.

Tatusov, R.L., Koonin, E.V., and Lipman, D.J. (1997). A genomic perspective on protein families. Science *278*, 631–637.

Tatusov, R.L., Natale, D.A., Garkavtsev, I.V., Tatusova, T.A., Shankavaram, U.T., Rao, B.S., Kiryutin, B., Galperin, M.Y., Fedorova, N.D., and Koonin, E.V. (2001). The COG database: new developments in phylogenetic classification of proteins from complete genomes. Nucleic Acids Res. *29*, 22–28.

Toledo-Arana, A., Dussurget, O., Nikitas, G., Sesto, N., Guet-Revillet, H., Balestrino, D., Loh, E., Gripenland, J., Tiensuu, T., Vaitkevicius, K., *et al.* (2009). The Listeria transcriptional landscape from saprophytism to virulence. Nature *459*, 950–956.

Wang, Z., Gerstein, M., and Snyder, M. (2009). RNA-Seq: a revolutionary tool for transcriptomics. Nat. Rev. Genet. *10*, 57–63.

Xie, G., Bonner, C., Brettin, T., Gottardo, R., Keyhani, N., and Jensen, R. (2003). Lateral gene transfer and ancient paralogy of operons containing redundant copies of tryptophan-pathway genes in Xylella species and in heterocystous cyanobacteria. Genome Biology *4*, R14.

Yada, T., Nakao, M., Totoki, Y., and Nakai, K. (1999). Modeling and predicting transcriptional units of *Escherichia coli* genes using hidden Markov models. Bioinformatics *15*, 987–993.

Yellaboina, S., Goyal, K., and Mande, S.C. (2007). Inferring genome-wide functional linkages in *E. coli* by combining improved genome context methods: comparison with high-throughput experimental data. Genome Res *17*, 527–535.

Yoder-Himes, D.R., Chain, P.S., Zhu, Y., Wurtzel, O., Rubin, E.M., Tiedje, J.M., and Sorek, R. (2009). Mapping the Burkholderia cenocepacia niche response via high-throughput sequencing. Proc. Natl. Acad. Sci. U.S.A. *106*, 3976–3981.

Zheng, Y., Roberts, R.J., and Kasif, S. (2002). Genomic functional annotation using co-evolution profiles of gene clusters. Genome Biol. 3, RESEARCH0060.

Small-molecule-mediated Signalling in Bacteria

5

Aswin Sai Narain Seshasayee and Nicholas M. Luscombe

Abstract

The ability to sense and respond to environmental cues is critical to the survival of unicellular organisms like bacteria. An important type of signals is represented by small molecules. In this review, following a general description of the prevalence of small-molecule-binding proteins in the prokaryotic kingdom, we discuss the following aspects of bacterial signalling mediated by small molecules: (a) the interplay between small-molecule signalling and metabolism, which involves local regulation of specific metabolic pathways, in addition to a more global integration of metabolic and other cellular functions; and (b) signal transduction via two second messenger small molecules – ppGpp and c-di-GMP – which initiate stringent response following nutrient starvation, and control switching between motile and adhesive states respectively. In conclusion, we briefly mention two other dimensions of small-molecule signalling: (a) the role of antibiotics in triggering transcriptional responses at sub-inhibitory concentrations; and (b) inter-kingdom signalling between bacteria and their mammalian hosts through hormones and quorum sensing signals.

Introduction

Bacteria, being unicellular organisms, are particularly exposed to the vagaries of their immediate microenvironment. Environmental adaptation is effected at two broad stages: (a) across evolutionary time-scales, bacterial genomes become streamlined and optimized to make appropriate use of their habitat; (b) over shorter time periods, changes in the milieu necessitate rapid reorganization of the phenotypic state, which is achieved through *signal transduction* networks. Optimal evolution of bacterial genomes is beyond the scope of this review. Signal transduction is initiated by the sensing of an environmental or a cellular cue. This is followed by the transmission of these signals largely to the transcriptional machinery which alters the gene expression programme of the cell, thus producing an apt response.

A large class of environmental cues is represented by small molecules. Though the definition of a small molecule may not always be clear, an acceptable description is that it is an organic molecule of molecular weight less than 1000 daltons. In this review, we make use of the following classes of molecules in describing small molecule signalling: (a) a wide range of organic molecules that are products of classical cellular metabolism (sugars, amino acids, fatty acids, nucleotides, cofactors, vitamins); (b) signal-transducing second messengers such as modified nucleotides; (c) antibiotics; and (d) a class of non-peptidic quorum-sensing molecules which promote communication across bacterial species and kingdoms of life. A common mechanism by which a macromolecule (mainly protein) responds to binding to a small-molecule is allostery, which is defined as a reversible change in protein activity effected by the binding of a small molecule to a site other than the activity centre; in enzymes, the activity centre is the catalytic site and in transcription factors, which modify transcriptional responses, this is the DNA-binding motif. In other cases, particularly in proteins related to motility and cell division, small-molecule-sensing leads to alterations in subcellular localization of the target protein (Duerig *et al.*, 2009; Shapiro *et al.*, 2009; Boehm *et al.*, 2010).

This review will discuss (mainly) genome-scale studies interrogating sensing of small-molecule signals and their interpretation via signal transduction networks. Particular emphasis will be laid on responses triggered by metabolic signals and modified nucleotides (a and b above). Transcriptional responses initiated by antibiotics at sublethal concentrations is a subject that has recently attracted the attention of genome-scale studies and will be briefly mentioned. From an anthropocentric perspective, interkingdom interactions between a human (or broadly mammalian) host and resident, and potentially disease-causing, bacteria are of importance; this has been the subject of recent reviews and will be remarked upon in the present context of small-molecule signals.

Small-molecule-binding proteins in bacteria

The availability of multiple genome sequences permitted studies in which protein sequences homologous to those of known function could be identified in organisms from diverse phylogenetic groups. In an attempt towards assessing the presence of small-molecule-mediated signalling across different kingdoms of life, an early comparative-genomic study by Anantharaman and colleagues surveyed 25 bacterial, archaeal and eukaryotic genomes for protein domains that can bind to small molecules (Anantharaman *et al.*, 2001). They restricted their analysis to ancient small-molecule-binding domains which are present in at least two of the three kingdoms of life: Bacteria, Archaea and Eukarya. First, they found that such ancient domains are relatively depleted in Eukarya which utilized a novel set of newer small-molecule-binding proteins; this might be explained by their argument that intracellular signal sensing is more prevalent in prokaryotes than in eukaryotes where signal transduction is mainly initiated at the cell surface. Secondly, they analysed the presence of 'output' functions in these proteins, which would give an idea of protein function, typically downstream of signal sensing. They found that many proteins contain only the small-molecule-binding domain and no other discrete functional domain; these single-domain proteins could be regulatory components of metabolic complexes. In a conceptually similar fashion, there are many proteins which contain domains that are phosphorylated by two-component histidine kinases, but no detectable output function; such 'single-domain response regulators' spatially separate signal receipt and output generation thus increasing the number of components involved in signal transduction (Jenal and Galperin, 2009). A second class of proteins contain an enzymatic or transporter domain in addition to the small-molecule-binding domain; once again, small molecule sensing could modulate catalysis or transport. A final class of proteins belong to the general class of 'two-headed' transcription factors, which regulate transcription initiation by binding to the DNA; this is exemplified by the *E. coli* Lac repressor first described by Jacob and Monod (Pardee *et al.*, 1959)where the DNA-binding activity of the protein is modulated by the sensing of lactose, a carbohydrate.

For the purpose of this article, we performed a preliminary analysis of over 92,000 proteins predicted to contain small-molecule-binding domains across 867 prokaryotic genomes obtained from the KEGG database (Kanehisa and Goto, 2000). Under half of all these proteins (~43%) are predicted transcription factors (Fig. 5.1). About a third contain at least one other partner domain; these include (a) proteins involved in second-messenger synthesis, (b) histidine kinases which on sensing a signal initiates a response by phosphorylating a receiver protein, and (c) other enzymes and transporters likely to be involved in metabolism. Finally, between 10% and 25% are likely to be single-domain small-molecule-binding proteins.

Small-molecule-sensing and bacterial transcriptional regulation

Transcription initiation is the first step in gene expression and arguably, at least in bacteria, the most important control point. Madan Babu and co-workers performed a detailed computational analysis of transcription factors in the best studied bacterial organism, *Escherichia coli K12*

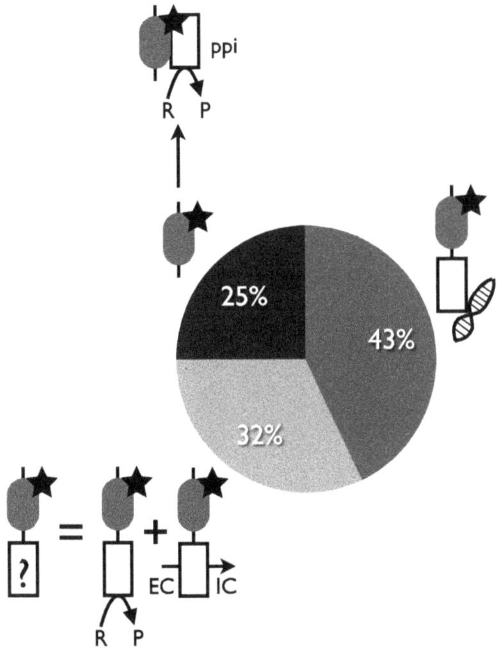

Figure 5.1 Small-molecule-binding proteins in prokaryotes. This Figure shows the approximate proportions of three types of small-molecule-binding proteins (with DNA-binding domains; with other domains such as those with enzymatic and transport functions; with no other detectable domain). The dataset comprises over 92,000 proteins identified by scanning protein sequences from 896 prokaryotic genomes using selected PFAM sequence models (http://pfam.sanger.ac.uk). As shown by Anantharaman and colleagues, these proportions will vary across different phylogenetic groups. The filled stars stand for small-molecules; small-molecule-binding domains are represented by the filled rounded rectangles. DNA-binding domains are shown in the form of unfilled rectangles in contact with a double helix. Unfilled rectangles marked with a '?' represent other output functions including, but probably not limited to, enzymatic (associated with a chemical transformation; R to P for reactant to product) and transport (arrow representing transport; EC–IC for extracellular–intracellular) domains. 'Protein–protein interaction' is abbreviated to 'ppi'.

(Madan Babu and Teichmann, 2003). An important result borne out of this study, in agreement with earlier investigations (Perez-Rueda, 2000), is that over than 85% of the ~270 transcription factors in this organism contain (or might contain) a domain that partners the DNA-binding domain (Fig. 5.2A). In most cases, this partner domain is one that is predicted to receive a signal either by sensing the signal directly or indirectly via phosphorlation by a sensory kinase (Ulrich *et al.*, 2005; Martínez-Antonio *et al.*, 2006). Remarkably, nearly half of all transcription factors contain an identifiable small-molecule-binding domain (Madan Babu and Teichmann, 2003); this proportion could be higher as a number of transcription factors, though with no detectable partner domain, have space to contain a small molecule-binding domain. Small-molecule-binding transcription factors in eukaryotes, though described (Sellick and Reece, 2005), are likely to be relatively rare.

Small molecule-sensing transcription factors and metabolism: a general association

We extended our analysis of the prevalence of small-molecule-binding domains to assess their occurrence in transcription factors from 867 prokaryotic genomes by assigning a set of DNA-binding domains to protein sequences. The proportion of transcription factors that contain a small-molecule-binding domain is variable across the set of genomes included in our analysis, to such an extent that the model organism *E. coli* appears to be an outlier which displays a strong preference for encoding such transcription factors. The numbers of both transcription factors (van Nimwegen, 2003; Ulrich *et al.*, 2005) and small molecule-binding proteins scale in a quadratic fashion with the total number of genes encoded in a genome. Given this scaling relationship, one would expect to see little association between the *proportion* of transcription factors that contain small-molecule-binding domains and the total number of genes in a genome; however, in contrast to this expectation, there is a significant overall correlation between the two.

Towards testing an intuitive relationship between small molecule signalling and metabolism, we first split the total gene content of each genome into 'enzymes' and 'non-enzymes'. Though the proportion of small-molecule-binding transcription factors shows a significant correlation with both these parameters, it is clear that it is more strongly associated with the numbers of enzymes than with non-enzymes

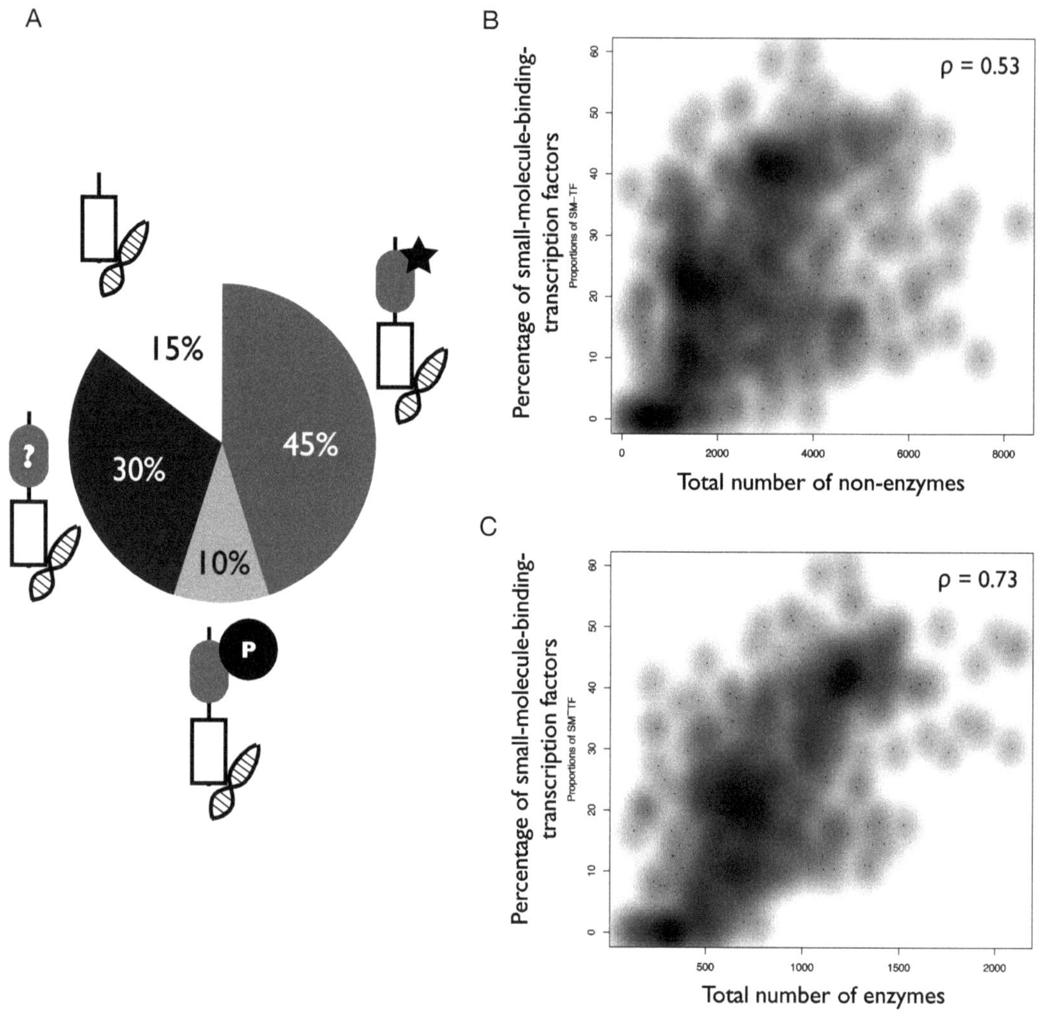

Figure 5.2 Small-molecule-binding transcription factors. (A) Classification of transcription factors in *E. coli* K12, as described by Madan Babu and Teichmann. Domain logos are as described in the legend for Fig. 5.1. The filled rounded rectangle with the '?' stands for possible domains not annotated. The dark circle with a 'P' indicates phosphorylation. Also shown are correlations between the proportion of transcription factors that contain a small-molecule-binding domain (y-axis) and the number of (B) non-enzyme proteins and (C) enzymes encoded in a genome. All proteins annotated with an Enzyme Commission number in KEGG were defined as enzymes. Small-molecule-binding domains and transcription factors were identified in 867 prokaryotic genomes using selected PFAM sequence models.

(Fig. 5.2B and C). Secondly, we made use of RegulonDB (Gama-Castro *et al.*, 2008), which is a rich source of regulatory network information – linking over 150 transcription factors to their corresponding target genes – for *E. coli* (see Chapter 9 by Martinez-Antonio). Nearly half of all genes targeted by small molecule-binding TFs are associated with metabolism and transport, whereas only under a third of the other genes are. These observations indicate the presence of a direct interaction between the metabolic and the transcriptional system in bacteria. In the next few sections, we discuss selected genome-scale studies illustrating such an interaction, while also presenting studies that demonstrate that the impact of small molecule-sensing on bacterial transcriptional control might be significantly more profound than anticipated.

Patterns underlying transcriptional regulation of metabolic networks: global and local regulation

Cellular metabolism can be represented in the form of metabolite- or protein-centric networks (Jeong *et al.*, 2000). In a metabolite–centric network, reactants and products of a reaction form the nodes of an edge. In a typical protein-centric network, two enzymes are connected if the small molecule product of the action of the first is a substrate to the second. Transcriptional regulation of such networks has been studied on a genome-wide scale in *E. coli* and the model eukaryote *Saccharomyces cerevisiae* using either transcriptional networks or gene expression measurements (Kharchenko *et al.*, 2005; Ihmels *et al.*, 2004; Notebaart *et al.*, 2008; Seshasayee *et al.*, 2009). Typical results include the observation that two enzymes connected to each other in the same pathway tend to be co-regulated; the further two enzyme are from each other in the metabolic network, in terms of either the number of metabolites separating the two (Kharchenko *et al.*, 2005)or the coupling between their metabolic fluxes (Notebaart *et al.*, 2008), the less likely that they will be co-regulated or co-expressed. At junctions where multiple reactions converge or diverge, more complex patterns of regulation emerge (Ihmels *et al.*, 2004; Seshasayee *et al.*, 2009); this is beyond the scope of this review.

An important observation that has emerged from decades of studying metabolic regulation in *E. coli* is that there are two broad classes of transcription factors: (a) specific factors that regulate genes belonging to a single metabolic pathway and (b) general factors which regulate genes belonging to different functions, but display a statistical enrichment for regulating genes from a single broad functional category (Seshasayee *et al.*, 2009). The latter category includes 'global' transcriptional regulators, which despite being only a few ($n = 7$ according to one definition) together regulate a majority of *E. coli* genes and respond to a multitude of different environmental and cellular signals (Martínez-Antonio and Collado-Vides, 2003). Specific transcription factors (a) have fewer binding sites, (b) are expressed at lower levels and (c) encoded physically closer, on the chromosome, to their target genes than general or global factors (Lozada-Chávez *et al.*, 2008; Menchaca-Mendez *et al.*, 2005; Hershberg *et al.*, 2005; Seshasayee *et al.*, 2009). General and specific transcription factors combine in different ways in regulating the three major functions of metabolism (Fig. 5.3A): (a) catabolic/degradation enzyme genes tend to be regulated by a single general and specific factor each; (b) anabolic/biosynthetic genes are regulated by a single factor, which may be general or specific and; (c) genes involved in energy metabolism are regulated by multiple general factors each (Seshasayee *et al.*, 2009). The distinctive regulation of catabolic enzymes is in part due to the phenomenon of catabolite repression which ensures that alternative carbon sources are not utilized when the preferred nutrient glucose is available (Görke and Stülke, 2008). The 'single-input' regulation of anabolism may enable 'just-in-time' transcription (Zaslaver *et al.*, 2004). Finally, energy metabolism is a central hub that should be modulated in response to a large variety of signals; this might be more efficiently achieved through general or global, rather than specific transcription factors (Martínez-Antonio and Collado-Vides, 2003).

Analysis of specific transcription factors is typically straightforward with little unexpected binding events, as illustrated by the genome-scale analysis of the binding of one such regulator, MelR, to the *E. coli* chromosome (Grainger *et al.*, 2004). However, general transcription factors present a different problem; though they display a statistical enrichment for targeting one functional category, they have numerous targets outside their preferred function, including non-metabolic targets (Martínez-Antonio and Collado-Vides, 2003) (Fig. 5.3B). In the following sections, we discuss genome-scale studies of two such transcription factors in *E. coli* – Lrp and Crp – both of which bind to small-molecules and might also contribute to defining the topology of the chromosome.

Lrp: the feast or famine global transcription factor

Lrp is a conserved bacterial transcription factor which, in *E. coli*, regulates genes involved in

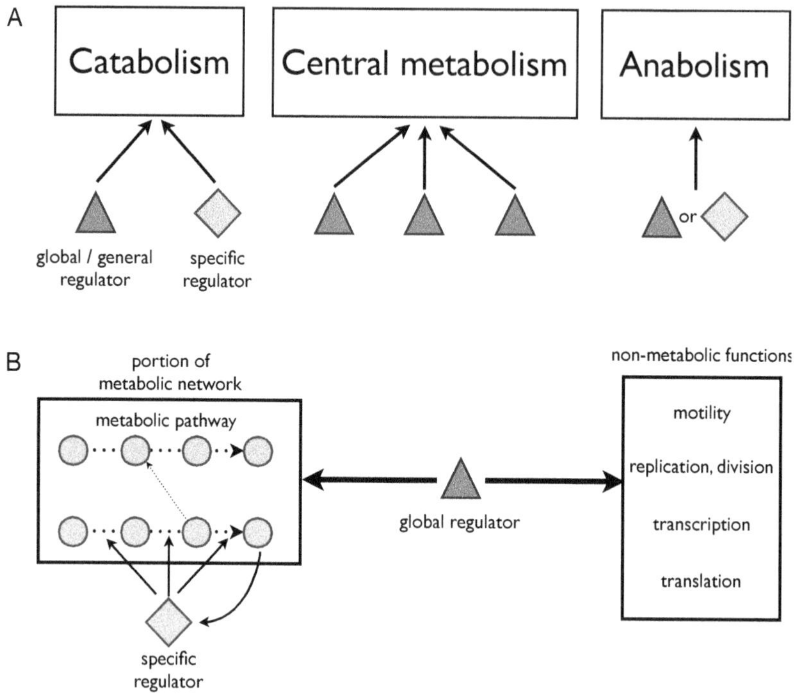

Figure 5.3 Regulation of metabolism. (A) General trends underlying regulation of the three general functions of metabolism: catabolism, central metabolism and anabolism. This is based on Seshasayee *et al.* (2009). (B) Specific transcription factors regulate a single metabolic pathway or functional category, whereas global regulators target multiple metabolic pathways and integrate them with non-metabolic functions. Solid arrows represent regulatory interactions; dotted arrows stand for metabolic transformations.

amino acid metabolism and transport, and non-metabolic functions such as pili biosynthesis. Lrp is present in about 250–700 molecules per cell, and its abundance varies inversely with growth rate: in rich medium, stationary phase cells contain about two-fold higher levels of Lrp than exponential phase cells (Chen *et al.*, 2001). Depending on the promoter that is regulated, the activity of Lrp can be enhanced, reduced or left unaffected by the binding of the amino acid leucine (Calvo and Matthews, 1994; Newman and Lin, 1995), which may act as a nutrient sensor.

A recent study interrogating the genome-wide binding of Lrp to the DNA identified just under ~140 Lrp-binding regions, thus expanding the catalogue of known Lrp targets by a factor of five (Cho *et al.*, 2008). By performing this study under three different conditions: (a) exponential phase with or (b) without exogenous leucine and (c) stationary phase, the authors were able to show that absence of leucine and stationary phase increase the number of Lrp-binding regions by 3- to 4-fold. Fewer than half of the genes regulated by Lrp are involved in small-molecule transport and metabolism. By integrating gene expression measurements in response to exogenous leucine and deletion of the *lrp* gene, the authors were able to identify three general classes of responses: (a) *independent response* where leucine has no effect on Lrp action; (b) *concerted response* in which leucine enhances the effect of Lrp and; (c) *reciprocal response* in which leucine antagonises the effect of Lrp. Though genes representative of each of these three classes of Lrp-regulated genes are presented in this work (e.g. concerted regulation of transporters of branched-chain amino acids; and reciprocal regulation of Arg, Ser, Ala and Pro transporters), the need for a given type of regulation for a gene function cannot be intuitively rationalized. However, one may speculate that this regulatory architecture, at the level of metabolism, would lead to homeostasis of amino acid levels.

Biochemical studies have shown that Lrp exists largely in two forms: octameric (Lrp_8) and hexadecameric (Lrp_{16}). Leucine binding favours the dissociation of Lrp to the octameric form (Lrp_8-leu) (Chen *et al.*, 2001). Therefore, at operons to which Lrp_8, Lrp_{16} and Lrp_8-leu have similar affinities, leucine has no effect on Lrp action. Similarly leucine potentiates the effect of Lrp at operons with particular affinity to Lrp_8-leu, whereas it overcomes the outcome of Lrp action at operons that specifically bind to Lrp_{16}/Lrp_8. Computer simulations making use of the above expectation showed that, in general, the sensitivity of an operon to growth conditions can be explained by its affinity to the three forms of Lrp and therefore, the impact of leucine on Lrp (Chen *et al.*, 2001).

Crp and transcriptional responses to carbon-source nutrition

Crp is the most prolific global transcription factor in *E. coli*, based on the information available in RegulonDB (Gama-Castro *et al.* 2008). Its activity is triggered by binding of the second messenger cyclic AMP (cAMP) in response to glucose starvation and other stresses. Its regulatory activity is most commonly described in the context of catabolite repression, the phenomenon in which alternative carbon sources are utilized only in the absence of glucose.

Liu and co-workers performed a systematic microarray study investigating the response of *E. coli* to different carbon sources of decreasing quality (Liu, 2005). They found that as the carbon source quality declines, the number of genes differentially expressed relative to growth on glucose increases; this happens in a hierarchical manner such that the set of genes upregulated in a given carbon source is a subset of that in one of lower quality. Overall, over 15% of differentially expressed genes are likely to be targets of Crp. Contrary to expectation based on strict metabolite feedback, genes for metabolizing better, though currently unavailable, carbon sources are induced. Despite the fact that the only variable is a single carbon source nutrient, there is an overall reorganization of the RNA polymerase–DNA interaction space as determined by fluorescence microscopy: whereas in glucose, there are distinct foci of transcriptional activity, these are lost in poorer carbon sources, leading to a more homogenous distribution of RNA polymerase.

Grainger and colleagues investigated the genome-wide binding of Crp to the *E. coli* chromosome using ChIP-chip (Grainger *et al.*, 2005). They found ~70 high-affinity binding regions, which is less than expected. Remarkably, they found that this relatively low enrichment could in part be due to the high background, which was attributed to weak binding events at low-affinity sites. Thus, the authors were able to propose that Crp is a chromosome-structuring protein and only a small minority of binding events directly affect transcription initiation, thus adding a new layer to the complexity of Crp-dependent gene regulation.

Small-molecule second-messengers: stringent response and the switch between motility and adhesion

A second messenger is a small molecule, which once produced in response to a signal, modulates effector proteins to effect a cellular response. Various nucleotides are famous for their role as second messengers in both prokaryotes and eukaryotes. The best studied of these are cAMP and cyclic GMP (cGMP). In *E. coli*, the role of cAMP is probably limited to its interaction with Crp. cGMP is uncommon in bacteria and is known only in photosynthetic cyanobacteria (Cadoret *et al.*, 2005). In this section, we restrict our discussion to two other second messenger nucleotides: (a) guanosine 3′,5′-bispyrophosphate (ppGpp), which is strongly associated with the stringent response, which involves down-regulation of macromolecule biosynthesis in response to nutrient starvation (Potrykus and Cashel, 2008) and (b) cyclic-di-GMP (c-di-GMP), which mainly regulates the switch between motile and adhesive phenotypes (Jenal and Malone, 2006). For the sake of completion, we mention here that cyclic-di-AMP has been recently described in bacteria in the context of response to DNA damage (Witte *et al.*, 2008).

ppGpp: an architect of stringent response

Stringent response is a form of bacterial stress response in which starvation of nitrogen and carbon sources leads to a global down-regulation of growth and division-associated genes (Potrykus and Cashel, 2008). This process is signalled by the abundance of uncharged tRNAs, which on binding to the ribosome, trigger synthesis of the alarmone ppGpp, from GTP and ATP, by the ribosome-associated enzyme RelA (Fig. 5.4A). A second enzyme SpoT has a less robust ppGpp synthesizing activity, but uniquely has the ability to hydrolyse the alarmone. ppGpp exerts its effects on transcription through several mechanisms: (a) binding to the secondary channel of RNA polymerase, aided by the protein DksA (Perederina *et al.*, 2004) (b) destabilizing the open complex (Barker *et al.*, 2001); (c) influencing the competition between the major and alternative sigma factors (Jishage *et al.*, 2002). In general, the role of ppGpp in down-regulating genes associated with translation, and stimulating transcription of several genes involved in amino acid biosynthesis has been studied. Moreover, there is a direct link between ppGpp and the global transcription factor Lrp: mutations in *relA* and *spoT* reduce the ability of *E. coli* to produce Lrp (Landgraf *et al.*, 1996).

Traxler and colleagues have performed two studies investigating the role of ppGpp in modulating transcription in *E. coli* following depletion of different nutrients. In the first, the experimental condition involved glucose–lactose diauxie, which generally involves transcriptional regulation of genes belonging to the RpoS (stationary phase sigma factor) and Crp regulatory networks (Traxler *et al.*, 2006). By investigating the global gene expression in the *relA* mutant strain, the authors showed that transcriptional response downstream of both RpoS and Crp were adversely affected: members of the RpoS regulon showed delayed induction; targets of Crp displayed reduced differential expression. This allowed the authors to

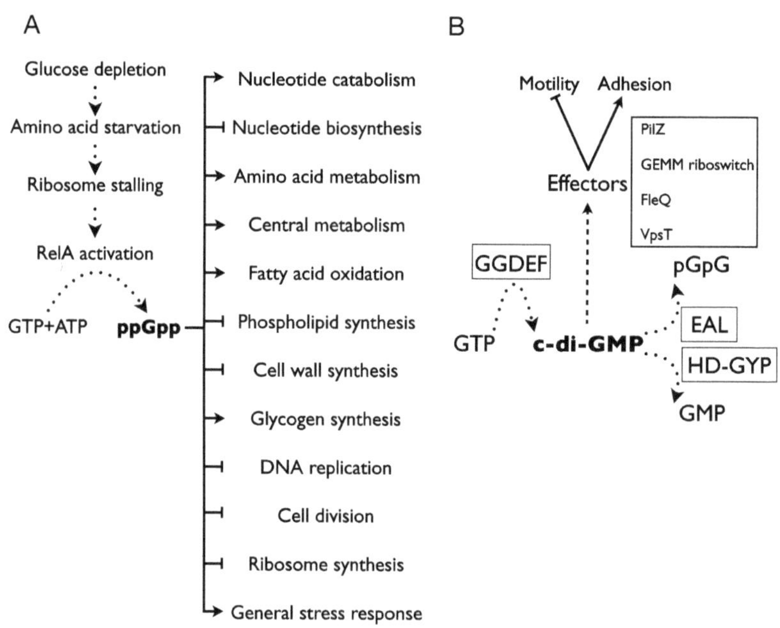

Figure 5.4 Regulation by ppGpp and c-di-GMP. (A) Signals and processes leading to the synthesis of ppGpp are represented by dotted arrow. The numerous functions regulated (directly or indirectly) by ppGpp are listed; the nature of regulation can be activation (solid arrows) or repression (solid lines with flat heads). This is based on the models presented in Traxler and colleagues (2006, 2008). (B) Enzymatic conversions involved in the turnover of c-di-GMP are shown in dotted arrows; GGDEF, EAL and HD-GYP are enzymatic protein domains. Motility is repressed and adhesion is activated by c-di-GMP. Various c-di-GMP effector molecules (PilZ, GEMM, FleQ and VpsT) are named in the box adjacent to 'Effectors'.

place ppGpp at the apex of the regulatory network governing transcriptional response to diauxie, where it regulates diverse processes including stress response, carbon scavenging and ribosome synthesis. In a second study, these authors examined the transcriptional response of *E. coli* to isoleucine starvation, which includes the wide-ranging restructuring of transcription of genes involved in central, amino acid, nucleotide and fatty acid metabolism (Traxler *et al.*, 2008). These responses were abolished in a *relA spoT* deletion strain (ppGpp^0). Further, the ppGpp^0 cells were larger and produced more macromolecules than the wild-type strain under isoleucine starvation. These global responses of *E. coli* to a deficiency in ppGpp synthesis suggest that 'ppGpp is the primary signal used by *E. coli* cells to adjust their reproductive potential to that defined by their nutritional environment' (Traxler *et al.*, 2008).

cyclic-di-GMP: a switch between motility and adhesion

Cyclic-di-GMP was first discovered in the late 1980s as a small-molecule regulator of cellulose biosynthesis in *Gluconacetobacter xylinus* and *Agrobacterium tumafaciens* (Ross *et al.* 1987; Amikam and Benziman 1989). Later, two enzymatic protein domains, called GGDEF and EAL, were implicated in the biosynthesis and hydrolysis of c-di-GMP respectively (Tal *et al.* 1998). A third domain called HD-GYP is less well studied and is also involved in c-di-GMP hydrolysis, albeit to a different end product when compared with the EAL domain (Ryan *et al.* 2006, 2009) (Fig. 5.4B). Functional characterization of these protein domains in various organisms, most notably *E. coli, Caulobacter crescentus, Vibrio cholerae, Salmonella enterica* serovar Typhimurium and *Pseudomonas aeruginosa*, identified fundamental roles for this small-molecule in regulating the complex decision to switch between motility and adhesion (Jenal and Malone, 2006; Hengge, 2009) (Fig. 5.4B). In addition, roles in the control of virulence (Tamayo *et al.*, 2007) and response to nutrient starvation have been pointed out (Kumar and Chatterji, 2008).

Sequence-based searches for proteins containing the GGDEF and EAL domains showed that this signalling system is nearly ubiquitous in Bacteria but absent in Archaea (Galperin *et al.*, 2001). Secondly, these proteins were found to be present in many copies in most genomes leading to speculation on their spatiotemporal regulation. Indeed, specific examples illustrate the intricate control of GGDEF and EAL gene expression. In *S. enterica* Typhimurium, the GGDEF-only protein AdrA is required for biofilm formation in rich media, whereas another GGDEF protein GcpA controls the same process under nutrient-deficient conditions (García *et al.*, 2004). In *E. coli*, only a few GGDEF and EAL proteins are expressed during the exponential phase of growth (Sommerfeldt *et al.*, 2009), and may operate in an additive fashion in adjusting cellular c-di-GMP levels (Boehm *et al.* 2010); many other proteins are specifically expressed during stationary phase (Sommerfeldt *et al.*, 2009; Weber *et al.*, 2006). Further, most of these proteins are associated with additional partner domains, most notably the small-molecule-sensing PAS domain, suggesting signal-dependent post-translational regulation (Galperin *et al.*, 2001). Distinct cellular localization of these proteins has also been demonstrated in specific cases, such as in the control of asymmetric cell division in *C. crescentus* (Paul *et al.*, 2004), and the gathering of the WspR protein in *P. aeruginosa* into distinct cytoplasmic foci in response to a surface-associated signal (Güvener and Harwood, 2007). Given this, it is clear that there is a lot of scope for systematic characterization of how these different c-di-GMP systems in a given bacterium are controlled, thus ensuring spatiotemporal separation of their activities (Hengge, 2009). In particular, the nature of signals that activate a given GGDEF or EAL protein remains largely unknown.

Genomic surveys also identified a large number of proteins which contain both GGDEF and EAL domains. These 'hybrid' proteins have been termed as a 'biochemical conundrum' (Ryan *et al.*, 2006). In fact, the first proteins implicated in c-di-GMP turnover are hybrid proteins. In many experimentally characterized proteins, only one of the two domains is catalytically active; the second domain might play the role of an allosteric regulator. However, in at least three recently studied proteins, both domains are catalytically active (Ferreira *et al.*, 2008; Kumar and Chatterji,

2008; Tarutina *et al.*, 2006). Our recent survey of catalytic site motifs in GGDEF and EAL domains suggests that over 85% of hybrid proteins might belong to one of two categories: (a) both domains are catalytically active and (b) only the EAL domain is active (Seshasayee *et al.*, 2010).

Though the enzymatic domains involved in the turnover of c-di-GMP are relatively well characterized, only recently have we started to identify downstream players that bind c-di-GMP and mediate its effects. One important protein domain, identified through bioinformatic searches, is called PilZ (Amikam and Galperin, 2006). This domain has been subsequently studied experimentally and its role as a c-di-GMP effector established (Pratt *et al.*, 2007; Benach *et al.*, 2007; Ryjenkov *et al.*, 2006). For example, the PilZ domain protein YcgR in *E. coli* acts downstream of an intricate c-di-GMP signalling network comprising five GGDEF and EAL proteins to control swimming velocity (Boehm *et al.*, 2010). In line with its role in c-di-GMP signalling, the phylogenetic distribution of PilZ closely mirrors that of GGDEF and EAL domains. However, to our knowledge, no PilZ protein has been associated with a transcription factor domain, thus currently ruling out the possibility of direct control of transcriptional initiation by c-di-GMP via PilZ. Other novel types of effector proteins, including those GGDEF domains that do not have catalytic activity but can bind c-di-GMP, are being discovered (Duerig *et al.*, 2009; Krasteva *et al.*, 2010; Hickman and Harwood, 2008). Interestingly, even degenerate GGDEF and EAL proteins, which might not retain any detectable relationship to c-di-GMP, have roles related to the control of motility and adhesion (Hengge, 2009).

Because of its structural similarity to nucleic acids, c-di-GMP was proposed to bind to RNA molecules (Jenal and Malone, 2006). This was experimentally verified later by the identification of riboswitches – regions in the 5′-UTR of certain mRNAs – that bind to c-di-GMP (Kulshina *et al.*, 2009; Sudarsan *et al.*, 2008; Smith *et al.*, 2009). This particular conserved RNA domain has been found in the mRNAs of GGDEF and EAL proteins as well as some other proteins whose activities respond to c-di-GMP levels. This is an unequivocal example for post-transcriptional regulation of gene expression by this second messenger. Observation that c-di-GMP-sensing riboswitches, unlike those that sense metabolic products, are found in phage genomes suggested that phages could sense 'physiological transformations' in the bacterial cell by monitoring c-di-GMP levels (Sudarsan, Lee, Weinberg, Moy, Kim, Link, and Breaker 2008).

In spite of the above studies, data on both signal-response and effect-mediation in c-di-GMP signalling lag behind those describing the enzymes involved in the molecule's turnover. We anticipate that further research will expand our knowledge of these aspects of this fundamental signalling mechanism.

Conclusion

Small-molecule signalling in complex ecological contexts: antibiotics and interkingdom signalling

In this review we have focused on two major, partially overlapping (for example, Crp can be discussed in the context of second messenger signalling because of the involvement of cAMP), themes in small-molecule signalling in bacteria: regulation of small-molecule metabolism and second-messenger signalling. We conclude by briefly describing two emerging areas in this context: antibiotics as signalling molecules and inter-kingdom signalling between bacteria and their mammalian hosts, which extends the popular phenomenon of intra- and inter-species quorum sensing.

Antibiotics are generally defined as synthetic or natural small molecules which are antagonistic to the growth of microorganisms. Recent evidence suggests that at sub-inhibitory concentrations, antibiotics act as signalling molecules which alter the transcriptional space of the target microorganism (Yim *et al.*, 2007). A range of cellular functions, spanning metabolic, adaptive and virulence capabilities, are targeted (Goh *et al.*, 2002). It is likely that these effects are tightly linked to quorum sensing, and this may blur the boundary between direct and indirect effects: (a) there are structural parallels between antibiotics and quorum signalling molecules and (b) antibiotic treatment also leads to up-regulation of quorum sensing systems (Goh *et al.*, 2002; Yim, Wang, and Davies 2007). Numerous observations illustrating the prevalence of antibiotic resistance and utilization among various naturally occurring microbial communities, and the hypothesis that there is an unexpectedly large pool of cryptic antibiotic resistance genes (Wright, 2007; Allen *et al.*, 2009; Sommer *et al.*, 2009) suggest that studies concerning regulatory functions of antibiotics are of fundamental importance to bacterial ecology.

Given the co-evolution of bacteria and their mammalian hosts for billions of years, it is not surprising that they can intercept each other's signals leading to what is called 'inter-kingdom signalling' (Pacheco and Sperandio, 2009). These signals are mainly agents of intercellular signalling: quorum sensing molecules which regulate several traits including virulence and motility in bacteria, and hormones in mammals. Studies describing such interactions have focused on the following: (a) control of host processes by bacterial quorum sensing molecules: an example is oxo-C12-homoserine (oxo-C12-HSL) lactone from *P. aeruginosa* which enters mammalian cells and induces apoptosis by influencing calcium signalling (Shiner *et al.*, 2006; Williams *et al.*, 2004); (b) disruption of bacterial quorum sensing by host enzymes: a class of mammalian esterases called paraoxanases can inactivate oxo-C12-HSL (Yang *et al.*, 2005); (c) sensing of mammalian hormones by bacteria: in the human colon, enterohaemorrhagic *E. coli* can sense the hormones adrenaline and noradrenaline and activate the expression of virulence-associated genes (Sperandio *et al.*, 2003). These investigations of cell signalling between bacteria and their mammalian hosts – through signalling molecules traditionally studied in the context of intra-kingdom signalling – represent a relatively new and exciting area of research.

Our review of small molecule-based gene regulation, we believe, has highlighted several genome-scale studies illustrating the complexities of bacterial signalling, especially in the context of integrating metabolic and non-metabolic cellular processes; for example, even the well-studied system of glucose-lactose diauxie involves induction of a large number of genes, well beyond an intuitive expectation of the up-regulation of the *lac* operon. However, we have unfortunately neglected the following popular and important areas of research: RNA-based gene regulation, especially riboswitches (Dambach and Winkler, 2009; Serganov, 2009; Garst and Batey, 2009; Serganov and Patel, 2009; Henkin, 2008), except in the context of c-di-GMP signalling, and intra-kingdom quorum sensing (Camilli and Bassler, 2006; Ng and Bassler, 2009; Boyer and Wisniewski-Dyé, 2009; Dickschat, 2010; Atkinson and Williams, 2009), which have been the subject of many recent reviews.

Chapter highlights

- Small-molecules are important regulators of bacterial physiology, with many regulatory proteins able to sense them.
- Metabolites establish direct feedback between transcriptional regulation and small-molecule metabolism; some metabolites can have global effects on the transcriptional states of a cell.
- Small-molecule second messengers – typically forms of nucleotides – control complex bacterial behaviour including starvation response and biofilm formation.
- Signalling mediated by small molecules could traverse kingdoms of life, for example, leading to interactions between a pathogenic bacterium and its mammalian host.

References

Al-Bahry, S.N., Mahmoud, I.Y., Al-Khaifi, A., Elshafie, A.E., and Al-Harthy, A. (2009). Viability of multiple antibiotic resistant bacteria in distribution lines of treated sewage effluent used for irrigation. Water Sci. Technol. *60*, 2939–2948.

Allen, H.K., Donato, J., Wang, H.H., Cloud-Hansen, K.A., Davies, J., and Handelsman, J. (2010). Call of the wild:

antibiotic resistance genes in natural environments. Nat. Rev. Microbiol. *8*, 251–259.
Amikam, D., and Benziman, M. (1989). Cyclic diguanylic acid and cellulose synthesis in *Agrobacterium tumefaciens*. J. Bacteriol. *171*, 6649–6655.
Amikam, D., and Galperin, M.Y. (2006). PilZ domain is part of the bacterial c-di-GMP binding protein. Bioinformatics *22*, 3–6.
Anantharaman, V., Koonin, E.V., and Aravind, L. (2001. Regulatory potential, phyletic distribution and evolution of ancient, intracellular small-molecule-binding domains. J. Mol. Biol. *307*, 1271–1292.
Atkinson, S., and Williams, P. (2009). Quorum sensing and social networking in the microbial world. Interface *6*, 959–978.
Barker, M.M., Gaal, T., Josaitis, C.A., and Gourse, R.L. (2001). Mechanism of regulation of transcription initiation by ppGpp. I. Effects of ppGpp on transcription initiation in vivo and *in vitro*. J. Mol. Biol. *305*, 673–688.
Benach, J., Swaminathan, S.S., Tamayo, R., Handelman, S.K., Folta-Stogniew, E., Ramos, J.E., Forouhar, F., Neely, H., Seetharaman, J., Camilli, A., Hunt, J.F., *et al.* (2007). The structural basis of cyclic diguanylate signal transduction by PilZ domains. EMBO J. *26*, 5153–5166.
Boehm, A., Kaiser, M., Li, H., Spangler, C., Kasper, C.A., Ackerman, M., Kaever, V., Sourjik, V., Roth, V., and Jenal, U. (2010). Second messenger-mediated adjustment of bacterial swimming velocity. Cell *14*, 107–116.
Boyer, M., and Wisniewski-Dyé, F. (2009). Cell–cell signalling in bacteria: not simply a matter of quorum. FEMS Microbiol. Ecol. *70*, 1–19.
Cadoret, J.C., Rousseau, B., Perewoska, I., Sicora, C., Cheregi, O., Vass, I., and Houmard, J. (2005). Cyclic nucleotides, the photosynthetic apparatus and response to a UV-B stress in the Cyanobacterium *Synechocystis* sp. Pcc 6803. J. Biol. Chem. *280*, 33935–33944.
Calvo, J.M., and Matthews, R.G. (1994). The leucine-responsive regulatory protein, a global regulator of metabolism in *Escherichia coli*. Microbiol. Rev. *58*, 466–490.
Camilli, A., and Bassler, B.L. (2006. Bacterial small-molecule signaling pathways. Science *311*, 1113–1116.
Chen, S., Hao, Z., Bieniek, E., and Calvo, J.M. (2001. Modulation of Lrp action in *Escherichia coli* by leucine: effects on non-specific binding of Lrp to DNA. J. Mol. Biol. *314*, 1067–7105.
Chen, S., Rosner, M.H., and Calvo, J.M. (2001. Leucine-regulated self-association of leucine-responsive regulatory protein (Lrp). from *Escherichia coli*. J. Mol. Biol. *312*, 625–635.
Cho, B., Barrett, C.L., Knight, E.M., Park, Y.S., and Palsson, B.Ø. (2008). Genome-scale reconstruction of the Lrp regulatory network in *Escherichia coli*. Proc. Natl. Acad. Sci. U.S.A. *105*, 19462–19467.
Dambach, M.D., and Winkler, W.C. (2009). Expanding roles for metabolite-sensing regulatory RNAs. Curr. Opin. Microbiol. *12*, 161–169.
Dickschat, J.S. (2010. Quorum sensing and bacterial biofilms. Natural Product Rep. *27*, 343–369.
Duerig, A., Abel, S., Folcher, M., Nicollier, M., Schwede, T., Amiot, N., Giese, B., and Jenal, U. (2009). Second messenger-mediated spatiotemporal control of protein degradation regulates bacterial cell cycle progression. Genes Dev. *23*, 93–104.
Ferreira, R.B., Antunes, L.C., Greenberg, E.P., and McCarter, L.L. (2008). *Vibrio parahaemolyticus* ScrC modulates cyclic dimeric GMP regulation of gene expression relevant to growth on surfaces. J. Bacteriol. *190*, 851–860.
Galperin, M.Y., Nikolskaya, A.N., and Koonin, E.V. (2001). Novel domains of the prokaryotic two-component signal transduction systems. FEMS Microbiol. Lett. (*203*, 11–21.
Gama-Castro, S., Jiménez-Jacinto, V., Peralta-Gil, M., Santos-Zavaleta, A., Peñaloza-Spinola, M.I., Contreras-Moreira, B., Segura-Salazar, J., Muñiz-Rascado, L., Martínez-Flores, I., Salgado, H., *et al.* (2008). RegulonDB (version 6.0): gene regulation model of *Escherichia coli* K-12 beyond transcription, active (experimental). annotated promoters and Textpresso navigation. Nucleic Acids Res. *36*, D120–D124.
García, B., Latasa, C., Solano, C., García-del Portillo, F., Gamazo, C., and Lasa, I. (2004). Role of the GGDEF protein family in Salmonella cellulose biosynthesis and biofilm formation. Mol. Microbiol. *54*, 264–277.
Garst, A.D., and Batey, R.T. (2009). A switch in time: detailing the life of a riboswitch. Biochim. Biophys. Acta *1789*, 584–591.
Goh, E., Yim, G., Tsui, W., McClure, J., Surette, M.G., and Davies, J. (2002). Transcriptional modulation of bacterial gene expression by subinhibitory concentrations of antibiotics. Proc. Natl. Acad. Sci. U.S.A. *99*, 17025–17030.
Grainger, D.C., Overton, T.W., Reppas, N., Wade, J.T., Tamai, E., Hobman, J.L., Constantinidou, C., Struhl, K., Church, G., Busby, S.J., *et al.* (2004). Genomic studies with *Escherichia coli* MelR protein: applications of chromatin immunoprecipitation and microarrays. J. Bacteriol. *186*, 6938–6943.
Grainger, D.C., Hurd, D., Harrison, M., Holdstock, J., and Busby, S.J. (2005). Studies of the distribution of *Escherichia coli* cAMP-receptor protein and RNA polymerase along the *E. coli* chromosome. Proc. Natl. Acad. Sci. U.S.A. *102*, 17693–17698.
Görke, B., and Stülke, J. (2008. Carbon catabolite repression in bacteria: many ways to make the most out of nutrients. Nat. Rev. Microbiol. *6*, 613–624.
Güvener, Z.T., and Harwood, C.S. (2007). Subcellular location characteristics of the *Pseudomonas aeruginosa* GGDEF protein, WspR, indicate that it produces cyclic-di-GMP in response to growth on surfaces. Mol. Microbiol. *66*, 1459–1473.
Hengge, R. (2009). Principles of c-di-GMP signalling in bacteria. Nat. Rev. Microbiol. *7*, 263–273.
Henkin, T.M. (2008). Riboswitch RNAs: using RNA to sense cellular metabolism. Genes Dev. *22*, 3383–3390.
Hershberg, R., Yeger-Lotem, E., and Margalit, H. (2005). Chromosomal organization is shaped by the

transcription regulatory network. Trends Genet. *21*, 138–142.

Hickman, J.W., and Harwood, C.S. (2008). Identification of FleQ from *Pseudomonas aeruginosa* as a c-di-GMP-responsive transcription factor. Mol. Microbiol. *69*, 376–389.

Ihmels, J., Levy, R., and Barkai, N. (2004). Principles of transcriptional control in the metabolic network of *Saccharomyces cerevisiae*. Nat. Biotechnol. *22*, 86–92.

Jenal, U., and Galperin, M.Y. (2009). Single domain response regulators: molecular switches with emerging roles in cell organization and dynamics. Curr. Opin. Microbiol. *12*, 152–160.

Jenal, U., and Malone, J. (2006). Mechanisms of cyclic-di-GMP signaling in bacteria. Annu. Rev. Genet. *40*, 385–407.

Jeong, H., Tombor, B., Albert, R., Oltvai, Z.N., and Barabási, A.L. (2000). The large-scale organization of metabolic networks. Nature *407*, 651–654.

Jishage, M., Kvint, K., Shingler, V., and Nyström, T. (2002). Regulation of sigma factor competition by the alarmone ppGpp. Genes Dev. *16*, 1260–1270.

Kanehisa, M., and Goto, S. (2000). KEGG: kyoto encyclopedia of genes and genomes. Nucl. Acids Res. *28*, 27–30.

Kharchenko, P., Church, G.M., and Vitkup, D. (2005). Expression dynamics of a cellular metabolic network. Mol. Systems Biol. *1*, 2005–0016.

Krasteva, P.V., Fong, J.C., Shikuma, N.J., Beyhan, S., Navarro, M.V., Yildiz, F.H., and Sondermann, H. (2010). *Vibrio cholerae* VpsT regulates matrix production and motility by directly sensing cyclic di-GMP. Science *327*, 866–868.

Kulshina, N., Baird, N.J., and Ferré-D'Amaré, A.R. (2009). Recognition of the bacterial second messenger cyclic diguanylate by its cognate riboswitch. Nat. Struct. Mol. Biol. *16*, 1212–1217.

Kumar, M., and Chatterji, D. (2008). Cyclic di-GMP: a second messenger required for long-term survival, but not for biofilm formation, in Mycobacterium smegmatis. Microbiology *154*, 2942–2955.

Landgraf, J.R., Wu, J., and Calvo, J.M. (1996). Effects of nutrition and growth rate on Lrp levels in *Escherichia coli*. J. Bacteriol. *178*, 6930–6936.

Liu, M. (2005). Global transcriptional programs reveal a carbon source foraging strategy by *Escherichia coli*. J. Biol. Chem. *280*, 15921–15927.

Lozada-Chávez, I., Angarica, V.E., Collado-Vides, J., and Contreras-Moreira, B. (2008). The role of DNA-binding specificity in the evolution of bacterial regulatory networks. J. Mol. Biol. *379*, 627–643.

Madan Babu, M., and Teichmann, S.A. (2003). Evolution of transcription factors and the gene regulatory network in *Escherichia coli*. Nucl. Acids Res. *31*, 1234–1244.

Martínez-Antonio, A., Janga, S.C., Salgado, H., and Collado-Vides, J. (2006). Internal-sensing machinery directs the activity of the regulatory network in *Escherichia coli*. Trends Microbiol. *14*, 22–27.

Martínez-Antonio, A., and Collado-Vides, J. (2003). Identifying global regulators in transcriptional regulatory networks in bacteria. Curr. Opin. Microbiol. *6*, 482–489.

Menchaca-Mendez, R., Janga, S.C., and Collado-Vides, J. (2005). The network of transcriptional interactions imposes linear constrains in the genome. Omics *9*, 139–145.

Newman, E.B., and Lin, R. (1995). Leucine-responsive regulatory protein: a global regulator of gene expression in *E. coli*. Annu. Rev. Microbiol. *49*, 747–775.

Ng, W., and Bassler, B.L. (2009). Bacterial quorum-sensing network architectures. Annu. Rev. Genet. *43*, 197–222.

Notebaart, R.A., Teusink, B., Siezen, R.J., and Papp, B. (2008). Co-regulation of metabolic genes is better explained by flux coupling than by network distance. PLoS Comp. Biol. *4*, e26.

Pacheco, A.R., and Sperandio, V. (2009). Inter-kingdom signaling: chemical language between bacteria and host. Curr. Opin. Microbiol. *12*, 192–198.

Pardee, A.B., Jacob, F., and Monod, J. (1959). The genetic control and cytoplasmic expression of 'Inducibility' in the synthesis of β-galactosidase by *E. coli*. J Mol. Biol. *1*, 165–178.

Paul, R., Weiser, S., Amiot, N.C., Chan, C., Schirmer, T., Giese, B., and Jenal, U. (2004). Cell cycle-dependent dynamic localization of a bacterial response regulator with a novel di-guanylate cyclase output domain. Genes Dev. *18*, 715–727.

Perederina, A., Svetlov, V., Vassylyeva, M.N., Tahirov, T.H., Yokoyama, S., Artsimovitch, I., and Vassylyev, D.G. (2004). Regulation through the secondary channel – structural framework for ppGpp-DksA synergism during transcription. Cell *118*, 297–309.

Perez-Rueda, E. (2000). The repertoire of DNA-binding transcriptional regulators in *Escherichia coli* K-12. Nucleic Acids Res. *28*, 1838–1847.

Potrykus, K., and Cashel, M. (2008. (p)ppGpp: still magical? Annu. Rev. Microbiol. *62*, 35–51.

Pratt, J.T., Tamayo, R., Tischler, A.D., and Camilli, A. (2007). PilZ domain proteins bind cyclic diguanylate and regulate diverse processes in *Vibrio cholerae*. J. Biol. Chem. *282*, 12860–12870.

Ross, P., *et al.* (1987). Regulation of cellulose synthesis in Acetobacter xylinum by cyclic diguanylic acid. Nature *325*, 279–281.

Ryan, R.P., *et al.* (2006). Cell–cell signaling in Xanthomonas campestris involves an HD-GYP domain protein that functions in cyclic di-GMP turnover. Proc. Natl. Acad. Sci. U.S.A. *103*, 6712–6717.

Ryan, R.P., Fouhy, Y., Lucey, J.F., and Dow, J.M. (2006). Cyclic di-GMP signaling in bacteria: recent advances and new puzzles. J. Bacteriol. *188*: 8327–8334.

Ryan, R.P., Lucey, J., O'Donovan, K., McCarthy, Y., Yang, L., Tolker-Nielsen, T., and Dow, J.M. (2009. HD-GYP domain proteins regulate biofilm formation and virulence in *Pseudomonas aeruginosa*. Env. Microbiol. *11*, 1126–1136.

Ryjenkov, D.A., Simm, R., Römling, U., and Gomelsky, M. (2006). The PilZ domain is a receptor for the second messenger c-di-GMP: the PilZ domain protein YcgR controls motility in enterobacteria. J. Biol. Chem. *281*, 30310–30314.

Römling, U. (2008). Great times for small molecules: c-di-AMP, a second messenger candidate in Bacteria and Archaea. Sci. Signaling *1*, pe39.

Sellick, C.A., and Reece, R.J. (2005). Eukaryotic transcription factors as direct nutrient sensors. Trends Biochem. Sci. *30*, 405–412.

Serganov, A., and Patel, D.J. (2009). Amino acid recognition and gene regulation by riboswitches. Biochim. Biophys. Acta *1789*, 592–611.

Serganov, A. (2009). The long and the short of riboswitches. Curr. Opin. Struct. Biol. *19*, 251–259.

Seshasayee, A.S., Fraser, G.M., Babu, M.M., and Luscombe, N.M. (2009). Principles of transcriptional regulation and evolution of the metabolic system in *E. coli*. Genome Res. *19*, 79–91.

Seshasayee, A.S., Fraser, G.M., and Luscombe, N.M. (2010). Comparative genomics of cyclic-di-GMP signalling in bacteria: post-translational regulation and catalytic activity. Nucleic Acids Res. *38*, 5970–5981.

Shapiro, L., McAdams, H.H., and Losick, R. (2009). Why and how bacteria localize proteins. Science *326*:, 1225–1228.

Shiner, E.K., Terentyev, D., Bryan, A., Sennoune, S., Martinez-Zaguilan, R., Li, G., Gyorke, S., Williams, S.C., and Rumbaugh, K.P. (2006). *Pseudomonas aeruginosa* autoinducer modulates host cell responses through calcium signalling. Cell. Microbiol. *8*, 1601–1610.

Smith, K.D., Lipchock, S.V., Ames, T.D., Wang, J., Breaker, R.R., and Strobel, S.A. (2009). Structural basis of ligand binding by a c-di-GMP riboswitch. Nat. Struct. Mol. Biol. *16*, 1218–1223.

Sommer, M.O., Dantas, G., and Church, G.M. (2009. Functional characterization of the antibiotic resistance reservoir in the human microflora. Science *325*, 1128–1131.

Sommerfeldt, N., Possling, A., Becker, G., Pesavento, C., Tschowri, N., and Hengge, R. (2009. Gene expression patterns and differential input into curli fimbriae regulation of all GGDEF/EAL domain proteins in *Escherichia coli*. Microbiology *155*, 1318–1331.

Sperandio, V., Torres, A.G., Jarvis, B., Nataro, J.P., and Kaper, J.B. (2003. Bacteria-host communication: the language of hormones. Proc. Natl. Acad. Sci. U.S.A. *100*, 8951–8956.

Sudarsan, N., Lee, E.R., Weinberg, Z., Moy, R.H., Kim, J.N., Link, K.H., and Breaker, R.R. (2008). Riboswitches in eubacteria sense the second messenger cyclic di-GMP. Science *321*, 411–413.

Tal, R., Wong, H.C., Calhoon, R., Gelfand, D., Fear, A.L., Volman, G., Mayer, R., Ross, P., Amikam, D., Weinhouse, H., *et al.* (1998). Three cdg operons control cellular turnover of cyclic di-GMP in *Acetobacter xylinum*: genetic organization and occurrence of conserved domains in isoenzymes. J. Bacteriol. *180*, 4416–4425.

Tamayo, R., Pratt, J.T., and Camilli, A. (2007). Roles of cyclic diguanylate in the regulation of bacterial pathogenesis. Annu. Rev. Microbiol. *61*, 131–148.

Tarutina, M., Ryjenkov, D.A., and Gomelsky, M. (2006). An unorthodox bacteriophytochrome from *Rhodobacter sphaeroides* involved in turnover of the second messenger c-di-GMP. J. Biol. Chem. *281*, 34751–34758.

Traxler, M.F., Chang, D., and Conway, T. (2006). Guanosine 3′,5′-bispyrophosphate coordinates global gene expression during glucose-lactose diauxie in *Escherichia coli*. Proc. Natl. Acad. Sci. U.S.A. *103*:, 2374–2379.

Traxler, M.F., Summers, S.M., Nguyen, H., Zacharia, V.M., Hightower, G.A., Smith, J.T., and Conway, T. (2008. The global, ppGpp-mediated stringent response to amino acid starvation in *Escherichia coli*. Mol. Microbiol. *68*, 1128–1148.

Ulrich, L.E., Koonin, E.V., and Zhulin, I.B. (2005). One-component systems dominate signal transduction in prokaryotes. Trends Microbiol. *13*, 52–56.

Weber, H., Pesavento, C., Possling, A., Tischendorf, G., and Hengge, R. (2006). Cyclic-di-GMP-mediated signalling within the sigma network of *Escherichia coli*. Mol. Microbiol. *62*, 1014–1034.

Williams, S.C., Patterson, E.K., Carty, N.L., Griswold, J.A., Hamood, A.N., and Rumbaugh, K.P. (2004). *Pseudomonas aeruginosa* autoinducer enters and functions in mammalian cells. J. Bacteriol. *186*, 2281–2287.

Witte, G., Hartung, S., Büttner, K., and Hopfner, K. (2008). Structural biochemistry of a bacterial checkpoint protein reveals diadenylate cyclase activity regulated by DNA recombination intermediates. Mol. Cell *30*, 167–718.

Wright, G.D. (2007). The antibiotic resistome: the nexus of chemical and genetic diversity. Nat. Rev. Microbiol. *5*, 175–186.

Yang, F., Wang, L., Wang, J., Dong, Y., Hu, J.Y., and Zhang, L. (2005). Quorum quenching enzyme activity is widely conserved in the sera of mammalian species. *FEBS letters* *579*, 3713–3177.

Yim, G., Wang, H.H., and Davies, J. (2007). Antibiotics as signalling molecules. Phil. Trans. Roy. Soc. Lond. B *362*, 1195–1(200.

Zaslaver, A., Mayo, A.E., Rosenberg, R., Bashkin, P., Sberro, H., Tsalyuk, M., Surette, M.G., and Alon, U. (2004). Just-in-time transcription program in metabolic pathways. Nat. Genet. *36*, 486–491.

van Nimwegen, E. (2003). Scaling laws in the functional content of genomes. Trends Genet. *19*, 479–484.

Transcriptional Circuits and Phenotypic Variation

6

Ákos T. Kovács and Oscar P. Kuipers

Abstract

By developing various survival strategies simultaneously, bacterial populations are well-prepared to meet harsh conditions. The Gram-positive model organism *B. subtilis* presents a superb example how noise, positive- and negative feedback loops and epigenetic inheritance influence developmental pathways. Noise in combination with a fast positive autoregulatory pathway creates the possibility to initiate natural competence in a subpopulation, followed by a slow negative feedback loop to escape from the competent state when needed. In contrast, sporulation is a one-way differentiation process that is accurately timed and regulated by a gradual increase of phosphorylated Spo0A levels in conjunction with a fine-tuned autoactivation through the phosphorelay in specific activated cells. Intertwinement of regulatory pathways depending on defined levels and activities of regulators can result in genetic logic AND gates, which ensure that appropriate pathways are only activated under specific conditions in a given subpopulation of cells. Finally, communication between subpopulations of cells within an isogenic culture aids and determines the development of complex microbial communities.

Heterogeneous response to fluctuating environmental conditions

Bacteria have developed various strategies and responses to deal with fluctuating environmental conditions. To optimize their chances to survive under harsh circumstances, bacteria employ various specific survival strategies. Commonly, they activate specific sets of genes in response to particular environmental conditions (Storz and Hengge-Aronis, 2009). However, development of specific responses in a rapidly changing environment can be costly and while the changes can take place fast, the reaction time of the bacterium can be limited. One solution bacteria apply to respond quickly to these stresses, is not only based on its genetic content, but also by preparing only part of the population for specific changeable conditions. Therefore, developing certain stress responses in part of the population can be advantageous: they can respond to stress conditions fast, while not all the cells in the population develop this phenotype being prepared for other changes. Heterogeneity in response to stress conditions could help a bacterium to increase the fitness of the species (Thattai and van Oudenaarden, 2004). Population heterogeneity can be caused by mutation, genetic rearrangements, such as phase variation, DNA methylation, cell cycle, variation in microenvironment or it can arise from phenotypic variation originating from feedback architectures of genetic networks (Avery, 2006; Smits *et al.*, 2006). Here, we will focus our attention on the development of phenotypic variability resulting from amplified noise in gene expression and discuss the importance of feedback loops for bistable gene transcription.

Contribution of noise to phenotypic heterogeneity and epigenetic inheritance

Random fluctuations in the biochemical reactions and concentrations of transcripts and proteins in cells are denoted as stochastic fluctuation or noise. Noise is believed to be a key determinant for the development of phenotypic variation, originating from intrinsic properties of transcription and translation. Depending on the biochemical network in which the produced protein is embedded, protein levels will show significant fluctuations from cell to cell. Noise has most influence when the number of molecules involved in the biochemical reaction is small, called the finite number principle (Raser and O'Shea, 2005; Kaern *et al.*, 2005). Noise comes in two ways. Intrinsic noise arises from the fluctuation in biochemical processes such as transcription and translation, whereas variation in the amount or states of other cellular components results indirectly in fluctuations in gene expression, which is called extrinsic noise (Elowitz *et al.*, 2002). By measuring cyan and yellow fluorescence controlled by identical promoter sequences, Elowitz and coworkers were able to discriminate between the contribution of intrinsic noise and extrinsic noise, affecting only one promoter or influencing both promoters. Low transcription rates are typically accompanied with increased intrinsic noise, while extrinsic noise peaks at intermediate transcript levels (Elowitz *et al.*, 2002). Intrinsic noise often fluctuates rapidly and is averaged out quickly in the cells. Therefore, it has in general less effect on the regulatory function of a gene, while extrinsic noise is presented during longer times, i.e. about one cell cycle (Rosenfeld *et al.*, 2005).

From a given promoter sequence, variable amounts of transcripts can be produced during transcription in cells of a clonal population, which is followed by translation that can amplify noise into protein level variation. The origin of noise was explored from both a theoretical and an experimental approach. In theoretical studies, the translation rate was shown to determine the extent of noise (Thattai and van Oudenaarden, 2001), while experimental studies showed examples for both cases, as increased noise was described with increasing translation efficiency in *Bacillus subtilis* (Ozbudak *et al.*, 2002), whereas generation of noise was dependent on transcriptional efficiency in *Saccharomyces cerevisiae* (Blake *et al.*, 2003). Essential and evolutionary conserved genes of *Escherichia coli* were shown to own promoters that confer low level of noise suggesting that natural selection can act to reduce gene expression noise (Silander *et al.*, 2012).

Under certain conditions noise generates phenotypic heterogeneity in an isogenic population of cells. A positive feedback loop can amplify noise and can convert it into a binary response, where a given gene is expressed at either high or low levels. This switch-like behaviour results in a bimodal expression of genes in genetically identical cells, a phenomenon commonly referred to as bistability (Fig. 6.1A), while in cases that more than two subpopulations evolve in a culture the output is called multistable. Theoretical modelling has suggested several factors that impact on the

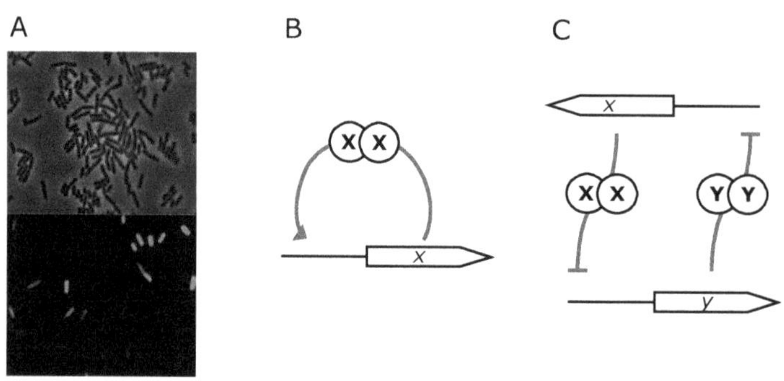

Figure 6.1 Bistable gene expression in bacteria. (A) Bright field (above) and fluorescent (below) microscopic image of *B. subtilis* strain harbouring a competence specific GFP reporter. Two subpopulations are visualized; one expressing the reporter (white cells), while others is not. (B) Positive feedback loop in combination with activation (here dimerization) or (C) double negative feedback loop (toggle switch) can be responsible for the development of multistationarity. Arrowheads indicate a positive effect, while perpendicular lines a negative effect.

development of phenotypic bi- and multistability. Principally, the system needs to show non-linear kinetics in addition to the presence of a feedback loop. This non-linearity can be achieved through oligomerization and cooperative DNA-binding or phosphorylation of the regulator, or a combination of these phenomena. Modelling further substantiated that an even number of negative feedback loops (Fig. 6.1C) or any number of positive feedback loops (Fig. 6.1B) can be responsible for the development of multistationarity (Angeli *et al.*, 2004). Subsequent experimental approaches confirmed the importance of both double-negative and positive feedback loops (for examples, see Smits see *et al.*, 2006), although the presence of a positive feedback loop does not necessarily mean that it results in multistationarity (Ferrell, 2002).

The impact of noise on multistable systems is controlled in several ways. The most obvious way is to increase the number of molecules in the system, so that addition or removal of single molecules has little effect on the properties of the system. However, this approach is energy consuming, hence nature utilizes other systems to control noise. A negative feedback loop is a feasible tool to ensure that the level of a given component is not increased, by which the effect of noise is tempered. Negative feedback loops are in fact overrepresented in *E. coli* (Thieffry *et al.*, 1998). Hysteresis can be another solution, when the switch from one state to another necessitates an unequal force compared to the reverse transition, exemplified by e.g. the direction of flagellar rotation in *Salmonella* chemotaxis (Bren and Eisenbach, 2001). Noise is amplified in case a longer signal cascade is present in the upstream regulatory pathway and therefore extrinsic noise has a bigger effect on gene expression variability than intrinsic noise (Hooshangi *et al.*, 2005; Pedraza and van, 2005). While noise has an important role in the development of bistable regulatory processes, environmental signals often also contribute to and modify the probability of noise-driven phenotypic heterogeneity. CRP (catabolite regulation protein) modulates the bistable expression of the *lac* operon in *E. coli* (Ozbudak *et al.*, 2004), while in *B. subtilis* the bistable expression of the sporulation regulator Spo0A and the competence regulator ComK are accurately controlled by a phosphorelay (Veening *et al.*, 2005; Chung *et al.*, 1994) and a proteolytic degradation system (Smits *et al.*, 2005; Maamar and Dubnau, 2005), respectively.

In general, noise is only beneficial for the cells, when it has no detrimental effect on the fitness of the organism. Therefore, production of essential proteins involves lower levels of noise compared to production of most other proteins (Fraser *et al.*, 2004). However, it can be advantageous, when a small proportion of the population switches to a phenotype, able to survive under an adverse environment that kills the majority of cells (Wolf *et al.*, 2005). Part of the population entering into a phenotypic state with reduced fitness anticipates future environmental assaults and can be considered as being a bet-hedging strategy. Bacterial persistence is a good example of bet-hedging (Lewis, 2007), where a certain part of the population enters into a dormant state, enabling these non-growing cells to survive harsh environments. Although these cells reduce their fitness under normal conditions, they present a possibility to survive a long-term exposure to unfavourable environments.

The cellular state can be transferred to the next generation, providing a case of epigenetic inheritance. A well-known example of epigenetic inheritance is site specific methylation (Casadesus and Low, 2006), where modification lasts over several cell division rounds. In the case of noise-based phenotypic heterogeneity, the state of the cell is depending on the state or/and level of certain proteins. Autophosphorylating kinases have previously been proposed to have the potential to store memory (Lisman, 1985). Using an artificial bistable gene regulatory circuit in *E. coli*, an autoregulatory system was shown to function as a memory device in this bacterium (Gardner *et al.*, 2000). Positive feedback helps to maintain and transfer the level of active transcriptional regulators into the new generation. After the feedback loop is activated, the positive feedback of the system will help to maintain a high level of the positive regulator, independently of the presence of the signal. The balance between degradation and growth rate will determine how long cells will maintain the relatively high level of protein needed to trigger positive autoregulation, and thus the duration of memory.

Phenotypic heterogeneity has been observed to occur in early days of *B. subtilis* research. Only part of the population has been shown to sporulate (Dawes and Thornley, 1970; Schaeffer *et al.*, 1965) or become competent (Nester and Stocker, 1963). New developments in single cell technology and the extensive research on the different heterogeneous processes made it possible to unravel how these systems are regulated and how they interact with each other. The heterogeneous nature of competence, sporulation, motility, biofilm formation and protease secretion of *B. subtilis* presents a convenient possibility to investigate the role of noise in the development of heterogeneity, how feedback loops influence the formation of bi- and multistable networks and finally how these regulatory systems interact with each other.

Noise defines the switching probability for bistable competence development

Natural competence is the ability of eubacteria to take up and incorporate genomic DNA from the environment (Lorenz and Wackernagel, 1994). It may help bacteria to survive under adverse conditions (Claverys *et al.*, 2006; Finkel and Kolter, 2001). Genes involved in natural competence are conserved in the *Bacillus* genus (Kovacs *et al.*, 2009) and commonly regulated by the ComK master regulator (van Sinderen *et al.*, 1994). ComK in addition to regulating hundreds of genes (Berka *et al.*, 2002; Hamoen *et al.*, 2002; Ogura *et al.*, 2002) also positively affects its own transcription (van Sinderen and Venema, 1994). As the high number of genes regulated by ComK is not solely connected to competence, the status of cells harbouring an activated ComK is called the K-state (Berka *et al.*, 2002). The transcription of *comK* is repressed by at least three different transcription factors: AbrB, Rok and CodY, while positively regulated by DegU in addition to ComK itself (Fig. 6.2) (Hamoen *et al.*, 2003). The MecA adaptor protein targets the ComK protein to the protease complex ClpCP, by which it is degraded (Turgay *et al.*, 1998). Quorum sensing pathways promote the production of the anti-adaptor protein ComS, which inhibits formation of the ComK/MecA/ClpCP complex and releases ComK for autostimulation (D'Souza *et al.*, 1994; Hamoen *et al.*, 1995; Prepiak and Dubnau, 2007; Solomon *et al.*, 1995). However, the increased ComK level and the concomitant competence development takes places only in part of the cellular population (Maamar and Dubnau, 2005; Smits *et al.*, 2005).

The presence of two subpopulations of *B. subtilis* cells within the population has been shown in early days by Nester and Stocker (1963), where penicillin G reduced the number of viable cells, but not competence, showing that the competent part of the population displays a biosynthetic latency. Competent cells were later separated by buoyant density centrifugation from the non-competent population (Hadden and Nester, 1968; Cahn and Fox, 1968). The fraction of competent cells expressed the *comK* gene and other late competent genes (Albano *et al.*, 1987; Hahn *et al.*, 1987, 1994). It was through single cell assays, e.g. using the *comK* promoter fused to the *gfp* reporter gene, that two populations of cells could be visualized in the cells (Haijema *et al.*, 2001) that allowed the in-depth examination of how two populations are developing during the K-state in *B. subtilis*. Two independent studies showed the importance of the ComK autoregulatory loop in the bistable *comK* expression (Maamar and Dubnau, 2005; Smits *et al.*, 2005). Deletion of separate components except the positive autoregulation by ComK itself still showed a bistable expression (Smits *et al.*, 2005). The opportunity to activate the autoregulatory circuit and therefore the probability that cells become competent, was shown to depend on the noisy expression of *comK* (intrinsic switching rate), that shows a burst of expression and reaches a maximum just before the level of ComK is increased enough to activate the autoregulatory loop and then decreased in stationary phase. Therefore, cells that do not switch to the K-state show a reduced level of *comK* expression after the transition point (Leisner *et al.*, 2007; Maamar *et al.*, 2007). Deletion of the *comK* gene to circumvent the presence of a positive autoregulatory loop did not alter the basal expression of *comK* compared to wild type non-competent cells (Maamar *et al.*, 2007). When either the repressor gene *rok* was deleted or the *comS* gene expression level was enhanced ectopically, the switching possibility was increased, most probably owing to the

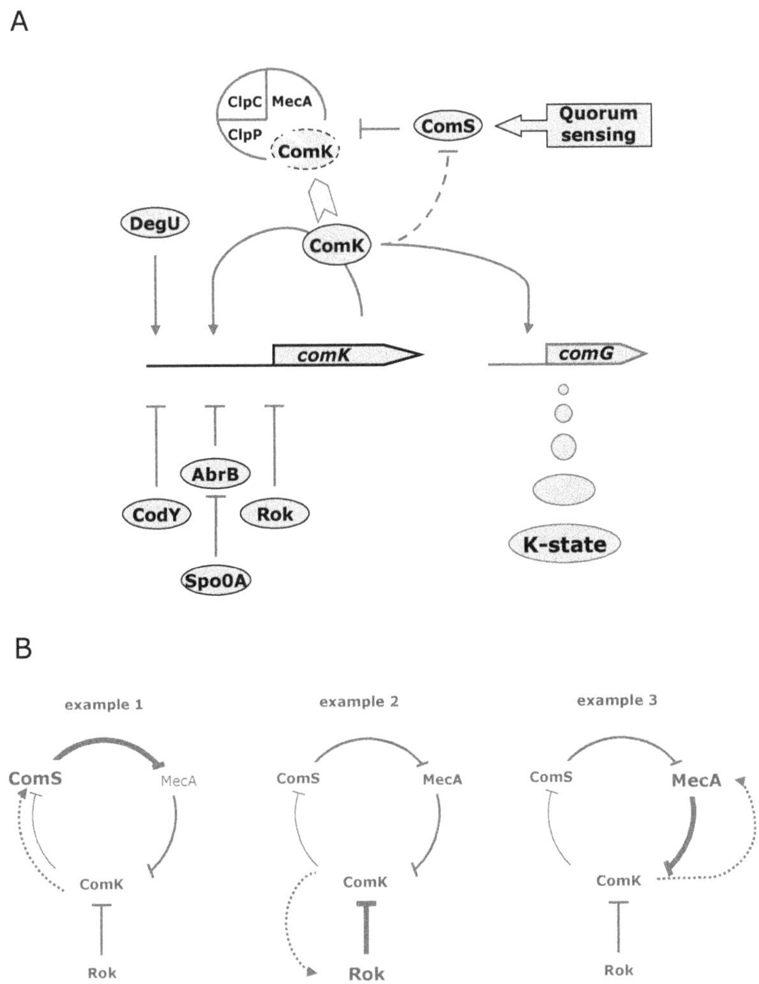

Figure 6.2 Interactions of regulators affect the regulation of competence development in *B. subtilis* (A) (Hamoen *et al.*, 2003). Arrowheads indicate a positive effect, while perpendicular lines a negative effect. Dashed line indicates the indirect affect of ComK on expression of *comS*. Chevron indicates the MecA mediated degradation of ComK protein. (B) Simplified model of synthetic competence circuits described in the text. Dotted lines indicate the introduced PcomG driven expression of *comS* (example 1), *rok* (example 2) and *mecA* (example 3). Note that effect of ComK on *comS* expression is a slow negative feedback loop.

increased basal expression of *comK* (Leisner *et al.*, 2007; Maamar *et al.*, 2007). Further, exchanging medium during the exponential growth phase by conditioned medium made from cells at high density (Magnuson *et al.*, 1994) the K-fraction raised to 36%, and an increase in basal *comK* expression and subsequent shutdown of expression was shifted to an earlier time interval (Leisner *et al.*, 2007). These studies suggest that temporal regulation of transcription controls the frequency of transitions to the K-state and the further decline of *comK* transcription during stationary phase defines a 'window of opportunity'. This window of opportunity is also affected by the cell fate choice of the bacteria, i.e. *B. subtilis* cells enter the competence or sporulation pathways (Kuchina *et al.*, 2011). The relative timing of these competing differentiation programmes determines cell fate outcomes. While initiation probability of competence is suggested to be independent from the state of the sporulation programme, the final commitment and development of the K-state is influenced by the sporulation process (e.g. asymmetric septum formation). On the other hand, Mirouze and coworkers suggested that the transcriptional bursts of the *spo0A* gene control developmental transitions, e.g. *comK* transcription and transitions to competence (Mirouze *et al.*, 2011).

The switching probability greatly depends on the level of ComK protein in the cells. The variation in ComK protein level originates from the intrinsic fluctuation in *comK* mRNA production and degradation. Reducing noise by increasing transcriptional efficiency (e.g. using a *rok* strain) and at the same time reducing translation (by altering the start codon of ComK) resulted in a lower population of cells initiating the K-state (Maamar *et al.*, 2007). Depletion of FtsW in *B. subtilis* cells results in elongated filamentous cells. These cells share cytoplasm and therefore the cellular content is averaged, which causes a reduction of noise in gene expression without affecting the mean concentration of cellular components. Thus, reducing noise decreases the probability of K-state initiation (Suel *et al.*, 2007).

Leisner *et al* suggested that shutdown of *comK* transcription is independent of the competence regulatory pathway as it was observed in non-competent cells, that lack a high level of ComK (Leisner *et al.*, 2007). However Suel *et al* (2006), using time-lapse microscopy showed that *comS* plays a crucial role in the escape from the K-state. By following expression of both *comG* and *comS* in the same cell, a negative correlation was observed between P*comS* and P*comG* activities (Suel *et al.*, 2006), which is consistent with the previously suggested negative regulation of *comS* by ComK (Hahn *et al.*, 1994). Under the given experimental set-up consecutive competence events were observed within a cell linage, indicating reinitiation of competence. Further analysis of cell lineages showed that the probability of initiating the K-state was similar both in cells previously initiating the K-state and in the ones not switching before. Further, cells were not significantly more or less likely to become competent if their sister became competent. If both sister cells became competent, the time interval, during which they stayed in the K-state was uncorrelated. This transient behaviour is caused by the combination of a fast positive and a slow negative feedback. The positive feedback is apparently important for the development of bistability, while the slow negative is necessary for the escape from competence. In this way ComS has a dual role in the regulatory pathway: on the one hand it is important to initiate competence by inhibiting MecA/ClpCP mediated degradation of ComK, but repression of *comS* expression is required to exit from the K-state, because a reduction in ComS level is concomitant with MecA-driven ComK degradation. By exchanging the negative feedback loop to a positive one (Fig. 6.2, example 1) most of the competent cells failed to escape from the K-state, which confirms the importance of the ComS-mediated feedback loop during reversion to the vegetative state. Increasing the level of ComS by expressing it from an inducible promoter also elongates the duration of competence, but the probability of initiation of the K-state is unchanged (Suel *et al.*, 2007). Rather, the probability of initiation of competence depends on the level of ComK. Increasing the level of ComK can produce an oscillatory regime, in which cells repeatedly go in and out of the competent state. Engineering the competence circuit helped to unravel the role of feedback loops and to modify different properties of the pathway. Introducing the *rok* repressor gene, expressed under control of P*comG* (Fig. 6.2, example 2), creates an additional negative feedback loop onto *comK*. Initiation of competence was unchanged in this strain, while the duration of the K-state was reduced, in agreement with the known role of negative feedback loops in such a circuit (Suel *et al.*, 2007). In other experiments the ComS-mediated negative feedback loop was changed by coupling *mecA* to the ComK-inducible *comG* promoter (Fig. 6.2, example 3) (Cagatay *et al.*, 2009). In this case, the MecA level was increased in response to an elevated ComK level, which in turns degrades ComK. This synthetic circuit showed similar excitable properties as in wild-type cells, although the distribution of duration times of the native circuit was broader than that of the synthetic circuit and although the synthetic cells were able to reconstitute the physiological function of competence, the frequency of DNA uptake was reduced. The probability of initiation and duration of the K-state greatly depends on growth conditions (Leisner *et al.*, 2008), but it also depends on the strains used for the study (Dubnau and Losick, 2006). Laboratory strains have been selected for an enhanced level of competence during domestication. Therefore, it is of great interest to compare natural isolates with laboratory strains and compare their competence regulatory circuits.

B. subtilis developmental pathways dependent on gradual activation of the transcription factors Spo0A and DegU

The development of dormant spores that are resistant to wide variety of stresses such as heat, cold, drought or UV-radiation are other examples of bistable phenotypes of *B. subtilis* (Dawes and Thornley, 1970). Since expression of more than 10% of all the genes in the genome is influenced during sporulation, initiation of sporulation is tightly controlled (Fawcett *et al.*, 2000). Both expression of Spo0A and its phosphorylation, and therefore its activity, are important for the development of an autoregulatory loop that results in bistable gene expression of the sporulation regulon (Fig. 6.3A) (Veening *et al.*, 2005; Chung *et al.*, 1994).

While competence development is a temporal phenotypic trait of *B. subtilis*, activation of the sporulation pathway in cells is unidirectional. Therefore, it is important to strictly control the definitive switching to enter sporulation during nutrient limitation. Cells in which the Spo0A autoregulatory circuit is activated (Spo0A-ON) produce a killing factor and a toxin that cause lysis of cells exhibiting a lower level of Spo0A~P (Gonzalez-Pastor *et al.*, 2003). Nutrients released from Spo0A-OFF cells delay sporulation and ensure that nutrient deprivation is not an initial decisive circumstance for the population. The strict control of the bistable Spo0A network, i.e. the presence of Spo0A-ON and Spo0A-OFF cells, is of crucial importance for the proper functioning of cannibalism during *B. subtilis* sporulation.

The autostimulatory loop of *spo0A*, in combination with the phosphorelay pathway

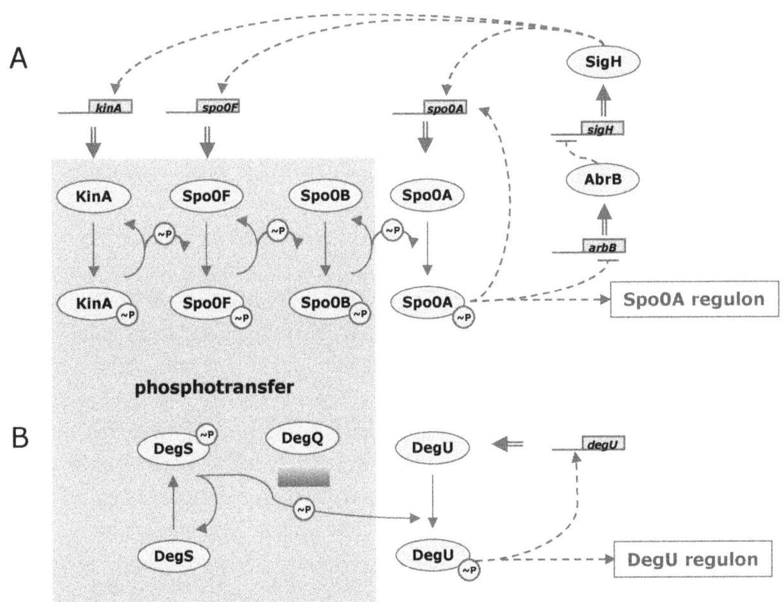

Figure 6.3 Phosphorylation pathways and feedback loops contribute to the heterogeneous gradual increase of the Spo0A (A) (Burbulys *et al.*, 1991; Fujita and Losick, 2005) and DegU (B) level (Kobayashi, 2007; Veening *et al.*, 2008a). Solid lines indicate phosphotransfer (~P), dashed lines transcriptional activation or repression (arrowheads indicate a positive effect, while perpendicular lines a negative effect.), while double lines symbolize transcription/translation.

modulated by different phosphatases, results in bistable expression of sporulation genes (Fujita and Losick, 2005; Veening *et al.*, 2005). Deletion of either the *rapA* or *spo0E* gene, both acting on the phosphorelay pathway, abolishes the bistable expression pattern, which could be restored by induction of heterologous Rap phosphates in the mutant strains (Veening *et al.*, 2005). The presence of Spo0A~P indirectly activates the expression of the stationary phase-specific sigma factor *sigH*, that in turn will increase the level of the phosphorelay components, generating a positive feedback loop (Fig. 6.3A) (Fujita *et al.*, 2005; Predich *et al.*, 1992). These feedback loops are not essential for the synthesis of the *spo0A*, but they contribute for the proper supply of the heterochronically expressed phosphorelay components for rising level of Spo0A~P (de Jong *et al.*, 2010; Chastanet *et al.*, 2010). Therefore heterogeneity originates from the highly dynamic and variable expression of the genes coding for the phosphorelay.

Fujita and Losick (2005) showed, that a certain threshold level of activated Spo0A is required to enter the sporulation pathway. Although the ratio between cells with activated Spo0A (Spo0A-ON) and cells with lower levels of Spo0A (Spo0A-OFF) depends on the cultivating conditions, a bistable output was observed under various conditions (Veening *et al.*, 2006). Genes regulated by Spo0A~P show a time-dependent expression, i.e. some genes are activated or repressed earlier (at lower levels of Spo0A~P), others are affected only at higher levels of Spo0A~P (Fujita *et al.*, 2005). The proper activation of genes displaying different threshold levels is achieved by the different binding affinity of Spo0A to the regulatory regions. Genes with high-affinity Spo0A-binding sites in their promoter regions are transcribed already at lower levels of Spo0A~P, while low-affinity sites are bound only at higher concentrations of Spo0A~P.

Competence and sporulation are events observed principally under dissimilar culture conditions, yet, owing to the noisy nature of the initiation of these pathways cells appear to initiate these developmental pathways in certain defined media sequentially (Veening *et al.*, 2006). As previously mentioned, the probability of competence initiation shows no correlation between sisters cells (Suel *et al.*, 2007). Unlike competence, time lapse microscopic examination of the sporulation process showed that the decision to sporulate occurs more than four cell divisions before activity of Spo0A can be detected in the cells using a *spoIIA–gfp* construct (Veening *et al.*, 2008c). It could be concluded, that the signal to activate Spo0A originates from a common ancestor, and that the cell fate inherited from the ancestor is therefore epigenetic. This epigenetic inheritance depends on the phosphorelay, which is important for both the induction of bistable sporulation and the propagation of the Spo0A activation signal.

The bistable loop coupled with phosphorylation-dependent activation was also observed in the case of the DegSU two-component system in *B. subtilis* (Fig. 6.3B) (Veening *et al.*, 2008a). Similar to Spo0A, a gradual increase in DegU~P determines the regulon and therefore the multicellular behaviour (Fig. 6.4). While low levels of DegU~P activate swarming behaviour of *B. subtilis* depending on its kinase DegS, a complex colony architecture is activated through DegS that is dependent on low levels of DegU~P. In case the level of DegU~P further increases, both swarming motility and complex colony architecture are inhibited, while exoprotease production is activated (Kobayashi, 2007; Verhamme *et al.*, 2007). DegU is also essential for proper activation of competence development, but in this case the unphosphorylated form of DegU facilitates ComK binding to its own promoter. Owing to the differences in the target genes, depending on the phosphorylation level of DegU, some of these genes are affected by a *degQ* mutation, while others are not (Kobayashi, 2007). DegQ stimulates phosphotransfer, and thereby modifies the level of active DegU~P. The affinity to bind to the promoter regions of target genes determines which genes are activated at a given level and phosphorylation state of DegU in the cells (Kobayashi, 2007). For example, DegU has higher affinity to bind to the promoter region of *flgB* involved in flagellum formation. This provides a molecular basis by which swarming motility can be activated prior to other developmental pathways.

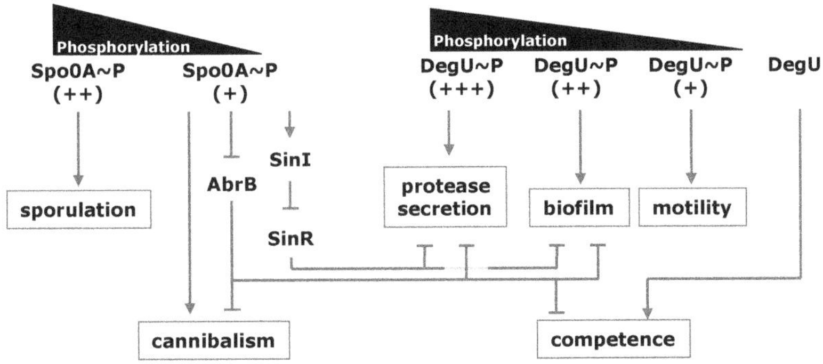

Figure 6.4 Gradual increase in the level of Spo0A and DegU defines which developmental pathways are activated in *B. subtilis* (Fujita *et al.*, 2005; Kobayashi, 2007; Verhamme *et al.*, 2007). Arrowheads indicate a positive effect, while perpendicular lines a negative effect.

The DegU regulon has been found to be upregulated in non-sporulating cells during sporulation initiating conditions (Veening *et al.*, 2008a), thus the DegU~P concentration is higher in non-sporulating cells, while a higher level of Spo0A~P is present in the sporulating cells. Non-sporulating cells contain lower levels of Spo0A~P (Fujita and Losick, 2005; Veening *et al.*, 2008c), that is needed for the expression of the *aprE* gene coding for subtilisin, an extracellular protease. Next to a high level of DegU~P, activation of *aprE* transcription necessitates the escape from AbrB and SinR repressors, which are both repressed directly and indirectly, respectively, by low levels of Spo0A~P (Veening *et al.*, 2008a). While DegS is expressed homogeneously, expression of *degU* is more heterogeneous compared to *degS* in stationary phase. To achieve a heterogeneous *degU* expression both the presence of a positive feedback within the DegU system and DegU~P autostimulation is required (Veening *et al.*, 2008a). This high level of DegU~P, arising from autostimulation is a prerequisite for the heterogeneous expression of *aprE*. Thus, two criteria must be met for the proper expression of *aprE* in vegetative cells. First, the Spo0A~P level must increase to a threshold level, but not too high, because that would initiate sporulation. Second, the level of DegU~P should be high, causing both the autoregulatory loop and the phosphorylation network to be active. The dual requirement of DegU~P and Spo0A~P activation presents a genetic logic-AND gate (Veening *et al.*, 2008b).

Intertwinement of regulatory networks in *B. subtilis* biofilms

Heterogeneous activation of Spo0A and DegU circuits results in proper activation of developmental pathways. Spo0A and DegU are both important transcription factors for the development of surface-associated architecturally complex communities of cells, known as biofilms (Abee *et al.*, 2011). Essential biofilm genes involved in the production of matrix protein (TasA derived from the *tap-A-sipW-tasA* operon) and exopolysaccharide (EPS produced from the *epsA-O* operon) are only expressed in a subpopulation of cells (Chai *et al.*, 2008). Interestingly, bistable expression of biofilm genes is achieved through the bistable antirepression of *sinR* that codes for the major repressor of biofilm gene expression. The small protein SinI binds to SinR and abolishes its action on biofilm genes. However, *sinI* is expressed only in a subpopulation of cells in a Spo0A~P-dependent manner (Chu *et al.*, 2008). In addition, the Spo0A~P-regulated AbrB also modulates expression by binding to the *tap-A* promoter region and also indirectly affects expression of the *eps* operon. In addition to SinI-SinR, a paralogous system, SlrA-Slr is involved in the regulation of the *eps* and *tap-A* operons (Kobayashi, 2008), but unlike *sinI*, expression of *slrA* is not bimodal in its expression (Chai *et al.*, 2009). SlrR also affects biofilm formation by repressing *sinR* and by forming a SinR–SlrR complex. It titrates SinR and prevents it from repressing *slrR* itself (Chai *et al.*, 2010).

Furthermore, this complex represses autolysin and motility genes. Thus, this epigenetic switch controls cell separation and helps the formation of long chains of cells that is a prerequisite for biofilm development.

Low Spo0A~P levels turn *sinI* expression ON and high levels turn *sinI* OFF and instead switch sporulation ON. Cells in which *sinI* and *sinR* were transplanted from their normal position near the chromosome replication terminus to positions near the origin and cells that harboured an extra copy of the genes were blocked in matrix production (Chai *et al.*, 2011). Thus, matrix gene expression is sensitive to the number of copies of *sinI* and *sinR*. Because cells at the start of sporulation have two chromosomes and matrix-producing cells one, chromosome copy number could contribute to cell-fate determination.

Interestingly, the heterogeneous expression of biofilm genes, as the motility and sporulation genes, localizes to distinct regions within the biofilm, and the localization and percentage of cell types expressing each set of genes is dynamic during development of the community (Vlamakis *et al.*, 2008). Sporulation depends upon functional expression of biofilm genes under biofilm-promoting conditions. Therefore, the pathways and the different heterogeneous phenotypes are connected and interdependent on each other. The initiation of biofilm formation not only depends on particular environmental conditions, but also requires the communication between cells expressing particular sets of genes (Lopez *et al.*, 2009). While all cells in a biofilm produce the ComX quorum sensing signal molecule, only few cells will respond to this signal and activate the expression of surfactin. Surfactin in its turn is sensed by another subpopulation of cells in the biofilm and activates the production of EPS and TasA. How these extracytoplasmic signals are sensed solely in a subpopulation of cells still remains a question, but we can imagine that heterogeneity at the level of activity of particular sensors can cause heterogeneous responses to a specific signal.

Concluding remarks

The *B. subtilis* developmental pathways mentioned present good examples of how bistable gene expression determines the eventual phenotype of a bacterium. Noise in transcription and proper autoregulatory circuits define the probability of switching to and escape from a developmental pathway of e.g. genetic competence. Bistability coupled with modulating protein activity results in a gradual increase of a regulator to determine temporal and spatial expression of genes. The intertwinement of regulatory pathways ensures that genes are expressed only in a subpopulation of cells, when the proper environmental signals are present.

Chapter highlights

Studies of bacterial developmental pathways have shown that:

- Noise can determine the possibility to activate a bistable circuit, while feedback loops can ensure the escape from or the fixation of a given developmental pathway.
- Both the levels and the activation states of regulators can accurately determine the size and specificity of a regulon.
- DNA-binding specificity, strength and distribution of binding sites determine which genes are transcribed upon a given level of a regulator.
- Sensing of external signals can differ from cell to cell and can result in heterogeneous activation of regulatory pathways.
- The proper ratio of active regulators and their interaction can be important for the development of specific pathways.
- Initiation of developmental pathways sometimes requires the preceding activation of other pathways. In some cases activation of one pathway excludes the stimulation of others in the same cell.

References

Abee, T., Kovács, Á.T., Kuipers, O.P., and van der Veen, S. (2011). Biofilm formation and dispersal in Gram-positive bacteria. Curr. Opin. Biotechnol. *22*, 172–179.

Albano, M., Hahn, J., and Dubnau, D. (1987). Expression of competence genes in *Bacillus subtilis*. J. Bacteriol. *169*, 3110–3117.

Angeli, D., Ferrell, J.E., Jr., and Sontag, E.D. (2004). Detection of multistability, bifurcations, and hysteresis in a large class of biological positive-feedback systems. Proc. Natl. Acad. Sci. U.S.A. *101*, 1822–1827.

Avery, S.V. (2006). Microbial cell individuality and the underlying sources of heterogeneity. Nat. Rev. Microbiol. *4*, 577–587.

Berka, R.M., Hahn, J., Albano, M., Draskovic, I., Persuh, M., Cui, X., Sloma, A., Widner, W., and Dubnau, D. (2002). Microarray analysis of the *Bacillus subtilis* K-state: genome-wide expression changes dependent on ComK. Mol. Microbiol. *43*, 1331–1345.

Blake, W.J., Kaern, M., Cantor, C.R., and Collins, J.J. (2003). Noise in eukaryotic gene expression. Nature *422*, 633–637.

Bren, A., and Eisenbach, M. (2001). Changing the direction of flagellar rotation in bacteria by modulating the ratio between the rotational states of the switch protein FliM. J. Mol. Biol. *312*, 699–709.

Burbulys, D., Trach, K.A., and Hoch, J.A. (1991). Initiation of sporulation in *B. subtilis* is controlled by a multicomponent phosphorelay. Cell *64*, 545–552.

Cagatay, T., Turcotte, M., Elowitz, M.B., Garcia-Ojalvo, J., and Suel, G.M. (2009). Architecture-dependent noise discriminates functionally analogous differentiation circuits. Cell *139*, 512–522.

Cahn, F.H., and Fox, M.S. (1968). Fractionation of transformable bacteria from ocompetent cultures of *Bacillus subtilis* on renografin gradients. J. Bacteriol. *95*, 867–875.

Casadesus, J., and Low, D. (2006). Epigenetic gene regulation in the bacterial world. Microbiol. Mol. Biol. Rev. *70*, 830–856.

Chai, Y., Chu, F., Kolter, R., and Losick, R. (2008). Bistability and biofilm formation in *Bacillus subtilis*. Mol. Microbiol. *67*, 254–263.

Chai, Y., Kolter, R., and Losick, R. (2009). Paralogous antirepressors acting on the master regulator for biofilm rormation in *Bacillus subtilis*. Mol. Microbiol. *74*, 876–887.

Chai, Y., Norman, T., Kolter, R., and Losick, R. (2010). An epigenetic switch governing daughter cell separation in *Bacillus subtilis*. Genes Dev. *24*, 754–765.

Chai, Y., Norman, T., Kolter, R., and Losick, R. (2011). Evidence that metabolism and chromosome copy number control mutually exclusive cell fates in *Bacillus subtilis*. EMBO J. *30*, 1402–1413.

Chastanet, A., Vitkup, D., Yuan, G.C., Norman, T.M., Liu, J.S., and Losick, R.M. (2010). Broadly heterogeneous activation of the master regulator for sporulation in *Bacillus subtilis*. Proc. Natl. Acad. Sci. U.S.A. *107*, 8486–8491.

Chu, F., Kearns, D.B., McLoon, A., Chai, Y., Kolter, R., and Losick, R. (2008). A novel regulatory protein governing biofilm formation in *Bacillus subtilis*. Mol. Microbiol. *68*, 1117–1127.

Chung, J.D., Stephanopoulos, G., Ireton, K., and Grossman, A.D. (1994). Gene expression in single cells of *Bacillus subtilis*: evidence that a threshold mechanism controls the initiation of sporulation. J. Bacteriol. *176*, 1977–1984.

Claverys, J.P., Prudhomme, M., and Martin, B. (2006). Induction of competence regulons as a general response to stress in Gram-positive bacteria. Annu. Rev. Microbiol. *60*, 451–475.

D'Souza, C., Nakano, M.M., and Zuber, P. (1994). Identification of *comS*, a gene of the *srfA* operon that regulates the establishment of genetic competence in *Bacillus subtilis*. Proc. Natl. Acad. Sci. U.S.A. *91*, 9397–9401.

Dawes, I.W., and Thornley, J.H. (1970). Sporulation in *Bacillus subtilis*. Theoretical and experimental studies in continuous culture systems. J. Gen. Microbiol. *62*, 49–66.

de Jong, I.G., Veening, J.W., and Kuipers, O.P. (2010). Heterochronic phosphorelay gene expression as a source of heterogeneity in *Bacillus subtilis* spore formation. J. Bacteriol. *192*, 2053–2067.

Dubnau, D., and Losick, R. (2006). Bistability in bacteria. Mol. Microbiol. *61*, 564–572.

Elowitz, M.B., Levine, A.J., Siggia, E.D., and Swain, P.S. (2002). Stochastic gene expression in a single cell. Science *297*, 1183–1186.

Fawcett, P., Eichenberger, P., Losick, R., and Youngman, P. (2000). The transcriptional profile of early to middle sporulation in *Bacillus subtilis*. Proc. Natl. Acad. Sci. U.S.A. *97*, 8063–8068.

Ferrell, J.E. Jr. (2002). Self-perpetuating states in signal transduction: positive feedback, double-negative feedback and bistability. Curr. Opin. Cell Biol. *14*, 140–148.

Finkel, S.E., and Kolter, R. (2001). DNA as a nutrient: novel role for bacterial competence gene homologs. J. Bacteriol. *183*, 6288–6293.

Fraser, H.B., Hirsh, A.E., Giaever, G., Kumm, J., and Eisen, M.B. (2004). Noise minimization in eukaryotic gene expression. PLoS. Biol. *2*, e137.

Fujita, M., Gonzalez-Pastor, J.E., and Losick, R. (2005). High- and low-threshold genes in the Spo0A regulon of *Bacillus subtilis*. J. Bacteriol. *187*, 1357–1368.

Fujita, M., and Losick, R. (2005). Evidence that entry into sporulation in *Bacillus subtilis* is governed by a gradual increase in the level and activity of the master regulator Spo0A. Genes Dev. *19*, 2236–2244.

Gardner, T.S., Cantor, C.R., and Collins, J.J. (2000). Construction of a genetic toggle switch in *Escherichia coli*. Nature *403*, 339–342.

Gonzalez-Pastor, J.E., Hobbs, E.C., and Losick, R. (2003). Cannibalism by sporulating bacteria. Science *301*, 510–513.

Hadden, C., and Nester, E.W. (1968). Purification of competent cells in the *Bacillus subtilis* transformation system. J. Bacteriol. *95*, 876–885.

Hahn, J., Albano, M., and Dubnau, D. (1987). Isolation and characterization of Tn917lac-generated competence mutants of *Bacillus subtilis*. J. Bacteriol. *169*, 3104–3109.

Hahn, J., Kong, L., and Dubnau, D. (1994). The regulation of competence transcription factor synthesis constitutes a critical control point in the regulation of competence in *Bacillus subtilis*. J. Bacteriol. *176*, 5753–5761.

Haijema, B.J., Hahn, J., Haynes, J., and Dubnau, D. (2001). A ComGA-dependent checkpoint limits growth during the escape from competence. Mol. Microbiol. *40*, 52–64.

Hamoen, L.W., Eshuis, H., Jongbloed, J., Venema, G., and van Sinderen, D. (1995). A small gene, designated *comS*, located within the coding region of the fourth amino acid-activation domain of *srfA*, is required for competence development in *Bacillus subtilis*. Mol. Microbiol. *15*, 55–63.

Hamoen, L.W., Smits, W.K., de, J.A., Holsappel, S., and Kuipers, O.P. (2002). Improving the predictive value of the competence transcription factor (ComK). binding site in *Bacillus subtilis* using a genomic approach. Nucleic Acids Res. *30*, 5517–5528.

Hamoen, L.W., Venema, G., and Kuipers, O.P. (2003). Controlling competence in *Bacillus subtilis*: shared use of regulators. Microbiology *149*, 9–17.

Hooshangi, S., Thiberge, S., and Weiss, R. (2005). Ultrasensitivity and noise propagation in a synthetic transcriptional cascade. Proc. Natl. Acad. Sci. U.S.A. *102*, 3581–3586.

Kaern, M., Elston, T.C., Blake, W.J., and Collins, J.J. (2005). Stochasticity in gene expression: from theories to phenotypes. Nat. Rev. Genet. *6*, 451–464.

Kobayashi, K. (2007). Gradual activation of the response regulator DegU controls serial expression of genes for flagellum formation and biofilm formation in *Bacillus subtilis*. Mol. Microbiol. *66*, 395–409.

Kobayashi, K. (2008). SlrR/SlrA controls the initiation of biofilm formation in *Bacillus subtilis*. Mol. Microbiol. *69*, 1399–1410.

Kovács, Á.T., Smits, W.K., Mironczuk, A.M., and Kuipers, O.P. (2009). Ubiquitous late competence genes in *Bacillus* species indicate the presence of functional DNA uptake machineries. Environ. Microbiol. *11*, 1911–1922.

Kuchina, A., Espinar, L., Cagatay, T., Balbin, A.O., Zhang, F., Alvarado, A., Garcia-Ojalvo, J., and Suel, G.M. (2011). Temporal competition between differentiation programs determines cell fate choice. Mol. Syst. Biol. 7, 557.

Leisner, M., Stingl, K., Frey, E., and Maier, B. (2008). Stochastic switching to competence. Curr. Opin. Microbiol. *11*, 553–559.

Leisner, M., Stingl, K., Radler, J.O., and Maier, B. (2007). Basal expression rate of *comK* sets a 'switching-window' into the K-state of *Bacillus subtilis*. Mol. Microbiol. *63*, 1806–1816.

Lewis, K. (2007). Persister cells, dormancy and infectious disease. Nat. Rev. Microbiol. *5*, 48–56.

Lisman, J.E. (1985). A mechanism for memory storage insensitive to molecular turnover: a bistable autophosphorylating kinase. Proc. Natl. Acad. Sci. U.S.A. *82*, 3055–3057.

Lopez, D., Vlamakis, H., Losick, R., and Kolter, R. (2009). Paracrine signaling in a bacterium. Genes Dev. *23*, 1631–1638.

Lorenz, M.G., and Wackernagel, W. (1994). Bacterial gene transfer by natural genetic transformation in the environment. Microbiol. Rev. *58*, 563–602.

Maamar, H., and Dubnau, D. (2005). Bistability in the *Bacillus subtilis* K-state (competence). system requires a positive feedback loop. Mol. Microbiol. *56*, 615–624.

Maamar, H., Raj, A., and Dubnau, D. (2007). Noise in gene expression determines cell fate in *Bacillus subtilis*. Science *317*, 526–529.

Magnuson, R., Solomon, J., and Grossman, A.D. (1994). Biochemical and genetic characterization of a competence pheromone from *B. subtilis*. Cell *77*, 207–216.

Mirouze, N., Prepiak, P., and Dubnau, D. (2011). Fluctuations in *spo0A* transcription control rare developmental transitions in *Bacillus subtilis*. PLoS Genet. 7, e1002048.

Nester, E.W., and Stocker, B.A. (1963). Biosynthetic latency in early stages of deoxyribonucleic acid transformation in *Bacillus subtilis*. J. Bacteriol. *86*, 785–796.

Ogura, M., Yamaguchi, H., Kobayashi, K., Ogasawara, N., Fujita, Y., and Tanaka, T. (2002). Whole-genome analysis of genes regulated by the *Bacillus subtilis* competence transcription factor ComK. J. Bacteriol. *184*, 2344–2351.

Ozbudak, E.M., Thattai, M., Kurtser, I., Grossman, A.D., and van, O.A. (2002). Regulation of noise in the expression of a single gene. Nat. Genet. *31*, 69–73.

Ozbudak, E.M., Thattai, M., Lim, H.N., Shraiman, B.I., and van, O.A. (2004). Multistability in the lactose utilization network of *Escherichia coli*. Nature *427*, 737–740.

Pedraza, J.M., and van, O.A. (2005). Noise propagation in gene networks. Science *307*, 1965–1969.

Predich, M., Nair, G., and Smith, I. (1992). *Bacillus subtilis* early sporulation genes *kinA*, *spo0F*, and *spo0A* are transcribed by the RNA polymerase containing sigma H. J. Bacteriol. *174*, 2771–2778.

Prepiak, P., and Dubnau, D. (2007). A peptide signal for adapter protein-mediated degradation by the AAA+ protease ClpCP. Mol. Cell *26*, 639–647.

Raser, J.M., and O'Shea, E.K. (2005). Noise in gene expression: origins, consequences, and control. Science *309*, 2010–2013.

Rosenfeld, N., Young, J.W., Alon, U., Swain, P.S., and Elowitz, M.B. (2005). Gene regulation at the single-cell level. Science *307*, 1962–1965.

Schaeffer, P., Millet, J., and Aubert, J.P. (1965). Catabolic repression of bacterial sporulation. Proc. Natl. Acad. Sci. U.S.A. *54*, 704–711.

Silander, O.K., Nikolic, N., Zaslaver, A., Bren, A., Kikoin, I., Alon, U., and Ackermann, M. (2012). A genome-wide analysis of promoter-mediated phenotypic noise in *Escherichia coli*. PLoS Genet. *8*, e1002443.

Smits, W.K., Eschevins, C.C., Susanna, K.A., Bron, S., Kuipers, O.P., and Hamoen, L.W. (2005). Stripping *Bacillus*: ComK auto-stimulation is responsible for the bistable response in competence development. Mol. Microbiol. *56*, 604–614.

Smits, W.K., Kuipers, O.P., and Veening, J.W. (2006). Phenotypic variation in bacteria: the role of feedback regulation. Nat. Rev. Microbiol. *4*, 259–271.

Solomon, J.M., Magnuson, R., Srivastava, A., and Grossman, A.D. (1995). Convergent sensing pathways mediate response to two extracellular competence factors in *Bacillus subtilis*. Genes Dev. *9*, 547–558.

Storz, G., and Hengge-Aronis, R. (2009). Bacterial stress responses. (Washington, DC: American Society for Microbiology Press).

Suel, G.M., Garcia-Ojalvo, J., Liberman, L.M., and Elowitz, M.B. (2006). An excitable gene regulatory circuit induces transient cellular differentiation. Nature *440*, 545–550.

Suel, G.M., Kulkarni, R.P., Dworkin, J., Garcia-Ojalvo, J., and Elowitz, M.B. (2007). Tunability and noise dependence in differentiation dynamics. Science *315*, 1716–1719.

Thattai, M., and van Oudenaarden, A. (2001). Intrinsic noise in gene regulatory networks. Proc. Natl. Acad. Sci. U.S.A. *98*, 8614–8619.

Thattai, M., and van Oudenaarden, A. (2004). Stochastic gene expression in fluctuating environments. Genetics *167*, 523–530.

Thieffry, D., Huerta, A.M., Perez-Rueda, E., and Collado-Vides, J. (1998). From specific gene regulation to genomic networks: a global analysis of transcriptional regulation in *Escherichia coli*. Bioessays *20*, 433–440.

Turgay, K., Hahn, J., Burghoorn, J., and Dubnau, D. (1998). Competence in *Bacillus subtilis* is controlled by regulated proteolysis of a transcription factor. EMBO J. *17*, 6730–6738.

van Sinderen, D., ten-Berge, A., Hayema, B.J., Hamoen, L., and Venema, G. (1994). Molecular cloning and sequence of *comK*, a gene required for genetic competence in *Bacillus subtilis*. Mol. Microbiol. *11*, 695–703.

van Sinderen, D., and Venema, G. (1994). *comK* acts as an autoregulatory control switch in the signal transduction route to competence in *Bacillus subtilis*. J. Bacteriol. *176*, 5762–5770.

Veening, J.W., Hamoen, L.W., and Kuipers, O.P. (2005). Phosphatases modulate the bistable sporulation gene expression pattern in *Bacillus subtilis*. Mol. Microbiol. *56*, 1481–1494.

Veening, J.W., Igoshin, O.A., Eijlander, R.T., Nijland, R., Hamoen, L.W., and Kuipers, O.P. (2008a). Transient heterogeneity in extracellular protease production by *Bacillus subtilis*. Mol. Syst. Biol. *4*, 184.

Veening, J.W., Smits, W.K., Hamoen, L.W., and Kuipers, O.P. (2006). Single cell analysis of gene expression patterns of competence development and initiation of sporulation in *Bacillus subtilis* grown on chemically defined media. J. Appl. Microbiol. *101*, 531–541.

Veening, J.W., Smits, W.K., and Kuipers, O.P. (2008b). Bistability, epigenetics, and bet-hedging in bacteria. Annu. Rev. Microbiol. *62*, 193–210.

Veening, J.W., Stewart, E.J., Berngruber, T.W., Taddei, F., Kuipers, O.P., and Hamoen, L.W. (2008c). Bet-hedging and epigenetic inheritance in bacterial cell development. Proc. Natl. Acad. Sci. U.S.A. *105*, 4393–4398.

Verhamme, D.T., Kiley, T.B., and Stanley-Wall, N.R. (2007). DegU co-ordinates multicellular behaviour exhibited by *Bacillus subtilis*. Mol. Microbiol. *65*, 554–568.

Vlamakis, H., Aguilar, C., Losick, R., and Kolter, R. (2008). Control of cell fate by the formation of an architecturally complex bacterial community. Genes Dev. *22*, 945–953.

Wolf, D.M., Vazirani, V.V., and Arkin, A.P. (2005). Diversity in times of adversity: probabilistic strategies in microbial survival games. J. Theor. Biol. *234*, 227–253.

Genomic Approaches to Reconstructing Transcriptional Networks

7

Stephen J.W. Busby and Stephen D. Minchin

Abstract

The traditional methods for discovering transcriptional regulatory networks in bacteria, based on genetics and biochemistry, are now being replaced by high throughput pan-genome methods. Experimental approaches include methods involving RNA or methods based on the direct observation of transcription factors. This chapter places the new methods in context and discusses their potential benefits and drawbacks.

Transcription regulatory networks

Gene expression in all bacteria is tightly controlled, with transcription initiation being the principal point of regulation for many genes. A typical bacterial genome encodes one RNA polymerase. Hence, all transcription is due to a single RNA polymerase (RNAP), which is a large multisubunit enzyme. To understand transcriptional regulation in bacteria, we need to understand how this RNAP is distributed between the different transcription units. We know that each transcription unit is controlled by a regulatory region containing at least one promoter, that is typically located just upstream of the first open reading frame. Such regulatory regions contain sequence elements that RNA polymerase interacts with, together with binding targets for different transcription factors (TFs) that are involved in regulation (Browning and Busby, 2004). The role of the different TFs at any promoter is to up- or down-regulate transcription initiation, and the activity of each TF is controlled by one or more signals. Because many signals evoke complicated transcriptional responses, many TFs interact at multiple promoter targets (Martínez-Antonio and Collado-Vides, 2003). Similarly, because regulation is sophisticated and the activity of most promoters is set by multiple input signals, most promoters are regulated by more than one TF (Barnard *et al.*, 2004). This results in a transcriptional network in which promoters are the nodes that integrate the activities of different TFs. In this chapter, we outline the different methods that are now available to elucidate such networks. The focus here is on experimental approaches, since computational approaches are dealt with in the following chapter, and special emphasis is placed on describing new whole-genome approaches and explaining their limitations.

Since the beginning of molecular biology, *Escherichia coli* K-12 has been the organism of choice for the study of transcription regulation, and it has been established that this regulation is due to a complex network of TFs and sigma factors that control the expression of ~1800 transcription units in response to changes in the environment (Salgado *et al.*, 2006). The *E. coli* genome encodes over 250 gene regulatory proteins that range from highly specific TFs such as the lactose operon repressor (Lac repressor), which controls a single transcription unit, through to global regulatory proteins, such as the cyclic AMP receptor protein (CRP), which controls scores of transcription units. In addition, the nucleoid-associated proteins, which are needed for maintaining chromosome folding and compaction, play important roles in transcriptional regulation (Dame, 2005). Many of these are present in large quantities that vary according to growth conditions, and they play key roles in up- or down-regulating specific promoters. Over 50 years, the network of *E. coli* gene regulatory proteins has been established by integrating information from studies on individual promoters and TFs. However, the arrival of whole genome sequences has catalysed the development of whole genome approaches to describing TF networks, and these have replaced strategies based on promoter-by-promoter studies. The importance of this switch is that the new methodologies can readily be applied to newly sequenced bacteria, many of which are not amenable to the methods used with *E. coli*. One of the aims of this chapter is to describe and contrast the different approaches.

Bottom-up methodologies

Some of these approaches are illustrated in Fig. 7.1: they usually start with a chosen specific promoter or factor. For a promoter, the transcript start would be located, followed by an investigation of how expression of the corresponding transcription unit varies in response to environmental change. Genetic approaches, exploiting a toolbox of genetic tricks, would then be used to select and characterize mutants in which the activity or regulation of the promoter or factor is altered (see Minchin and Busby, 2009). Such approaches can identify *trans*-acting factors that up- or down-regulate expression of a target gene. Promoters can be found from *cis*-acting mutations that identify key elements such as −10 and −35 hexamers. Binding sites for repressors can be defined by *cis*-acting mutations that led to constitutive expression. Similarly, activator binding sites can be postulated from the location of upstream *cis*-acting mutations that lead to loss of expression. Although, genetic analysis, identifying *cis*-acting elements and *trans*-acting factors, has led to models for the regulation of scores promoter regions, it is important to realize the limitations. First genetic analysis can only suggest, rather than prove, a model. Second, some regulatory regions are complex, sometimes with multiple promoters, or many factors that each have but small effects. Additionally, in some cases, genetic approaches are hindered by redundancy whereby the effects of one factor are masked by another.

Ideally, information from genetic experiments is confirmed with *in vitro* biochemical studies using purified proteins and cloned regulatory regions. At its best, biochemistry can give unique access to a wealth of quantitative and mechanistic information that cannot be accessed by genetics. The simplest applications concern binding of proteins to regulatory region DNA, and an arsenal of techniques is available, ranging from filter binding,

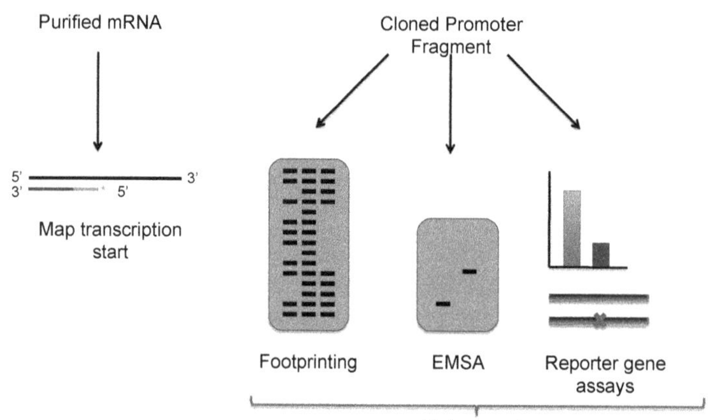

Figure 7.1 Bottom-up methodologies. Several approaches can be exploited to study the regulation of individual genes. Often the transcription start site is mapped by primer extension of purified mRNA. Once the promoter has been cloned it can be analysed using footprinting, electromobility shift assays (EMSA) and reporter gene assays. Further information can be obtained from mutational analysis. A colour version of this figure is located in the plate section at the back of the book.

through electromobility shift assays (EMSAs) and footprinting, to atomic force microscopy and cryo-electron microscopy. These studies can give information on binding target locations, binding affinities, individual rate constants, and how they are affected by different conditions or specific ligands. It is possible to recapitulate both factor-dependent and factor-independent transcription initiation *in vitro* and both closed and open complexes can be monitored, together with the location of the different binding factors. The occupation of a promoter by RNA polymerase can be measured by filter binding assays, footprinting and EMSAs. The formation of transcriptionally competent complexes can be followed in parallel by footprinting with potassium permanganate to detect DNA unwinding in the open complex, or by measuring the formation of an abortive or full-length transcript (examples of this approach can be found in Browning *et al.*, 2009).

Amazingly, for nearly 30 years, up to about 1996, this combination of genetic and biochemistry, based on case-by-case studies, mainly, though not exclusively, with *E. coli*, was the sole available strategy for investigating transcriptional regulation in bacteria. However, as data accumulated (for example in Regulon DB), bioinformatics became more useful, with consensus targets being established for promoters and TFs. The drawback of these 'bottom-up' approaches is that they are time-consuming to perform, and, amazingly, even after decades of effort less than 50% of the regulatory regions of *E. coli* K-12 have been subjected to this type of analysis. For this reason, 'top-down' whole genome approaches have become increasingly exploited and the principal aim of this chapter is to review their application.

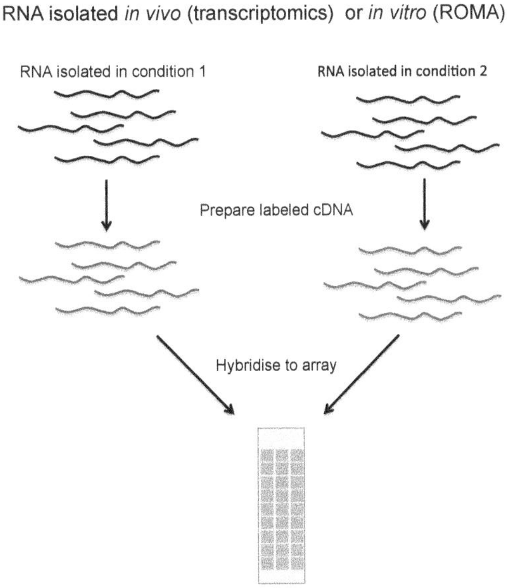

Figure 7.2 Transcriptomics and ROMA. mRNA levels can be determined on a genomic scale. RNA is purified either from cells (transcriptomics) or from an *in vitro* run-off transcription assays using genomic DNA as template (ROMA). Labelled cDNA is prepared and hybridized to a microarray to quantify different sequences, and hence different starting mRNAs. Two samples can be compared directly, for example, wild-type versus mutant or cells growing in different conditions. A colour version of this figure is located in the plate section at the back of the book.

Transcript profiling

Transcriptomics can be used to analyse the total complement of RNA molecules made by a bacterium in any particular environment (Rhodius *et al.*, 2002). The RNA is converted into cDNA and then high-density microarrays or high throughput sequencing runs are used to deduce its sequence composition (Fig. 7.2). The simplest application to the discovery of TF networks is to compare RNAs made in genetic backgrounds with and without the TF under examination. Although this results in a list of transcripts whose expression is affected by the TF, the list is unlikely to be complete since this approach suffers from a number of complications. First, the effects of deletion of many TFs on many RNAs may be due to indirect effects, and this is especially problematic with global TFs, such as CRP. For example the TF may control the expression of other TFs. More subtly, certain gene products, whose expression is affected by deletion of the TF in question, may affect metabolism, and this will, in turn, affect the activity of other TFs. The problems arise because this type of transcriptomics measures all transcripts and even more complications arise if growth rate is affected. Other problems can arise because certain genes whose transcription is regulated by the factor are poorly expressed in the conditions of the experiment, or, because, other TFs are involved, the effects of the TF deletion are too small to be detected.

One way to sidestep some of these problems is to switch the focus to study changes in growth conditions. Such an experiment might compare the transcriptome in two different growth conditions either with or without the TF in question. This approach identifies changes that depend on the TF, and, often more important, can show that some changes are not dependent on a particular TF. This may then lead the way to identifying another relevant TF. Another useful and easy strategy is to measure effects of overproducing a TF on the transcriptome of cells. This is especially useful to identify promoters whose expression is highly condition-dependent, since just a small incremental TF-dependent activity can be sufficient to identify a gene as a target for that TF (for a full discussion of problems and strategies, see Rhodius and Larose, 2003).

A further problem with transcriptomics methodologies is that the measured level of any transcript depends on both its synthesis and degradation, and any alteration in the turnover of a specific transcript may distort results. This is not an easy problem to resolve and it is ignored by most investigators, who make the assumption that any distortion of their results owing to messenger turnover will be minor.

ROMA: transcriptomics *in vitro*

Many of the shortcomings of in *vivo* transcript profiling come from the complications of the intact cell (such as messenger turnover) and the technique of ROMA (run-off microarray analysis) was developed to counter these (reviewed in Maclellan *et al.*, 2009). ROMA exploits the ability of purified bacterial RNAP to locate many promoters and initiate transcription in a simple *in vitro* system consisting of DNA, buffer and nucleoside triphosphates. Furthermore, for many promoters, regulation by purified TFs can be recapitulated in this system. Thus, in a typical ROMA experiment, microarrays will be used to analyse total RNA made after purified RNAP is incubated with total genomic DNA from the bacterium under study and nucleoside triphosphates. The experiment will be run with and without the TF in question and targets for the TF will be deduced from the RNA species whose synthesis is repressed or activated by the TF (Fig. 7.2). ROMA experiments are also especially useful for identifying promoters that are recognized by specific sigma factors and these experiments are performed by comparing the transcripts made by purified RNA polymerase that has been reconstituted with different sigmas. Thus, in principle, in a single experiment, ROMA can define the regulon for a TF or the 'simulant' for a sigma factor. Thus, working with CRP, 190 targets could be identified in the *Escherichia coli* K-12 genome, and bioinformatic analysis that exploited the known DNA consensus for CRP binding was used to locate many CRP-dependent promoters (Zheng *et al.*, 2004). Although seductively simple, the ROMA method does have some caveats. First, there may be some promoters where TFs simply fail to function *in vitro*. This may be because the templates used in ROMA are, by necessity, linear rather than supercoiled, but, also, could well be because certain TF transactions simply cannot be reproduced in a plastic tube. A good example of this is transcription activation by CRP at promoters where the DNA site for CRP is located further than 70 base pairs upstream of the transcript start. A second caveat is that transcripts whose initiation is dependent on two activators, or an activator as well as an alternative sigma factor, are likely to be missed. Similarly a transcript with two promoters, only one of which is activator-dependent, would be missed. Finally, false positive results can arise because the artificially high RNAP availability of *in vitro* transcription leads to adventitious activation at promoters with little or no role *in vivo*.

It should be clear that neither transcript profiling nor ROMA are likely to give a complete description of the activity of any particular TF or sigma factor, and yet, carefully handled, both can give valuable information. Both methods depend on identifying changes in specific transcripts that are the downstream consequences of the action of the TF or sigma factor. Because of the intrinsic problems with both methodologies, investigators have now sought to develop direct experimental approaches to catalogue TF binding targets that are independent of their effects. The two principal methods, which are discussed here, are Genomic SELEX and Chromatin Immunoprecipitation (ChIP).

Genomic SELEX

SELEX is the systematic evolution of ligands by exponential enrichment. It was originally developed as a way of selecting the targets for a specific DNA binding protein from amongst a completely random mix of base sequences generated by synthetic chemistry. The particular protein under study is purified (usually exploiting an affinity tag) and native polyacrylamide gel electrophoresis is then used to enrich DNA sequences that bind to the protein. Repeated rounds of target amplification and electrophoresis are then used to select for the tightest binding sequences. The principle of genomic SELEX is very similar, save that the starting material is random 200–300 base pair DNA fragments from total genomic DNA of the bacterium under study (Fig. 7.3). A typical genomic SELEX experiment with a purified bacterial TF might result in the selection of dozens of different fragments to which the protein binds, and these can be identified either by base sequencing or by analysis on a high density microarray. These fragments are likely to contain the strongest DNA sites for the TF which can then be identified by searching for common sequence motifs in the selected fragments. From the location of these targets, it will then be possible to identify genes whose expression may be regulated by the TF. Although genomic SELEX is simple and direct, there are a number of potential problems. One is that weaker binding targets, or targets where binding requires a ligand, covalent modification of the TF, or interaction with another TF, are likely to be missed. Although it may be possible to address this by re-running the experiment in different conditions, this is unlikely to result in a full repertoire of binding targets. However because this method picks targets on the criterion of binding in the absence of any function, some of the selected targets may play no role in transcription regulation. Such targets may well be *bona fide* binding sites for TFs with binding at these targets playing alternative roles such as structuring the bacterial chromosome. Alternatively, the binding may simply be a vestige of evolution. In any case, the existence of such sites creates a complication if the

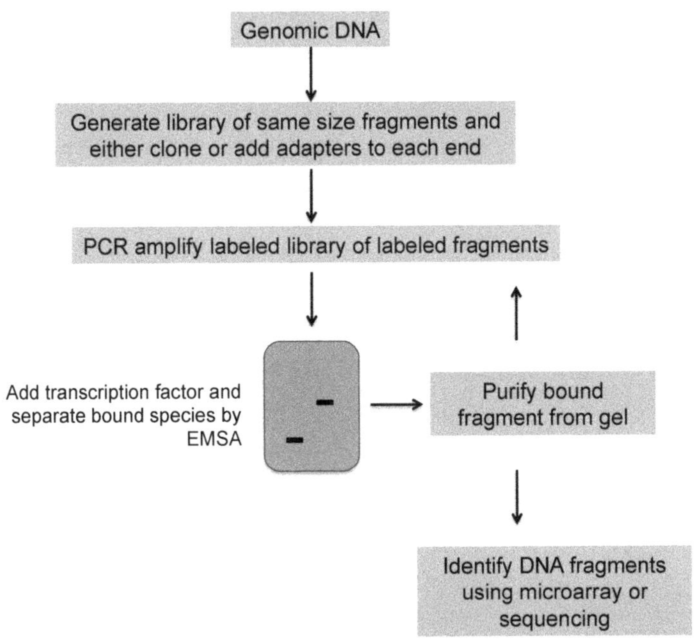

Figure 7.3 Genomic SELEX. A library of DNA fragments is generated from genomic DNA and these are either cloned into a vector or adapters added. PCR, using primers complementary to vector or the adaptor sequences, is used to amplify and label the DNA. Several rounds of EMSA, purification of bound fragments and PCR amplification can be used to select for genomic binding sites for the transcription factor. The purified fragments are analysed by microarray or direct sequencing.

sole aim of the experiment of an experiment with a TF is to establish its regulon. Examples of the use of genomic SELEX can be found in Shimada *et al.* (2009).

Chromatin immunoprecipitation

ChIP can be regarded as the *in vivo* equivalent of genomic SELEX, since it identifies protein–DNA interactions *in vivo* directly, independent of the biological consequences of binding. Briefly, bacterial cells are exposed to formaldehyde, thereby instantly crosslinking DNA binding proteins to the chromosome. After cell lysis and shearing of chromosomal DNA by sonication, the protein of interest is immunoprecipitated with specific antibodies, together with cross-linked DNA fragments. After reversal of the crosslinks and purification, the immunoprecipitated DNA is analysed in order to detect enrichment of the sequences bound by the protein of interest (Fig. 7.4). Using ChIP in conjunction with DNA microarray analysis (ChIP-on-chip) permits DNA binding to be measured on a chromosome-wide scale. The most straightforward use of ChIP in bacterial systems is in the location of transcription factors (Wade *et al.*, 2007; Grainger *et al.*, 2009). Following on from the global analysis of the *C. crescentus* CtrA regulon and the *B. subtilis* Spo0A and CodY regulators, ChIP-on-chip has now been applied to many other bacterial systems, including several pathogens. For instance, the *H. pylori* Fur protein has been studied and found to bind at about 200 genomic loci in an iron-dependent manner, supporting the idea that this protein acts as a pleiotropic regulator. The previously uncharacterised BlaI transcription factor from *M. tuberculosis* was recently shown to regulate five DNA loci including the *blaI* gene itself, and others involved in resistance to β-lactam antibiotics. Unexpectedly, BlaI was found to bind upstream of the operon encoding ATP synthase, suggesting links between cell wall damage and ATP production. In the case of the *Salmonella enterica* PhoP transcription regulator, ChIP experiments were used to understand the hierarchy with which different promoters are 'served' as concentrations of the trigger ligand, magnesium ions, change. Studies performed in *E. coli* K-12 represent the paradigm for ChIP-on-chip studies of transcription factors. The distribution of ArcA, CRP, FNR, LexA, Lrp, MelR, NsrR and RutR has been determined. These studies show that some factors recognize single binding sites (as in the case of MelR) whilst others have more complex distributions (for example, CRP and LexA).

The great advantage of ChIP is that it is simple and flexible, it directly confirms binding *in vivo*, and can easily be used to measure changes in response to environmental change or as cells pass through different growth phases (Grainger and Busby, 2008). However, ChIP studies with *Escherichia coli* have yielded some surprises. Thus, CRP binds throughout the *E. coli* chromosome to nearly 1000 sites and, at many targets, appears to have no effect on transcription (Grainger *et al.*, 2005). With RutR, most of its 20 binding sites mapped within coding regions, suggesting that it may play some other, as yet undiscovered, role (Shimada *et al.*, 2008). It is plausible that the binding of transcription factors to specific sites without any function is a by-product of evolution, perhaps due to inadequate 'purging' after horizontal gene transfer.

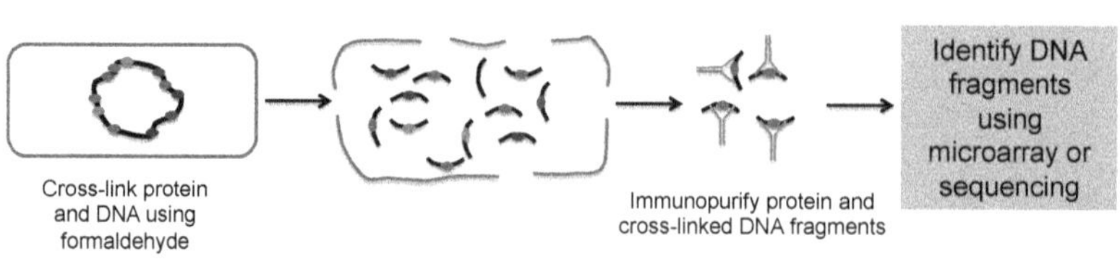

Figure 7.4 Chromatin immunoprecipitation (ChIP-ChIP and ChIP-Seq). Cells are grown and transcription factors cross-linked to their DNA binding sites. An antibody specific for the transcription factor of interest is used to purify the transcription factor and the cross-linked target DNA. The purified DNA target can be analysed using microarrays or by direct sequencing. A colour version of this figure is located in the plate section at the back of the book.

A special application of ChIP is to study the distribution of RNA polymerase across a bacterial chromosome and this can be exploited to identify transcription units, to study different sigma factors or to investigate the regulon of any TF (Grainger *et al.* 2005; Reppas *et al.*, 2006; Wade *et al.*, 2006). For example, instead of measuring the effects of deleting a particular TF on the composition of the transcriptome, it is possible directly to measure the effects on RNAP distribution across the chromosome. Another application of ChIP directed to RNAP polymerase is to measure its redistribution in response to environmental change. For example, the response of *E. coli* to rifampicin was described in terms of re-localization of RNA polymerase to promoter sequences, and a similar redistribution was observed as *E. coli* enters stationary phase (Herring *et al.*, 2005). Thus, measurements of changes in the distribution of RNA polymerase (RNA polymerase-omics), may eventually supplant RNA-based strategies (transcript-omics) for the study of genome-wide transcriptional responses. Unexpectedly, ChIP studies on RNAP in *E. coli* have identified many locations where RNAP binds without producing any transcript (Reppas *et al.*, 2006). This underscores the power of ChIP strategies that measure binding in the absence of its function. Hence ChIP detects targets for non-transcribing DNA-bound RNAP that would be missed by strategies based on measurement of RNA.

DNA sampling

The aim of a ChIP experiment is to identify all of the binding locations on a chromosome for one specified protein. DNA sampling is a complementary protocol, aimed to identify all of the proteins that are bound at one specified chromosomal location (Butala *et al.*, 2009). Briefly, the DNA segment to be 'sampled' is cloned into a low copy number plasmid, adjacent to a short array of binding targets for the Lac repressor, sandwiched between two recognition sequences for the yeast homing endonuclease, I-SceI. The plasmid is transformed into a bacterial host that carries a second plasmid encoding I-SceI under the control of an arabinose-inducible promoter. Induction of I-SceI expression triggers excision of a DNA fragment carrying the target DNA segment, together with the Lac repressor binding targets. Since I-SceI recognizes an 18-base pair target sequence, excision is specific for the plasmid and the host chromosome is unaffected. In the sampling protocol, the fragment is immunopurified by exploiting a host that has been modified by recombineering so that the Lac repressor is fused to an epitope tag. By using such a host, it is possible to isolate the fragment, together with bound proteins, using magnetic beads to which antibodies directed to the epitope tag have been immobilized. Gel electrophoresis and mass spectrometry are then used to identify the different bound proteins. This protocol, illustrated in Fig. 7.5, allows the comparison of proteins bound at a particular DNA location in cells grown in different conditions (e.g. the changes in bound RNA polymerase before and after induction of a promoter). DNA sampling is similar to ChIP in that it exploits an affinity reagent to pull down a protein (Lac repressor) that is bound to a DNA target. In DNA sampling, target sites for the Lac repressor are engineered to be close to the DNA target that is to be sampled and the emphasis is on the analysis of co-purified proteins rather than analysis of the DNA target. To date, DNA sampling has been applied only to *E. coli* K-12, but, in principle, it should be applicable in a variety of bacterial species (Grainger *et al.*, 2009).

The main limitation of DNA sampling is that it is cumbersome to execute. It also has some technical limitations, such as the identification of gene regulatory proteins whose binding to the sampled target is weak or ligand-dependent, although it may be possible to develop crosslinking strategies to remedy this. Additionally, with the development of more sensitive detection methods and the application of novel imaging technologies, DNA sampling may eventually be applicable on a whole genome scale.

Perspective

A TF network for any bacterium can be proposed if the results from experiments involving biochemistry and genetics, transcript profiling, genomic SELEX, chromatin immunoprecipitation and DNA sampling are pooled together. The information from each data set can be enhanced

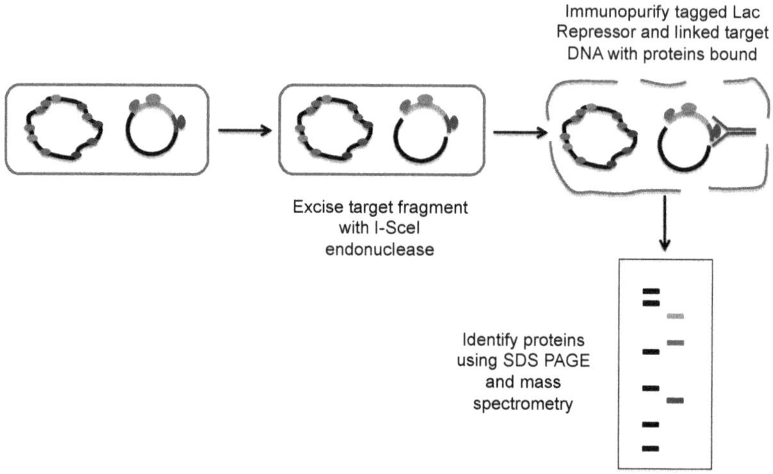

Figure 7.5 DNA sampling. The target DNA sequence is cloned into a vector adjacent to binding sites for the Lac Repressor and flanked by I-SceI recognition sites. During the assay, the target DNA is excised by I-SceI endonuclease from the vector, *in vivo*, in cells containing Lac repressor that is tagged with the FLAG epitope. The protein–DNA complex is then purified using antibodies specific for the tagged Lac Repressor. Following SDS PAGE, the proteins are identified by mass spectrometry. A colour version of this figure is located in the plate section at the back of the book.

by bioinformatics, and, as more and more experimental data has accumulated, the contribution of bioinformatics and the potential for computational approaches has grown (see the following chapter). The simplest application is to locate likely DNA sites for different TFs in DNA segments that have been identified by the different approaches. We can now aspire to obtaining a complete description of the location and role of all the different transcription factors. This is likely to be realized for *E. coli* K-12 in the next few years (see Cho *et al.*, 2009) and this will provide the essential base for constructing regulatory networks and understanding the extent of different regulons. Historically, transcription regulation in bacteria has been studied by genetics, with deductions being made from intelligent thinking based on phenotypes. The arrival of bacterial genome sequences prompted the adoption of whole genome methodologies such as transcriptomics, which have transformed our aspirations. This has now been followed by the adoption of powerful experimental strategies founded on direct measurements, which provide new yet different insights. These experimental approaches can be applied not only in axenic cultures but also in more complex environments, including natural hosts and situations where many bacterial species are growing together.

Chapter highlights

- Traditional 'bottom-up' approaches focussed on individual regulatory regions are being replaced by 'top-down' pan-genome methodologies.
- Transcriptome-based approaches, combined with bioinformatics, provide a powerful tool to understand regulatory circuitry in previously unexplored bacteria.
- Chromatin immunoprecipitation and DNA sampling provide complementary approaches to the direct study of transcription factors and RNA polymerase in live bacteria, independent of their function.
- Genomic SELEX and ROMA permit pan-genome studies of transcription factors and RNA polymerase *in vitro*.

Acknowledgements

Work from the authors' laboratory is supported by a Wellcome Trust Programme Grant to S.J.W.B. We are grateful to Akira Ishihama, David Grainger and David Lee for helpful comments and access to unpublished data.

References

Barnard, A., Wolfe, A., and Busby, S. (2004). Regulation at complex bacterial promoters: how bacteria use different promoter organizations to produce different regulatory outcomes. Curr. Opin. Microbiol. *7*, 102–108.

Browning, D.F., and Busby, S.J. (2004). The regulation of bacterial transcription initiation. Nature Rev. Microbiol. *2*, 57–65.

Browning, D., Savery, N., Kolb, A., and Busby. S. (2009). Assays for transcription factor activity. Methods Mol. Biol. *543*, 369–387.

Butala, M., Busby, S., and Lee, D. (2009). DNA sampling: a method for probing protein binding at specific loci on bacterial chromosomes. Nucl. Acids Res. doi:10.1093/nar/gkp043.

Cho, B.-K.,. Zengler, K., Qiu, Y., Park, Y.S., Knight, E., Barrett, C., Gao, Y., and Palsson, B. (2009). Nature Biotechnol. *27*, 1043–1049.

Dame, R.T. (2005). The role of nucleoid-associated proteins in the organization and compaction of bacterial chromatin. Mol. Microbiol. *56*, 858–870.

Grainger, D., Hurd, D., Harrison, M., Holdstock, J., and Busby, S. (2005). Studies of the distribution of *Escherichia coli* cAMP receptor protein and RNA polymerase along the *E. coli* chromosome. Proc. Natl. Acad. Sci. U.S.A. *102*, 17693–17698.

Grainger, D., and Busby, S. (2008). Global regulators of transcription in *Escherichia coli*: mechanisms of action and methods for study. Adv. Appl. Microbiol. *65*, 93–113.

Grainger, D., Lee, D., and Busby, S. (2009). Direct methods for studying transcription regulatory proteins and RNA polymerase in bacteria. Curr. Opin. Microbiol. *12*, 531–535.

Herring, C.D., Raffaelle, M., Allen, T.E., Kanin, E.I., Landick, R., Ansari, A.Z., and Palsson, B.Ø. (2005). Immobilization of *Escherichia coli* RNA polymerase and location of binding sites by use of chromatin immunoprecipitation and microarrays. J. Bacteriol. *187*, 6166–6174.

Maclellan, S.R., Eiamphungporn, W., and Helmann, J.D. (2009). ROMA: an *in vitro* approach to defining target genes for transcription regulators. Methods *47*, 73–77.

Martínez-Antonio, A., and Collado-Vides J. (2003). Identifying global regulators in transcriptional regulatory networks in bacteria. Curr. Opin. Microbiol. *6*, 482–489.

Minchin, S.D., and Busby, S.J. (2009). Analysis of mechanisms of activation and repression at bacterial promoters. Methods *47*, 6–12.

Reppas, N.B., Wade, J.T., Church, G.M., and Struhl, K. (2006). The transition between transcriptional initiation and elongation in *E. coli* is highly variable and often rate limiting. Mol. Cell. *24*, 747–757.

Rhodius, V., Van Dyk, T.K., Gross, C., and LaRossa, R.A. (2002). Impact of genomic technologies on studies of bacterial gene expression. Annu. Rev. Microbiol. *256*, 599–624.

Rhodius, V.A., and LaRossa, R.A. (2003). Uses and pitfalls of microarrays for studying transcriptional regulation. Curr. Opin. Microbiol. *6*, 114–119.

Salgado, H., Santos-Zavaleta, A., Gama-Castro, S., Peralta-Gil, M., Peñaloza-Spínola, M.I., Martínez-Antonio, A., Karp, P.D., and Collado-Vides, J. (2006). The comprehensive updated regulatory network of *Escherichia coli* K-12. BMC Bioinformatics *7*, 5.

Shimada, T., Ishihama, A., Busby, S.,and Grainger, D. (2008). The *Escherichia coli* RutR transcription factor binds at targets within genes as well as intergenic regions. Nucl. Acids Res. *36*, 3950–3955.

Shimada, T., Yamamoto, K., and Ishihama, A. (2009). Involvement of the leucine response transcription factor LeuO in regulation of the genes for sulfa drug efflux. J. Bacteriol. *191*, 4562–4571.

Wade, J., Roa, D., Grainger, D., Hurd, D., Busby, S., Struhl, K., and Nudler, E. (2006). Extensive functional overlap between sigma factors in *Escherichia coli*. Nat. Struct. Mol. Biol. *13*, 806–814.

Wade, J.T., Struhl, K., Busby, S.J., and Grainger, D.C. (2007). Genomic analysis of protein–DNA interactions in bacteria: insights into transcription and chromosome organization. Mol. Microbiol. *65*, 21–26.

Zheng, D., Constantinidou, C., Hobman, J.L., and Minchin, S.D. (2004). Identification of the CRP regulon using *in vitro* and in vivo transcriptional profiling. Nucleic Acids Res. *32*, 5874–5893.

Structure and Evolution of Transcriptional Regulatory Networks

8

Guilhem Chalancon and M. Madan Babu

Abstract

Regulation of gene expression is primarily mediated by proteins called transcription factors (TFs), which recognize and bind specific nucleotide sequences and affect transcription of nearby genes. Over the last years, considerable information has been accumulated on regulatory interactions between the TFs and their regulated target genes (TGs) in various model prokaryotic systems such as *Escherichia coli* and *Bacillus subtilis*. This has permitted researchers to model the transcriptional regulatory system of an organism as a network, wherein TFs or TGs are represented as nodes and regulatory interactions are denoted as directed links. Representation of this information as a network has provided us with a robust conceptual framework to investigate this system, and work in the last decade has uncovered several fundamental general principles pertaining to its structure and evolution. In this chapter, we first introduce the concept of transcriptional regulatory networks. We then discuss our current understanding of the structure of transcriptional regulatory networks. Specifically, we discuss the local and global structure of such networks. We then discuss the various forces that influence network evolution such as gene duplication, horizontal gene transfer, and gene loss. In particular, we discuss how the transcriptional regulatory network evolves across organisms that live in different environments. Finally, we conclude by discussing major challenges for future research and highlighting how the new understanding can have implications for biotechnology and medicine and can be exploited in applications such as microbial engineering and synthetic biology.

Introduction

The ability to coordinate and bring about changes in gene expression in response to environmental variation is crucial for the maintenance of cellular homeostasis. Among all the regulatory processes modulating the synthesis of a gene product, regulation of transcription is essential, because this is the first step in a series of events that give rise to a protein. Such alterations in the expression level of particular genes eventually trigger phenotypic changes in response to the environment, thereby permitting the organism to adapt to the new environment.

Regulation of transcription is mediated through proteins called transcription factors (TFs). TFs are DNA binding proteins that bind to specific regions, the cis-regulatory elements, in the promoter regions of certain genes and eventually influence gene expression. In addition to a DNA binding domain (DBD) that recognizes the DNA, most TFs also contain an additional regulatory domain (e.g. a small molecule binding domain, enzymatic domain, etc.) that responds to the signal (e.g. a small molecule). The affinity of the DBD to bind a specific DNA sequence can be modulated through the state of the regulatory domain (e.g. a ligand binding to a regulatory domain). The regulatory domain itself is influenced by the presence or absence of a signal in the internal or the external environment. For example, in a simple free-living organism such as *E. coli*, studies have estimated the presence of ~320 TFs and over 80% of them have been shown to also contain a regulatory domain in addition to a DBD (Madan Babu and Teichmann, 2003).

The binding of a TF to a promoter region can

either result in an increased or decreased transcription of the regulated target gene (TG). In addition to exerting their effect independently, TFs can also affect gene expression in a combinatorial manner. More specifically, TFs regulate the initiation of transcription through different strategies operating on the transcriptional machinery. In bacteria, we can roughly distinguish two classes of mechanisms for repression: the binding of TFs can block the RNA polymerase by steric hindrance, or can recruit co-repressors that decrease the affinity of the holoenzyme (α2ββ'ω) for the promoter region. Similarly, activation can either be achieved through the binding of the TF, which increases the local concentration of the holoenzyme at the promoter region, or by the subsequent recruitment of a co-activator. See Browning and Busby (2004) for a more detailed description of the other mechanisms of activation and repression of transcription in bacteria.

The affinity of transcription factor DBDs for promoters is sequence dependent. Therefore, genes containing identical or similar DNA sequences (cis-regulatory elements) in their promoter region are susceptible to be targeted and regulated by the same TF. Moreover, the unit of the prokaryotic genome organization comprises of operons, which consist of a collection of genes that are adjacent to each other, placed under the control of a single promoter, and that give rise to a polycistronic transcript (e.g. mRNA molecule which can have independent translation initiation sites for the generation of multiple protein products that are encoded in the same transcript (Davies and Jacob, 1968). As a consequence, genes belonging to the same operon can be regulated at once, by one single TF. Because the genes contained in operons tend to have similar biological functions, this organization is considered to facilitate the coordinated regulation of gene expression (Osbourn and Field, 2009).

The expression pattern of a TF itself is extremely dynamic and dependent on stress. In *E. coli*, a key response to stress is the general stress response, which triggers the transcription of genes required for survival during starvation. This response is induced by growth-rate reduction, which is a consequence of nutrient limitation or starvation. It can also be induced by acidic pH, rapid variations in temperature, or in osmolarity (Weber *et al.*, 2006).

Modulators of the general stress response include TFs and subunits of the RNA polymerase such as σ factors. Particularly σ38, also called RpoS, controls the expression of ~10% of the genome in case of starvation (Foster, 2007; Weber *et al.*, 2005). RpoS is structurally very similar to σ70, which is largely expressed in rapidly growing cells, but controls the transcription of distinct set of genes that decrease the growth rate but target DNA protection and repair. This example highlights the importance of transcriptional regulation for survival.

Concept of transcriptional networks

A fast, precise, and global regulation of transcription is essential for cell survival in changing environments. This regulation is mostly controlled by TFs, which are expressed or regulated differentially depending on environmental conditions and which specifically target promoter regions. This knowledge results from decades of detailed investigations that focused on specific cases of prokaryotic gene regulation, mostly performed in *E. coli*. However, deciphering general rules governing transcription regulation at the genome level in bacterial organisms has become an achievable goal in recent years.

As one would imagine, myriad TFs not only bind to promoter sequences with combinatorial effects on the transcription of downstream genes, but those interactions are also highly dynamic. This dynamic allows cells to coordinate elaborate responses to external and internal stimuli, but is a major challenge for understanding transcriptional regulation in its global nature. The availability of sequenced genomes as from the late 1990s undoubtedly changed the scenario. It has now become possible to collect and analyse large amounts of information (in hundreds, then thousands) of bacterial species, allowing annotations and predictions of TF binding sites. Simultaneously, the development of genome-scale high-throughput experiments detecting protein–DNA interactions became possible. For instance, chromatin-immunoprecipitation and protein- DNA microarrays played a central role in the identification of new protein–DNA interactions (Grainger *et al.*, 2005; Grainger *et al.*, 2009; Molle *et al.*, 2003).

The understanding of the diverse nature of information on TFs and their regulated targets (see Table 8.1) was facilitated by the adoption of network theory, which permitted uncovering patterns in gene regulation on a genomic scale (Babu *et al.*, 2004; Milo *et al.*, 2002; Thieffry *et al.*, 1998). The investigation of interactions between TFs and their TGs as a network provided a general framework to identify general principles that govern such complex systems. Formally, transcriptional regulatory networks (TRNs) are modelled as directed graphs that are composed of vertices or nodes that are connected by directed edges. In this case, vertices denote both TFs and their TGs. Directed edges, which connect a TF to its TG, represent a regulatory interaction. Such an object can be studied with a set of analytic tools derived from network theory (Babu *et al.*, 2004; Barabasi and Oltvai, 2004). Consequently, during the past decade, such approaches have facilitated detailed investigations into the structure, the dynamics, and the evolution of the regulation of transcription at the genome level.

In this chapter, we first discuss the main characteristics of the structure of prokaryotic transcriptional regulatory networks. In the second part, we discuss about the various forces that influence their evolution. Finally we discuss how the understanding gained is being exploited in biotechnology and medicine.

Structure of transcriptional networks

TRNs have a complex and hierarchical structure and can be investigated at several levels of organization (Babu *et al.*, 2004) (Fig. 8.1). At the most basic level, the network is made up of basic units: a TF, its TG, and the cis-regulatory element through which it regulates the expression of the TG (Fig. 8.1A). At the local level of organization, these basic units are arranged into recurrent wiring patterns called network motifs, which appear frequently throughout the network (Fig. 8.1B). The network motifs have been shown to perform specific information processing task, the details of which have been discussed in great detail (Alon, 2007). The global level of organization involves the set of all known regulatory interactions among the TFs and the TGs in an organism (Fig. 8.1C). In particular, it was shown that the presence of a few TFs characterizes TRNs, which were referred to as global regulators because they control the expression of a large number of genes.

It should be noted that much of the work on bacterial regulatory networks has focused on *E. coli* for which data are most abundant. Although much of our understanding of TRNs has been obtained by investigating the *E. coli* network, work on the *B. subtilis*, *Corynebacterium*, and *Saccharomyces cerevisiae* networks and the TRNs from other organisms have shown that the general principles of organization are largely the same. Currently, there over 2500 regulatory interactions in *E. coli*, which are available through the RegulonDB database (Gama-Castro *et al.*, 2008). For a comprehensive list of databases providing information about known and inferred TRNs, please see Table 8.1.

Local network structures

At a local level, TRNs have been shown to contain small recurrent patterns of interconnections whose number of occurrence is substantially higher than what is expected by chance when compared with random networks of identical size. These structures, which were first defined by (Shen-Orr *et al.*, 2002), are known as network motifs (Alon, 2007). Milo *et al.* (2002) and Lee *et al.* (2002) discovered three overrepresented network motifs in the *E. coli* and yeast transcriptional regulatory network (Fig. 8.1B). These three motifs are referred to as (i) feed forward motifs (FFMs), (ii) single input modules (SIMs), and (iii) multiple input modules (MIMs). Several subsequent works have shown that each motif possesses distinct kinetic properties with respect to the control of TG expression (Alon, 2007).

FFMs

In FFMs, a top-level TF regulates a TG and an intermediate TF, which also regulates the same TG. One should note that, because the top and the intermediate TFs can either be activators or repressors, four combinations are possible in

Table 8.1 Databases and computer programs for investigating transcriptional regulatory networks. Adapted from Babu MM (Babu, 2008) and Janky *et al.* (Janky *et al.*, 2009)

Databases containing regulatory information	Description	Website
RegTransBase	TF-binding sites and regulatory interactions	http://regtransbase.lbl.gov/cgi-bin/regtransbase?page=main
ORegAnno	An open access database for gene regulatory element and polymorphism annotation	http://www.oreganno.org/
STRING	Genome context and SMART (simple modular architecture research tool), domain assignment	http://smart.embl-heidelberg.de/
RegulonDB	Database of TFs and binding sites for *E. coli*	http://regulondb.ccg.unam.mx/
DBTBS	Database of TFs and binding sites for *B. subtilis*	http://dbtbs.hgc.jp/
Coryneregnet	Database of regulatory network for several microbes	http://www.coryneregnet.de/
Prodoric	Prokaryotic database of gene regulation	http://www.prodoric.de/
TractorDB	Predicted TF-binding sites in gamma proteobacterial genomes	http://www.tractor.lncc.br/
Microbes Online	Domain assignment, expression data, evolutionary relationships and operon structure	http://www.microbesonline.org/
BacTregulators	Database of transcription factors in bacteria and archaea	http://www.bactregulators.org/
DBD	Database of predicted transcription factors of over 700 completely sequenced genomes based on SCOP DNA binding domains	http://dbd.mrc-lmb.cam.ac.uk/DBD/index.cgi?Home
RegPrecise	Database of curated genomic inference of regulons in prokaryotic genomes	http://regprecise.lbl.gov/RegPrecise/
Transfac	Transcription factor database	http://www.biobase-international.com/pages/index.php?id=transfac
ArchaeaTF	Archaeal transcription factor database	http://bioinformatics.zj.cn/archaeatf/Homepage.php
Tools for analysis of transcription regulation	**Description**	**Website**
Vista	Tools for comparative analysis of genomic sequences	http://genome.lbl.gov/vista/index.shtml
RSAT	A very powerful platform for regulatory sequence analysis	http://rsat.ulb.ac.be/rsat/
Webmotifs	motif discovery, scoring, analysis, and visualization using different programs	http://fraenkel.mit.edu/webmotifs/finalout.html
seqVISTA	Platform for binding site discovery	http://zlab.bu.edu/SeqVISTA/index.htm
Weblogo	Visualizing binding site information	http://weblogo.berkeley.edu/
Enologos	Logo visualization	http://biodev.hgen.pitt.edu/cgi-bin/enologos/enologos.cgi
Network visualization	**Description**	**Website**
Biolayout	Visualization	http://cgg.ebi.ac.uk/services/biolayout/
Cytoscape	Visualization and analysis	http://www.cytoscape.org/
GraphViz	Visualization	http://www.graphviz.org/

Network analysis	Description	Website
H3Viewer	Visualization	http://graphics.stanford.edu/~munzner/h3/
Neat	Visualization and analysis	http://rsat.ulb.ac.be/rsat/index_neat.html
Netminer	Visualization and analysis (Commercial)	http://www.netminer.com/
Osprey	Visualization and analysis	http://biodata.mshri.on.ca/osprey/index.html
Pajek	Visualization and analysis	http://vlado.fmf.uni-lj.si/pub/networks/pajek/
Visant	Visualization and analysis	http://visant.bu.edu/
Yed	Visualization and analysis	http://www.yworks.com/
Mfinder	Network motif finder	http://www.weizmann.ac.il/mcb/UriAlon/groupNetworkMotifSW.html
FanMod	Network motif finder	http://www.minet.uni-jena.de/~wernicke/motifs/
Clique finder	Identification of cliques	http://topnet.gersteinlab.org/clique/
MCode	Identification of densely connected subnetwork	http://baderlab.org/Software/MCODE
Cytoscape	Several plugins in cytoscape allows advanced analysis of network topology	http://www.cytoscape.org/
Vanted	Analysis of network with experimental data	http://vanted.ipk-gatersleben.de/
Biotapestry	Drawing, analysis and visualization	http://www.biotapestry.org/
TYNA/Topnet	Network analysis	http://tyna.gersteinlab.org/tyna/
NCT	Network comparison toolkit	http://chianti.ucsd.edu/nct/
iGraph	Network analysis and visualization	http://igraph.sourceforge.net/

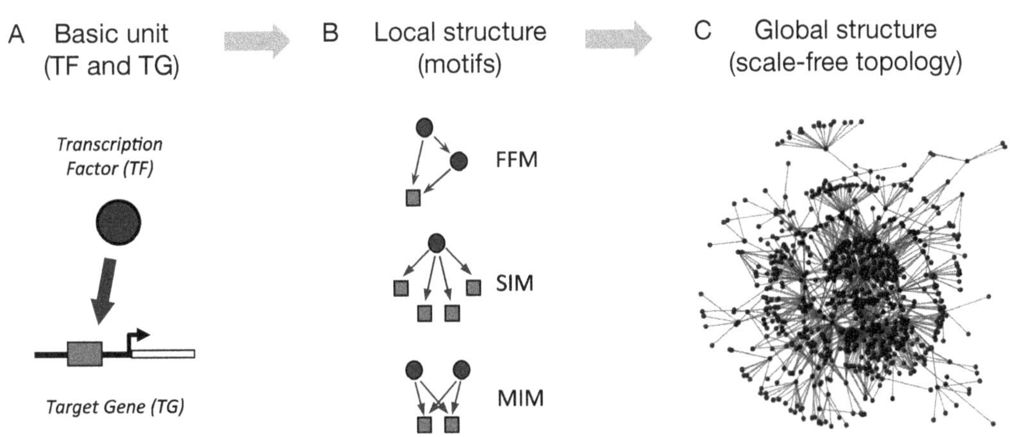

Figure 8.1 Structure of transcriptional regulatory network (A) The basic unit consists of a transcription factor (TF) which recognized specific regulatory sequence upstream of its target gene (TG) (B) At the local level, the basic units assemble to form network motifs: the feed-forward motif (FFM), c (SIM) and multiple input motif (MIM). (C) At the global level, transcriptional regulatory networks display a scale-free topology, which is characterized by the presence of a few TFs (hubs or global regulators) that regulate many genes and many TFs that regulate a few genes.

response to two possible inputs (that is activation or repression of the top-level TF), resulting in eight distinct cases. However, two particular combinations are prevalent in the *E. coli* TRN (Mangan and Alon, 2003). In the most recurrent FFM, both TFs are activators. This pattern ensures that the TG is only transcribed when a persistent signal activates the top-level TF, because expression of the TG relies on the activation of the two TFs. This configuration prevents fluctuating concentrations of the top-level TF from regulating the downstream TG, thereby filtering stochastic variation or noise in the input signal. Noticeably, the second most frequent FFM in the *E. coli* TRNs is comprised of TFs acting in an opposing manner: the intermediate-level TF is a repressor whereas the top-level one is an activator. This pattern is referred to as an incoherent FFM (Mangan *et al.*, 2006) and possesses a pulse-like dynamic in the expression of the TG: the top-level TF activates the expression of the TG until a response threshold activates the intermediate TF. At that point, the expression of the TG is inhibited.

SIMs

In SIMs, a single TF regulates a group of TGs simultaneously, therefore allowing a coordinated regulation of those set of genes. However, the concentration of TF necessary to activate the regulated genes varies depending on their promoter strength. Therefore, an SIM can show a rather subtle behaviour because the TF concentration changes with time. Such a motif can set a temporal order in the pattern of expression of individual TGs. Such patterns have been indeed observed experimentally in several metabolic pathway genes (Zaslaver *et al.*, 2004) and in the flagellar biogenesis pathway (Kalir *et al.*, 2001).

MIMs

In this type of motif, multiple TFs regulate the expression of numerous TGs. Consequently, distinct signals can be integrated in the motif, providing distinct ways of regulating gene expression. Consistently, MIMs provide a flexible regulation of their TGs in a combinatorial manner that is very likely to confer a fitness advantage under different environmental conditions.

Global structure

The global level of organization of TRNs has been extensively studied by several groups. It has been shown that TRNs display a 'scale-free' like topology (Babu *et al.*, 2004; Madan Babu and Teichmann, 2003; Thieffry *et al.*, 1998). Such a topology is characterized by the presence of a few TFs (referred to as global regulators) that regulate a strikingly large number of TGs and a vast majority of TFs (called fine-tuners) that regulate a small number of TGs. An analysis of the *E. coli* transcriptional network has defined global regulators as the top 20% of the TFs with the highest number of regulated TGs. An investigation of the function of the global regulators showed that they are TFs involved in carbon degradation (Mlc and Lrp), redox status sensing (ArcA, NarL, and Fnr), ion transport regulation (Fur), environmental sensors (CspA and Crp), and nucleoid associated proteins (Hns, Ihf, and Fis). It has been proposed that the global regulators contribute to the robustness of the gene regulatory system, where robustness is defined as the ability of the TRN to remain functional while its structure is significantly perturbed (Barabasi and Albert, 1999; Kitano, 2004). In addition to the above mentioned topology, recent studies have also shown that the TRN of *E. coli* and that of other organisms display extensive combinatorial regulation (Balaji *et al.*, 2007; Janga *et al.*, 2007b) and tend to possesses a multilayer hierarchical (i.e. a serial cascade of TFs) structure without feedback regulation at the transcription level (Cosentino Lagomarsino *et al.*, 2007; Jothi *et al.*, 2009; Ma *et al.*, 2004; Martinez-Antonio *et al.*, 2008; Yu and Gerstein, 2006).

Dynamic nature of transcriptional networks

The maintenance of cellular homeostasis and the successful adaptation to environmental changes are challenges that microorganisms face all the time and both rely on the rapid integration of external and internal stimuli via changes in gene expression. Unsurprisingly, the capacity of the transcriptional regulatory machinery to quickly bring about changes in the gene expression pattern reflects the highly dynamic dimension of TRNs. Cells must respond to change in temperature and pH, nutrient or toxins concentrations, etc. Consistently, active parts of the TRN change over time. In addition to sequence-specific TFs that respond to distinct signals, nucleoid-like architectural proteins have been shown to affect the local chromosome structure and influence the availability of specific sites on the DNA. Such chromosomal dynamics have been shown to influence the expression of several genes (Marr *et al.*, 2008). In this sense, knowledge on the topological properties of regulatory network, though informative, is not sufficient to explain this fundamental function. Accordingly, a change in regulatory network topology across different conditions and the impact of architectural proteins such as Hns, Fis, etc., have gained considerable attention and are directions of current intense research (Balaji *et al.*, 2007; Berger *et al.*, 2010; Dillon and Dorman, 2010; Dorman, 2009a; Janga *et al.*, 2007a; Luijsterburg *et al.*, 2006; Luijsterburg *et al.*, 2008; Martinez-Antonio *et al.*, 2008). In addition to architectural proteins, secondary messenger molecules such as cyclic di-GMP, (p)ppGpp, riboswitches, and small regulatory RNAs can affect gene expression dynamics. Their prevalence and impact on gene regulation at the genome level and how they tune the transcriptional response is another intense area of research (Hengge, 2009; Montange and Batey, 2008; Pesavento and Hengge, 2009; Schirmer and Jenal, 2009; Sharma *et al.*, 2010; Storz *et al.*, 2005; Waters and Storz, 2009)

Evolution of transcriptional networks

The increasing availability of completely sequenced genomes and the development of high-throughput experiments have facilitated extensive investigation of gene phylogenies for all protein families from hundreds of prokaryotic organisms. This has allowed us to gain insights into the intricate interplay of evolutionary forces that drive the evolution of TRNs. In this part of the chapter, we will first provide a short overview of the major mechanisms of gene evolution and then discuss the role these evolutionary forces have in shaping the prokaryotic regulatory networks.

Mechanisms for the evolution of gene regulatory networks

Mutations in the genome of an organism contribute to the evolution of TRNs. Such mutations, which fall on a spectrum, may affect just a single or few bases (e.g. single nucleotide substitutions) or may result in the generation of a large chunk of genetic material (e.g. duplication, repeat element expansion by transposons or horizontal transfer). Accordingly, such events may have a range of outcomes. For instance, they can affect regulatory interactions either (i) at the *cis* level, by mutating TF-binding sites or incorporate cis-regulatory elements upstream of genes during repeat element expansion or (ii) at the *trans* level, through the modification or generation of new DBDs that may recognize a different DNA sequence or may respond to a different ligand. Most of these mutations are likely to either be deleterious or cause disruption of an existing regulatory interaction. Evolution of the TRNs, on the other hand, consists of addition of new nodes (TFs and TGs) and new edges (regulatory interactions). As we will see in the following sections, gain of genes is crucial for those two aspects. As illustrated in Fig. 8.2, gene gain is driven in prokaryotes either by gene duplication (Brenner *et al.*, 1995; Chothia and Gough, 2009; Teichmann *et al.*, 1998) or by horizontal gene transfer (Koonin *et al.*, 2001; Kunin *et al.*, 2005). Although these two processes intrinsically add new nodes in TRNs, more importantly, they increase the evolvability of such networks by facilitating gain and the rewiring of regulatory interactions (Babu *et al.*, 2004; Gelfand, 2006; Janga and Collado-Vides, 2007; McAdams *et al.*, 2004; Perez and Groisman, 2009a). This point is well illustrated by a recent work that showed that artificial incorporation of new regulatory interactions into *E. coli* is rarely a barrier for evolution and even contributes to the fitness under various selection pressures (Isalan *et al.*, 2008). In this section, we only consider gene duplication, loss, and horizontal gene transfer. We do not explicitly address evolution of new interactions through repeat element expansion, which is another mechanism that may influence network evolution (Marino-Ramirez *et al.*, 2005).

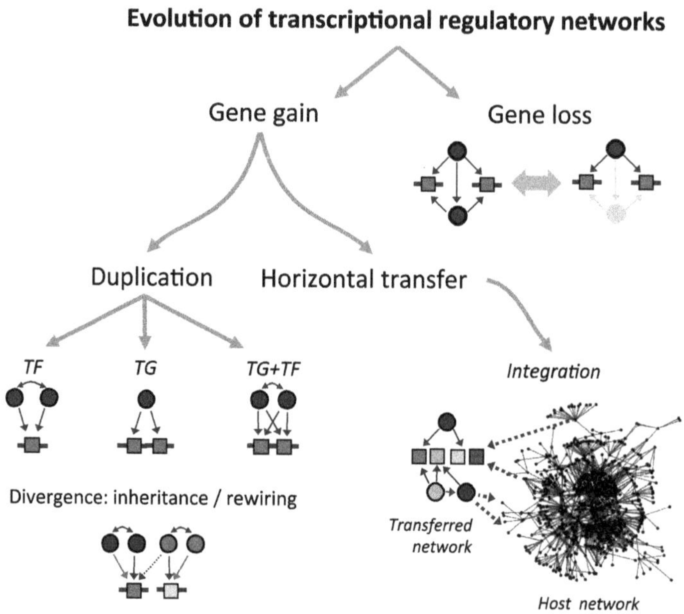

Figure 8.2 The major evolutionary forces that drive transcriptional regulatory network evolution.

Impact of gene duplication on transcriptional regulatory network evolution

Evolution by gene duplication involves the generation of a second copy of the genomic segment harbouring a gene, thereby resulting in the emergence of two identical copies of the same gene in a genome. Following duplication, one of the copies retains the ancestral function and the other copy may diverge under a relaxed selection pressure until it acquires a new function (neo-functionalization). Alternatively, the two copies may share a part of the function of the ancestral copy (sub-functionalization) or the second copy may become degenerate (Lynch and Conery, 2000). In a simplistic scenario, three different cases (Fig. 8.2) must be considered: whether the duplicated segment contains either a TF or TG, or both (Madan Babu and Teichmann, 2003; Teichmann and Babu, 2004). As a consequence of this event, gene duplication will double the quantity of regulatory interactions in addition to the number of genes involved. In each case, the fate of those shared interactions (i.e. their maintenance or removal during evolution) is of crucial importance to understand the evolution of TRNs.

Through a systematic analysis of the TRN of the prokaryote *E. coli* and the unicellular eukaryote *S. cerevisiae*, Teichmann and Babu (2004) found that more than two-thirds of the interactions have evolved as a consequence of gene duplication. They also observed that more than one-half of the known regulatory interactions were inherited from ancestral TFs or TGs after duplication with the rest of the regulatory interactions having been rewired and gained during divergence after gene duplication (Madan Babu and Teichmann, 2003; Teichmann and Babu, 2004). The authors also noticed that only a small fraction of the genes and the regulatory interactions have evolved as a consequence of gene recombination or innovation (Teichmann and Babu, 2004).

An obvious question that arises given the vast amount of gene duplication during the evolution of transcription networks is if this has had any significant role in the generation of the network motifs or of the global topology of the network. In the same study (Teichmann and Babu, 2004), the authors investigated the individual network motifs and demonstrated that although the individual genes in the network motifs may have evolved as a consequence of gene duplication, the interactions have either been gained or have evolved as a consequence of rewiring. Conant and

Wagner (2003) also observed the same trend by investigating yeast and the *E. coli* network. These studies together demonstrate that network motifs have evolved independently (i.e. convergent evolution) multiple times, possibly because they contribute to fitness by tuning the expression level of genes in a way that maximizes fitness. This is supported by the observation from experimental evolution studies, where *E. coli* was found to optimize its expression level of a protein that maximizes growth rate and therefore its fitness (Dekel and Alon, 2005). An investigation of the global structure of the TRN by Teichmann and Babu (2004) showed that the scale-free structure is not a direct consequence of gene duplication. Although this observation is consistent with the possibility that the scale-free structure could have evolved because of selection, there are other possible mechanisms, which are non-adaptive (e.g. neutral evolution), that may also give rise to the same structure (Lynch, 2007).

Taken together, these studies show that gene duplication has played a key role in the evolution of the network component and losses and gains of regulatory interactions. In addition, they have contributed to the growth of the TRN through the inheritance of regulatory interactions, gain, and through rewiring, thereby fuelling network evolution.

Horizontal gene transfer: getting connected

In eukaryotes, gene duplication and loss are believed to be the major source of genome diversification. However, in prokaryotes, horizontal gene transfer (HGT) of genetic material also represents a substantial source of genetic novelty (Koonin *et al.*, 2001; Lerat *et al.*, 2005). Interestingly, the uptake of foreign genes is often biased towards the acquisition of traits that directly contribute to fitness, such as virulence, symbiosis, or resistance to toxins (Becq *et al.*, 2007; Nakamura *et al.*, 2004; Sorek *et al.*, 2007). Thus, although understanding the role of HGT is of particular importance in prokaryotic evolution, it also has implications for understanding how they contribute to network evolution and adaptation of organisms to new environments (Ahmed *et al.*, 2008; Juhas *et al.*, 2009).

HGT requires the physical incorporation of foreign DNA into the receiver organism, its integration into the host regulatory network, and eventually its selection through the bacterial population (i.e. its fixation). The incorporation of DNA during HGT is driven by three distinct mechanisms referred to as conjugation, transduction, and transformation. The molecular mechanisms of these processes have been extensively studied and are beyond the scope of this chapter (see Chen *et al.*, 2005). Here, we discuss the regulatory constraints and mechanisms that shape the integration of new genes in TRNs. When a segment of DNA is horizontally transferred into an individual, the immediate impact on fitness of the imported genes is indeed crucial for the adaptation and survival of the individual in a bacterial population and during changing environments. However, how the gene gets integrated into the chromosome over the long run and how it integrates into an existing regulatory network is only now being understood in detail (Dorman, 2007, 2009b; Lercher and Pal, 2008; Navarre *et al.*, 2007; Stoebel *et al.*, 2008) (Fig. 8.2).

If the transferred segment is transcriptionally active, an imported gene must be successfully translated and folded in a non-lethal protein. In such cases, its protein expression level must be adequately regulated. This implies the need for a tighter transcriptional regulation and, thus, a proper recognition of its promoter region and TF binding sites by the resident transcriptional network, or requires a horizontally transferred TF that came along with the segment. Therefore, the probability of integrating a transferred gene into a network is expected to generally decrease with phylogenetic distance (Sorek *et al.*, 2007). It has been observed in *E. coli* K-12 that genes in K-loops, known to be hot spots of HGT, are poorly translated (Taoka *et al.*, 2004). Taoka and colleagues notably provided evidence that most of the recently acquired foreign genes in *E. coli* K-12 are generally not translated in laboratory conditions, suggesting that their expression may not be directly contribute to fitness (i.e. growth) in log-phase culture. In another study, Sorek *et al.* (2007) show that genes that failed to be horizontally transferred are those that are generally highly expressed. Thus, viability and successful synthesis of newly acquired genes alone are unlikely to be sufficient conditions for fixation. A balance between fitness benefits and cost in synthesis of the new gene is therefore necessary for the survival and competitiveness of the individual harbouring the transferred gene in a mixed bacterial population.

How can the cell find a strategy to favour such balance? Interestingly, several recent reports have suggested that it might be important, as a first step, to silence the transferred gene. The transferred gene can then be subsequently expressed (through anti-silencing mechanisms) (Stoebel *et al.*, 2008) when the benefit of its expression is higher than the cost of its synthesis. This is likely to tip the balance in the population, favouring the emergence of individuals who harbour the transferred gene. For example, it was observed that nucleoid-associated proteins, such as Hns, contribute to silencing the transcriptional activation of recently acquired genes, providing a 'stealth function' that minimizes the cost on fitness of their expression, thus facilitating their transmission (Doyle *et al.*, 2007; Stoebel *et al.*, 2008). Consistently, Navarre *et al.* (2006) demonstrated that Hns selectively silences horizontally acquired genes by targeting sequences with GC content lower than the resident genome in Salmonella. In addition to these studies, Perez and Groisman (Perez and Groisman, 2009b) have suggested that mutations in orthologous TFs and in their dependent promoters in different organisms may allow bacterial TFs to incorporate newly acquired genes into ancestral regulatory circuits and yet retain control of the core members of a regulon.

Taken together, these studies have begun to help us understand the role of HGT in network evolution and better appreciate various aspects of laterally acquired genes that contribute to their increased likelihood to be successfully integrated into existing regulatory networks.

Evolution of networks across organisms

Although the previously mentioned studies have provided insights into how networks evolve in an organism, it is of fundamental interest to understand how TRNs evolve across species. In other words, are interactions between TFs and TGs sufficiently conserved to be able to predict a regulatory interaction in an organism from a closely related one? This question is important because less information is available on the transcriptional networks of many prokaryotes, as most of the experimental studies performed over the past decades have been focused on model organisms such as *E. coli* and *B. subtilis*. Approaches used to address the problem of the inference of TRN from other prokaryotes can broadly be grouped into two categories, depending on whether we focus on orthology or on sequence similarity of TF binding sites (Babu, 2008; Janky *et al.*, 2009; Venancio and Aravind, 2009). The first category of methods exploits the assumption that orthologous TFs regulate orthologous TGs in distinct genomes. The latter exploits the assumption that identical binding sites upstream of two genes in closely related species imply similar regulatory interactions with orthologous TFs. Overall, these methods, in addition to methods discussed in the introduction, have provides us with a deeper insight into the evolution of TRNs across organisms.

Recent studies that have investigated over 150 completely sequenced genomes have shown that TFs are less conserved across genomes than their TGs (Lozada-Chavez *et al.*, 2006; Madan Babu *et al.*, 2006), suggesting a greater evolvability of TFs. Noticeably, it was observed that global regulators do not differ from other TFs in terms of sequence conservation. Another study, by Hershberg and Margalit, showed that the mode of regulation (activation or repression) exerted by TFs has an effect on their evolution. Repressors were found to coevolve tightly with their TGs. In contrast, activators were found to be lost independently of their targets. These results suggest that prokaryote organisms rapidly evolve their own set of transcriptional regulators and are therefore able to rewire regulation interaction in a very flexible way (Hershberg and Margalit, 2006). These observations are also supported by a study by Isalan *et al.* (2008) that shows that artificial incorporation of new regulatory interactions into *E. coli* is rarely a barrier for evolution and, in fact, contributes to the fitness under various selection pressures.

An analysis of the local structure revealed that motifs are not conserved as whole units and that individual interactions within a motif may be lost or retained. Given the functional importance of network motifs, these results may seem surprising at a first glance because one would have expected that closely related species will conserve local network structures. However, a careful analysis by Madan Babu and co-workers (Madan Babu *et al.*, 2006) showed that organisms with similar lifestyles tend to conserve similar interactions and similar motifs. In fact, it was noticed that losing or gaining interactions can result in embedding orthologous genes in different motif contexts (Fig. 8.2C). Thus, this result is more meaningful when one considers the environment in which an organism lives. This trend appeared to be statistically significant and the study identified interesting examples (Madan Babu *et al.*, 2006). For instance, in *E. coli*, it was observed that the fumarate reductase genes FrdB and FrdC are under the control of the TFs Fnr and NarL in a FFM. These enzymes, which convert fumarate to succinate under anaerobic conditions to derive energy, are therefore only expressed when both Fnr and NarL are active – that is, only under a persistent signal for lack of oxygen. Consistently, *E. coli* faces alternations of aerobic and anaerobic phases over long periods, which makes it important to induce fumarate reductases only when the bacteria is likely to stay in an anaerobic environment for extended periods. In contrast, *Haemophilus influenzae* is a pathogen that faces strong redox fluctuations during host infection. Interestingly, contrary to what happens in *E. coli*, NarL is lost and the expression of FrdB or FrdC only depends on Fnr. Therefore the fumarate reductases are regulated in a simpler manner (through an SIM) in this pathogen, which again seems relevant given its environmental lifestyle. Interestingly, this FFM found in *E. coli* is also conserved in distantly related organisms that have similar lifestyles, such as *Bordetella pertussis* (beta-proteobacterium) and *Desulfitobacterium hafniense* (firmicute).

At the level of the global structure, it was observed that global regulatory hubs are not preferentially more conserved than other TFs. It was found that the condition-specific global regulatory hubs are the ones that may be lost more easily. This observation lends support to an idea that orthologous TFs may contribute to different fitness to organisms living in different environments and, hence, completely different TFs may emerge as global regulators. Consistent with this, an analysis of the *E. coli* and the *B. subtilis* network revealed that, although the global topology was similar, very different proteins emerged as global hubs. This observation again points to the importance of the environment in shaping network structure (Madan Babu *et al.*, 2006).

Taken together, these observations highlight an important principle: TRNs are extremely plastic, evolve rapidly, and adapt to the environment by tinkering individual interactions (Lozada-Chavez *et al.*, 2006; Madan Babu *et al.*, 2006; Price *et al.*, 2007). More specifically, the specific principles can be summarized as follows (Fig. 8.3): at the level of network components, TFs evolve more rapidly than their TGs, allowing organisms to evolve their own set of regulators in line with their environment. Besides, both at the basic and at the local structure level, organisms with similar lifestyles tend to possess similar regulatory interactions. Finally, at the level of the global structure, conservation of TFs is independent of their connectivity (i.e. the number of TGs), whereas the environment, again, seems to be the major force driving gain and loss of TF and regulatory interactions.

Outlook

In this chapter, we introduced the concept of TRNs and discussed how representing the transcriptional regulatory system of an organism as a network could provide us with a better understanding of the complexity of gene regulation on a genomic scale. Specifically, we discussed research in the last decade and highlight general principles of network structure and evolution. Finally, we discuss major challenges and important directions for future research and describe how our understanding of the structure and evolution of gene networks are already being exploited in different ways.

Quantitative modelling of gene networks

Although experimental advances in sequencing are providing us with an avalanche of information

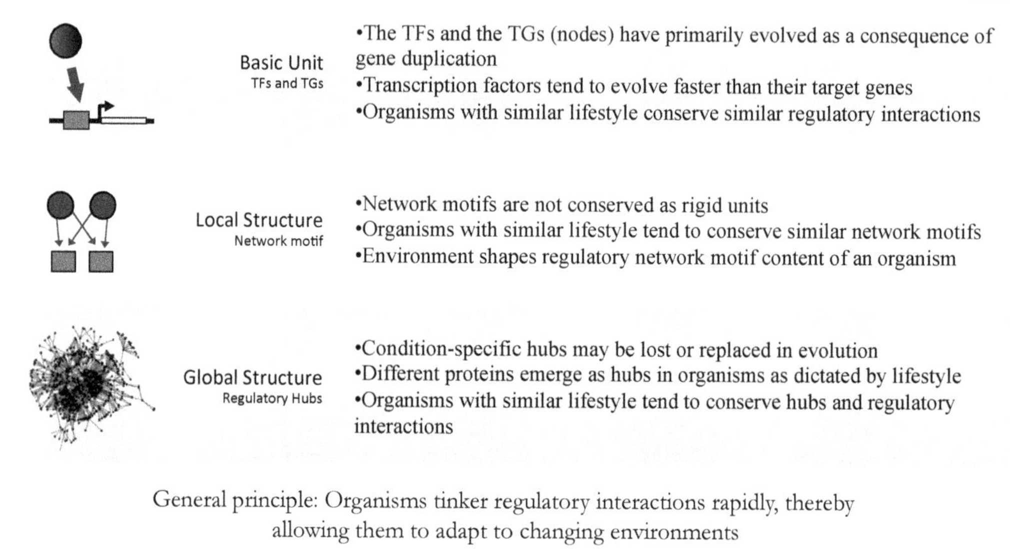

Figure 8.3 General principles of evolution at three distinct levels of network organization.

about the repertoire of genes and their expression levels across different conditions from diverse microbes and microbial communities, one of the fundamental challenges for the future is to develop conceptual and computational frameworks to integrate all these data to quantitatively model how individual genes are regulated within a cell in different context, such as stress, during infection, in the presence of a particular food source, etc. In this direction, computational and experimental approaches that model regulation of individual genes at high resolution (Ronen *et al.*, 2002; Zaslaver *et al.*, 2006) or the changes in the structure of entire regulatory network of an organism (Luscombe *et al.*, 2004; Martinez-Antonio *et al.*, 2008) are already being investigated. A key advance would be to investigate different biological systems such as DNA damage response, stress response, etc., from diverse organisms; develop new methods for investigating network dynamics; and uncover general principles through comparative analysis.

Natural variation and network evolution

The ability to sequence different strains of the same species or different individuals from the same population is providing us with a wealth of information about natural variation in the genomic sequences of different organisms [e.g. *Mycobacterium leprae* (Monot *et al.*, 2009), *E. coli* (Ooka *et al.*, 2009; Studier *et al.*, 2009)]. Such variation might involve single nucleotide changes (Brochet *et al.*, 2008) or structural alterations such as insertion and deletion of sequences through transposable elements and HGT (Brzuszkiewicz *et al.*, 2006). These events affect not only protein coding regions, but also inter-genic regions and, hence, may influence the expression of relevant genes. For example, it was recently shown that the gain of a regulatory interaction through mutations in the promoter region of Salmonella enterica serovar Typhimurium strains allowed the regulation of a virulence gene. This feature conferred a fitness advantage to those strains and permitted them to adapt better to the host environment (Osborne *et al.*, 2009). Given the fluid nature of bacterial genomes, another important future direction would be to understand natural variation in gene circuits within distinct populations of the same species. Such an understanding can provide fundamental insights into the emergence of pathogens (Brzuszkiewicz *et al.*, 2006)and has implications for human health and disease (Ahmed *et al.*, 2008).

Noise and gene networks

Non-genetic cell-to-cell variation in gene expression (i.e. noise) has been another exciting area that has recently gained attention (Losick and Desplan, 2008; Raj and van Oudenaarden, 2008). Such stochastic variation in a cell population can be beneficial where phenotypic diversity is advantageous but detrimental if homogeneity and fidelity in cellular behaviour is required. Recent work in this direction has shown that different circuits have the potential to either amplify or buffer noise (Losick and Desplan, 2008; Raj and van Oudenaarden, 2008). For instance, it was recently shown that, while seemingly different alternative circuits can provide similar patterns of outputs in gene expression, the impact of fluctuations in protein levels was shown to be an important determinant of why some circuits were selected in evolution (Cagatay *et al.*, 2009). An important challenge in this direction would be to understand the interplay between network structure and the noise level of individual genes in such networks. In this direction, a study by Jothi *et al.* (2009) has shown that TFs that are in the top of the hierarchy generally tend to show higher cell-to-cell variation in their expression level. Based on this and other observations, it was proposed that the interplay between network organization and TF dynamics could permit differential utilization of the same underlying network by distinct members of a clonal cell population. Gaining a better understanding of how gene circuits could influence stochasticity in gene expression and will have a significant impact in understanding phenomenon such as (i) bacterial persistence or adaptive resistance (Balaban *et al.*, 2004; Jayaraman, 2008), (ii) differential cell-fate outcome in response to the same uniform stimulus (Maamar *et al.*, 2007), (iii) phenotypic variability in fluctuating environments (Acar *et al.*, 2008), and (iv) cellular differentiation and development (Suel *et al.*, 2006, 2007).

Engineering gene circuits

Another major challenge would be to exploit the knowledge gained about regulatory networks to engineer gene circuits with defined properties (e.g. tunable circuits; An and Chin, 2009) for different applications. In this context, several groups have made important contributions and synthetic gene circuits are already being exploited in medicine (e.g. engineering interactions between bacterial and human cells) (Anderson *et al.*, 2006; Steidler *et al.*, 2000); bioenergy (e.g. production of fatty-acid derived fuels) (Steen *et al.*, 2010); bioremediation (e.g. to harness the concentration gradient of metals) (Xu and Lavan, 2008); laboratory applications (e.g. creation of bacterial strains resistant to specific antibiotics for selection experiments) (Dantas *et al.*, 2008; Martinez, 2008); and biotechnology (e.g. for the production of proteins) (Alper *et al.*, 2005). For a more detailed and current account of synthetic biology and engineering of gene circuits, the reader is directed to the following reviews by Chin (Chin, 2006), Kiel *et al.* (Kiel *et al.*, 2010), and Lu *et al.* (Lu *et al.*, 2009).

This is truly an exciting time for experimental and computational biologists who aim to understand gene regulatory networks. Especially, with the advances in computing and genomic technologies, we foresee the availability of more extensive and detailed maps of transcriptional regulation and other mechanisms of regulation (e.g. riboswitches and small RNAs) in a number of microorganisms. The availability of such information will fuel research that addresses fundamental questions linking different types of regulation (Leonard *et al.*, 2008; Purnick and Weiss, 2009). All these advancements collectively have the potential to transform our understanding of gene regulation in the near future.

Chapter highlights

- Information on transcriptional regulation can be represented as a network where nodes denote transcription factors or target genes and edges denote transcriptional regulatory interactions.
- Transcriptional networks can be investigated at the local level and the global level. At the local level, they are made up of small patterns of interconnections called network motifs. At the global level, they are characterized by the presence of hubs, which are TFs that regulate several genes.
- Bacterial transcriptional networks evolve by gene duplication, gene loss and horizontal gene transfer. During the course of evolution, regulatory interactions can also be re-wired by mutations in *cis*-regulatory elements or in the DNA binding domain of TFs.
- A comparison of diverse microbial genomes reveals that bacterial transcriptional networks evolve rapidly and organisms tinker regulatory interactions to efficiently adapt to changing environments.
- Representing gene regulation on a genomic scale provides a framework for integrating multiple types of data to model gene networks, study the impact of natural variation and engineer gene circuits.
- With the advent of next generation sequencing, future studies on transcriptional networks from pathogens and other relevant microbes will have significant implications for biotechnology and medicine.

Acknowledgements

The authors would like to thank the Medical Research Council, UK, for funding their research. G.C. thanks the ENS Cachan for financial support. We thank ASM press for kindly providing us the permission to reproduce this chapter. This chapter was originally published as Chalancon G and Madan Babu M, Structure and Evolution of Transcriptional Regulatory Networks (Chapter 1) in Bacterial Stress Responses, 2nd edn (eds Gisela Storz and Regine Hengge), 2011, ASM Press, Washington DC, USA.

References

Acar, M., Mettetal, J.T., and van Oudenaarden, A. (2008). Stochastic switching as a survival strategy in fluctuating environments. Nat. Genet. *40*, 471–475.

Ahmed, N., Dobrindt, U., Hacker, J., and Hasnain, S.E. (2008). Genomic fluidity and pathogenic bacteria: applications in diagnostics, epidemiology and intervention. Nat. Rev. Microbiol. *6*, 387–394.

Alon, U. (2007). Network motifs: theory and experimental approaches. Nat. Rev. Genet. *8*, 450–461.

Alper, H., Fischer, C., Nevoigt, E., and Stephanopoulos, G. (2005). Tuning genetic control through promoter engineering. Proc. Natl. Acad. Sci. U.S.A. *102*, 12678–12683.

An, W., and Chin, J.W. (2009). Synthesis of orthogonal transcription–translation networks. Proc. Natl. Acad. Sci. U.S.A. *106*, 8477–8482.

Anderson, J.C., Clarke, E.J., Arkin, A.P., and Voigt, C.A. (2006). Environmentally controlled invasion of cancer cells by engineered bacteria. J. Mol. Biol. *355*, 619–627.

Babu, M.M. (2008). Computational approaches to study transcriptional regulation. Biochem. Soc. Trans. *36*, 758–765.

Babu, M.M., Luscombe, N.M., Aravind, L., Gerstein, M., and Teichmann, S.A. (2004). Structure and evolution of transcriptional regulatory networks. Curr. Opin. Struct. Biol. *14*, 283–291.

Balaban, N.Q., Merrin, J., Chait, R., Kowalik, L., and Leibler, S. (2004). Bacterial persistence as a phenotypic switch. Science *305*, 1622–1625.

Balaji, S., Babu, M.M., and Aravind, L. (2007). Interplay between network structures, regulatory modes and sensing mechanisms of transcription factors in the transcriptional regulatory network of *E. coli*. J. Mol. Biol. *372*, 1108–1122.

Barabasi, A.L., and Albert, R. (1999). Emergence of scaling in random networks. Science *286*, 509–512.

Barabasi, A.L., and Oltvai, Z.N. (2004). Network biology: understanding the cell's functional organization. Nat. Rev. Genet. *5*, 101–113.

Becq, J., Gutierrez, M.C., Rosas-Magallanes, V., Rauzier, J., Gicquel, B., Neyrolles, O., and Deschavanne, P. (2007). Contribution of horizontally acquired genomic islands to the evolution of the tubercle bacilli. Mol. Biol. Evol. *24*, 1861–1871.

Berger, M., Farcas, A., Geertz, M., Zhelyazkova, P., Brix, K., Travers, A., and Muskhelishvili, G. (2010). Coordination of genomic structure and transcription by the main bacterial nucleoid-associated protein HU. EMBO Rep. *11*, 59–64.

Brenner, S.E., Hubbard, T., Murzin, A., and Chothia, C. (1995). Gene duplications in H. influenzae. Nature *378*, 140.

Brochet, M., Rusniok, C., Couve, E., Dramsi, S., Poyart, C., Trieu-Cuot, P., Kunst, F., and Glaser, P. (2008). Shaping a bacterial genome by large chromosomal replacements, the evolutionary history of

Streptococcus agalactiae. Proc. Natl. Acad. Sci. U.S.A. *105*, 15961–15966.

Browning, D.F., and Busby, S.J. (2004). The regulation of bacterial transcription initiation. Nat. Rev. Microbiol. *2*, 57–65.

Brzuszkiewicz, E., Bruggemann, H., Liesegang, H., Emmerth, M., Olschlager, T., Nagy, G., Albermann, K., Wagner, C., Buchrieser, C., Emody, L., *et al.* (2006). How to become a uropathogen: comparative genomic analysis of extraintestinal pathogenic *Escherichia coli* strains. Proc. Natl. Acad. Sci. U.S.A. *103*, 12879–12884.

Cagatay, T., Turcotte, M., Elowitz, M.B., Garcia-Ojalvo, J., and Suel, G.M. (2009). Architecture-dependent noise discriminates functionally analogous differentiation circuits. Cell *139*, 512–522.

Chen, I., Christie, P.J., and Dubnau, D. (2005). The ins and outs of DNA transfer in bacteria. Science *310*, 1456–1460.

Chin, J.W. (2006). Modular approaches to expanding the functions of living matter. Nat Chem Biol 2, 304–311.

Chothia, C., and Gough, J. (2009). Genomic and structural aspects of protein evolution. Biochem J *419*, 15–28.

Conant, G.C., and Wagner, A. (2003). Convergent evolution of gene circuits. Nat. Genet. *34*, 264–266.

Cosentino Lagomarsino, M., Jona, P., Bassetti, B., and Isambert, H. (2007). Hierarchy and feedback in the evolution of the *Escherichia coli* transcription network. Proc. Natl. Acad. Sci. U.S.A. *104*, 5516–5520.

Dantas, G., Sommer, M.O., Oluwasegun, R.D., and Church, G.M. (2008). Bacteria subsisting on antibiotics. Science *320*, 100–103.

Davies, J., and Jacob, F. (1968). Genetic mapping of the regulator and operator genes of the lac operon. J. Mol. Biol. *36*, 413–417.

Dekel, E., and Alon, U. (2005). Optimality and evolutionary tuning of the expression level of a protein. Nature *436*, 588–592.

Dillon, S.C., and Dorman, C.J. (2010). Bacterial nucleoid-associated proteins, nucleoid structure and gene expression. Nat. Rev. Microbiol. *8*, 185–195.

Dorman, C.J. (2007). H-NS, the genome sentinel. Nat. Rev. Microbiol. *5*, 157–161.

Dorman, C.J. (2009a). Nucleoid-associated proteins and bacterial physiology. Adv Appl Microbiol *67*, 47–64.

Dorman, C.J. (2009b). Regulatory integration of horizontally-transferred genes in bacteria. Front Biosci *14*, 4103–4112.

Doyle, M., Fookes, M., Ivens, A., Mangan, M.W., Wain, J., and Dorman, C.J. (2007). An H-NS-like stealth protein aids horizontal DNA transmission in bacteria. Science *315*, 251–252.

Foster, P.L. (2007). Stress-induced mutagenesis in bacteria. Crit. Rev. Biochem. Mol. Biol. *42*, 373–397.

Gama-Castro, S., Jimenez-Jacinto, V., Peralta-Gil, M., Santos-Zavaleta, A., Penaloza-Spinola, M.I., Contreras-Moreira, B., Segura-Salazar, J., Muniz-Rascado, L., Martinez-Flores, I., Salgado, H., *et al.* (2008). RegulonDB (version 6.0): gene regulation model of *Escherichia coli* K-12 beyond transcription, active (experimental). annotated promoters and Textpresso navigation. Nucleic Acids Res. *36*, D120–124.

Gelfand, M.S. (2006). Evolution of transcriptional regulatory networks in microbial genomes. Curr. Opin. Struct. Biol. *16*, 420–429.

Grainger, D.C., Hurd, D., Harrison, M., Holdstock, J., and Busby, S.J. (2005). Studies of the distribution of *Escherichia coli* cAMP-receptor protein and RNA polymerase along the *E. coli* chromosome. Proc. Natl. Acad. Sci. U.S.A. *102*, 17693–17698.

Grainger, D.C., Lee, D.J., and Busby, S.J. (2009). Direct methods for studying transcription regulatory proteins and RNA polymerase in bacteria. Curr. Opin. Microbiol. *12*, 531–535.

Hengge, R. (2009). Principles of c-di-GMP signalling in bacteria. Nat. Rev. Microbiol. *7*, 263–273.

Hershberg, R., and Margalit, H. (2006). Co-evolution of transcription factors and their targets depends on mode of regulation. Genome Biol. *7*, R62.

Isalan, M., Lemerle, C., Michalodimitrakis, K., Horn, C., Beltrao, P., Raineri, E., Garriga-Canut, M., and Serrano, L. (2008). Evolvability and hierarchy in rewired bacterial gene networks. Nature *452*, 840–845.

Janga, S.C., and Collado-Vides, J. (2007). Structure and evolution of gene regulatory networks in microbial genomes. Res. Microbiol. *158*, 787–794.

Janga, S.C., Salgado, H., Collado-Vides, J., and Martinez-Antonio, A. (2007a). Internal versus external effector and transcription factor gene pairs differ in their relative chromosomal position in *Escherichia coli*. J. Mol. Biol. *368*, 263–272.

Janga, S.C., Salgado, H., Martinez-Antonio, A., and Collado-Vides, J. (2007b). Coordination logic of the sensing machinery in the transcriptional regulatory network of *Escherichia coli*. Nucleic Acids Res. *35*, 6963–6972.

Janky, R., Helden, J., and Babu, M.M. (2009). Investigating transcriptional regulation: from analysis of complex networks to discovery of cis-regulatory elements. Methods *48*, 277–286.

Jayaraman, R. (2008). Bacterial persistence: some new insights into an old phenomenon. J. Biosci. 33, 795–805.

Jothi, R., Balaji, S., Wuster, A., Grochow, J.A., Gsponer, J., Przytycka, T.M., Aravind, L., and Babu, M.M. (2009). Genomic analysis reveals a tight link between transcription factor dynamics and regulatory network architecture. Mol. Syst. Biol. *5*, 294.

Juhas, M., van der Meer, J.R., Gaillard, M., Harding, R.M., Hood, D.W., and Crook, D.W. (2009). Genomic islands: tools of bacterial horizontal gene transfer and evolution. FEMS Microbiol. Rev. *33*, 376–393.

Kalir, S., McClure, J., Pabbaraju, K., Southward, C., Ronen, M., Leibler, S., Surette, M.G., and Alon, U. (2001). Ordering genes in a flagella pathway by analysis of expression kinetics from living bacteria. Science *292*, 2080–2083.

Kiel, C., Yus, E., and Serrano, L. (2010). Engineering signal transduction pathways. Cell *140*, 33–47.

Kitano, H. (2004). Biological robustness. Nat. Rev. Genet. *5*, 826–837.

Koonin, E.V., Makarova, K.S., and Aravind, L. (2001). Horizontal gene transfer in prokaryotes: quantification and classification. Annu. Rev. Microbiol. *55*, 709–742.

Kunin, V., Goldovsky, L., Darzentas, N., and Ouzounis, C.A. (2005). The net of life: reconstructing the microbial phylogenetic network. Genome Res. *15*, 954–959.

Lee, T.I., Rinaldi, N.J., Robert, F., Odom, D.T., Bar-Joseph, Z., Gerber, G.K., Hannett, N.M., Harbison, C.T., Thompson, C.M., Simon, I., *et al.* (2002). Transcriptional regulatory networks in *Saccharomyces cerevisiae.* Science *298*, 799–804.

Leonard, E., Nielsen, D., Solomon, K., and Prather, K.J. (2008). Engineering microbes with synthetic biology frameworks. Trends Biotechnol. *26*, 674–681.

Lerat, E., Daubin, V., Ochman, H., and Moran, N.A. (2005). Evolutionary origins of genomic repertoires in bacteria. PLoS Biol. *3*, e130.

Lercher, M.J., and Pal, C. (2008). Integration of horizontally transferred genes into regulatory interaction networks takes many million years. Mol. Biol. Evol. *25*, 559–567.

Losick, R., and Desplan, C. (2008). Stochasticity and cell fate. Science *320*, 65–68.

Lozada-Chavez, I., Janga, S.C., and Collado-Vides, J. (2006). Bacterial regulatory networks are extremely flexible in evolution. Nucleic Acids Res. *34*, 3434–3445.

Lu, T.K., Khalil, A.S., and Collins, J.J. (2009). Next-generation synthetic gene networks. Nat. Biotechnol. *27*, 1139–1150.

Luijsterburg, M.S., Noom, M.C., Wuite, G.J., and Dame, R.T. (2006). The architectural role of nucleoid-associated proteins in the organization of bacterial chromatin: a molecular perspective. J. Struct. Biol. *156*, 262–272.

Luijsterburg, M.S., White, M.F., van Driel, R., and Dame, R.T. (2008). The major architects of chromatin: architectural proteins in bacteria, archaea and eukaryotes. Crit. Rev. Biochem. Mol. Biol. *43*, 393–418.

Luscombe, N.M., Babu, M.M., Yu, H., Snyder, M., Teichmann, S.A., and Gerstein, M. (2004). Genomic analysis of regulatory network dynamics reveals large topological changes. Nature *431*, 308–312.

Lynch, M. (2007). The evolution of genetic networks by non-adaptive processes. Nat. Rev. Genet. *8*, 803–813.

Lynch, M., and Conery, J.S. (2000). The evolutionary fate and consequences of duplicate genes. Science *290*, 1151–1155.

Ma, H.W., Kumar, B., Ditges, U., Gunzer, F., Buer, J., and Zeng, A.P. (2004). An extended transcriptional regulatory network of *Escherichia coli* and analysis of its hierarchical structure and network motifs. Nucleic Acids Res. 32, 6643–6649.

Maamar, H., Raj, A., and Dubnau, D. (2007). Noise in gene expression determines cell fate in *Bacillus subtilis.* Science *317*, 526–529.

Madan Babu, M., and Teichmann, S.A. (2003). Evolution of transcription factors and the gene regulatory network in *Escherichia coli.* Nucleic Acids Res. *31*, 1234–1244.

Madan Babu, M., Teichmann, S.A., and Aravind, L. (2006). Evolutionary dynamics of prokaryotic transcriptional regulatory networks. J. Mol. Biol. *358*, 614–633.

Mangan, S., and Alon, U. (2003). Structure and function of the feed-forward loop network motif. Proc. Natl. Acad. Sci. U.S.A. *100*, 11980–11985.

Mangan, S., Itzkovitz, S., Zaslaver, A., and Alon, U. (2006). The incoherent feed-forward loop accelerates the response-time of the gal system of *Escherichia coli.* J. Mol. Biol. *356*, 1073–1081.

Marino-Ramirez, L., Lewis, K.C., Landsman, D., and Jordan, I.K. (2005). Transposable elements donate lineage-specific regulatory sequences to host genomes. Cytogenet. Genome Res. *110*, 333–341.

Marr, C., Geertz, M., Hutt, M.T., and Muskhelishvili, G. (2008). Dissecting the logical types of network control in gene expression profiles. BMC Syst. Biol. *2*, 18.

Martinez, J.L. (2008). Antibiotics and antibiotic resistance genes in natural environments. Science *321*, 365–367.

Martinez-Antonio, A., Janga, S.C., and Thieffry, D. (2008). Functional organisation of *Escherichia coli* transcriptional regulatory network. J. Mol. Biol. *381*, 238–247.

McAdams, H.H., Srinivasan, B., and Arkin, A.P. (2004). The evolution of genetic regulatory systems in bacteria. Nat. Rev. Genet. *5*, 169–178.

Milo, R., Shen-Orr, S., Itzkovitz, S., Kashtan, N., Chklovskii, D., and Alon, U. (2002). Network motifs: simple building blocks of complex networks. Science *298*, 824–827.

Molle, V., Nakaura, Y., Shivers, R.P., Yamaguchi, H., Losick, R., Fujita, Y., and Sonenshein, A.L. (2003). Additional targets of the *Bacillus subtilis* global regulator CodY identified by chromatin immunoprecipitation and genome-wide transcript analysis. J. Bacteriol. *185*, 1911–1922.

Monot, M., Honore, N., Garnier, T., Zidane, N., Sherafi, D., Paniz-Mondolfi, A., Matsuoka, M., Taylor, G.M., Donoghue, H.D., Bouwman, A., *et al.* (2009). Comparative genomic and phylogeographic analysis of Mycobacterium leprae. Nat. Genet. *41*, 1282–1289.

Montange, R.K., and Batey, R.T. (2008). Riboswitches: emerging themes in RNA structure and function. Annu. Rev. Biophys. *37*, 117–133.

Nakamura, Y., Itoh, T., Matsuda, H., and Gojobori, T. (2004). Biased biological functions of horizontally transferred genes in prokaryotic genomes. Nat. Genet. *36*, 760–766.

Navarre, W.W., McClelland, M., Libby, S.J., and Fang, F.C. (2007). Silencing of xenogeneic DNA by H-NS-facilitation of lateral gene transfer in bacteria by a defense system that recognizes foreign DNA. Genes Dev. *21*, 1456–1471.

Navarre, W.W., Porwollik, S., Wang, Y., McClelland, M., Rosen, H., Libby, S.J., and Fang, F.C. (2006). Selective silencing of foreign DNA with low GC content by the H-NS protein in Salmonella. Science *313*, 236–238.

Ooka, T., Ogura, Y., Asadulghani, M., Ohnishi, M., Nakayama, K., Terajima, J., Watanabe, H., and Hayashi, T. (2009). Inference of the impact of insertion sequence (IS). elements on bacterial genome diversification through analysis of small-size structural polymorphisms in *Escherichia coli* O157 genomes. Genome Res *19*, 1809–1816.

Osborne, S.E., Walthers, D., Tomljenovic, A.M., Mulder, D.T., Silphaduang, U., Duong, N., Lowden, M.J., Wickham, M.E., Waller, R.F., Kenney, L.J., *et al.* (2009). Pathogenic adaptation of intracellular bacteria by rewiring a cis-regulatory input function. Proc. Natl. Acad. Sci. U.S.A. *106*, 3982–3987.

Osbourn, A.E., and Field, B. (2009). Operons. Cell Mol Life Sci *66*, 3755–3775.

Perez, J.C., and Groisman, E.A. (2009a). Evolution of transcriptional regulatory circuits in bacteria. Cell *138*, 233–244.

Perez, J.C., and Groisman, E.A. (2009b). Transcription factor function and promoter architecture govern the evolution of bacterial regulons. Proc. Natl. Acad. Sci. U.S.A. *106*, 4319–4324.

Pesavento, C., and Hengge, R. (2009). Bacterial nucleotide-based second messengers. Curr. Opin. Microbiol. *12*, 170–176.

Price, M.N., Dehal, P.S., and Arkin, A.P. (2007). Orthologous transcription factors in bacteria have different functions and regulate different genes. PLoS Comput Biol 3, 1739–1750.

Purnick, P.E., and Weiss, R. (2009). The second wave of synthetic biology: from modules to systems. Nat. Rev. Mol. Cell Biol. *10*, 410–422.

Raj, A., and van Oudenaarden, A. (2008). Nature, nurture, or chance: stochastic gene expression and its consequences. Cell *135*, 216–226.

Ronen, M., Rosenberg, R., Shraiman, B.I., and Alon, U. (2002). Assigning numbers to the arrows: parameterizing a gene regulation network by using accurate expression kinetics. Proc. Natl. Acad. Sci. U.S.A. *99*, 10555–10560.

Schirmer, T., and Jenal, U. (2009). Structural and mechanistic determinants of c-di-GMP signalling. Nat. Rev. Microbiol. *7*, 724–735.

Sharma, C.M., Hoffmann, S., Darfeuille, F., Reignier, J., Findeiss, S., Sittka, A., Chabas, S., Reiche, K., Hackermuller, J., Reinhardt, R., *et al.* (2010). The primary transcriptome of the major human pathogen *Helicobacter pylori*. Nature *464*, 250–255.

Shen-Orr, S.S., Milo, R., Mangan, S., and Alon, U. (2002). Network motifs in the transcriptional regulation network of *Escherichia coli*. Nat. Genet. *31*, 64–68.

Sorek, R., Zhu, Y., Creevey, C.J., Francino, M.P., Bork, P., and Rubin, E.M. (2007). Genome-wide experimental determination of barriers to horizontal gene transfer. Science *318*, 1449–1452.

Steen, E.J., Kang, Y., Bokinsky, G., Hu, Z., Schirmer, A., McClure, A., Del Cardayre, S.B., and Keasling, J.D. (2010). Microbial production of fatty-acid-derived fuels and chemicals from plant biomass. Nature *463*, 559–562.

Steidler, L., Hans, W., Schotte, L., Neirynck, S., Obermeier, F., Falk, W., Fiers, W., and Remaut, E. (2000). Treatment of murine colitis by *Lactococcus lactis* secreting interleukin-10. Science *289*, 1352–1355.

Stoebel, D.M., Free, A., and Dorman, C.J. (2008). Anti-silencing: overcoming H-NS-mediated repression of transcription in Gram-negative enteric bacteria. Microbiology *154*, 2533–2545.

Storz, G., Altuvia, S., and Wassarman, K.M. (2005). An abundance of RNA regulators. Annu. Rev. Biochem. *74*, 199–217.

Studier, F.W., Daegelen, P., Lenski, R.E., Maslov, S., and Kim, J.F. (2009). Understanding the differences between genome sequences of *Escherichia coli* B strains REL606 and BL21(DE3). and comparison of the *E. coli* B and K-12 genomes. J. Mol. Biol. *394*, 653–680.

Suel, G.M., Garcia-Ojalvo, J., Liberman, L.M., and Elowitz, M.B. (2006). An excitable gene regulatory circuit induces transient cellular differentiation. Nature *440*, 545–550.

Suel, G.M., Kulkarni, R.P., Dworkin, J., Garcia-Ojalvo, J., and Elowitz, M.B. (2007). Tunability and noise dependence in differentiation dynamics. Science *315*, 1716–1719.

Taoka, M., Yamauchi, Y., Shinkawa, T., Kaji, H., Motohashi, W., Nakayama, H., Takahashi, N., and Isobe, T. (2004). Only a small subset of the horizontally transferred chromosomal genes in *Escherichia coli* are translated into proteins. Mol. Cell Proteomics 3, 780–787.

Teichmann, S.A., and Babu, M.M. (2004). Gene regulatory network growth by duplication. Nat. Genet. *36*, 492–496.

Teichmann, S.A., Park, J., and Chothia, C. (1998). Structural assignments to the *Mycoplasma genitalium* proteins show extensive gene duplications and domain rearrangements. Proc. Natl. Acad. Sci. U.S.A. *95*, 14658–14663.

Thieffry, D., Huerta, A.M., Perez-Rueda, E., and Collado-Vides, J. (1998). From specific gene regulation to genomic networks: a global analysis of transcriptional regulation in *Escherichia coli*. Bioessays *20*, 433–440.

Venancio, T.M., and Aravind, L. (2009). Reconstructing prokaryotic transcriptional regulatory networks: lessons from actinobacteria. J Biol *8*, 29.

Waters, L.S., and Storz, G. (2009). Regulatory RNAs in bacteria. Cell *136*, 615–628.

Weber, H., Pesavento, C., Possling, A., Tischendorf, G., and Hengge, R. (2006). Cyclic-di-GMP-mediated signalling within the sigma network of *Escherichia coli*. Mol. Microbiol. *62* 1014–1034.

Weber, H., Polen, T., Heuveling, J., Wendisch, V.F., and Hengge, R. (2005). Genome-wide analysis of the general stress response network in *Escherichia coli*: sigmaS-dependent genes, promoters, and sigma factor selectivity. J. Bacteriol. *187*, 1591–1603.

Xu, J., and Lavan, D.A. (2008). Designing artificial cells to harness the biological ion concentration gradient. Nat. Nanotechnol. *3*, 666–670.

Yu, H., and Gerstein, M. (2006). Genomic analysis of the hierarchical structure of regulatory networks. Proc. Natl. Acad. Sci. U.S.A. *103*, 14724–14731.

Zaslaver, A., Bren, A., Ronen, M., Itzkovitz, S., Kikoin, I., Shavit, S., Liebermeister, W., Surette, M.G., and Alon, U. (2006). A comprehensive library of fluorescent transcriptional reporters for *Escherichia coli*. Nat. Methods *3*, 623–628.

Zaslaver, A., Mayo, A.E., Rosenberg, R., Bashkin, P., Sberro, H., Tsalyuk, M., Surette, M.G., and Alon, U. (2004). Just-in-time transcription program in metabolic pathways. Nat. Genet. *36*, 486–491.

Operation of the Gene Regulatory Network in *Escherichia coli*

9

Agustino Martínez-Antonio

Abstract

Transcription factors function as sensory systems acting at the core of genetic regulatory switches. The transcriptional regulatory network in *Escherichia coli* can be studied as the integration of the whole of these genetic sensory systems. The operation of this regulatory system affect the expression of genes by interacting with the DNA at the promoter regions of transcription units. In this chapter I present the advances of what we know about the mechanistic logic for the operation of the regulatory programme in *E. coli*. It is proposed that for a better understanding on the operation of the regulatory network it should be considered the globalism of transcription factors, the signal perceived by each, their co-regulating activity, the genome position of regulatory and target genes, and cellular concentration of the regulatory proteins, among others.

Escherichia coli as a biological model system

Escherichia coli (abbreviated *E. coli*) is a Gram-negative, facultative anaerobic and non-sporulating bacterium. Cells are typically rod-shaped and are about 2 μm long and 0.5 μm in diameter, with a cell volume of 0.6–0.7 μm^3. *E. coli* was discovered by the German paediatrician and bacteriologist Theodor Escherich, in 1885 and it is now classified as part of the Enterobacteriaceae family of gamma-proteobacteria (Savageau, 1983; Blattner *et al.*, 1997). Optimal growth of *E. coli* occurs at 37°C but some laboratory strains can multiply at temperatures up to 49°C. This bacterium is commonly found in the lower intestine of warm-blooded organisms. *E. coli* normally colonizes an infant's gastrointestinal tract within 40 hours of birth, arriving with food, water or from the individuals handling the child. Most *E. coli* strains are harmless, but some, such as serotype O157:H7, can cause serious food poisoning in humans (Ohnishi *et al.*, 1999). The harmless strains are part of the normal flora of the gut, and they can beneficial their hosts by assisting food assimilation, providing some vitamins, and preventing the establishment of pathogenic bacteria within the intestine (Marteau, 2006; Seksik *et al.*, 2008).

Cultivated strains (e.g. *E. coli* K12) are well adapted to the laboratory environment, and, unlike wild type strains, these strains have lost their ability to thrive in the intestine. In 1946, Joshua Lederberg and Edward Tatum first described the phenomenon known as bacterial conjugation (Lederberg and Tatum, 1946; Lederberg and Tatum, 1953), which became an integral part of the first experiments performed for the understanding of phage genetics. These findings helped early researchers, such as Seymour Benzer, who used *E. coli* and phage T4, to understand the topography of gene structure (Miller *et al.*, 2003). After extensive studies in phages, molecular biologists centred their investigations in bacteria, particularly *E. coli*, which became their first-choice model. Thanks to theses studies in bacteria, such *E. coli*, it was possible to developed many of the modern fundaments of molecular biology and the refinement of several tools in the post-genomic era. Nowadays, we know the DNA sequence from more than 40 genomes of *E. coli* strains, most of them pertaining to pathogens (Gordon *et al.*,

2005; Wirth *et al.*, 2006). The *E. coli* pangenome (the total repertoire of genes in all the sequenced genomes of the genus) is thought to be as large as those encountered in humans (Rasko *et al.*, 2008; Vieira *et al.*, 2011). Without any doubt, *E. coli* continues to be the best choice as a model organism for driving pioneering studies, like it is the case of the new field of synthetic biology; where researches take their functional genetic elements – along with those of their phages – as biological parts list, or bio-bricks, and use them to construct new arrangements of genetic circuits (Arkin, 2001; Benner and Sismour, 2005; Keasling, 2008).

The regulatory genome of *E. coli*

Initially, bacteria were considered merely as a biochemical reactor inside a bag. However, conform we gain knowledge about its molecular components we realized that there is an orchestrated machinery operating behind its functioning and that the elements of this machinery are printed in the genome. We are aware now that the blueprints to sustain multiple forms and life-styles are written in the genomes, and more specifically into the DNA strands. The DNA in genomes has evolved to encrypt a sophisticated, flexible and dynamic biological code that by now is only partially understood; even in the most studied biological models such as *E. coli*. The genomic DNA has the planes and regulatory directives about how cell should function. This includes how to produce the molecular machinery to read and execute this genetic regulatory code, the machinery to sense and regulate gene expression and constructors and recyclers of energy and cellular building blocks, among others. We can study genomes as those physical entities that preserve the integrity of long-term memory in an organism and at the same time are flexible enough as to update their codified information to respond to changes in their living conditions. *E. coli* K-12, the best studied strain and to which I will refer through this entire chapter, has around 4605 open reading frames (ORF) arranged in a 4.6 Mbp genome size (Blattner *et al.*, 1997) and these genes are organized in 3386 transcription units (Gama-Castro *et al.*, 2011). Depending on the environmental conditions, bacteria can express a limited repertoire of genes with different functions and following different patterns. From a transcriptional point of view, the transcription unit is the operative component of DNA for gene expression as it is delimited by promoter and terminator regions, which correspond to the start and terminus of transcription, respectively. A transcription unit can encode one (monocistronic) or more (polycistronic) genes. For the purpose of gene expression the regulatory region, at the beginning of the transcription unit, is the relevant zone for the turning *off* or *on* of gene expression. This DNA region is defined as the promoter zone where a sigma factor – bound to a RNA polymerase – can bind and start the transcription of encoding DNA to RNA. The activity of the RNA polymerase holoenzyme is further modulated by transcriptional regulatory proteins – the transcription factors – which also are bound around this promoter region and act as activators or repressors of gene expression by assisting or obstructing the bound -and the activity- of RNA polymerase. In this way, transcription and sigma factors compose the regulatory machinery that controls gene expression at the transcriptional -promoter- level. The natural balance among the components of the regulatory machinery (transcription units, transcription factors and sigma) in free-living bacterial genomes differ in order of magnitude; i.e. thousands, hundreds and dozens, respectively (Pérez-Rueda *et al.*, 2009). Since genomes are the repositories of heredity information about how to transit through life, to perform these tasks, genomes must have a sensing and an auto-regulatory capability.

Transcription and sigma factors

Although most of the proteins with DNA-binding properties could be engaged in activities of gene regulation, these activities are made by transcription and sigma factors. Out of the total genes encoded in *E. coli*, around 300 of them are predicted (by computational analyses of predicted proteins with DNA-binding domains) to encode for transcription factors and seven for sigma factors (σ), (these two class of genes correspond to

~7% of the total genes encoded), (Pérez-Rueda and Collado-Vides, 2000; Madan Babu and Teichmann, 2003). Nevertheless, From the total predicted transcription factors, exists experimental evidence for only 176 (60%) showing that they can bound different DNA-sites within the regulatory regions of transcription units, and for the seven sigma factors (σ), there is experimental evidence of their promoter sites; these findings are extensively documented in RegulonDB. Since the very first studies in the regulatory machinery, transcription factors were recognized as two-headed or domain-proteins (Jacob and Monod, 1961; Jacob, 1971), one for (information) sensing and another for (transcriptional) response. The coupling of DNA-regulatory activity depending of environmental signals make the transcription factors an adequate regulatory switches found at the core of any genetic regulatory circuit (Gann, 2001; Ptashne, 2004). The variety in the repertoire of families of transcription factors in bacterial genomes is related to the richness of information from the surrounding environment that the cell has to process: e. g. the regulatory repertoire in genomes of free-living bacteria is richer that the repertoire of bacteria residing in intracellular environments or more uniform conditions (Molina and van Nimwegen, 2009; Pérez-Rueda *et al.*, 2009; Grilli *et al.*, 2012).

The transcriptional regulatory network of *E. coli*

Genetic regulatory networks are reconstructed from experimental evidence that indicates that the product of one gene (a regulatory protein) controls its own (self-regulation) or additional genes (encoding for regulatory or non-regulatory products). The result of putting all these genes and their interactions is a network graph where vertices symbolize the genetic components (genes, transcription units, operons, etc.) and edges symbolize their regulatory interactions. This network can now be studied employing graph theory tools, and interpreted it from a biological point of view. In this way the genetic components of each organism can -in principle- be represented within a network from which we can compute unravel the complexity in order to characterize it and compare the functional structures among different biological systems. The study of gene regulatory networks might help scientist to put into context of the whole network the regulatory interactions between a transcription factor and their targeted genes. This network context can stimulate the consideration of additional elements that impact the activity of a gene, which are difficult to appreciate outside this network context. Here I will consider the regulatory network formed by transcription factors and their target genes (Fig. 9.1 and Table 9.1). The genetic network of *E. coli* contains 1565 genes (nodes) and 3531 regulatory interactions (edges). Of the total number of nodes, 176 correspond to transcription factors and the rest to structural and RNA genes. Taking into account all possible genes, this entire network involves around one third of the total genes annotated in *E. coli*. Based on these numbers, it is necessary to continue the efforts for finding DNA-binding sites for all the known transcription factors as there is a large fraction without identified targets. It is not easy to calculate which proportion of the total of regulatory interactions in *E. coli* is representing the 3531 interactions between transcription factors and known target genes up to date. All biological networks described so far, have a node-degree distribution nearest to a power-law topology, which means there are only a few highly connected genes, while most of the others poorly connected (Newman *et al.*, 2006; Galan-Vasquez *et al.*, 2011). Since transcription factors are the main actors in a genetic regulatory network onward I will describe the operation of the regulatory network centring my analysis on the subnetwork of these regulatory proteins (Fig. 9.2). The subnetwork analysis, comprising the regulatory interactions only among transcription factors is a small window of the regulatory machinery wherewith it is possible to defined the behaviour of specific cellular processes

Self-repression is the dominating activity of transcription factors

Transcription factors can have regulatory effect over a large number of genes and it is common that they are self-regulated as an efficient form to

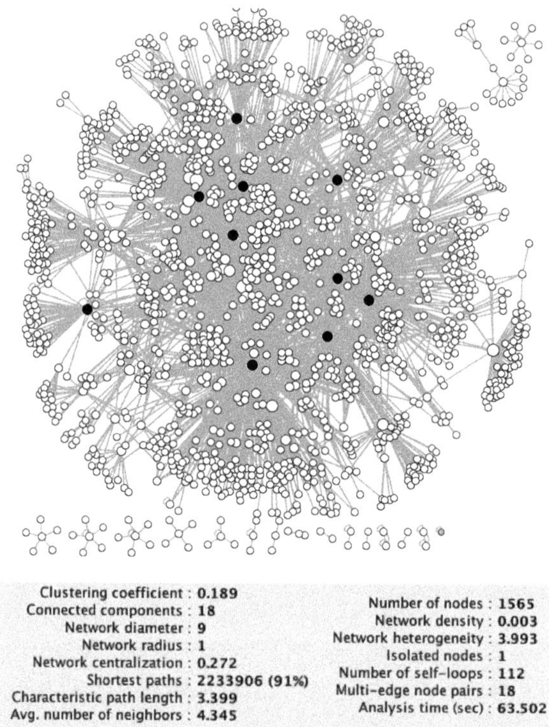

Figure 9.1 *Escherichia coli* transcriptional regulatory network. This network corresponds to the regulonDB version 7.0 draw with Cytoscape programme. Nodes represent genes and arcs the regulatory interactions among them. Black nodes represent the most global regulators, the rest are ordered in agreements to their degree-connectivity. The table inside this figure shows the statistics of this network obtained with the Network Analyser of Cytoscape (Shannon *et al.*, 2003).

Table 9.1 Statistics in operon organization and regulation in *E. coli*

Operons	Numbers	Names
Biggest operon (with larger number of genes)	16 genes	*mraZ-rsm*H-*fts*LI-*mur*EF-*mra*Y-*mur*D-*fts*W-*mur*GC-*ddl*B-*fts*QAZ-*lpx*C
Operon with larger number of promoters (and therefore transcription units)	12 promoters	*mraZ-rsm*H-*fts*LI-*mur*EF-*mra*Y-*mur*D-*fts*W-*mur*GC-*ddl*B-*fts*QAZ-*lpx*C
Operon with larger number of terminators	Five terminators	*rho*L-*rho*
Operon regulated by the larger number of TFs	12 transcription factors	*gad*AXW
Operon with larger number of DNA-binding sites for TFs	21 DNA-binding sites	*glp*TQ
Operon with more promoters for different sigma factors	Four different sigma factors	*rpo*H and *clp*PX-*lon*

respond quickly to changes in the environmental conditions (Thomas, 1991; Alon, 2007). There are three forms in which a regulatory gene can self-regulates: activation, repression or both. From the topological analysis it is possible to observe that self-repression dominates in this kind of self-regulation on transcription factors. This make biological sense since negative feedbacks are necessary to keep homeostasis and this is a function that most of transcription factors perform in *E. coli* and *B. subtilis*. However, in *Pseudomonas aeruginosa* self-activation seems to be dominating

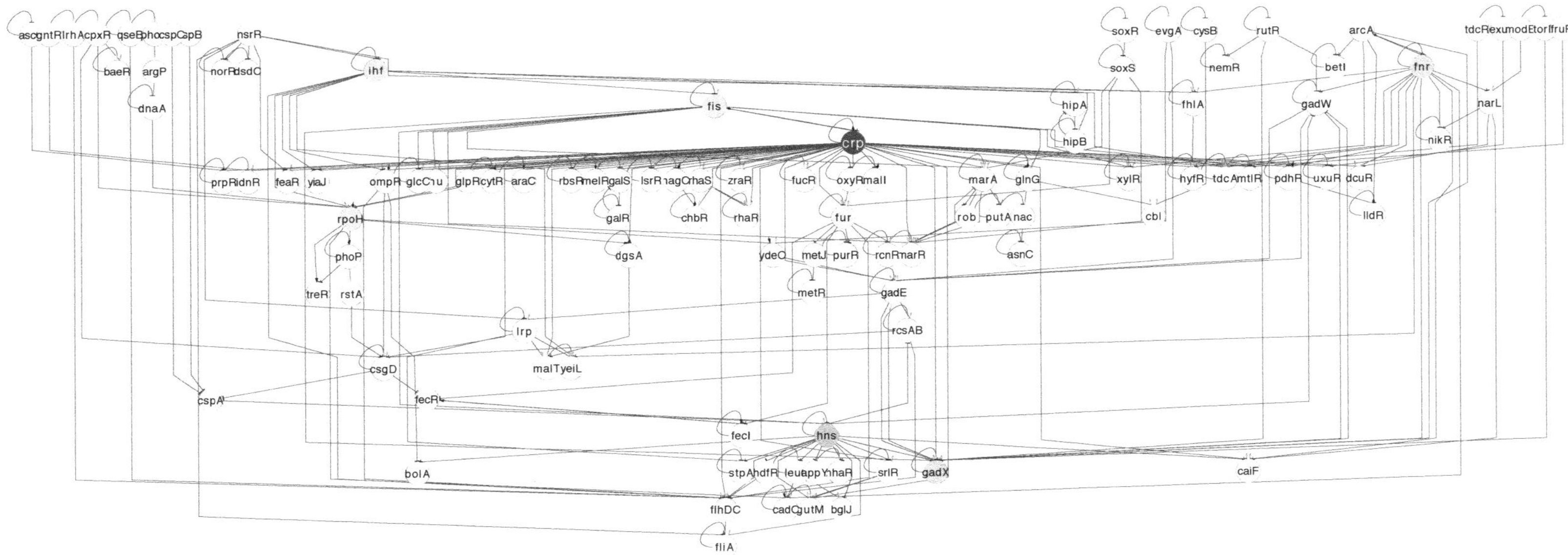

Figure 9.2 Hierarchical network organization of transcription factors in *E. coli*. The subnetwork represents the connected regulatory hardware of *E. coli*. Nodes represent TF. This figure was drawn with the aid of cytoscape software (Shannon *et al.*, 2003).

(Galán-Vasquez *et al.*, 2011). It is pending to characterize extensively the regulatory network in this last bacterium in order to know if this is a general feature of the regulatory network in this organism, or it is a feature of the network for the control of specific cellular process such as pathogenic and virulence factors; it is possible that there is a bias in this conclusion as these are the most characterized interactions of the network in *P. aeruginosa* up to date. The study of pathogeny and virulence in *P. aeruginosa* is in sharp contrast with the other two organisms, where most of the characterized networks are related to diverse physiological aspects including central metabolism.

The regulatory flux in the network is propagated by sequential activation

There are three forms in which transcription factors affect the expression of a target gene: activation, repression or both (activation and repression). One interesting observation in the execution of the regulatory programme of *E. coli* is that, in spite of the fact that most transcription factors are negatively auto-regulated, the regulation from one transcription factor to the next one inside the network is mostly by activation. This is also true for other bacteria such as *B. subtilis* and *P. aeruginosa* (Galan-Vasquez *et al.*, 2012). The fact that activation dominates the regulation of many genes might be a necessary condition for the correct execution of large transcriptional programmes since the sequential activation of several genes is needed. When several transcription factors are involved in a regulatory pathway imply that for their execution several environmental signal should be check out on the way to proceed from the beginning to the end. This also implies the existence of regulatory paths into the network that are prone to run defined transcriptional programmes. These aspects need more study and characterization into the transcriptional networks. It is noteworthy to differentiate between regulatory networks and regulatory pathways. In the former concept, this entity consists in the connectivity among transcription factors, whereas the latest concept includes not only the transcription factors interactions but all the interactions or different mechanisms of regulation, i.e. phosphorylation, allosteric inhibition etc. In RegulonDB, an endeavour on this direction is the proposal of Gensor Units (Gama-Castro *et al.*, 2011). We can speculate that the sequential order in which the regulatory software should be executed, as encoded in the genome, influences the way the network is structured. In other words, the regulatory network topology might be an evolutionary physical structure that facilitates the adequate interactions of their molecular components, following the scripts encoded in the genome. The structure of regulatory interactions might be then, responsible for the transition from genotype to phenotype in this bacterium (Martínez-Antonio *et al.*, 2008). Similar topological regulatory networks could be working in other bacteria.

The regulatory network operates with different hierarchies (path lengths)

A biological path in a network, besides their mathematical interpretation, functionally describes the regulatory hierarchical cascade. It means that the regulatory information, which starts in a sensory protein or directly by a transcription factor, can go through different transcription factors until it ends in a non-regulatory gene. It has been thought that prokaryotes have short regulatory cascades since they lack of developmental programmes where long regulatory paths are normally required. The average length of paths is of 3.19 in the regulatory network of *E. coli* while the maximal path length is 14 steps; as it will be described below. A close analysis on the network structure gives us an insight about the role of short and long regulatory cascades.

Short regulatory cascades for the catabolism of sugars

The best example of short regulatory cascades is the regulatory machinery used for the uptake of carbon sources in *E. coli*. This is a fan-like structure constituted by CRP (cAMP receptor protein) and many local transcriptional repressors of catabolic operons for each carbon source. In some cases local repressors are assisted by additional transcription factors which in each case (the different use of an alternative carbon source) the length do not exceed three steps. The global regulator, CRP, is normally co-regulating together with one of the numerous specific- catabolic transcriptional repressors. When glucose is present in the milieu, there are low levels of cAMP and the repressors are mostly set on the operators regions of their regulated catabolic operons silencing them. The absence of glucose increase the levels of cAMP and it binds to CRP, in this conformation CRP also binds to the DNA on the operators sites of catabolic operons activating its transcription. If there is an alternative sugar in the milieu it eventually is transported inside the cell and binds to the repressor. The repressor in this conformation relief their site on the operator region increasing the transcription rate. The multiple alternative switches for the use of alternative carbon sources require the persistent absence of glucose and the presence of others alternative sugars. The coupled activity of CRP and one of the local regulators activates the machinery that directs the uptake and the first catabolic steps of one of the alternative carbon sources until their derivative products are taken over by the general streams of the glycolysis or pentose pathways (Martínez-Antonio *et al.*, 2008).

Long regulatory cascades for developmental processes

Conversely, the longest transcriptional cascades are those regulating physiological processes needed for development in bacteria, such as the regulatory cascades directing flagella and biofilm formation (Martínez-Antonio *et al.*, 2008). If we consider the time taken for the transcriptional programme to run along the different networks it is natural to expect that their execution in short cascades will be quicker and therefore could result in a more uniform response by the bacterial population. In contrast, the execution of regulatory programmes along large transcriptional cascades means that longer times will be required for the cell to accomplish them. Considering that, in addition, each transcription factor in the cascade is auto-regulated, it could be expected that longer transcriptional programmes would give place to a more noise-responding population. This might result in a heterogeneity of bacterial phenotypes, as it has been observed in bacterial populations during flagella and biofilm formation whose transcriptional programmes include the longest paths in *E. coli* (Sauer *et al.*, 2002; Gaynor *et al.*, 2004; Allegrucci *et al.*, 2006). The structure of large regulatory cascades make sense since some of bacterial endogenous programmes are considered as developmental process, which implies that for their completion many checking points (conditions) should go over and possibly more than one cellular phenotype should be apt in each case.

Global transcription factors in the regulatory network

The most highly connected nodes in a regulatory network (HUBs) correspond to regulatory genes, which they have high connectivity degree: this is mainly because they are regulating too many target genes – *output* degree (Shen-Orr *et al.*, 2002; Bhardwaj *et al.*, 2010). Although, the high number of regulated genes could be a common criterion for considering a given transcription factor as a true global regulator, and for being as true global regulators, these transcription factors have to fulfil additional operative criteria such as: (i) the capability to regulate a large number of other transcription factors; (ii) exhibit co-regulation with various other transcription factors; (iii) their target genes should have promoters recognized by more than one kind of sigma factor; (iv) the products of their regulated genes should fall into different functional classes; (v) the transcription factors should be active in different growth conditions, and; (vi) these transcription factors commonly belong to evolutionary protein families with few paralogues (Martinez-Antonio and Collado-Vides, 2003). Some of the above criterions can be computational described; for instance, the calculated G coefficient for each transcription factor, indicates the relative global activity of a transcription factor inside the regulatory network (Galan-Vasquez *et al.*, 2011):

$$G = \frac{1}{4}\left(\frac{TFR}{N_{TF} + N_{SF} - 1} + \frac{GR}{N_G} + \frac{SF}{N_{SF}} + \frac{CR}{N_{TF} - 1}\right) \qquad (9.1)$$

where N_{TF} indicates the total number of known transcription factors contained in the network, N_G is the number of non-regulatory genes, and N_{SF} is the number of sigma factors used by the promoters of genes in the whole network. Additionally, TFR and GR represent the number of transcription factors and non-regulatory genes regulated by each transcription factors, respectively. SF represents the distinct sigma factors used by the promoters of genes regulated by each transcription factor and CR represents the number of transcription factors each transcription factor co-regulates with. The G value for the most global regulators in *E. coli* are given in Table 9.2.

Global transcription factors are commonly set at the highest hierarchies in genetic regulatory networks and normally have the property of auto-regulation in a dual way (i.e. they auto-activate

Table 9.2 Hubs and highly regulated target genes in the regulatory network of *E. coli*

Most global regulators (G value)	Name	Number of regulated genes
CRP (0.461)	Cyclic AMP receptor protein, also known as catabolite activator protein (CAP)	440 genes in 128 regulons
H-NS (0.186)	Histone-like nucleoid-associated protein	286 genes in 44 regulons
FNR (0.282)	Fumarate and nitrate reductase regulatory protein	284 genes in 69 regulons
FIS (0.235)	Factor for inversion stimulation	225 genes in 48 regulons
IHF (0.286)	Integration host factor	223 genes in 60 regulons
ArcA (0.169)	Aerobic respiration regulatory protein	160 genes in 48 regulons
Lrp (0.141)	Leucine-responsive regulatory protein	97 genes in 26 regulons
Most highly regulated transcription units	**Transcription factors regulating**	**Encoded gene(s)**
*flh*DC	9: CRP, Fur, H-NS, HdfR, IHF, LrhA, OmpR, QseB, RcsAB	Heterodimer master regulator of flagella synthesis
*sod*A	8: ArcA, CRP, FNR, Fur, IHF, MarA, Rob, SoxS	Superoxide dismutase, Mn. Alleviate oxidative stress
*nir*BCD	8: CRP, FNR, Fis, FruR, H-NS, IHF, NarL, NarP	Large and small subunits of nitrite reductase and nitrite transporter
*mic*F	8: H-NS, HU, IHF, Lrp, MarA, OmpR, Rob, SoxS	Antisense negative regulator of OmpF abundance
*gad*AX	8: ArcA, CRP, FNR, GadE, GadW, GadX, H-NS, TorR	Regulators of glutamate decarboxylase synthesis
*omp*F	7: CRP, CpxR, EnvY, Fur, IHF, Lrp, OmpR	Outer membrane porin for secretion of toxic compounds
*nrf*ABCDEFG	7: FNR, Fis, FlhDC, IHF, NarL, NarP, NsrR	Nitrite reductase, formate-dependent, cytochrome c
*glt*BDF	7: ArgR, CRP, FNR, GadE, IHF, Lrp, Nac	Large and small subunits of glutamate synthase
*dcu*B-*fum*B	7: ArcA, CRP, DcuR, FNR, Fis, Fur, NarL	C4-dicarboxylate antiporter and fumarase B
*nap*FDAGHBC-*ccm*ABCDEFGH	6: FNR, FlhDC, IscR, ModE, NarL, NarP	Proteins with predicted roles in electron transfer to periplasmic nitrate reductase
*mar*RAB	6: CRP, Fis, MarA, MarR, Rob, SoxS	Regulators of weak acids and antibiotics resistance systems

and auto-repress their own transcription). This type of dual operation guarantees that the protein level of these important regulators will fluctuate between certain levels but will never fall to zero (Thomas, 1973; Savageau, 1974). In agreement with this prediction, the mRNA and protein levels for global regulators are normally higher than the rest of transcription factors in the network, which makes biological sense since they should be functional most of the time (see below). In *E. coli*, these global regulators might be divided into two main groups; The group of global transcription factors controlling the global metabolism and, the group of transcription factors known as nucleoid-associated proteins (NAPs). In the first group, we find regulators for controlling carbon uptake (CRP), respiration mode (FNR anaerobic and ArcA aerobic) and stringent response due to the lack of important amino acids (Lrp, leucine-responsive protein). The common characteristic in this group of global regulators is they use small-molecules as effector signals to switch their regulatory activity. In the second group, we can find FIS, H-NS and IHF, whose maximal protein abundance, in each case, has been associated to different points in a growth curve (Ali Azam *et al.*, 1999). The nucleiod-associated proteins have DNA-bending and bridging properties and is considered they exert an analogue type of regulation on gene expression (Marr *et al.*, 2008), this is, exerting their regulation by structuring the bacterial nucleoid into different forms (see below). Physically, it is difficult to test this hypothesis. Moreover, the extreme dynamism of nucleoids makes almost impossible to define their structures, even more, the identification of physical limits of the proposed chromosomal loops (Savageau, 1983; Scolari *et al.*, 2011; Fritsche *et al.*, 2012; Sobetzko *et al.*, 2012).

The most regulated transcription units in the network

In contrast to global regulators, the most highly regulated transcription units (nodes with the highest input degree) encode for products that define metabolic and adaptive capabilities of this bacterium: such as motility, response to various stresses (*sodA*, encoding for a superoxide dismutase), and nitrogen metabolism (*nirBCD*, encoding large subunit of nitrate reductase). It is interesting that two transcription units encoding for transcription factors are among the most highly regulated ones (*flhCD*, encoding for the heterodimer master regulator of flagella synthesis and *micF*, a sRNA that regulates to *ompF*). Moreover, these genes are not the most highly conserved in bacteria, and probably are more specific to the *Escherichia* genus and their tight regulation might be responsible to maintain the fitness level required for the distinctive lifestyle of this bacterium including the gut of animals.

Transcription factors as sensory regulatory switches

From the first studies performed on transcription factors, they were recognized as two-headed molecules as they have, usually, two functional domains: one for DNA-binding and the other one for sensing small molecule effectors or for interaction with additional proteins. The existence of the last domain is very important in transcription factors as it provides the 'switch' character to these regulatory proteins (Jacob, 1971). About three-quarters of the *E. coli* transcription factors were identified as two-domain proteins (Madan Babu and Teichmann, 2003). The binding of specific effectors, normally one for each transcription factor, induces conformational changes that switch their regulatory activity (from an active to an inactive state or vice versa), (Jacob and Monod, 1961; Browning and Busby, 2004; Wall *et al.*, 2004).

Signal effectors of transcription factors

The signal effectors for the proper activation-deactivation of transcription factors can have a highly diverse nature: osmotic pressure, light, temperature, organic compounds, waste products, metal ions, etc. These effectors can be sensed direct or indirectly by the transcription factor either inside, or on the periphery, of the cell (i.e. in the periplasm). Signal effectors may have different sources, for instance, small-molecules, such metabolites, can play a signal effector role and these molecules are the product of enzymatic reaction, respiration, or metabolic recycling. On the other hand, signals sensed in the cell periphery are mainly of two types: (i) organic molecules transported into the cell, which serve to supply energy, or as precursors of the cellular building blocks, and (ii) physicochemical conditions in the milieu (heat, osmotic stress, etc.). In general, all the signal effectors produced inside the cell and transported from the milieu are sensed by the so-called 'one-component' sensory systems (Ulrich *et al.*, 2005). This type of system mainly consists of transcription factors which have a protein domain where the effector binds. On the other hand, physicochemical conditions surrounding the cell are normally sensed by the 'two-component' sensory systems (Gao and Stock, 2009; Casino *et al.*, 2010). In this case, the transcription factors involved differ in their operative domains, therefore, in the system performs two different types of proteins.. The sensory part of the system is located in a sensory protein which should be located in the periplasm, and it is capable to perceive exogenous conditions, *in situ*. These sensory proteins normally have additional domains, as those named PAS, which are habilitated with chemical structures to perceive photons, electrons, protons, etc. Once the signal is perceived by these sensory proteins, they are auto-phosphorylated and then they transfer the phosphate group to their partners, the so-called response regulators. These partner proteins have DNA-binding domains to exert the transcriptional regulation. These two-component sensory proteins have sometimes additional partners to strength or modulate the activity of the sensory domain.

The regulatory network as a sensory system for changes in exogenous and endogenous conditions

Transcription factors with the capability of sensing signal effectors produced inside the cell might be considered as sensors of 'internal or endogenous' conditions; while those sensing transported molecules and milieu conditions are considered as sensors of 'external or exogenous conditions'. Some transcription factors, however, can sense signal effectors produced either inside or outside the bacteria; these were considered as 'hybrid' sensory systems, which typically correspond to transcription factors able to recognize amino acids; either, synthesized inside the cell or transported from the milieu (Martínez-Antonio *et al.*, 2006).

Owing to the fact that each of these sensory systems is represented by a transcription factor, the transcriptional network could also be considered as a network of transcriptional sensory systems. The logic behind the coordinated function of sensory systems seems to be simple to understand in some cases and complex in others. For instance, it makes biological sense that a co-regulatory activity should exist among the transcription factors that control the transport of molecules from the milieu with those that control the expression of enzymes needed to degrade these metabolites inside the cell; as it really happens. On the other hand, it is more complicated to understand the co-regulatory activity of two-component systems (required for environmental perception) with the nucleoid-associated proteins; For example, it has been observed that environmental conditions and nucleoid structure (i.e. their regulatory actors) are factors that coordinately influence the expression of genes in *E. coli* (Martínez-Antonio *et al.*, 2006).

The operative mechanism of the regulatory programme

The main actors in a transcriptional regulatory network are the transcription factors (and sometimes sigma factors), which are represented in a networks graph as nodes. These elements constitute the regulatory machinery responsible for the transcriptional *switching* acting in the promoter region of transcription units and leading to either, activation or repression of genes. However, each transcription factor connected to the network repertoire has its own distinctiveness characterized by (i) a wide heterogeneity in their number of target genes; (ii) the type of signals perceived (of endogenous or exogenous origin) and, (iii) their co-regulatory ability to work with different sigma factors and other transcription factors. Taking into account all these features in the catalogue of transcription factors then, the question is: What is the operative logic harmonizing the operation of the different cellular regulatory programmes that have to be in agreement with the environment conditions and the cellular necessities?. A careful analysis of data generated by experimentalists might offer the following suggestions about the operative logic. First, it is known empirically that transcription factors have a wide range of target genes, driving their classification as global or local regulators;. The number of genes regulated by global or local transcription factors can range from hundreds of genes to a less than a dozen, respectively, within the same transcription unit. The decreasing global activity of transcription factors grossly correlates with their cellular concentrations (Isalan *et al.*, 2008) and with the increased specificity for their target DNA-binding sites (Lozada-Chávez *et al.*, 2008; Janga *et al.*, 2009). Second, even if a co-regulatory activity is sometimes observed among global regulators, it is more common to observe a co-regulation among global and local regulators, and hardly any in a local-local co-regulation. Third, local regulators tend to be encoded proximal to their target genes in the linear DNA molecule. In fact, local transcription factors are frequently encoded in divergent orientation or in the same operon within the genetic context of their regulated genes; which not only facilitates regulation but also the horizontal transfer of these functional genetic modules among bacteria. Fourth, it is known that nucleoid-associated proteins can wholly re-shape the bacterial nucleoid and that these proteins are preferentially expressed at different time-points in a growing population. The nucleoid-associated proteins are abundant small proteins of less than 20 kDa, present in several thousand of copies per cell. These small proteins bind cooperatively over the DNA, bending, or bridging, and reshaping the whole DNA structure. The hypothesis states that these different conformations acquired by the nucleoid influence the performance of global transcriptional programmes, and it is considered as an analogue form of gene regulation. This analogue regulation is complemented with the more precise – digital – regulation exerted by local transcription factors; from which switching activity is driven by specific signal effectors that fine-tune the expression of genes in response to precise changing conditions (Marr *et al.*, 2008; Martínez-Antonio *et al.*, 2009). Functionally, this hypothesis is supported by different findings where most of cases co-regulation is observed between nucleoid-associated proteins and some transcription factors binding effector signals. Concerning the physical chemistry of the regulation at the transcription initiation site, it makes sense that local regulators are produced very close to their respective DNA-binding sites as they are produced in very low quantities (one or two dozen of molecules per cell and many of them expressed less than one time per cell generation) (Kolesov *et al.*, 2007; Taniguchi *et al.*, 2010; Wang *et al.*, 2011). If they were produced far of their targets DNA-binding sites probably will never meet their targets. Conversely, the relatively abundant global regulators, normally present at a few of thousands copies per cell, can easily diffuse and reach their target DNA-binding sites. In this way, we can visualize the complementary activity of nucleoid-associated proteins, global–local regulators, and transcription factors for exogenous and endogenous conditions in the genetic regulatory network (Janga *et al.*, 2009).

Evolutionary conservation of the bacterial transcriptional machinery

Bacterial genomes sizes have different coding capacity that varies from 120 until to almost 10,000 open reading frames. Under this context, *E. coli* has a medium-size genome. It is natural to assume that as genomes increase in size not all the type of genes should increase in the same proportion. In large genomes, the genes that normally are enriched correspond to those encoding for regulatory and for secondary metabolism functions (Molina and van Nimwegen, 2009; Grilli *et al.*, 2012). Considering the genes encoding for the transcriptional machinery; larger genomes encode more transcription factors per gene than the smaller ones and this trend applies also for sigma factors. Moreover, not all the evolutionary families of transcription and sigma factors increase uniformly as genomes sizes do; only some evolutionary families expand their members as genomes become larger. In the case of transcription factors families, their expansion provide the bacteria with the abilities to contend with environment changes, as exemplified by cell–cell communication (LuxR), sensing of exogenous signals (OmpR), use of exogenous metabolites (GntR and AraC), and response to toxic compounds and antibiotics (TetR). In the same logic, the family of sigma factors that is more expanded corresponds to the one contributing with extra-cytoplasmic functions (ECF), which normally allow bacteria to adapt and successfully compete in diverse exogenous conditions (Pérez-Rueda *et al.*, 2009). The above suggests that, as environmental conditions become increasingly variable, signal integration and regulation of gene expression will necessarily require a more complex coordination to enable a rapid response and adaptation of bacteria (Cases *et al.*, 2003; Molina and van Nimwegen, 2009). This implies that the repertoire of gene interactions will increase concomitantly with an increase in gene number in larger genomes. To understand regulatory networks mechanisms as the evolutionary mean for flexibility and adjustment in each bacterium, we should consider the three main biological mechanisms driving evolution in regulation: (i) gene duplication; (ii) rewiring of edges by mutation/selection of TF/DNA interactions and, (iii) horizontal gene transfer. Consistent with these observations, the transcriptional sensory machinery for exogenous signals in *E. coli* is less conserved than the sensory systems required for endogenous conditions (Madan Babu *et al.*, 2006; Salgado *et al.*, 2007; Seshasayee *et al.*, 2009). These same apply even for the case of hybrid-sensing transcription factors where the genes encoding for the synthesizing enzymes are more conserved that those for the respective transporters of the amino acids.

Future directions

I would like to mention that this chapter gathers a personal conception derived from the studies performed in the last 10 years to elucidate the way the regulatory network operated in *E. coli*. It is possible that some important topics raised in the research of many of my colleagues were not covered or not discussed enough; I apologize with them for that. My personal perspective is that we have advanced in the way the transcription and sigma factors are organized in bacteria, being *E. coli* the best understood model. These kind of regulatory factors are directly interacting with the DNA but we know that much of gene regulation is also exerted by additional modes and actors such as those by small RNAs, by metabolites acting directly over the mRNA, by protein-protein and RNA–protein interactions, among others.

It is pending to integrate all these components in our current drawing chart of the regulatory network in order to have a full understandable panorama on the network for gene regulation in bacteria. Another relevant aspect that is worth to mention is that, the full regulatory networks are commonly constructed from available pairwise interactions data, nevertheless, when we study the structure of these networks, it is possible to observe that many cellular functions are the result of several circuits and regulatory cascades operating inside the network in coordinated fashion. This aspect draws our attention for experimental designs in order to validate and learn from the dynamics and function of the regulatory elements as a mean for genetic circuits engineering.

Chapter highlights

- The regulatory network of *E. coli* discussed here includes 1531 genes and 3421 regulatory interactions (around one third of their total genes). This network shows a power-law distribution with a few global regulators and most of genes poorly connected.
- There is a small set of proteins (<6) know as NAPs, which are in a high cellular concentrations. These proteins reshape the nucleoid structure and influence the running of global transcriptional programs, this mode of regulation is proposed as an analogue regulation type.
- The binding of specific signal effectors assists the activity of most of TFs in *E. coli*. These effectors switch and tune the activity of TFs. To this type of regulation depending of environmental signals is named the digital – more precise – regulation type.
- The integration of regulatory programs have place in the promoter region of TUs where it is common to observe co-regulation among global and local TFs as well as of TFs sensing exogenous and endogenous conditions.
- The mechanistic logic to understand the harmonious operation of regulatory programs in the network should consider the globality of TFs, their signal perceived, co-regulation, genome position, and cellular concentration.

Acknowledgements

The author thanks Roberto Olivares-Hernández for their critical reading of the ms, Edgardo Galán-Vásquez and Beatriz C. Luna for their assistance with the figures and Heladia Salgado for providing some data from RegulonDB. This work was supported by CONACYT through the grants 102854 and 103686.

References

Ali Azam, T., Iwata, A., Nishimura, A., Ueda, S., and Ishihama, A. (1999). Growth phase-dependent variation in protein composition of the *Escherichia coli* nucleoid. J. Bacteriol. *181*, 6361–6370.

Allegrucci, M., Hu, F.Z., Shen, K., Hayes, J., Ehrlich, G.D., Post, J.C., and Sauer, K. (2006). Phenotypic characterization of *Streptococcus pneumoniae* biofilm development. J. Bacteriol. *188*, 2325–2335.

Alon, U. (2007). Network motifs: theory and experimental approaches. Nat. Rev. Genet. *8*, 450–461.

Arkin, A.P. (2001). Synthetic cell biology. Curr. Opin. Biotechnol. *12*, 638–644.

Benner, S.A., and Sismour, A.M. (2005). Synthetic biology. Nat. Rev. Genet. *6*, 533–543.

Bhardwaj, N., Kim, P.M., and Gerstein, M.B. (2010). Rewiring of transcriptional regulatory networks: hierarchy, rather than connectivity, better reflects the importance of regulators. Sci Signal *3*, ra79.

Blattner, F.R., Plunkett, G., Bloch, C.A., Perna, N.T., Burland, V., Riley, M., Collado-Vides, J., Glasner, J.D., Rode, C.K., Mayhew, G.F., *et al.* (1997). The complete genome sequence of *Escherichia coli* K-12. Science *277*, 1453.

Browning, D.F., and Busby, S.J. (2004). The regulation of bacterial transcription initiation. Nat. Rev. Microbiol. *2*, 57–65.

Cases, I., de Lorenzo, V., and Ouzounis, C.A. (2003). Transcription regulation and environmental adaptation in bacteria. Trends Microbiol. *11*, 248–253.

Casino, P., Rubio, V., and Marina, A. (2010). The mechanism of signal transduction by two-component systems. Curr. Opin. Struct. Biol. *20*, 763–771.

Fritsche, M., Li, S., Heermann, D.W., and Wiggins, P.A. (2012). A model for *Escherichia coli* chromosome packaging supports transcription factor-induced DNA domain formation. Nucleic Acids Res. *40*, 972–980.

Galan-Vasquez, E., Luna, B., and Martínez-Antonio, A. (2011). The ç of *Pseudomonas aeruginosa*. Microbial Informatics Experimentation *1*, 3.

Galan-Vasquez, E., Luna, B., and Martínez-Antonio, A. (2012). Description of gene regulatory networks in three bacterial models: *Escherichia coli, Bacillus subtilis* and *Pseudomonas aeruginosa*. Recent Res. Dev. Microbiol. *12*: 19–39.

Gama-Castro, S., Salgado, H., Peralta-Gil, M., Santos-Zavaleta, A., Muñiz-Rascado, L., Solano-Lira, H., Jimenez-Jacinto, V., Weiss, V., García-Sotelo, J.S., López-Fuentes, A., *et al.* (2011). RegulonDB version 7.0: transcriptional regulation of *Escherichia coli* K-12 integrated within genetic sensory response units (Gensor Units). Nucleic Acids Res. *39*, D98–105.

Gann, M.P., and A. (2001). Genes and Signals (Cold Spring Harbor Laboratory Press).

Gao, R., and Stock, A.M. (2009). Biological insights from structures of two-component proteins. Annu. Rev. Microbiol. *63*, 133–154.

Gaynor, E.C., Cawthraw, S., Manning, G., MacKichan, J.K., Falkow, S., and Newell, D.G. (2004). The genome-sequenced variant of *Campylobacter jejuni* NCTC 11168 and the original clonal clinical isolate differ markedly in colonization, gene expression, and virulence-associated phenotypes. J. Bacteriol. *186*, 503–517.

Gordon, D.M., Stern, S.E., and Collignon, P.J. (2005). Influence of the age and sex of human hosts on the distribution of *Escherichia coli* ECOR groups and virulence traits. Microbiology *151*, 15–23.

Grilli, J., Bassetti, B., Maslov, S., and Cosentino Lagomarsino, M. (2012). Joint scaling laws in functional and evolutionary categories in prokaryotic genomes. Nucleic Acids Res. *40*, 530–540.

Isalan, M., Lemerle, C., Michalodimitrakis, K., Horn, C., Beltrao, P., Raineri, E., Garriga-Canut, M., and Serrano, L. (2008). Evolvability and hierarchy in rewired bacterial gene networks. Nature *452*, 840–845.

Jacob, F. (1970). La logique du vivant: Une histoire de l'heredite. Publisher Gallimard, Paris.

Jacob, F., and Monod, J. (1961). Genetic regulatory mechanisms in the synthesis of proteins+. J. Mol. Biol. *3*, 318–356.

Janga, S.C., Salgado, H., and Martínez-Antonio, A. (2009). Transcriptional regulation shapes the organization of genes on bacterial chromosomes. Nucleic Acids Res. *37*, 3680–3688.

Keasling, J.D. (2008). Synthetic biology for synthetic chemistry. ACS Chem. Biol. *3*, 64–76.

Kolesov, G., Wunderlich, Z., Laikova, O.N., Gelfand, M.S., and Mirny, L.A. (2007). How gene order is influenced by the biophysics of transcription regulation. Proc. Natl. Acad. Sci. U.S.A. *104*, 13948–13953.

Lederberg, J., and Tatum, E.L. (1946). Gene recombination in *Escherichia coli*. Nature *158*, 558.

Lederberg, J., and Tatum, E.L. (1953). Sex in bacteria; genetic studies, 1945–1952. Science *118*, 169–175.

Lozada-Chávez, I., Angarica, V.E., Collado-Vides, J., and Contreras-Moreira, B. (2008). The role of DNA-binding specificity in the evolution of bacterial regulatory networks. J. Mol. Biol. *379*, 627–643.

Madan Babu, M., and Teichmann, S.A. (2003). Evolution of transcription factors and the gene regulatory network in *Escherichia coli*. Nucleic Acids Res. *31*, 1234–1244.

Madan Babu, M., Teichmann, S.A., and Aravind, L. (2006). Evolutionary dynamics of prokaryotic transcriptional regulatory networks. J. Mol. Biol. *358*, 614–633.

Marr, C., Geertz, M., Hütt, M.-T., and Muskhelishvili, G. (2008). Dissecting the logical types of network control in gene expression profiles. BMC Syst Biol 2, 18.

Marteau, P. (2006). Probiotics, prebiotics, synbiotics: ecological treatment for inflammatory bowel disease? Gut 55, 1692–1693.

Martinez-Antonio, A., and Collado-Vides, J. (2003). Identifying global regulators in transcriptional regulatory networks in bacteria. Curr. Opin. Microbiol. *6*, 482–489.

Martínez-Antonio, A., Janga, S.C., Salgado, H., and Collado-Vides, J. (2006). Internal-sensing machinery directs the activity of the regulatory network in *Escherichia coli*. Trends Microbiol. *14*, 22–27.

Martínez-Antonio, A., Janga, S.C., and Thieffry, D. (2008). Functional organisation of *Escherichia coli* transcriptional regulatory network. J. Mol. Biol. *381*, 238–247.

Martínez-Antonio, A., Medina-Rivera, A., and Collado-Vides, J. (2009). Structural and functional map of a bacterial nucleoid. Genome Biol. *10*, 247.

Miller, E.S., Kutter, E., Mosig, G., Arisaka, F., Kunisawa, T., and Rüger, W. (2003). Bacteriophage T4 genome. Microbiol. Mol. Biol. Rev. *67*, 86–156.

Molina, N., and van Nimwegen, E. (2009). Scaling laws in functional genome content across prokaryotic clades and lifestyles. Trends Genet. *25*, 243–247.

Newman, M.E.J., Barabasi, A.L., and Watts, D.J. (2006). The Structure and Dynamics of Networks (Princeton University Press).

Ohnishi, M., Tanaka, C., Kuhara, S., Ishii, K., Hattori, M., Kurokawa, K., Yasunaga, T., Makino, K., Shinagawa, H., Murata, T., *et al.* (1999). Chromosome of the enterohemorrhagic *Escherichia coli* O157:H7; comparative analysis with K-12 MG1655 revealed the acquisition of a large amount of foreign DNAs. DNA Res. *6*, 361–368.

Pérez-Rueda, E., and Collado-Vides, J. (2000). The repertoire of DNA-binding transcriptional regulators in *Escherichia coli* K-12. Nucleic Acids Res. *28*, 1838–1847.

Pérez-Rueda, E., Janga, S.C., and Martínez-Antonio, A. (2009). Scaling relationship in the gene content of transcriptional machinery in bacteria. Mol. BioSyst. *5*, 1494–1501.

Ptashne, M. (2004). A Genetic Switch, 3rd edn: Phage Lambda Revisited (Cold Spring Harbor Laboratory Press).

Rasko, D.A., Rosovitz, M.J., Myers, G.S.A., Mongodin, E.F., Fricke, W.F., Gajer, P., Crabtree, J., Sebaihia, M., Thomson, N.R., Chaudhuri, R., *et al.* (2008). The pangenome structure of *Escherichia coli*: comparative genomic analysis of *E. coli* commensal and pathogenic isolates. J. Bacteriol. *190*, 6881–6893.

Salgado, H., Martínez-Antonio, A., and Janga, S.C. (2007). Conservation of transcriptional sensing systems in prokaryotes: a perspective from *Escherichia coli*. FEBS Letters *581*, 3499–3506.

Sauer, K., Camper, A.K., Ehrlich, G.D., Costerton, J.W., and Davies, D.G. (2002). *Pseudomonas aeruginosa* displays multiple phenotypes during development as a biofilm. J. Bacteriol. *184*, 1140–1154.

Savageau, M.A. (1974). Comparison of classical and autogenous systems of regulation in inducible operons. Nature 252, 546–549.

Savageau, M.A. (1983). *Escherichia coli* Habitats, Cell Types, and Molecular Mechanisms of Gene Control. The American Naturalist *122*, 732–744.

Scolari, V.F., Bassetti, B., Sclavi, B., and Lagomarsino, M.C. (2011). Gene clusters reflecting macrodomain structure respond to nucleoid perturbations. Mol Biosyst 7, 878–888.

Seksik, P., Dray, X., Sokol, H., and Marteau, P. (2008). Is there any place for alimentary probiotics, prebiotics or synbiotics, for patients with inflammatory bowel disease? Mol Nutr Food Res 52, 906–912.

Seshasayee, A.S.N., Fraser, G.M., Babu, M.M., and Luscombe, N.M. (2009). Principles of transcriptional regulation and evolution of the metabolic system in *E. coli*. Genome Res. *19*, 79–91.

Shannon, P., Markiel, A., Ozier, O., Baliga, N.S., Wang, J.T., Ramage, D., Amin, N., Schwikowski, B., and Ideker, T. (2003). Cytoscape: a software environment for integrated models of biomolecular interaction networks. Genome Res *13*, 2498–2504.

Shen-Orr, S.S., Milo, R., Mangan, S., and Alon, U. (2002). Network motifs in the transcriptional regulation network of *Escherichia coli*. Nat. Genet. *31*, 64–68.

Sobetzko, P., Travers, A., and Muskhelishvili, G. (2012). Gene order and chromosome dynamics coordinate spatiotemporal gene expression during the bacterial growth cycle. Proc. Natl. Acad. Sci. U.S.A. *109*, E42–50.

Taniguchi, Y., Choi, P.J., Li, G.-W., Chen, H., Babu, M., Hearn, J., Emili, A., and Xie, X.S. (2010). Quantifying *E. coli* proteome and transcriptome with single-molecule sensitivity in single cells. Science *329*, 533–538.

Thomas, R. (1973). Boolean formalization of genetic control circuits. J. Theoret. Biol. *42*, 563–585.

Thomas, R. (1991). Regulatory networks seen as asynchronous automata: a logical description. J. Theoret. Biol. *153*, 1–23.

Ulrich, L.E., Koonin, E.V., and Zhulin, I.B. (2005). One-component systems dominate signal transduction in prokaryotes. Trends Microbiol. *13*, 52–56.

Vieira, G., Sabarly, V., Bourguignon, P.-Y., Durot, M., Le Fèvre, F., Mornico, D., Vallenet, D., Bouvet, O., Denamur, E., Schachter, V., *et al.* (2011). Core and panmetabolism in *Escherichia coli*. J. Bacteriol. *193*, 1461–1472.

Wall, M.E., Hlavacek, W.S., and Savageau, M.A. (2004). Design of gene circuits: lessons from bacteria. Nat. Rev. Genet. *5*, 34–42.

Wang, W., Li, G.-W., Chen, C., Xie, X.S., and Zhuang, X. (2011). Chromosome organization by a nucleoid-associated protein in live bacteria. Science *333*, 1445–1449.

Wirth, T., Falush, D., Lan, R., Colles, F., Mensa, P., Wieler, L.H., Karch, H., Reeves, P.R., Maiden, M.C.J., Ochman, H., *et al.* (2006). Sex and virulence in *Escherichia coli*: an evolutionary perspective. Mol. Microbiol. *60*, 1136–1151.

Bacillus subtilis Transcriptional Network

10

Yuko Makita and Kenta Nakai

Abstract

Bacillus subtilis is a soil living bacterium, long known as a representative of the low G+C group of Gram-positive bacteria in contrast to *Escherichia coli*, a representative of Gram-negative bacteria. Its genome contains about 4177 protein-coding genes and 178 RNA genes. The number of sigma factors is 18, which is much larger than 7 in *E. coli*. Although there are very few theoretical studies of its global transcriptional network, our preliminary analysis based on experimentally validated data stored in the DBTBS database suggests that it is a typical scale-free network largely governed by a small number of hub transcription factors like that of many other model organisms. However, some of the results of the network motif analysis in *E. coli* were not confirmed in *B. subtilis*. Although this is likely to be due to the small sample size in *B. subtilis*, a more cautious approach might be necessary to perform network motif analyses of bacterial gene regulatory networks, which are not so large, in general. As a specific subnetwork, the gene regulatory network for sporulation, which is regarded as a model of eukaryotic differentiation process, is described. In the description of the network, not only transcriptional regulation but also post-translational regulation as well as genome recombination are necessary.

General features of *Bacillus subtilis*

Historically, all bacteria were simply classified into two groups: Gram-positive bacteria and Gram-negative bacteria. *B. subtilis* had been regarded as a typical representative (i.e. model organism) of Gram-positive bacteria while *Escherichia coli* that of Gram-negative ones (for comprehensive reviews of *B. subtilis*, see (Sonenshein *et al.*, 2001). Nowadays, the classification of bacteria has become more complicated but *B. subtilis* is still a representative of the low G+C group of Gram-positive bacteria, which roughly corresponds to the Firmicutes phylum. Thus, *B. subtilis* is a relative to the *Staphylococcus* genus, such as *S. aureus*, and is also distantly related to various important pathogenic bacteria, such as *Streptococcus pneumoniae* and *Clostridium botulinum*. Thus, the great accumulation of knowledge on *B. subtilis* should be useful for understanding as well as overcoming the diseases caused by those bacteria.

B. subtilis is a rod-shaped bacterium, living in soil. One of the main reasons why it is so useful as a model organism is in its ability to import exogenous DNA molecules and to use them as a source of genetic information (transformation), the phenomenon known as competence. Owing to this ability, this bacterium can be used as a large cloning vector (Itaya *et al.*, 2008). Another characteristics of *B. subtilis* is spore formation, which is induced with starvation and other stresses, enabling this bacterium to survive under extremely harsh conditions, e.g. boiling water. The molecular mechanism of spore formation has been extensively studied as a model of cell differentiation (see below).

The complete genome of *B. subtilis* was first sequenced by an international consortium of more than 30 laboratories in 1997 (Kunst *et al.*, 1997) (its publication was faster that of *E. coli*) and was recently re-sequenced with new sequencing technology to correct previous sequencing errors (Barbe *et al.*, 2009). The genome is circular and about 4.2 Mbp in length. As a member of the low G+C Gram-positive bacteria, its average G+C content is 43.5%. The genome is predicted to encode 4177 protein-coding genes (CDSs) and 178 RNA genes (according to the GenBank entry: AL009126.3). Those numbers are comparable to those of *E. coli* (4146 and 176 in strain K-12 subst. MG1655). To clarify which of these genes are real and are expressed in what conditions is an important issue. Comparison of the genome sequence with that of close relatives (comparative genomics) is an effective way for illustrating evolutionarily conserved regions and then for predicting genes. By the end of 2011, 39 genome sequences had been determined in the *Bacillus* genus (299 in the entire Firmicutes). With the progress of the Genomic Encyclopedia of Bacteria and Archaea project, we can expect that much more comprehensive comparison will become possible, soon (Wu *et al.*, 2009). Another way is to conduct systematic experiments using microarrays, for example. In *B. subtilis*, a number of so-called transcriptome analyses as well as proteome ones have been reported (Kocabaş *et al.* 2009; Noirot *et al.*, 2004; Völker *et al.*, 2005). According to a recently published analysis using tiling arrays of 269 samples under 104 conditions, only 186 (4.7%) CDSs were not expressed under any conditions (Nicolas *et al.*, 2012).

The transcriptional apparatus of *B. subtilis*

Sigma factors

Like in *E. coli*, the regulation of transcription in *B. subtilis* consists of two levels: the first level is achieved by the modulation of promoter selectivity with the alternative use of multiple sigma factors with the RNA polymerase complex. The second level is mediated by the use of hundreds of transcription factors.

There are 18 sigma factors in *B. subtilis*. This number is much larger than seven in *E. coli*. Each of those sigma factors recognizes a pair of specific DNA motifs located about –35/–10 base pairs from the transcription start site except σ^L, which recognizes motifs located –24/–12 base pairs from the TSS (Helmann *et al.*, 2001). However, most of the genes (about 4000 genes) are directly or indirectly regulated by the principal sigma factor: σ^A. In fact, a systematic knockout analysis of each gene coding sigma factors showed that only *sigA*, which encodes σ^A, is an essential gene in all sigma factors (Kobayashi *et al.*, 2003). As a secondary sigma factor, σ^B regulates about 200 genes, which are induced upon general stress responses through the increased binding of σ^B to RNA polymerase. In the spore formation, five sporulation sigma factors, σ^E, σ^F, σ^G, (σ^H,) and σ^K, are involved (see below). There is also a distinct, i.e. lower homology, class of sigma factors, called extracytoplasmic function (ECF) sigma factors. The *B. subtilis* genome encodes seven such genes. It is believed that those ECF factors typically control functions related to the cell surface, such as signalling from the outside to the inside of the cell. Interestingly, sigA, on ECF sigma factor, sigW, and its anti-sigma factor, rsiW, were highly expressed under all 104 conditions (Nicolas *et al.*, 2012).

To understand the network of gene regulation, it is important to know how the expression of those sigma factors is regulated. An interesting example is the involvement of post-translational process in the regulation of σ^E and σ^K, which will be described later. Another important mechanism is the use of factors that bind and inhibit sigma factors, which are called anti-sigma factors. This type of regulation will also appear in the section of sporulation regulation.

Transcription factors

The second stage of gene regulation is mediated by a number of transcription factors, each of which regulates various genes positively and/or negatively. In bacteria including *B. subtilis*, majority of these DNA-binding transcription factors are characterized with the typical helix–turn–helix DNA-binding domain. Thus, with the use of sequence motifs corresponding to this domain or other DNA-binding domains, it is possible to predict which ORF encodes a transcription factor or not. According to such a study of Pérez-Rueda's group, *B. subtilis* is predicted to have 237 transcription factors, more than half of which are further supported by experimental evidences (Moreno-Campuzano *et al.*, 2006). For transcription factors with the helix–turn–helix motif, some correlation has been observed between the motif position in their amino acid sequences and their function: repressors tend to have the motif in their N-terminal region while activators in their C-terminal region. Using this tendency as well as literature survey, it was predicted that 59% of *B. subtilis* transcription factors are repressors, 17% are activators, and 17% are dual regulators. The tendency that the majority of the encoded transcription factors are repressors seems to hold for the entire Firmicutes (Moreno-Campuzano *et al.*, 2006); if we focus on only experimentally validated 141 transcription factors, 48.9% of them are repressors, 30.5% are activators, and 20.6% are dual regulators. However, if we focus on hub regulators that regulate more than 20 experimentally validated genes, seven out of nine are dual regulators and the rest two are repressors (Table 10.1). In an extreme case, a transcription factor can switch its mode of regulation depending on its concentration: in lower levels, GerE activates the transcription of *cotD* with σ^K and RNA polymerase *in vitro* whilst it represses the *cotD* expression in higher levels (Ichikawa *et al.*, 1999).

Transcriptional regulatory databases in *B. subtilis*

Bacterial transcription regulatory systems evolve rapidly: transcription factors are suggested to evolve faster than regulated genes (Lozada-Chávez *et al.*, 2006; Madan Babu *et al.*, 2007) and regulon/operon structures are not conserved even between closely related bacteria (Price *et al.*, 2007). For example, only 35 transcription factors, including four sigma factors, have bi-directional best hits (BBHs) in *B. subtilis* against 159 *E. coli* factors (Price *et al.*, 2007). Further, only 4 of the 35 BBHs can be assigned as evolutionarily orthologous from their analysis using phylogenetic trees and functional conservation. It is therefore important to construct reference databases on a wider variety of model organisms. Here, we introduce two databases of transcriptional regulatory information in *B. subtilis*.

DBTBS (http://dbtbs.hgc.jp) is a reference-based database on transcriptional regulation in *B. subtilis*, summarizing the experimentally characterized transcription factors, their recognition sequences and the genes they regulate (Ishii *et al.* 2001; Makita *et al.*, 2004; Sierro *et al.*, 2008). At present (June 2012), the number of *B. subtilis* transcriptional regulations annotated in the database is 1523. In addition, 773 experimentally

Table 10.1 Top 10 of TFs that have many regulatory operons

TF	No. of known regulating operons (genes)	Regulatory direction	Function
Spo0A	62 (138)	Dual	Central role in the initiation of sporulation
CcpA	45 (146)	Dual	Transcriptional regulator mediating carbon catabolite repression
AbrB	34 (80)	Dual	Transcriptional pleiotropic regulator of transition state genes
LexA	34 (53)	Negative	Transcriptional repressor of the SOS regulon (DNA-damage inducible genes)
SpoIIID	32 (68)	Dual	Regulation of genes controlled by mother cell-specific sigma factors E and K
ComK	27 (70)	Dual	Competence transcription factor
TnrA	27 (68)	Dual	Transcriptional pleiotropic regulator involved in global nitrogen regulation
Fur	23 (45)	Negative	Transcriptional repressor of iron uptake
GerE	22 (41)	Dual	Transcriptional regulator required for the expression of late spore coat genes
DegU	19 (64)	Dual	Pleiotropic regulator involved in various post-exponential phase responses
PhoP	19 (36)	Dual	Two-component response regulator involved in phosphate regulation

validated transcription start sites, 463 terminators, and 141 transcription factors are included. These data were extracted from 998 references. For comparative analyses of potential regulatory regions in the genome of related bacteria, DBTBS shows the conservation of the upstream regulatory sequences between homologous genes in 40 Gram-positive bacterial species, as well as the presence of overrepresented hexameric motifs that may have regulatory functions.

PRODORIC (http://www.prodoric.de) supports a large compilation of gene regulatory data including transcription factor binding sites, promoter structures and gene expression patterns. The whole dataset is manually curated and relies on published results extracted from the scientific literature. 734 binding sites are annotated for 698 genes in *B. subtilis*. Seventy-one position weight matrices are collected. In the latest version, PRODORIC is improved by the implementation of various new tools for the elucidation of gene regulatory interactions (Grote *et al.*, 2009). In addition to these databases, some useful gene annotation databases are listed in Table 10.2.

The transcriptional network of *B. subtilis*

Since there are no extensive analyses on the transcriptional network in *B. subtilis*, here we describe our preliminary unpublished results based on DBTBS, following pioneering studies done in *E. coli* or *Saccharomyces cerevisiae*.

A global view of the network

Transcriptional network is an integrated relationship between transcription factors and regulated genes. In Fig. 10.1, such a network is drawn based on the data stored in DBTBS with the graphviz software (http://www.graphviz.org/). This graph contains 686 links (or edges) in total (sigma factors were not included); of course, this network is imperfect, missing unknown number of uncharacterized regulatory relationship. Nevertheless, we can expect that some global features of the entire *B. subtilis* network are obtainable because this bacterium has a rich history of investigation by many researchers.

A standard way to characterize a given network is to see to which of the three typical network models, 'random', 'scale-free', or 'hierarchical', it belongs (or which is most the similar) (Barabási *et al.*, 2004). In the random network model, where its nodes are connected at random, most of the nodes have approximately the same (average) number of links. Looking at the network in Fig. 10.1, it is not likely that this network fits well with the random network model: most transcription factors regulate only a few genes while there are several hub regulators that interact with many genes. In fact, these features are characteristic to the scale-free network. Mathematically, a network is scale-free if the distribution of the connectivity (or the degree, which means the number of nodes linked to a given node) follows the power-law. As shown in Fig. 10.2, the transcription network of *B. subtilis* is in fact a scale-free network. The most important property of scale-free networks is its robustness against random mutations. For example, even if 80% of randomly selected nodes are damaged, the remaining 20% could still keep the compact cluster because random mutations affect the vast number of small degree nodes mainly, not disrupting the network integrity (Albert *et al.*, 2000); of course,

Table 10.2 Useful databases for the study of *B. subtilis*

Database	URL	Category	Reference
DBTBS	http://dbtbs.hgc.jp	Transcriptional regulations	Sierro *et al.* (2008)
PRODORIC	http://prodoric.tu-bs.de/	Transcriptional regulations	Grote *et al.* (2009)
SubtiList	http://genolist.pasteur.fr/SubtiList/	Gene annotations	Moszer *et al.* (2002)
SubtiWiki	http://subtiwiki.uni-goettingen.de/	Gene annotations	Mäder *et al.* (2012)
BSORF	http://bacillus.genome.jp/	Gene annotations and microarray data	–
BsubCyc	http://bsubcyc.org/	Gene annotations and metabolic and transcriptional networks	Karp *et al.* (2011)

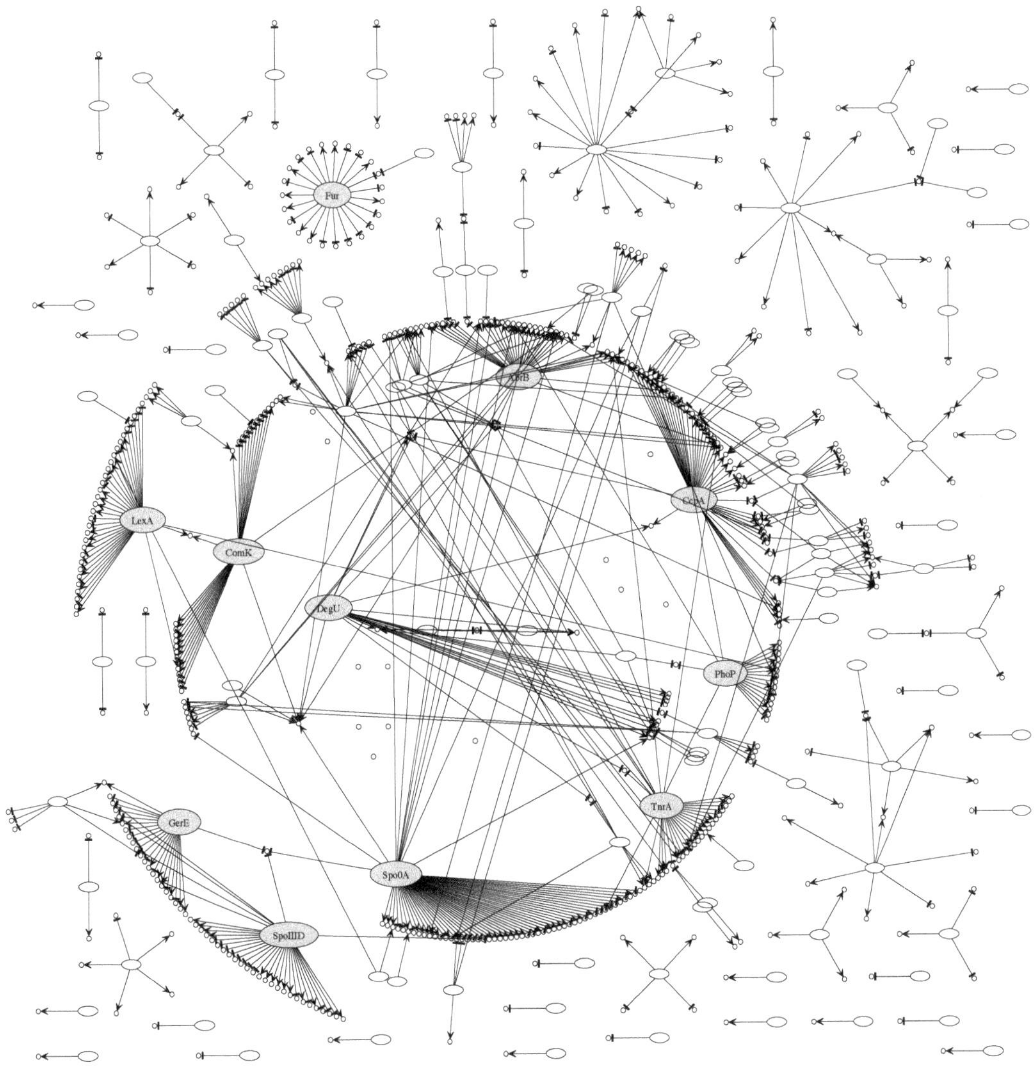

Figure 10.1 The global transcriptional network in *B. subtilis*. Top 10 transcription factors that control many different regulatory operons are shown in grey ovals. Small white ovals represent the other transcription factors and small dots represent regulated operons. Arrows represent positive regulations and tees are negative regulations. See text for further details.

on the other hand, any mutation on a single hub regulator would be a fatal disaster. As pointed out in *E. coli* and *S. cerevisiae* (Barabási *et al.*, 2000), if we divide the nodes into transcription factors and regulated genes, the transcriptional network shows asymmetry between them: if we look at the network from transcription factors, which means how many different regulated genes interact with a transcription factor, the network is typically scale-free; on the other hand, if we look at it from the side of regulated genes, which means how many different factors interact with a regulated gene, it is not scale-free (Table 10.3). This tendency can be understood from the hierarchy between the two classes of genes.

Network motif analyses

Another standard way for analysing complex networks is to characterize them using network motifs. Network motifs, which are subnetwork

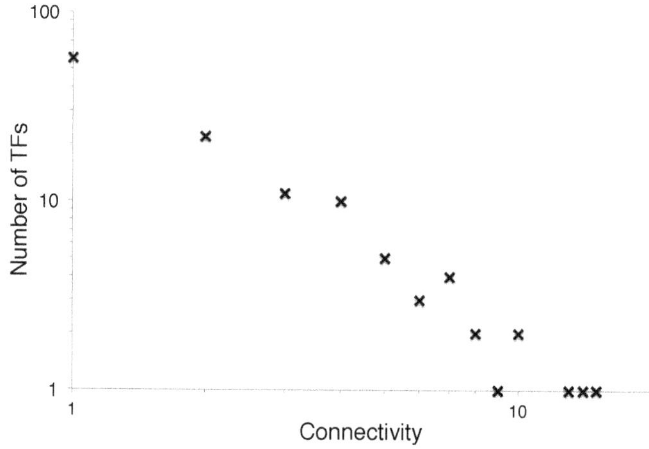

Figure 10.2 Log-log plot for the number of transcription factors and the network connectivity (i. e., the number of regulated genes for each factor).

Table 10.3 Top 10 genes that are regulated by many different transcription factors

RG	No. of known regulators	Function	Functional category	
comK	7	Competence transcription factor	3.5.2	Transcription regulation
aprE	5	Extracellular alkaline serine protease (subtilisin E)	2.2	Metabolism of amino acids and related molecules
nprE	5	Extracellular neutral metalloprotease	2.2	Metabolism of amino acids and related molecules
rok	5	Repressor of comK	1.10	Transformation/competence
alsS	4	Alpha-acetolactate synthase	2.1.1	Specific pathways
citB	4	Aconitate hydratase	2.1.3	TCA cycle
ldh	4	L-Lactate dehydrogenase	1.4	Membrane bioenergetics (electron transport chain and ATP synthase)
nasD	5	Assimilatory nitrite reductase (subunit)	2.2	Metabolism of amino acids and related molecules
phoP	4	Two-component response regulator involved in phosphate regulation	3.5.2	Transcription regulation
sacB	4	Levansucrase (EC 2.4.1.10)	2.1.1	Specific pathways

The IDs in the Functional category column are those defined in SubtiList.

patterns that frequently recur in many different parts of a network, are classified into three categories: the feedforward loop (FFL), the single input module (SIM), and the dense overlapping regulons (DOR) (Shen-Orr *et al.*, 2002).

In the FFL motif, a general transcription factor regulates a specific transcription factor and both of them jointly regulate an effector operon. As a typical example in *B. subtilis*, a global transcription factor, CcpA, which regulates the carbon catabolite repression, represses a specific transcriptional regulator, AcoR, which is an activator of the acetoin dehydrogenase operon (acoABCL). What is important here is that CcpA also represses acoABCL directly, thus configuring the FFL motif. In DBTBS, which only contains experimentally validated relationships, there are 28 examples of the FFL motif. These 28 examples are further classified into two types. In cases when both the direct and indirect effects of the general transcription factor on the regulated operon are in the same direction (positively or negatively), they are called the coherent FFLs; otherwise, they are called the incoherent FFLs. Previous analyses in *E. coli* showed that 85% of its FFLs are the coherent type (Shen-Orr *et al.*, 2002, Milo *et al.*, 2002). In *B. subtilis*, coherent FFLs share 53.6% (15/28). Although coherent FFLs still constitute the majority, this tendency may not be significant with this sample size.

The single-input module (SIM) is defined as a set of operons that are regulated solely by a single transcription with the same direction (all positive or all negative). For example, a transcriptional repressor, CtsR, regulates six class III stress operons; a zinc-specific repressor, Zur, regulates three operons. From the network motif analysis in *E. coli*, it was revealed that the majority of transcription factors in SIMs are autoregulatory, i.e. the transcription factor is regulated by itself. More specifically, although 50% of transcription factors are autoregulated in the complete dataset in *E. coli*, 70% of them are in SIMs (Shen-Orr *et al.*, 2002). In *B. subtilis*, however, only one transcription factor out of five SIMs is autoregulatory although 36.2% (51/141) of the entire transcription factors that have experimental evidence are engaged in autoregulation. Since the data size is still small, it is dangerous to say anything definite but the discrepancies observed in between *E. coli* and *B. subtilis* seem to imply the potential limitation of previous studies that were also based on relatively small data sizes.

The dense overlapping regulation (DOR) is a layer of overlapping interactions between operons and a group of input transcription factors. DOR is also called the multiple input module (MIM). In *E. coli*, sets of genes regulated by different transcription factors are much more overlapping than expected at random. The same tendency is confirmed in *B. subtilis*.

These network motif analyses have been applied to evolutionary dynamics analyses of transcriptional regulatory networks (Babu *et al.*, 2006, Jothi *et al.*, 2009) as well as evolutionary analyses of network motifs caused by gene duplications (Teichmann *et al.*, 2004, Babu *et al.*, 2004) using the data of several organisms but such studies have not been performed in *B. subtilis* yet.

The gene regulatory network for sporulation

Sporulation in *B. subtilis*

When *B. subtilis* detects relatively long-range stress, such as starvation or increased cell concentration, it starts to form a spore, which phenomenon is not observed in *E. coli*. As a model of cell differentiation, where gene expression must be regulated specific to each stage or cell type, the molecular mechanism of this sporulation process has been extensively studied. In the last section of this chapter, we introduce its outline emphasizing the transcriptional network, where five sigma factors and tens of transcription factors constitute a cascade, which means a successive one-way flow. Those regulations are more conserved then gene presence, while the opposite is true for non-sporulation genes (De Hoon *et al.*, 2010) The entire process is classified into seven stages (Errington, 2003) and it takes about 7 hours for a spore coat is formed (24 hours for the release of the mature spore). Refer to Fig. 10.3, which is a schematic representation of the entire regulatory network, when reading the following (rather complicated) explanation.

The asymmetrical cell division

First of all, various extracellular or intracellular signals are detected by one of five sensor proteins, which are the so-called two-component histidine kinases (KinA, KinB, and KinE). These kinases phosphorylate a secondary-messenger protein, Spo0F and its phosphoryl group is then relayed to a phosphotransferase, Spo0B, and subsequently to a master transcription regulator, Spo0A (Fig. 10.3). Spo0A directly/indirectly controls the transcription of over 500 genes (Molle *et al.*, 2003). The 3D structures of Spo0A (Zhao *et al.*, 2002) as well as Spo0F and Spo0B (Varughese *et al.*, 2006) are solved. The *spo0A* gene is regulated by both σ^A-dependent and σ^H-dependent promoters. The *sigH* (*spo0H*) gene, which encodes σ^H, has a σ^A-dependent promoter and is also repressed by AbrB. The activated (phosphorylated) form of Spo0A (Spo0A-P) represses the expression of the *abrB* gene. Accumulation of Spo0A-P causes asymmetric cell division that produces two types of cells with distinct fates. The smaller progeny

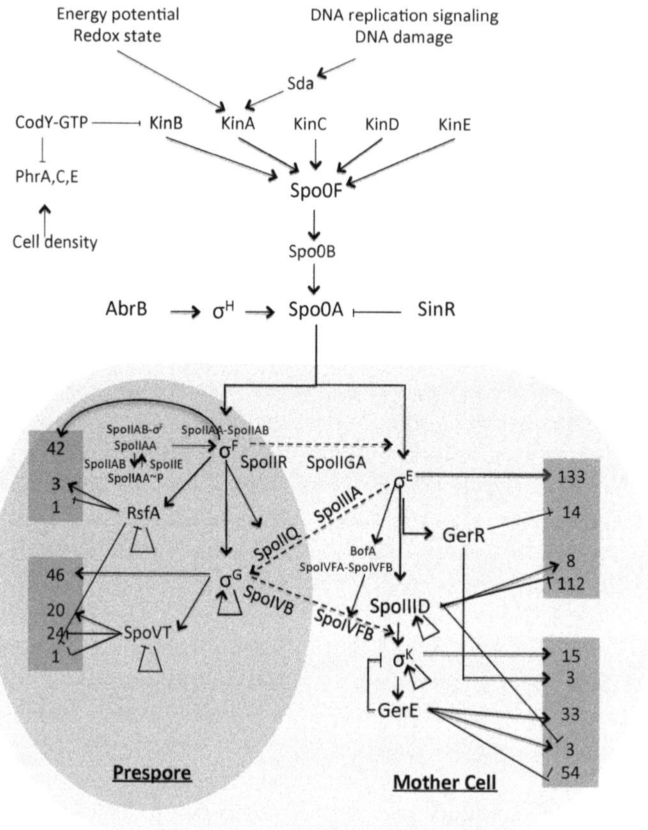

Figure 10.3 The sporulation cascade in *B. subtilis*. The diagram represents the main transcriptional regulators in sporulation (adapted from Piggot *et al.*, 2004, Eichenberger *et al.*, 2004, Paredes *et al.*, 2005, Wang *et al.*, 2006, Kroos, 2007). The numbers in dark rectangles represent the number of genes regulated by each sigma factor or an associated transcription factor. These numbers are taken from Eichenberger *et al.* (2004), Wang *et al.* (2006), Sierro *et al.* (2008).

cell is called the prespore (it becomes the forespore in the next step), which develops into the spore, while the bigger one is called the mother cell, which helps the spore formation but will lyse in the end (programmed cell death).

The activation of σ^F in early prespore

Spo0A-P activates the expression of three operons: the *spoIIA* operon, the *spoIIE* gene, and the *spoIIG* operon. The *spoIIA* operon contains three genes: *spoIIAA*, *spoIIAB* and *spoIIAC* (*sigF*). The *spoIIAC* gene encodes σ^F; the SpoIIAB protein is an anti-sigma factor that binds to σ^F, repressing its activity; in addition, in higher ATP concentration, SpoIIAB phosphorylates SpoIIAA, which works otherwise an anti-anti-sigma-factor. The *spoIIE* gene encodes a membrane-bound phosphatase that dephosphorylates SpoIIAA-P. From a still controversial reason, the ADP concentration is elevated specifically in the prespore, SpoIIAA-P is dephosphorylated by SpoIIE, and then binds to SpoIIAB, which releases and activates σ^F in prespore. σ^F regulates around 50 genes, including *spoIIIG*, which encodes σ^G.

The activation of σ^E in early mother cell

The *spoIIG* operon, which is activated by Spo0A-P, contains two genes: one of them, *spoIIGB* (*sigE*), encodes the precursor of σ^E (pro σ^E) while the other, *spoIIGA*, encodes a membrane-bound protease that can activate σ^E. In the prespore, σ^F activates the expression of *spoIIR*, which encodes a secretary protein that is secreted into the space between the mother cell and the prespore. After secreted, SpoIIR activates SpoIIGA, which then activates σ^E specifically in the mother cell. σ^E then activates about 260 sporulation-specific genes, including a coat proteins regulator GerR and a hub regulator SpoIIID.

The activation of σ^G in middle to late forespore

Like other sporulation-specific sigma factors, σ^G is first expressed as an inactive form. The mechanism of its activation is unknown but the involvement of SpoIIIAA-SpoIIIAH and SpoIIQ, which are expressed in the mother cell under the control of σ^E, is suggested. If so, it makes sense that the cascade of sigma factor activation is successive and is always communicated across the two types of the cells. σ^G regulates about 80 genes, including a σ^G-dependent transcription factor, SpoVT.

The formation of *sigK* and the activation of its product in late mother cell

The last sigma factor that is activated in the cascade is strictly regulated in multiple ways and one of them is rather unique: the *sigK* gene is newly created specifically in the late mother cell. Namely, the gene for *sigK* is separated into two parts (*spoIVCB* and *spoIIIC*) with an interval of 48 kb. This interval is excised by a specific recombinase, SpoIVCA, whose gene is dependent on σ^E and SpoIIID. This ensures that the *sigK* gene is only created in the mother cell and that this change is not inherited. Both *spoIVCB* and *spoIIIC* are also regulated by σ^E. Again, the *sigK* gene encodes a precursor protein, pro σ^K. Pro σ^K is processed into its active form by a protease, SpoIVFB, that is integrated into the outer membrane of the forespore. Interestingly, this protease is also inactive until it is activated by the SpoIVFA and BofA proteins, whose synthesis is indirectly regulated by σ^E and which are integrated into the outer membrane of the forespore after the completion of the forespore formation. More accurately, there is another σ^G-dependent protein, SpoIVB, which goes through the inner membrane of the forespore and inactivates SpoIVFA and BofA, resulting to the activation of SpoIVFB. σ^K regulates about 110 genes, including a typical global regulator, GerE. Many of these σ^K-dependent genes code genes for the formation of the spore-coat, which is very resistant to external conditions.

Chapter highlights

- *Bacillus subtilis* has been regarded as a model organism of the low G+C group of Gram-positive bacteria, or the Firmicutes, while *Escherichia coli* that of Gram-negative ones. Thus, they are ideal for comparative analyses.
- Although the genome size of *B. subtilis* is similar to that of *E. coli*, *B. subtilis* has more than twice of sigma factors than *E. coli*.
- As databases of *B. subtilis* transcription, DBTBS and PRODORIC are easy to use.
- The transcriptional network of *B. subtilis* is scale-free like that of *E. coli*. However, their results of the network motif analysis are inconsistent probably owing to their small sample size.
- The sporulation mechanism of *B. subtilis* has been extensively studied as a model of cell differentiation (stage-specific and cell type-specific gene expression). In its mechanism, four specific sigma factors and a few important transcription factors constitute a regulatory cascade.
- The regulatory cascade of sporulation includes extensive post-translational regulations and a developmental genome recombination. The existence of many intracellular communication ensures the stepwise ongoing of the entire process.

Acknowledgements

We would like to thank Drs Hiromu Takamatsu and Ashwini Patil for critically reading the manuscript.

References

Albert, R., Jeong, H., and Barabasi, A. (2000). Error and attack tolerance of complex networks. Nature *406*, 378–382.

Babu, M., Luscombe, N., Aravind, L., Gerstein, M., and Teichmann, S. (2004). Structure and evolution of transcriptional regulatory networks. Curr. Opin. Struct. Biol. *14*, 283–291.

Barabási, A., and Oltvai, Z. (2004). Network biology: understanding the cell's functional organization. Nat. Rev. Genet. *5*, 101–113.

Barbe, V., Cruveiller, S., Kunst, F., Lenoble, P., Meurice, G., Sekowska, A., Vallenet, D., Wang, T., Moszer, I., Médigue, C., *et al.* (2009). From a consortium sequence to a unified sequence: the *Bacillus subtilis* 168 reference genome a decade later. Microbiology *155*, 1758–1775.

De Hoon, M.J., Eichenberger, P., and Vitkup, D. (2010). Hierarchical evolution of the bacterial sporulation network. Curr Biol. *20*, R735–45.

Eichenberger, P., Fujita, M., Jensen, S., Conlon, E., Rudner, D., Wang, S., Ferguson, C., Haga, K., Sato, T., Liu, J., *et al.* (2004). The program of gene transcription for a single differentiating cell type during sporulation in *Bacillus subtilis*. PLoS Biol. *2*, e328.

Grote, A., Klein, J., Retter, I., Haddad, I., Behling, S., Bunk, B., Biegler, I., Yarmolinetz, S., Jahn, D., and Münch, R. (2009). PRODORIC (release 2009): a database and tool platform for the analysis of gene regulation in prokaryotes. Nucleic Acids Res. *37*, D61–65.

Errington, J. (2003). Regulation of endospore formation in *Bacillus subtilis*. Nat. Rev. Microbiol. *1*, 117–126.

Helmann, J.D., and Moran, C.P., Jr. (2001). RNA polymerase and sigma factors. In *Bacillus subtilis* and its Closest Relatives: From Genes to Cells, Sonenshein, A.L. , Hoch, J.A. and Losick, R. , eds. (American Society for Microbiology), pp. 289–312.

Ichikawa, H., Halberg, R., and Kroos, L. (1999). Negative regulation by the *Bacillus subtilis* GerE protein. J. Biol. Chem. *274*, 8322–8327.

Ishii, T., Yoshida, K., Terai, G., Fujita, Y., and Nakai, K. (2001). DBTBS: a database of *Bacillus subtilis* promoters and transcription factors. Nucleic Acids Res. *29*, 278–280.

Itaya, M., Fujita, K., Kuroki, A., and Tsuge, K. (2008). Bottom-up genome assembly using the *Bacillus subtilis* genome vector. Nat Methods *5*, 41–43.

Jothi, R., Balaji, S., Wuster, A., Grochow, J., Gsponer, J., Przytycka, T., Aravind, L., and Babu, M. (2009). Genomic analysis reveals a tight link between transcription factor dynamics and regulatory network architecture. Mol Syst Biol *5*, 294.

Karp, P.D., Keseler, I.M., Altman, T., Caspi, R., Fulcher, C.A., Subhraveti, P., Kothari, A., Krummenacker, M., Latendresse, M., *et al.* (2011). BioCyc: microbial genomes and cellular networks. Microbe *6*,176–182.

Kobayashi, K., Ehrlich, S., Albertini, A., Amati, G., Andersen, K., Arnaud, M., Asai, K., Ashikaga, S., Aymerich, S., Bessieres, P., *et al.* (2003). Essential *Bacillus subtilis* genes. Proc. Natl. Acad. Sci. U.S.A. *100*, 4678–4683.

Kocabaş, P., Calik, P., Calik, G., and Ozdamar, T. (2009). Microarray Studies in *Bacillus subtilis*. Biotechnol. J. *4*, 1012–1027.

Kroos, L. (2007). The *Bacillus* and *Myxococcus* developmental networks and their transcriptional regulators. Annu. Rev. Genet. *41*, 13–39.

Kunst, F., Ogasawara, N., Moszer, I., Albertini, A., Alloni, G., Azevedo, V., Bertero, M., Bessières, P., Bolotin, A., Borchert, S., *et al.* (1997). The complete genome sequence of the Gram-positive bacterium *Bacillus subtilis*. Nature *390*, 249–256.

Lozada-Chávez, I., Janga, S., and Collado-Vides, J. (2006). Bacterial regulatory networks are extremely flexible in evolution. Nucleic Acids Res. *34*, 3434–3445.

Madan Babu, M., Balaji, S., and Aravind, L. (2007). General trends in the evolution of prokaryotic transcriptional regulatory networks. Genome Dyn. *3*, 66–80.

Madan Babu, M., Teichmann, S., and Aravind, L. (2006). Evolutionary dynamics of prokaryotic transcriptional regulatory networks. J. Mol. Biol. *358*, 614–633.

Mäder, U., Schmeisky, A.G., Flórez, L.A., and Stülke, J. (2012). SubtiWiki – a comprehensive community resource for the model organism *Bacillus subtilis*. Nucleic Acids Res. 40 (Database issue): D1278–D1287.

Makita, Y., Nakao, M., Ogasawara, N., and Nakai, K. (2004). DBTBS: database of transcriptional regulation in *Bacillus subtilis* and its contribution to comparative genomics. Nucleic Acids Res. *32*, D75–77.

Milo, R., Shen-Orr, S., Itzkovitz, S., Kashtan, N., Chklovskii, D., and Alon, U. (2002). Network motifs: simple building blocks of complex networks. Science *298*, 824–827.

Molle, V., Fujita, M., Jensen, S.T., Eichenberger, P., González-Pastor, J.E., Liu, J.S., and Losick, R. (2003). The Spo0A regulon of *Bacillus subtilis*. Mol. Microbiol. *50*, 1683–1701.

Moreno-Campuzano, S., Janga, S., and Pérez-Rueda, E. (2006). Identification and analysis of DNA-binding transcription factors in *Bacillus subtilis* and other Firmicutes--a genomic approach. BMC Genomics *7*, 147.

Moszer, I., Jones, L., Moreira, S., Fabry, C., and Danchin, A. (2002). *subtilist*: the reference database for the *Bacillus subtilis* genome. Nucleic Acids Res. *30*, 62–65.

Nicolas, P., Mäder, U., Dervyn, E., Rochat, T., Leduc, A., Pigeonneau, N., Bidnenko, E., Marchadier, E., Hoebeke, M., Aymerich, S., *et al.* (2012). Condition-dependent transcriptome reveals high-level regulatory architecture in *Bacillus subtilis*. Science *335*, 1103–1106.

Noirot, P., and Noirot-Gros, M. (2004). Protein interaction networks in bacteria. Curr. Opin. Microbiol. *7*, 505–512.

Paredes, C., Alsaker, K., and Papoutsakis, E. (2005). A comparative genomic view of clostridial sporulation and physiology. Nat. Rev. Microbiol. 3, 969–978.

Piggot, P., and Hilbert, D. (2004). Sporulation of *Bacillus subtilis*. Curr. Opin. Microbiol. 7, 579–586.

Price, M., Dehal, P., and Arkin, A. (2007). Orthologous transcription factors in bacteria have different functions and regulate different genes. PLoS Comput Biol 3, 1739–1750.

Rasmussen, S., Nielsen, H., and Jarmer, H. (2009). The transcriptionally active regions in the genome of *Bacillus subtilis*. Mol 73 Microbiol 1043–1057.

Shen-Orr, S., Milo, R., Mangan, S., and Alon, U. (2002). Network motifs in the transcriptional regulation network of *Escherichia coli*. Nat. Genet. *31*, 64–68.

Sierro, N., Makita, Y., de Hoon, M., and Nakai, K. (2008). DBTBS: a database of transcriptional regulation in *Bacillus subtilis* containing upstream intergenic conservation information. Nucleic Acids Res. *36*, D93–96.

Sonenshein, A.L., Hoch, J.A., and Losick, R. (2001). *Bacillus Subtilis* and Its Closest Relatives: From Genes to Cells. (Amer Society for Microbiology).

Teichmann, S., and Babu, M. (2004). Gene regulatory network growth by duplication. Nat. Genet. *36*, 492–496.

Varughese, K., Tsigelny, I., and Zhao, H. (2006). The crystal structure of beryllofluoride SpoOF in complex with the phosphotransferase SpoOB represents a phosphotransfer pretransition state. J. Bacteriol. *188*, 4970–4977.

Völker, U., and Hecker, M. (2005). From genomics via proteomics to cellular physiology of the Gram-positive model organism *Bacillus subtilis*. Cell 7 Microbiol 1077–1085.

Wang, S., Setlow, B., Conlon, E., Lyon, J., Imamura, D., Sato, T., Setlow, P., Losick, R., and Eichenberger, P. (2006). The forespore line of gene expression in *Bacillus subtilis*. J Mol Biol, 358, 16–37.

Wu, D., Hugenholtz, P., Mavromatis, K., Pukall, R., Dalin, E., Ivanova, N., Kunin, V., Goodwin, L., Wu, M., Tindall, B., *et al.* (2009). A phylogeny-driven genomic encyclopaedia of Bacteria and Archaea. Nature *462*, 1056–1060.

Zhao, H., Msadek, T., Zapf, J., Madhusudan, Hoch, J., and Varughese, K. (2002). DNA complexed structure of the key transcription factor initiating development in sporulating bacteria. *10* Structure 1041–1050.

Helicobacter pylori Transcriptional Network

11

Alberto Danielli and Vincenzo Scarlato

Abstract

The human gastric pathogen *Helicobacter pylori* appears to enroll only 17 transcriptional regulators to transduce environmental signals into coordinated output expression of the genome. We show that the low number of transcriptional regulators, together with the large body of molecular tools, makes *H. pylori* an appealing model organism to characterize transcriptional network structures involved in virulence regulation and host–pathogen interactions.

In particular we provide evidence that the regulators are wired in a shallow transcriptional regulatory network (TRN), which orchestrates the key physiological responses needed to colonise the gastric niche: heat and stress response, motility and chemotaxis, acid acclimation and metal ion homeostasis. Interestingly, long regulatory cascades are absent, and rather than having a plethora of specialized regulators, the TRN of *H. pylori* appears to transduce separate environmental inputs by using different combinations of a small set of regulators. It is not tailored to adapt to many environmental stimuli, and is apparently unable to react to metabolic signals encountered outside of the gastric niche. On the other hand, the predominance of negative regulatory interactions suggests that this architecture of the TRN evolved to quickly respond to changing conditions in the gastric niche in order to maintain homeostasis. Metal-responsive regulators such as NikR and Fur appear to have a very important role in this TRN, forming a central regulatory hub, with regulatory interaction feeding into all other sub-network circuits.

Basic biology

Helicobacter pylori is a Gram-negative, flagellated, microaerophilic and spirally shaped bacterium, belonging to the class of γ-proteobacteria, which colonizes the human gastric niche. Pioneering work by Robin Warren and Barry Marshall (Marshall and Warren, 1983) enabled *H. pylori* infections to be correlated with inflammation of the gastric epithelium and gastritis in over 50% of the human population, worldwide. *H. pylori* is a very successful pathogen, and in untreated infections it can persist in the host for the entire duration of life (de Reuse and Bereswill, 2007). The bacterium can even lurk asymptomatically for years, contributing in the long term to the insurgence of more severe diseases such as peptic ulcer, dietary allergies and malignant gastric tumours, ranking this microorganism a class 1 carcinogen (Parsonnet *et al.*, 1994; Bouvard *et al.*, 2009).

To establish infection and survive in the stomach, *H. pylori* must traverse the gastric mucous layer and gain close contact with the epithelial cells of the stomach (Scott *et al.*, 2007), tolerating wide fluctuations of the stomach pH (Sachs *et al.*, 2003). The regulated expression of a dedicated set of acid acclimation genes permits to keep acidity of the bacterial cell close to neutrality. Metallo-enzymes and metal ion sensors are involved in this process and also participate to other essential functions such as respiration and detoxification (Bury-Mone *et al.*, 2004; van Vliet *et al.*, 2004b). For example, a correct metal ion homeostasis is at the basis of the activation mechanism of the Ni^{2+}-dependent urease and the [Ni^{2+}–Fe^{2+}] hydrogenase enzymes (Mehta *et al.*, 2003). These are central players in the infectious process: urease allows buffering of the acidic environment by conversion of urea to ammonium and bicarbonate (Tsuda *et al.*, 1994; Scott *et al.*, 2002), while hydrogenase allows infection through the energy-yielding consumption of hydrogen (Maier *et al.*, 1996, Olson and Maier, 2002). In addition, deletion of most flagellar and motility genes severely impairs the ability of the bacterium to colonize the gastric niche, resulting in attenuated infections (Eaton *et al.*, 1996; Josenhans and Suerbaum, 2002), possibly because of failure to counter gastric shedding forces, or move in response to favourable or noxious gradients (Schreiber *et al.*, 2004).

H. pylori as model for pathogen TRNs

The infectious capacity of *H. pylori* is dependent on the coordinated expression in time and space of virulence factors and housekeeping genes, which enable to withstand the stresses imposed by the harsh acidic environment and to counteract host's responses. These signals are processed into output expression of the genome by the transcriptional regulatory network (TRN) of the bacterium (Fig. 11.1), which, in analogy to other bacteria controls the global expression of the genetic repertoire in response to changes in the environment (Balazsi and Oltvai, 2005; Seshasayee *et al.*, 2006). Prokaryotic TRNs are organized as multilayered, hierarchical structures (Babu *et al.*, 2009), characterized by local subnetworks, or regulatory modules, which contribute to respond with particular response dynamics to an environmental signal. Specifically, origons have been described as local subnetwork modules of the TRN that originate at a specific class of sensory transcription factor (TF), which roots the input node of the origon and is responsible for the transduction a distinct intracellular or extracellular signal (Balazsi *et al.*, 2005). Network motifs, on the other hand, represent specific patterns of interconnection between regulators and target genes (Alon, 2007). Based on their design they can be classified in different types, implementing distinctive circuit logic to respond to diverse biological stimuli.

With the plethora of systemic and molecular data available and the effortlessness of genetics, *Escherichia coli* served as model organism to describe the first detailed prokaryotic TRN to date (Balaji *et al.*, 2007; Salgado *et al.*, 2012), while TRNs of pathogenic bacterial species have been described with less detail. In addition, many bacteria employ hundreds of TFs, making difficult the global dissection of the TRN. On the contrary, *H. pylori* has an unusual low number of regulators (Scarlato *et al.*, 2001), many of which have been described both at the molecular and biochemical level. In addition, transcriptome and ChIP-chip analyses are available for several regulators. The available molecular and '-omic' data describe with different degrees of detail nearly 80% of the annotated regulators, a value that goes beyond that of the model organism *E. coli*, in which only 50% of regulators have an assigned function (Seshasayee *et al.*, 2006). *H. pylori*, therefore, represents a very appealing species to study in general how bacterial TRNs are assembled, and more in specific how a TRN for a human pathogen is wired up.

Association studies with *H. pylori* have been brilliantly exploited to study human population dynamics (Falush *et al.*, 2003; Linz *et al.*, 2007). This reflects a very stringent dependence of *H. pylori* with the host. Accordingly, it is accountable to expect that *H. pylori* lacks a regulatory apparatus capable to respond to stresses encountered outside of the stomach (Babu *et al.*, 2006). On the other hand, given its success as a life-lasting pathogen, the regulatory apparatus can be considered as paradigmatic example of TRN tailored to guarantee prolonged infection in the host (Danielli *et al.*, 2010). Because the wiring architecture of regulatory motifs encompassing orthologous TFs does not evolve rigidly, but appears to be conserved in distantly related species if they share similar living environments (Babu *et al.*, 2007), identification of specific features of the transcriptional network of *H. pylori* may provide findings relevant for the dissection of transcriptional networks of other infectious agents. The purpose of the following paragraphs is to illustrate known regulatory interactions that characterize the TRN of *H. pylori* and to demonstrate how in this bacterium the need for coordinated responses is achieved by extensive tinkering and horizontal wiring of few key transcription factors.

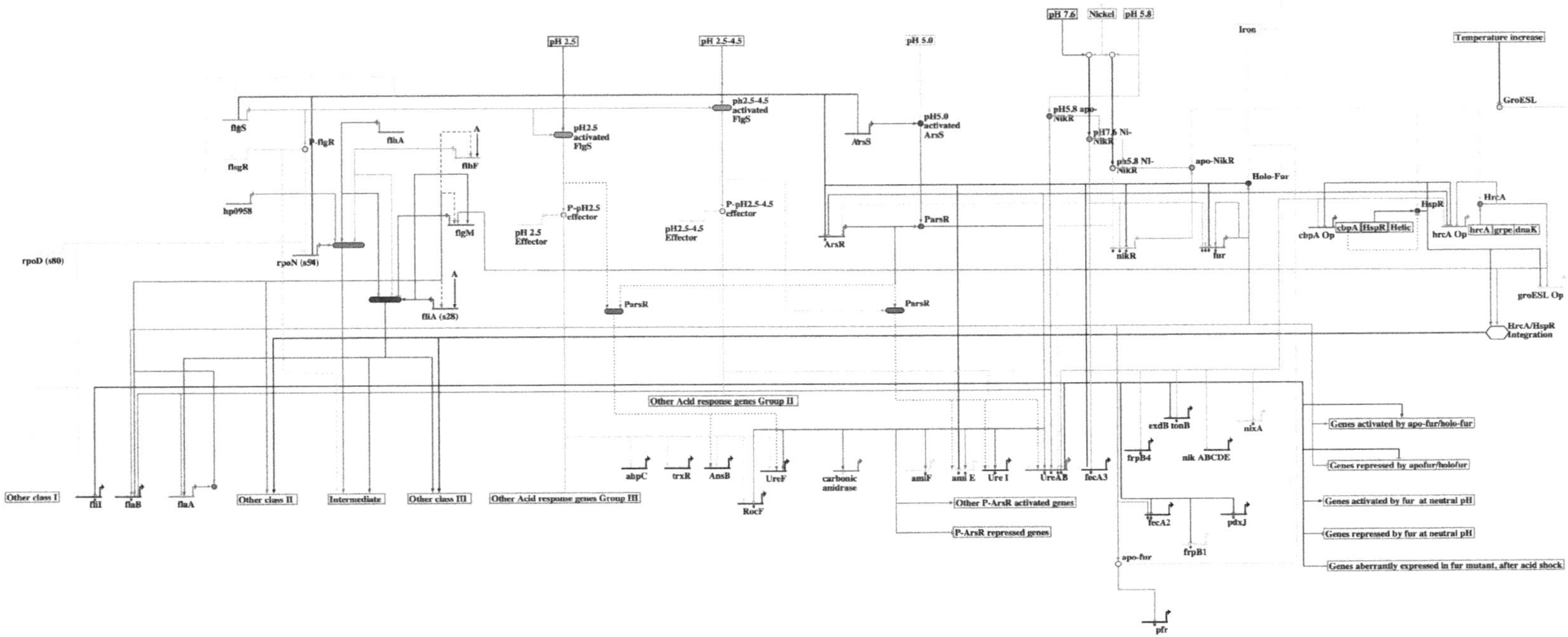

Figure 11.1 The *H. pylori* TRN. Overall view of the regulatory connections in the TRN, modelled using BioTapestry (Longabaugh *et al*., 2009; Danielli *et al*., 2010). Three layers are recognizable: a top layer containing environmental signals; a middle layer, with TFs and associated proteins; a bottom layer with effector genes. Question marks are used for uncertain regulatory connections. Dashed lines are used when interactions are inferred but not formally demonstrated. Each gene is indicated with a line topped by a bent arrow, representing the transcriptional start site (TSS). Regulatory connections can be positive (arrows), negative (barred lines), or neutral (simple lines, used when the result of the interaction is unknown). Direct TF–DNA interactions are marked by a dot. Proteins are represented with ovals and a line emerging from an oval symbolizes the interaction between a specific protein and its target(s). Protein–protein interactions are represented as lines emerging from a TSS and connecting to a specific oval. Some genes are represented as separated entities even if belonging to the same operon, while in the case of the *hrcA* and *cbpA* operons, ORFs belonging to the same operon are represented as boxes downstream of a common TSS.

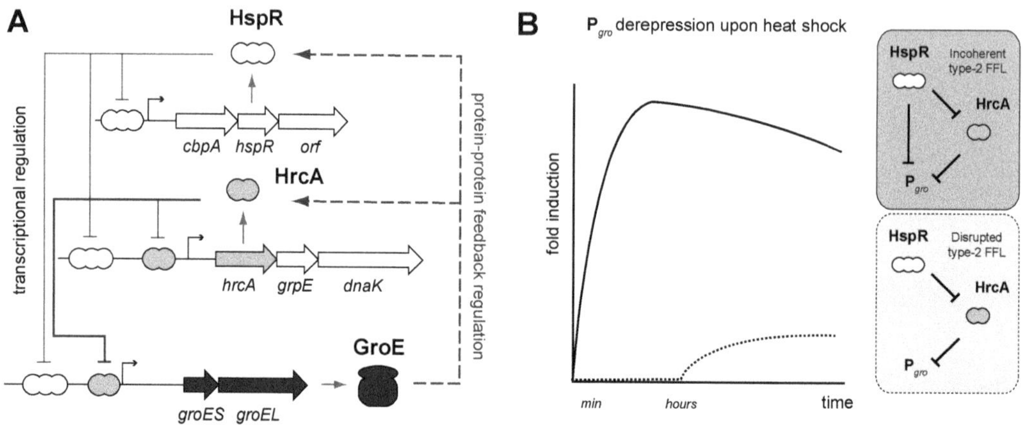

Figure 11.2 The heat shock origon. (A) Schematic drawing of the regulatory interactions that control expression of the three main heat shock operons in *H. pylori*. Protein-protein feedback regulation mediated by the GroE system is indicated by a dashed line. Arrowheads represent positive effects, hammerheads negative regulation. (B) The *H. pylori* heat shock response circuit conforms to an incoherent type-2 FFL motif with AND logic, as both HrcA and HspR can respond to the same stress signals. In addition, because GroESL has a positive effect on the DNA-binding activity of the regulators, the circuit is modulated under steady-state conditions by chaperone concentration feedback. Incoherent FFLs significantly accelerate the response kinetics of a simple regulation cascade at the same steady state input levels (Mangan and Alon, 2003; Mangan *et al*., 2006). Accordingly, in wild-type *H. pylori* strains the P_{gro} promoter is rapidly derepressed (< 120 s) upon heat shock (Spohn *et al*., 2002), while in a P_{gro} mutant lacking the HspR binding site, transcriptional derepression is observed only 60 minutes after temperature up-shift (Spohn *et al*., 2004), as expected for a type-2 FFL motif.

The *H. pylori* transcriptional apparatus

H. pylori is characterized by a reduced genome size (~1600kb), and the genome sequences of four clinical isolates confirm a conserved and limited repertoire of TFs and regulators (Tomb *et al.*, 1997; Alm *et al.*, 1999; Oh *et al.*, 2006; Baltrus *et al.*, 2009). The reduced number of regulators is a landmark of small parasitic genomes, where transcription factors have been lost owing to absence of selective pressure (Babu *et al.*, 2006), and well reflects the reductive evolution of this pathogen, which is constrained to the gastric habitat, where the presence of other competitive microorganisms is virtually absent (Josenhans *et al.*, 2007). The *H. pylori* transcriptional apparatus enrols only 17 bona fide TFs, which transduce the key physiological responses needed to colonize the gastric niche: (1) heat and stress response, (2) motility and chemotaxis, (3) acid acclimation and (4) metal ion homeostasis (Table 11.1). The list includes three sigma factors, including the vegetative σ^{80} and two alternative sigma factors, σ^{54} and σ^{28}, which are involved in regulation of flagellar genes, four complete two-component systems [ArsRS for acid acclimation, CheAY/Y2 (Foynes *et al.*, 2000) and FlgRS for motility and flagellar gene regulation, and CrdRS involved in copper resistance], two orphan response regulators HP1043 and HP1021 with unknown function, and four transcriptional regulators HspR/HrcA and Fur/NikR, respectively involved in stress response and metal ion homeostasis. The number of regulatory targets varies consistently between regulators (Table 11.1). Fur and NikR are pleiotropic regulators targeting many genes belonging to different functional categories. Also ArsR appears to have a moderately extended regulon (Pflock *et al.* 2006a; Wen *et al.*, 2006), while other regulators such as HspR, HrcA, FliA, RpoN, have a more restricted set of target genes, with functionally similar roles (Niehus *et al.*, 2004; Roncarati *et al.*, 2007a; Danielli *et al.*, 2010).

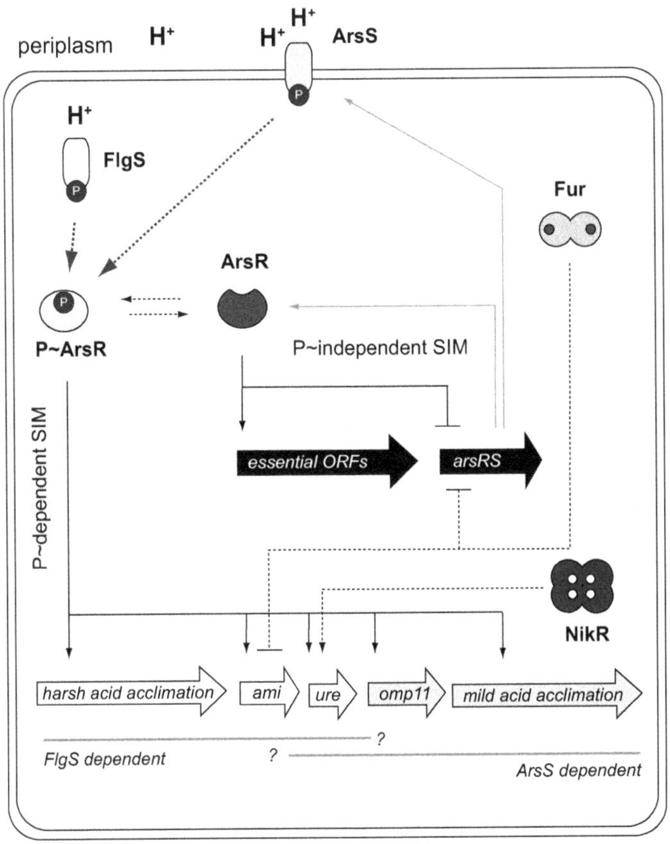

Figure 11.3 The acid acclimation origon. ArsR roots the sensing of pH-responsive genes through a bipartite regulon, controlling a different set of target genes through two separate SIM motifs, according to its phosphorylation status. Target genes can be subdivided in three main classes according to the sensory kinase that activates ArsR. A first cluster of genes encompasses P~ArsR-dependent target operons regulated in a phosphorylation-dependent manner, upon mild acidification of the periplasm through ArsS signalling (grey block arrows; *omp11*, *carbonic anhydrase*, *ureAB*, *hypA*, and others). A second cluster of genes contains target operons that are regulated by more harsh acidic conditions, triggering also intracellular cytoplasmic pH decrease to different extents and therefore likely overlapping with the ArsS-dependent targets (white block arrows). Their regulation is P~ArsR-dependent and phosphorylation of the regulator similarly promotes DNA binding to their promoters. However, they are not deregulated in *arsS* deletion mutants, and may therefore rely on a different (cytoplasmic) acid signal transducer to promote P~transfer needed to activate ArsR, possibly the aforementioned FlgS histidine kinase. The third group includes genes that are targeted by unphosphorylated ArsR (*arsRS* operon, and others), and whose regulation is not necessarily pH-dependent. The latter group contains genes of unknown function that appear to be essential for viability (black block arrows).

As described in next paragraphs, it appears that different master regulators are required to root the origons involved in responses central for gastric infection. Within origons, different network motifs are evidenced, which account for distinct regulatory logic of the circuits.

Origons required for the colonization

The heat shock and stress response origon

The heat shock origon of *H. pylori* is responsible for sensing stress signals, such as misfolded proteins generating from heat shock and osmotic

Table 11.1 Transcription factors and regulators in *H. pylori*

TF	Gene	Description	Function	Origon	Motif	Target genes	Target TFs
σ^{80}	*rpoD* (*HP0088*)	Vegetative sigma factor	housekeeping	–	–	~1600	all except *fliA?* (upstream RpoN consensus)
σ^{54}	*rpoN* (*HP0714*)	Alternative sigma factor	Flagellar regulation *Class ii* genes Basal body, hook	FLAG	SIM	26/26	*fliA* Based on upstream RpoN consensus sequence
σ^{28}	*fliA* (*HP1032*)	Alternative sigma factor	Flagellar regulation *Class iii* genes Late structures	FLAG	SIM	6/6	–
FlgR	*flgR* (*HP0703*)	NtrC-like response regulator	Flagellar regulation RpoN-dependent regulation	FLAG	SIM	21/32	*fliA* Based on upstream RpoN consensus sequence
FlgS	*flgS* (*atoS*) (*HP0244*)	NtrB-like cytoplasmic histidine kinase	Flagellar regulation Acid acclimation	FLAG	P~	-/123	P~FlgR P~ArsR (based on genetic evidence)
CheA	*cheA* (*HP0392*)	Histidine kinase	Chemotaxis	FLAG		NA	P~CheY P~CheY2
CheY	*cheY* (*HP1067*)	Response regulator	Chemotaxis	FLAG	P~	NA	–
CheY2	(*HP0392*)	CheY-like receiver domain fused to CheA	Chemotaxis P~sink	FLAG	P~	NA	–
ArsR	*arsR* (*HP0166*)	ompR-like response regulator	Acid acclimation	pH	SIM	60/63	*arsR* Essential gene
ArsS	*arsS* (*HP0165*)	Transmembrane histidine kinase	Acid acclimation Periplasmic acid sensor	pH	P~	–/142	P~ArsR
HP1021	(*HP1021*)	Atypical orphan response regulator	Acetone metabolism	NA	NA	NA	NA Essential gene
HP1043	(*HP1043*)	Atypical orphan response regulator	Growth	NA	NA	NA	*HP1043* Essential gene
CrdR	*crdR* (*HP1364*)	Response regulator	Copper resistance	NA	NA	NA	NA
CrdS	*crdS* (*HP1365*)	Histidine kinase	Copper resistance	NA	P~		P~CrdR
Fur	*Fur* (*HP1027*)	Ferric uptake regulator	Pleiotropic Metal ion homeostasis	METAL	BM MCL FFL	80/191	*fur*, *nikR*, *arsR*, *rpoN**, *flgR** *flgS**, *hrcA**
NikR	*nikR* (*HP1338*)	*E. coli* NikR homologue	Pleiotropic Nickel ion homeostasis	METAL	BM MCL FFL	15/45	*nikR*, *fur*, *hrcA*, *hspR*
HrcA	*hrcA* (*HP0111*)	*B. subtilis hrcA* repressor homologue Membrane associated	Heat shock Stress response	HS	FFL	5/36	*hrcA*
HspR	*hspR* (*HP1025*)	*Streptomyces* spp. *hspR* repressor homologue	Heat shock Stress response	HS	FFL	8/24	*hspR*, *hrcA*

TF, transcription factor; motif, type of local network structure or interconnection in the main origon; target genes: number of genes belonging respectively to the regulon (direct regulation), or the transcriptome (indirect regulation) of the TF (sourced from Danielli *et al.*, 2010); SIM, single input motif; BM, bifan input motif; MCL, multicomponent loop; FFL, feedforward loop; P~, phospho-transfer activation. FLAG, flagellar biosynthesis and motility origon; pH, acid acclimation origon; METAL, metal-ion homeostasis origon; HS, heat shock and stress response origon. *Evidence gathered from ChIP-chip experiments. Note that many TFs, especially repressors, are autoregulated, and that Fur and NikR are wired in a multicomponent feedback loop (MCL), which is rarely found in bacteria.

stresses. While many Gram-negative bacteria employ specialized sigma factors (σ^{32}) to activate the transcriptional responses upon heat shock, *H. pylori* has evolved an opposite strategy, which resembles the heat shock regulation of Gram-positive bacteria (Naberhaus, 1999): it implements two repressors with homology respectively to *Bacillus subtilis* HrcA (Schulz and Schumann, 1996) and *Streptomyces* spp. HspR (Servant and Mazodier, 2001). HspR and HrcA regulate three main heat shock operons including the *groESL* chaperone operon (Fig. 11.2A). All three operons are responsive to heat shock and are induced by the presence of misfolded proteins or stress signals (Homuth *et al.*, 2000; Spohn *et al.*, 2002).

HspR represents the master regulator of the origon and represses all three operons, thereby autoregulating negatively its own synthesis and controlling the expression of the intermediary regulator HrcA (Spohn and Scarlato, 1999b). On the contrary, HrcA represses only its own *hrc* operon and the *groESL* operon (Roncarati *et al.*, 2007a). The regulation of the latter requires both HrcA and HspR in a dual repression mechanism: both TFs are required for correct repression of the *hrc* and *gro* operons (Spohn *et al.*, 2004) (Fig. 11.2A). Interestingly, the DNA-binding activity of both repressors is enhanced by the product of the *groESL* target genes, indicating a positive protein-protein feedback loop exerted by the GroE chaperonin system on HrcA and HspR (Roncarati *et al.*, 2007a). This renders the mechanism similar to the *B. subtilis* 'titration model', which postulates that upon heat shock, GroESL is titrated away by misfolded polypeptides, thereby relieving HrcA repression (Mogk *et al.*, 1997; Reischl *et al.*, 2002).

Despite the functional interdependency, and flanking binding sites in the promoters of co-regulated genes, the binding of HrcA and HspR to DNA is not cooperative, making unclear the requirement of two repressors. A coherent explanation can be drawn from the network motif found in the heat shock origon. In fact, TFs and genes targets are wired in a shining example of an incoherent type-2 feedforward loop (FFL), which greatly accelerates the response kinetics of GroESL expression in response to heat shock (Fig. 11.2B). This has very important implications, as the altered homeostasis of chaperones appears to indirectly influence virulence gene expression as well as the assembly/expression of the flagellar apparatus (Roncarati *et al.*, 2007a). Incoherent type-2 FFLs are rare motifs (Mangan and Alon, 2003), thus the heat shock FFL motif of *H. pylori* may reflect evolutionary adaptation of the bacterium for prolonged infections in the host. In this light, it is interesting to recall that a similar dual heat shock regulation mode has been reported in *Staphylococcus aureus* (CstR/HrcA) (Chastanet *et al.*, 2003).

In conclusion, the heat shock response of *H. pylori* is regulated by a plethora of feedforward and feedback mechanisms, including the autoregulation of both HspR and HrcA, the positive feedback of GroESL on their DNA-binding activity, and membrane-association of HrcA (Roncarati *et al.*, 2007b), which may confer additional sensory potential to the circuit.

The flagellar biosynthesis origon

The genome of *H. pylori* encompasses ~40 unclustered motility genes, often scattered within multicistronic housekeeping gene operons (Tomb *et al.*, 1997; Niehus *et al.*, 2004). In analogy to other bacteria, flagellar genes are typically positively regulated and hierarchically organized in three main classes according to their activating TF (Josenhans and Suerbaum, 2002; Macnab, 2003): *class I* genes are transcribed by the vegetative σ^{80} RNAP, and comprise mostly flagellar regulatory genes (such as *flgR*, *flgS*, *rpoN*, as well as *flhA*); *class II* includes specific targets of the alternative σ^{54} factor (middle expression), encoding components of the flagellar basal body and hook, including *flaB* flagellin and *fliA*, encoding the alternative σ^{28} factor; finally, *class III* genes encode late flagellar structures, such as *flaA* flagellin, transcribed by the σ^{28}-containing RNAP. The flagellar biosynthesis module is a rather complex origon employing a remarkable fraction of dedicated TFs (7 out of 17), and including post-transcriptional as well as pleiotropic regulatory effects, such as polarity and DNA supercoiling (Spohn and Scarlato, 1999a; Ye *et al.*, 2007).

A master regulator with homology to the enterobacterial FlhDC system is absent. Conversely, the flagellar biosynthesis origon adopts a short σ regulatory cascade ($\sigma^{80} > \sigma^{54} > \sigma^{28}$) (Niehus *et al.*, 2004), in which each σ-factor feeds into the downstream flagellar gene class through a Single Input Motif (SIM) (Danielli *et al.*, 2010). SIMs frequently occur in systems expressing gene products that form assemblies with controlled stoichiometry (Shen-Orr *et al.*, 2002), and exhibit time-shifted dynamics (Yu *et al.*, 2003; Zaslaver *et al.*, 2004). The 3-noded σ-regulatory chain of SIMs thus suggests sequential expression of early, middle, and late flagellar components. The role of FlgR, a NtrC-like response regulator belonging to the FlgRS two component systems, appears to be that of enhancer dependent activator of σ^{54} transcription (Spohn and Scarlato, 1999a; Pereira *et al.*, 2006). A putative type-1 coherent FFL with OR logic may contribute to sustain flagellar gene expression in the presence of discontinuous triggering signals (Mangan *et al.*, 2003; Mangan *et al.*, 2006), paralleling observations made in *E. coli* (Kalir *et al.*, 2005). In fact, several *intermediate class* genes, under the transcriptional control of the σ^{54}-containing RNAP, may on the appearance of FliA (σ^{28}) also be transcribed by the σ^{28}-containing RNAP (Niehus *et al.*, 2004). However, formal demonstration of the dynamic responses governed by such a putative FFL motif has yet to be provided.

In addition to TF–DNA interactions controlling transcription, several accessory proteins modulate the output of the flagellar circuit through post-transcriptional feedback mechanisms. FlgM, transcribed as *class* 2 gene, encodes an anti-sigma factor involved in FliA feedback inhibition of *class* 3 gene transcription (e.g. *flaA*) (Colland *et al.*, 2001; Josenhans *et al.*, 2002; Rust *et al.*, 2009), while components of the basal body, FlhA and FlhF, positively modulate transcription of both *class II* (middle) and *class III* (late) flagellar genes (Niehus *et al.*, 2004). Finally, the HP0958 protein described as a RpoN chaperone (Pereira and Hoover, 2005), has been also associated with the decay of the *flaA* RNA transcript (Douillard *et al.*, 2008). Despite the complexity of the flagellar origon, it is evident that, protein–protein interactions of accessory factors with TFs are fundamental to modulate the output of circuit, in analogy with the HspR-HrcA origon, controlled by GroESL protein level feedback.

The acid acclimation origon

The subnetwork controlling transcription of acid responsive genes is rooted by the autoregulated acid response regulator ArsR (Pflock *et al.*, 2005, Pflock *et al.*, 2006a). It is encoded by a multicistronic operon coding also for the cognate transmembrane ArsS histidine kinase. It has been proposed that ArsS senses acidification of the periplasm, through changes in protonation of histidine residues (pK_a ~6.0) encompassed in its extracytosolic sensory domain (Pflock *et al.*, 2004). This triggers autophosphorylation of ArsS and the transfer of a phosphoryl group to the receiver domain of ArsR, thereby promoting its DNA-binding activity towards a specific set of acid responsive promoters (Pflock *et al.*, 2005; Pflock *et al.*, 2006a; Wen *et al.*, 2006).

Knockout mutants of *arsR* have not been obtained, possibly because the protein is essential for growth of *H. pylori*. However, strains carrying a point mutation that disrupts only the ArsR phosphoacceptor site, as well as *arsS* deletion mutants, are viable (Forsyth *et al.*, 2002; Pflock *et al.*, 2004; Wen *et al.*, 2006). This genetic evidence suggests that ArsR regulates alternative gene targets according to its phosphorylation status. Consistently, protein-DNA binding studies provided evidence for DNA elements recognized by either phosphorylated as well as unphosphorylated ArsR (Wen *et al.*, 2006), some of which characterized by distinctive features, such as different extension of the footprint and nucleotide sequence (Dietz *et al.*, 2002). Additional complexity is provided by the position of the binding site in the target promoter, according to which ArsR can act as activator or repressor of transcription (Pflock *et al.*, 2006b). Finally, recent work identified FlgS, the NtrB-like cytoplasmic histidine kinase of the flagellar origon, as being involved in the response to cytosolic pH decrease, suggesting a possible and unusual interlink between the flagellar circuit with the acid acclimation module controlled by ArsR (Wen *et al.*, 2009).

Thus, although the bulk of transcriptome and DNA-binding analyses conducted at different pH conditions show a certain degree of variation, the available data substantiate the hypothesis of a bipartite ArsR regulon. According

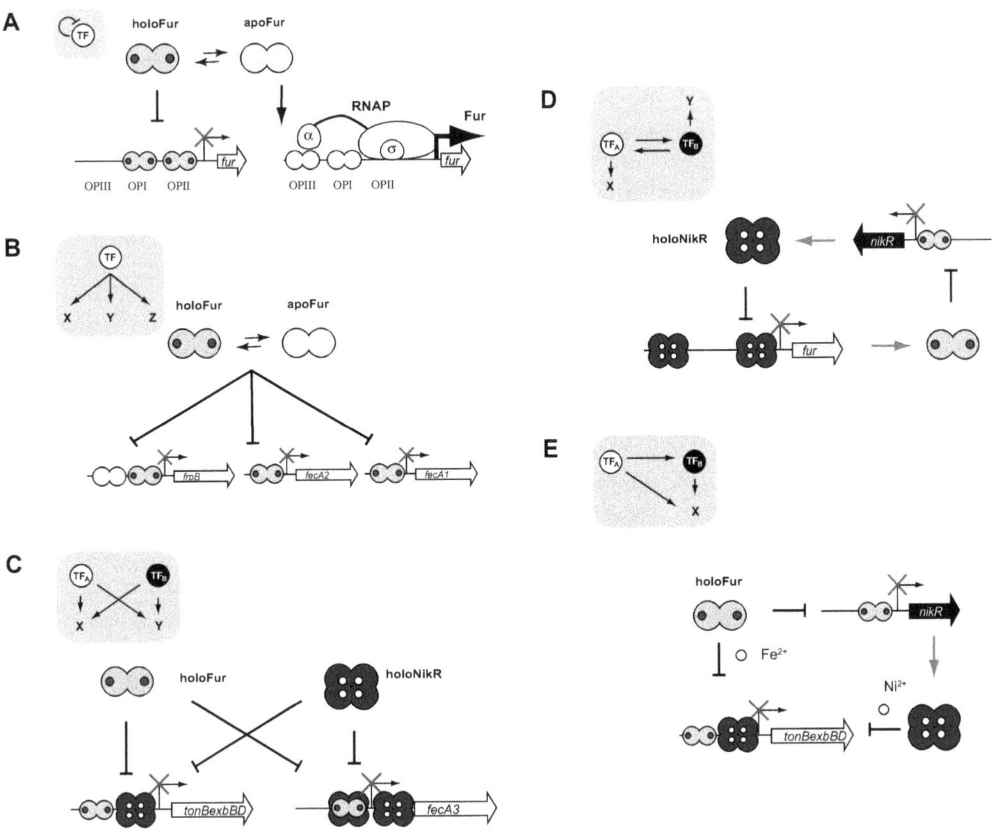

Figure 11.4 Network motifs in the metal ion homeostasis module. (A) Autoregulation; this motif frequently involves repressors of transcription, controlling multiple gene targets through a Single Input Motif (SIM). Both NikR and Fur are autoregulated. The latter involves binding to three operators, recognized with different affinities in the presence or absence of the iron co-factor. (B) SIMs originate from a single TF that feeds directly into set of target operons (X, Y, Z); they frequently occur in circuits expressing protein complexes or belonging to related metabolic pathways. (C) Bifan Motif (BM); multiple operons (X, Y) are regulated in concomitance by two transcription factors (TF_A, TF_B). These symmetrical motifs occur when multiple responses need to be triggered in answer to shared stimuli. For example the *exbBD-tonB* system needs to be expressed in iron replete conditions to foster metal-ion uptake mediated by *fecA* metal-dicitrate transporters. (D) Multicomponent loops (MCLs) are frequently found in developmental and differentiative programmes of eukaryotes, but to date rarely described in prokaryotic TRNs. Fur and NikR constitute a clear example. (E) Feed Forward Loops (FFL) are asymmetric three node motifs, occurring with high frequency in prokaryotic TRNs, where TFs feeding into the same target genes exhibit a different hierarchy (see also legend to Fig. 11.2). According to the net signs feeding into the target gene FFLs can be coherent (same sign) or incoherent (opposite sign), negative or positive. They enable the system to respond with different dynamics, at the same steady state of the triggering signal, thus conferring different response kinetics (Alon, 2007).

to the phosphorylation status, ArsR controls transcription of different sets of genes through two distinct SIMs (Fig. 11.3). Moreover, regulators belonging to other origons such as FlgS and the metallo-regulators Fur and NikR also intimately contribute to the regulation of acid acclimation genes.

The metal homeostasis origon

H. pylori infections have been associated with disorders in iron metabolism of the host, particularly in adolescents and pregnant women (Muhsen and Cohen, 2008). This is linked to the expression of haemolysins, metal-scavengers, transporters, siderophores, and other factors, which allow *H.*

pylori to compete with the host for essential metal ions (Carpenter *et al.*, 2009b). On the other hand, high intracellular concentrations of these ions are toxic. *H. pylori* implements three systems to control expression of genes encoding metal-trafficking proteins and metallo-enzymes: a homologue of the *E. coli* NikR regulator (Chivers and Sauer, 2000; Contreras *et al.*, 2003), the ferric uptake regulator Fur (Bereswill *et al.*, 1998; Bereswill *et al.*, 1999) and the CrdRS two-component system (Waidner *et al.*, 2005). Whereas the CrdRS system appears to be only involved in copper-resistance to date (Pflock *et al.*, 2007b), NikR and Fur constitute key virulence TFs of the network, as both *fur* and *nikR* mutants are impaired in competitive colonization experiments in animal models (Bury-Mone *et al.*, 2004, Gancz *et al.*, 2006). These regulators act as selective metal sensors (Giedroc and Arunkumar, 2007), correlating the intracellular concentration of a specific metal ion to the expression of genes involved not only in metal trafficking, but also in acid acclimation, stress response, and motility.

NikR is an extensively studied metal-dependent regulator, which controls the transcription of genes coding for Ni-enzymes and Ni-trafficking proteins in response to varying intracellular concentrations of Ni^{2+} (Abraham *et al.*, 2006; Dosanjh and Michel, 2006; Zambelli *et al.*, 2008). In *H. pylori*, NikR regulation is pleiotropic, controlling genes encoding nickel uptake and detoxification systems, virulence and acid acclimation factors such as urease, outer membrane proteins, and other regulators, including HrcA, HspR, and Fur (van Vliet *et al.*, 2002a; Contreras *et al.*, 2003). According to the position of the operator elements, NikR can activate or repress transcription in a Ni^{2+}-responsive fashion (Ernst *et al.*, 2005c). Binding to proximal operators, overlapping the transcriptional start site and/or the –10 box, represses the transcription of *fecA3* and *frpB4*, encoding metal transport systems (Davis *et al* 2006; Ernst *et al.*, 2006; Danielli *et al.*, 2009), the *nixA* permease gene (Wolfram *et al.*, 2006), and the *exbBD-tonB* system (Bury-Mone *et al.*, 2004; Ernst *et al.*, 2006), responsible for energizing the outer membrane transport complexes (Schaurer *et al.*, 2007) (Fig. 11.4C). Conversely, binding to elements upstream of the core promoter, appears to induce expression of the *ureAB* operon in response to Ni^{2+} (van Vliet *et al.*, 2002a; Bahlawane *et al.*, 2010) (Fig. 11.3). Finally, NikR is negatively autoregulated (Contreras *et al.*, 2003) and binds to the *fur* promoter, contributing to its transcriptional repression at physiological iron conditions (Delany *et al.*, 2005).

Fur, on the other hand, is involved in Fe^{2+} homeostasis, important to limit excessive concentrations of ferric ions, which may induce oxidative stresses through Haber–Weiss and Fenton reactions (Hantke, 2001). Fur target genes are transcriptionally repressed by iron, derepressed in *fur* mutants and in response to iron chelation, and bound by the Fur protein in the core promoter region. These regulatory targets encompass *fecA1*, *fecA2*, and *frpB1* genes involved in Fe^{2+} uptake (Delany *et al.*, 2001a; van Vliet *et al.*, 2002b; Danielli *et al.*, 2009) (Fig. 11.4B), as well as genes important for detoxification and redox control (Ernst *et al.*, 2005a; Ernst *et al.*, 2005b; Danielli *et al.*, 2006; Alamuri *et al.*, 2006; Carpenter *et al.*, 2009a). Their regulation conforms to the classic Fur repression paradigm, which postulates Fe^{2+} as co-repressor (de Lorenzo *et al.*, 1987). Oddly, in *H. pylori* iron appears also to induce transcription of Fur repressed genes. In fact, Fur directly regulates *pfr* (Bereswill *et al.*, 2000) and *sodB* (Ernst *et al.*, 2005b), encoding respectively a bacterio-ferritin-like protein and superoxide dismutase, which need to be promptly derepressed under Fe^{2+}-replete conditions. Instead as co-repressor, iron acts as inducer of transcription of these genes. This is accomplished by so called apo-operators, which are bound by Fur with higher affinity in the absence of iron ions (Delany *et al.*, 2001b; Carpenter *et al.*, 2009b). Thereby, the same steady state signal (presence or absence of intracellular Fe^{2+}) is processed by Fur into two opposite transcriptional outputs, one for apo-Fur and one for holo-Fur repressed genes. Operators recognized with different affinities by holo- and/or apo-Fur have also been shown to mediate the auto-repression mechanism of the *fur* gene itself (Delany *et al.*, 2002a, 2003) (Fig. 11.4A). In addition, Fur may also have a positive effect on transcription, as suggested by the existence of operons that are targeted by the protein and down-regulated in *fur* knockout strains (Ernst *et al.*, 2005a; Danielli *et al.*, 2006). These genes encode proteins important for chemotaxis and motility, interactions with the host, and redox equilibrium (*oorDABC*). The latter is an interesting case, as it codes for an essential ferredoxin-like protein complex (Hughes *et al.*, 1998), which may be involved in de novo acquisition of antibiotic resistance to nitrimidazoles (Gerritis *et al.*, 2006). Although Fur-dependent transcriptional activation has been reported in *Neisseria meningitidis* (Delany *et al.*, 2004), in *H. pylori* the postulated molecular mechanism of activation has not been elucidated yet, leaving the question open on how Fur can function as true transcriptional activator (Alamuri *et al.*, 2006).

Network motifs in the metal homeostasis regulatory module

While initial sensing of Ni^{2+} and Fe^{2+} is selectively transduced by respectively NikR and Fur, the output of their circuits is modulated by a goldmine of regulatory motifs and feedback control mechanisms. In addition to autoregulation (Fig. 11.4A), both TFs feed into their target genes via SIMs (Fig. 11.4B), in which multiple target operons are regulated at once by each regulator. However, it has been shown that a restricted number of targets genes is under the control of both Fur and NikR (Bury-Mone *et al.*, 2004; Delany *et al.*, 2005; Danielli *et al.*, 2006). This superimposes a symmetric bifan motif to the metal homeostasis origon (BM; Fig. 11.4C). Given the central role of metal ions, this is not surprising. Metal ions are in fact needed to cofactor distinct metallo-enzymes, but compete at the same time for similar chaperones and transporters, and may therefore trigger coordinated responses. In addition, NikR and Fur are also wired in a symmetrical multicomponent loop (MCL) (Fig. 11.4D), as both metalloregulators bind to each other promoter regions, contributing to regulate reciprocally their own transcription.

Because Fur can bind specific promoters also in the apo-form, and NikR appears to bind DNA also in the absence of Ni^{2+} under acidic conditions (Li and Zamble, 2009), the BM and MCL motifs in their subnetwork give the possibility to switch to a FFL motif as soon as the concentrations of a specific metal ion, or pH, reach threshold levels to cofactor the DNA binding activity of either regulator (Fig. 11.4E, example with *exbBD*). This grants a great flexibility to the regulatory response, which combines cis-elements, such as the coexistence and position of Fur and NikR operators, with the modulation of their DNA binding affinity in response to different input signals. In addition, the response includes feedback control of metal availability, tuned by metal-trafficking complexes whose expression is regulated by the same circuit.

Centrality of the metallo-regulators in the *H. pylori* TRN

The centrality of the metal ion homeostasis origon in the pathogenomics of *H. pylori* is reflected by elevated number of direct regulatory targets of this subnetwork (Table 11.1) and further strengthened by a multitude of direct regulatory links with TFs and target genes encompassed in all other origons (Fig. 11.5). In fact, NikR is directly involved in the up-regulation of urease and other putative acid acclimation genes in response to acidic pH (van Vliet *et al.*, 2002a, 2004a, 2004b). Similarly to NikR, *fur* deletion mutants are impaired in acid acclimation (Bijlsma *et al.*, 2002; Gancz *et al.*, 2006). Consistently, Fur regulates genes coding for amidase (*ami*), involved in acid acclimation (van Vliet *et al.*, 2003), and Fur targeting of the *arsRS* operon has been reported in ChIP-chip experiments (Danielli *et al.*, 2006), substantiating a hierarchically important role of Fur in the regulation of acid acclimation (Fig. 11.5).

Fur also appears to be involved in a switch in gene expression during the bacterial growth-phase. In fact, the growth-phase dependent regulation of iron-responsive genes (Merrell *et al.*, 2003; Danielli *et al.*, 2009) is likely to be directly linked to Fur, as the growth-phase-dependent regulation of the proteome is lost in *fur* knockout mutants (Choi *et al.*, 2009) and increased intracellular Fur protein concentrations have been observed in stationary growth phase (Danielli *et al.*, 2006).

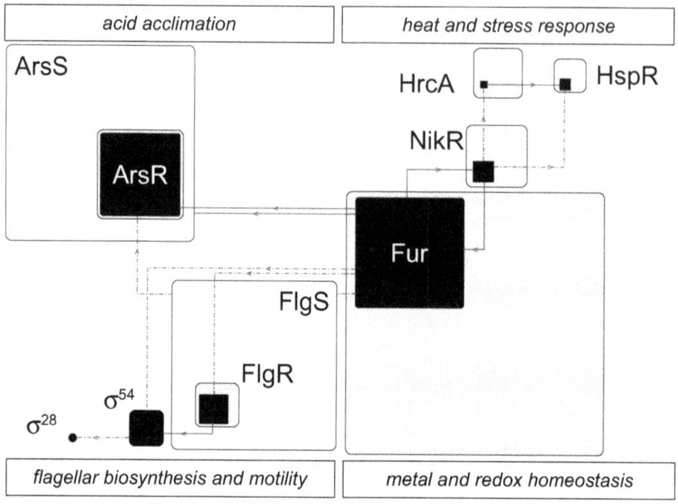

Figure 11.5 Centrality of metallo-regulators in the *H. pylori* TRN. Squares symbolize the number of known gene targets for each TF. Grey squares represent the extent of the transcriptome (genes de-regulated in knockout mutants of the regulator). Black squares represent the extent of the regulon, defined as the set of genes regulated via protein–DNA interaction at the promoter level. The size of each square is proportional to the number of target genes (see also Table 11.1). The extent of the transcriptome of ArsS and FlgS histidine kinases is depicted by white boxes. Text boxes represent gene categories dedicated to physiological responses important for *H. pylori* colonization. The TFs mediating these responses are positioned next to each text box. Full lines emanating form the black boxes denote direct regulatory control. Dotted lines indicate indirect evidence gathered by ChIP-chip experiments or consensus motif searches, which need to be confirmed by EMSA or footprinting analyses. The CheAY/Y2 two component system involved in motility is not depicted. Note that direct regulatory connections emanate from the metal homeostasis module, directly or via intermediary regulators, to all origons and most annotated TFs. According to its central position, Fur can be considered as *H. pylori* regulatory hub or global regulator.

The metal homeostasis origon is also linked to the heat shock and stress response origon controlled by HspR and HrcA. In fact, several operons coding for heat shock and stress response factors are deregulated in *nikR* deletion mutants (Contreras *et al.*, 2003). Interestingly, a search for NikR operators using a recently identified TRWYA-n15-TRWYA consensus motif (Stoof *et al.*, 2010), pinpoints two binding sites upstream of the *hspR* and *hrcA* encoding operons, suggesting that their transcription may be directly subjected to NikR control. Finally, both Fur and NikR have been associated to the control of flagellar biosynthesis and motility, through ChIP-chip and transcriptome analyses (Ernst *et al.*, 2005a, Danielli *et al.*, 2006). This association certainly deserves further investigation, as it may also involve Fur-dependent regulation of flagellar regulators such as RpoN, FlgS, FlgR (Danielli *et al.*, 2006).

A shallow TRN wired to maintain homeostasis

Together the data provide evidence for extensive intra- and inter-origon wiring between TFs, generating a shallow network, with low level of hierarchy and extensive horizontal connections. Feedback to this circuit is provided through interaction of modulatory proteins with TFs and cofactor availability. With the exception of the flagellar regulatory chain, long regulatory cascades of sequentially activated TFs, frequently associated with bacterial differentiation pathways (e.g. biofilm formation and sporulation) (Martinez-Antonio *et al.*, 2008), are absent. Conversely, separate environmental inputs are often processed by different combinations of small sets of TFs or associated proteins. For example, the flagellar FlgS histidine kinase may foster the crosstalk between signal transduction pathways that control motility and acid acclimation in the network. Fur and

NikR participate clearly as pleiotropic regulators with comprehensive connections to all origons through direct control or intermediary regulators (Fig. 11.5). Given the absence of Crp, Fnr, H-NS and Fis global regulators, Fur appears as a particularly interesting case as it may function as regulatory hub, with hundreds of genomic binding sites (Danielli *et al.*, 2006), many of which have not been linked to a direct regulatory function yet. Another peculiar feature of this network is the reduced number of response regulators and the widespread use of negative regulatory interactions that may confer the possibility to respond with fast response kinetics upon a specific triggering signal. The shallow architecture organization of the *H. pylori* TRN explicitly points to a network wired up to maintain homeostasis for prolonged life-lasting infections of the stomach: built to lurk (Danielli *et al.*, 2010). Owing to the paucity of sensor TFs the network is not tailored to adapt to many environmental stimuli, and apparently not flexible to react to metabolic signals encountered outside of the gastric niche. Possibly, the shallow network design has emerged by reductive evolution, selecting for transcriptional interactions needed to respond only to gastric inputs.

Certainly, important aspects of the TRN remain to be elucidated. The proposed consensus binding motifs remain to be validated for many *H. pylori* TFs. The responses to oxidative and osmotic stresses are just beginning to be understood (Wang *et al.*, 2006), and essential orphan response regulators (Delany *et al.*, 2002b; Pflock *et al.*, 2007a; Müller *et al.*, 2007), for which knockout mutant strains are unavailable (HP1043 and HP1021), may contribute substantially to tune the network. In addition, the role of antisense regulation and small RNAs are just beginning to emerge (Xiao *et al.*, 2009a,b; Sharma *et al.*, 2010; Wen *et al.*, 2011), and will possibly reserve interesting revelations in the near future. For example, in the human pathogen *Mycoplasma pneumoniae* the extensive use of non-coding RNAs permits to mount complex regulatory responses despite the low number (10) of regulators (Güell *et al.*, 2009). This adds to the concept that, rather than genetic model systems with many regulators, infectious agents with simple genomes may be excellent model organisms in which to attempt dissection of pathogenic regulatory networks at the systems level.

In this light *H. pylori* represents an ideal organism as the low number of TFs permits to combine systems biology approaches with indispensable in depth molecular analyses for a remarkably wide fraction of network regulators.

Chapter highlights

- The regulatory network of *Helicobacter pylori* lists a remarkably reduced number of transcriptional regulators, organized in four main origons.
- This network exhibits a shallow design, characterized by extensive horizontal wiring between regulators and by the absence of long vertical regulatory cascades.
- Two metallo-regulators form a central regulatory hub, with direct regulatory connections emanating to all origons and to most other transcription factors.
- Frequent occurrence of feedback control, in conjunction with the prevailing use of negative regulations over positive ones, clearly indicates that the network is wired to maintain homeostasis.
- The limited number of regulators and their shallow horizontal wiring suggest that the circuit is poorly adapted to respond to specific signals encountered outside the gastric niche: this well reflects the successful reductive evolution of *H. pylori* as obligate human pathogen.

Acknowledgements

We thank present and past members of the lab for discussions and daily commitment. This work was partially supported by funding from the Italian Ministry for Research (PRIN) and Fondazione del Monte di Bologna (FDM) to A.D., and from the University of Bologna (ex60% and Strategic Project) to V.S.

References

Abraham, L.O., Li, Y., and Zamble, D.B. (2006). The metal- and DNA-binding activities of *Helicobacter pylori* NikR. J. Inorg. Biochem. *100*, 1005–1014.

Alamuri, P., Mehta, N., Burk, A., and Maier, R.J. (2006). Regulation of the *Helicobacter pylori* Fe–S cluster synthesis protein NifS by iron, oxidative stress conditions, and fur. J. Bacteriol. *188*, 5325–5330.

Alm, R.A., Ling, L.S., Moir, D.T., King, B.L., Brown, E.D., Doig, P.C., Smith, D.R., Noonan, B., Guild, B.C., deJonge, B.L., *et al.* (1999). Genomic-sequence comparison of two unrelated isolates of the human gastric pathogen *Helicobacter pylori.* Nature *397*, 176–180.

Alon, U. (2007). Network motifs, theory and experimental approaches. Nat. Rev. Genet. *8*, 450–461.

Babu, M.M., Balaji, S., and Aravind, L. (2007). General trends in the evolution of prokaryotic transcriptional regulatory networks. Genome Dyn. *3*, 66–80.

Babu, M.M., Lang, B., and Aravind, L. (2009). Methods to reconstruct and compare transcriptional regulatory networks. Methods. Mol. Biol. *541*, 163–180.

Babu, M.M., Teichmann, S.A., and Aravind, L. (2006). Evolutionary dynamics of prokaryotic transcriptional regulatory networks. J. Mol. Biol. *358*, 614–633.

Bahlawane, C., Dian, C., Muller, C., Round, A.,Fauquant, C., Schauer, K., de Reuse, H., Terradot, L., and Michaud-Soret, I. (2010). Structural and mechanistic insights into *Helicobacter pylori* NikR activation. Nucleic Acids Res. *38*, 3016–3018.

Balaji, S., Babu, M.M., and Aravind, L. (2007). Interplay between network structures, regulatory modes and sensing mechanisms of transcription factors in the transcriptional regulatory network of *E. coli.* J. Mol. Biol. *372*, 1108–1122.

Balazsi, G., and Oltvai, Z.N. (2005). Sensing your surroundings, how transcription–regulatory networks of the cell discern environmental signals. Sci. STKE *2005*, pe20.

Balazsi, G., Barabasi, A.L., and Oltvai, Z.N. (2005). Topological units of environmental signal Processing in the transcriptional regulatory network of *Escherichia coli.* Proc. Natl. Acad. Sci. U S A *102*, 7841–7846.

Baltrus, D.A., Amieva, M.R., Covacci, A., Lowe, T.M., Merrell, D.S., Ottemann, K.M., Stein, M., Salama, N.R., and Guillemin, K.J. (2009). The complete genome sequence of *Helicobacter pylori* strain G27. J. Bacteriol. *191*, 447–448.

Bereswill, S., Greiner, S., van Vliet, A.H., Waidner, B., Fassbinder, F., Schiltz, E., Kusters, J.G., and Kist, M. (2000). Regulation of ferritin-mediated cytoplasmic iron storage by the ferric uptake regulator homolog (Fur). of *Helicobacter pylori.* J. Bacteriol. *182*, 5948–5953.

Bereswill, S., Lichte, F., Greiner, S., Waidner, B., Fassbinder, F., and Kist, M. (1999). The ferric uptake regulator (Fur). homologue of *Helicobacter pylori*, functional analysis of the coding gene and controlled production of the recombinant protein in *Escherichia coli.* Med. Microbiol. Immunol (Berl). *188*, 31–40.

Bereswill, S., Lichte, F., Vey, T., Fassbinder, F., and Kist, M. (1998). Cloning and characterization of the fur gene from *Helicobacter pylori.* FEMS Microbiol. Lett. *159*, 193–200.

Bijlsma, J.J., Waidner, B., van Vliet, A.H., Hughes, N.J., Häg, S., Bereswill, S., Kelly, D.J., Vandenbroucke-Grauls, C.M., Kist, M., and Kusters, J.G. (2002). The *Helicobacter pylori* homologue of the ferric uptake regulator is involved in acid resistance. Infect. Immun. *70*, 606–611.

Bouvard, V., Baan, R., Straif, K., Grosse, Y., Secretan, B., El Ghissassi, F., Benbrahim-Tallaa, L., Guha, N., Freeman, C., Galichet, L., Cogliano, V., and WHO International Agency for Research on Cancer Monograph Working Group (2009). A review of human carcinogens--Part B: biological agents. Lancet Oncol. *10*, 321–322.

Bury-Mone, S., Thiberge, J.M., Contreras, M., Maitournam, A., Labigne, A., and de Reuse, H. (2004). Responsiveness to acidity via metal ion regulators mediates virulence in the gastric pathogen *Helicobacter pylori.* Mol. Microbiol. *53*, 623–638.

Carpenter, B.M., Gancz, H., Gonzalez-Nieves, R.P., West, A.L., Whitmire, J.M., Michel, S.L., and Merrell, D.S. (2009a). A single nucleotide change affects fur-dependent regulation of sodB in *H. pylori.* PLoS *4* One 5369.

Carpenter, B.M., Whitmire, J.M., and Merrell, D.S. (2009b). This is not your mother's repressor, the complex role of fur in pathogenesis. Infect. Immun. *77*, 2590–2601.

Chastanet, A., Fert, J., and Msadek, T. (2003). Comparative genomics reveal novel heat shock regulatory mechanisms in *Staphylococcus aureus* and other Gram-positive bacteria. Mol. Microbiol. *47*, 1061–1073.

Chivers, P.T., and Sauer, R.T. (2000). Regulation of high affinity nickel uptake in bacteria. Ni^{2+}-dependent interaction of NikR with wild-type and mutant operator sites. J. Biol. Chem. *275*, 19735–19741.

Choi, Y.W., Park, S.A., Lee, H.W., and Lee, N.G. (2009). Alteration of growth-phase-dependent protein regulation by a fur mutation in *Helicobacter pylori.* FEMS Microbiol. Lett. *294*, 102–110.

Colland, F., Rain, J.C., Gounon, P., Labigne, A., Legrain, P., and de Reuse, H. (2001). Identification of the *Helicobacter pylori* anti-sigma28 factor. Mol. Microbiol. *41*, 477–487.

Contreras, M., Thiberge, J.M., Mandrand-Berthelot, M.A., and Labigne, A. (2003). Characterization of the roles of NikR, a nickel-responsive pleiotropic autoregulator of *Helicobacter pylori.* Mol. Microbiol. *49*, 947–963.

Danielli, A., Amore, G., and Scarlato, V. (2010). Built shallow to maintain homeostasis and persistent infection, insight into the transcriptional regulatory network of the gastric human pathogen *Helicobacter pylori.* PLoS Pathog. *10*, e1000938.

Danielli, A., Romagnoli, S., Roncarati, D., Costantino, L., Delany, I., and Scarlato, V. (2009). Growth phase and metal-dependent transcriptional regulation of the fecA genes in *Helicobacter pylori.* J. Bacteriol. *191*, 3717–3725.

Danielli, A., Roncarati, D., Delany, I., Chiarini, V., Rappuoli, R., and Scarlato, V. (2006). In vivo dissection of the *Helicobacter pylori* Fur regulatory circuit by genome-wide location analysis. J. Bacteriol. *188*, 4654–4662.

Davis, G.S., Flannery, E.L., and Mobley, H.L. (2006). *Helicobacter pylori* HP1512 is a nickel-responsive NikR-regulated outer membrane protein. Infect. Immun. *74*, 6811–6820.

de Lorenzo, V., Wee, S., Herrero, M., and Neilands, J.B. (1987). Operator sequences of the aerobactin operon of plasmid ColV-K30 binding the ferric uptake regulation (fur). repressor. J. Bacteriol. *169*, 2624–2630.

de Reuse, H., and Bereswill, S. (2007). Ten years after the first *Helicobacter pylori* genome, comparative and functional genomics provide new insights in the variability and adaptability of a persistent pathogen. FEMS Immunol. Med. Microbiol. *50*, 165–176.

Delany, I., Ieva, R., Soragni, A., Hilleringmann, M., Rappuoli, R., and Scarlato, V. (2005). *In vitro* analysis of protein–operator interactions of the NikR and fur metal-responsive regulators of coregulated genes in *Helicobacter pylori*. J. Bacteriol. *187*, 7703–7715.

Delany, I., Pacheco, A.B., Spohn, G., Rappuoli, R., and Scarlato, V. (2001a). Iron-dependent transcription of the frpB gene of *Helicobacter pylori* is controlled by the Fur repressor protein. J. Bacteriol. *183*, 4932–4937.

Delany, I., Rappuoli, R., and Scarlato, V. (2004). Fur functions as an activator and as a repressor of putative virulence genes in *Neisseria meningitidis*. Mol. Microbiol. *52*, 1081–1090.

Delany, I., Spohn, G., Pacheco, A.B., Ieva, R., Alaimo, C., Rappuoli, R., and Scarlato, V. (2002a). Autoregulation of *Helicobacter pylori* Fur revealed by functional analysis of the iron-binding site. Mol. Microbiol. *46*, 1107–1122.

Delany, I., Spohn, G., Rappuoli, R., and Scarlato, V. (2001b). The Fur repressor controls transcription of iron-activated and -repressed genes in *Helicobacter pylori*. Mol. Microbiol. *42*, 1297–1309.

Delany, I., Spohn, G., Rappuoli, R., and Scarlato, V. (2003). An anti-repression Fur operator upstream of the promoter is required for iron-mediated transcriptional autoregulation in *Helicobacter pylori*. Mol. Microbiol. *50*, 1329–1338.

Delany, I., Spohn, G., Rappuoli, and R., and Scarlato, V. (2002b). Growth phase-dependent regulation of target gene promoters for binding of the essential orphan response regulator HP1043 of *Helicobacter pylori*. J. Bacteriol. *184*, 4800–4810.

Dietz, P., Gerlach, G., and Beier, D. (2002). Identification of target genes regulated by the two-component system HP166-HP165 of *Helicobacter pylori*. J. Bacteriol. *184*, 350–362.

Dosanjh, N.S., and Michel, S.L. (2006). Microbial nickel metalloregulation, NikRs for nickel ions. Curr Opin Chem. Biol. *10*, 123–130.

Douillard, F.P., Ryan, K.A., Caly, D.L., Hinds, J., Witney, A.A., Husain, S.E., and O'Toole, P.W. (2008). Posttranscriptional regulation of flagellin synthesis in *Helicobacter pylori* by the RpoN chaperone HP0958. J. Bacteriol. *190*, 7975–7984.

Eaton, K.A., Suerbaum, S., Josenhans, C., and Krakowka, S. (1996). Colonization of gnotobiotic piglets by *Helicobacter pylori* deficient in two flagellin genes. Infect. Immun. *64*, 2445–2448.

Ernst, F.D., Bereswill, S., Waidner. B., Stoof, J., Mäder, U., Kusters, J.G., Kuipers, E.J., Kist, M., van Vliet, A.H., and Homuth, G. (2005a). Transcriptional profiling of *Helicobacter pylori* Fur- and iron-regulated gene expression. Microbiology *151*, 533–546.

Ernst, F.D., Homuth, G., Stoof, J., Mäder, U., Waidner, B., Kuipers, E.J., Kist, M., Kusters, J.G., Bereswill, S., and van Vliet, A.H. (2005b). Iron-responsive regulation of the *Helicobacter pylori* iron-cofactored superoxide dismutase SodB is mediated by Fur. J. Bacteriol. *187*, 3687–3692.

Ernst, F.D., Kuipers, E.J., Heijens, A., Sarwari, R., Stoof, J., Penn, C.W., Kusters, J.G., and van Vliet, A.H. (2005c). The nickel-responsive regulator NikR controls activation and repression of gene transcription in *Helicobacter pylori*. Infect. Immun. 73, 7252–7258.

Ernst, F.D., Stoof, J., Horrevoets, W.M., Kuipers, E.J., Kusters, J.G., and van Vliet, A.H. (2006). NikR-mediates nickel-responsive transcriptional repression of the *Helicobacter pylori* outer membrane proteins FecA3 (HP1400). and FrpB4 (HP1512). Infect. Immun. *74*, 6821–6828.

Falush, D., Wirth, T., Linz, B., Pritchard, J.K., Stephens, M., Kidd, M., Blaser, M.J., Graham, D.Y., Vacher, S., Perez-Perez, G.I., Yamaoka, Y., Mégraud, F., Otto, K., Reichard, U., Katzowitsch, E., Wang, X., Achtman, M., and Suerbaum, S. (2003). Traces of human migrations in *Helicobacter pylori* populations. Science 7, 1582–1585.

Forsyth, M.H., Cao, P., Garcia, P.P., Hall, J.D., and Cover, T.L. (2002). Genome-wide transcriptional profiling in a histidine kinase mutant of *Helicobacter pylori* identifies members of a regulon. J. Bacteriol. *184*, 4630–4635.

Foynes, S., Dorrell, N., Ward, S.J., Stabler, R.A., McColm, A.A., Rycroft, A.N., and Wren, B.W. (2000). *Helicobacter pylori* possesses two CheY response regulators and a histidine kinase sensor, CheA, which are essential for chemotaxis and colonization of the gastric mucosa. Infect. Immun. *68*, 2016–2023.

Gancz, H., Censini, S., and Merrell, D.S. (2006). Iron and pH homeostasis intersect at the level of Fur regulation in the gastric pathogen *Helicobacter pylori*. Infect. Immun. *74*, 602–614.

Gerrits, M.M., van Vliet, A.H., Kuipers, E.J., and Kusters, J.G. (2006). *Helicobacter pylori* and antimicrobial resistance, molecular mechanisms and clinical implications. Lancet Infect. Dis. *6*, 699–709.

Giedroc, D.P., and Arunkumar, A.I. (2007). Metal sensor proteins: nature's metalloregulated allosteric switches. Dalton Trans., 3107–3120.

Güell, M., van Noort, V., Yus, E., Chen, W.H., Leigh-Bell, J., Michalodimitrakis, K., Yamada, T., Arumugam, M., Doerks, T., Kühner, S., Rode, M., Suyama, M., Schmidt, S., Gavin, A.C., Bork, P., and Serrano, L. (2009). Transcriptome complexity in a genome-reduced bacterium. Science *326*, 1268–1271.

Hantke, K. (2001). Iron and metal regulation in bacteria. Curr. Opin. Microbiol. *4*, 172–177.

Homuth, G., Domm, S., Kleiner, D., and Schumann, W. (2000). Transcriptional analysis of major heat shock genes of *Helicobacter pylori*. J. Bacteriol. *182*, 4257–4263.

Hughes, N.J., Clayton, C.L., Chalk, P.A., and Kelly, D.J. (1998). *Helicobacter pylori porCDAB* and *oorDABC* genes encode distinct pyruvate: flavodoxin and 2-oxoglutarate: acceptor oxidoreductases which mediate electron transport to NADP. J. Bacteriol. *180*, 1119–1128.

Josenhans, C., Beier, D., Linz, B., Meyer, T.F., and Suerbaum, S. (2007). Pathogenomics of *Helicobacter*. Int. J. Med. Microbiol. *297*, 589–600.

Josenhans, C., Niehus, E., Amersbach, S., Hörster, A., Betz, C., Drescher, B., Hughes, K.T., and Suerbaum, S. (2002). Functional characterization of the antagonistic flagellar late regulators FliA and FlgM of *Helicobacter pylori* and their effects on the *H. pylori* transcriptome. Mol. Microbiol. *43*, 307–322.

Josenhans, C., and Suerbaum, S. (2002). The role of motility as a virulence factor in bacteria. Int. J. Med. Microbiol. *291*, 605–614.

Kalir, S., Mangan, S., and Alon, U. (2005). A coherent feed-forward loop with a SUM input function prolongs flagella expression in *Escherichia coli*. Mol. Syst Biol. *1*, 2005.0006.

Li, Y., and Zamble, D. (2009). The pH-responsive DNA-binding activity of *Helicobacter pylori* NikR. Biochemistry *48*, 2486–2496.

Linz, B., Balloux, F., Moodley, Y., Manica, A., Liu, H., Roumagnac, P., Falush, D., Stamer, C., Prugnolle, F., van der Merwe, S.W., Yamaoka, Y., Graham, D.Y., Perez-Trallero, E., Wadstrom, T., Suerbaum, S., and Achtman, M. (2007). An African origin for the intimate association between humans and *Helicobacter pylori*. Nature *445*, 915–918.

Longabaugh, W.J., Davidson, E.H., and Bolouri, H. (2009). Visualization, documentation, analysis, and communication of large-scale gene regulatory networks. Biochim. Biophys. Acta *1789*, 363–374.

Macnab, R.M. (2003). How bacteria assemble flagella. Annu. Rev. Microbiol. *57*, 77–100.

Maier, R.J., Fu, C., Gilbert, J., Moshiri, F., Olson, J., and Plaut, A.G. (1996). Hydrogen uptake hydrogenase in *Helicobacter pylori*. FEMS Microbiol. Lett. *141*, 71–76.

Mangan, S., and Alon, U. (2003). Structure and function of the feed-forward loop network motif. Proc. Natl. Acad. Sci. U.S.A. *100*, 11980–11985.

Mangan, S., Itzkovitz, S., Zaslaver, A., and Alon, U. (2006). The incoherent feed-forward loop accelerates the response-time of the gal system of *Escherichia coli*. J. Mol. Biol. *10*, 1073–1081.

Mangan, S., Zaslaver, A., and Alon, U. (2003). The coherent feedforward loop serves as a sign-sensitive delay element in transcription networks. J. Mol. Biol. *21*, 197–204.

Marshall, B.J., and Warren, J.R. (1983). Unidentified curved bacilli in the stomach of patients with gastritis and peptic ulceration. Lancet *1*, 1311–1315.

Martinez-Antonio, A., Janga, S.C., and Thieffry, D. (2008). Functional organisation of *Escherichia coli* transcriptional regulatory network. J. Mol. Biol. *381*, 238–247.

Mehta, N., Olson, J.W., and Maier, R.J. (2003). Characterization of *Helicobacter pylori* nickel metabolism accessory proteins needed for maturation of both urease and hydrogenase. J. Bacteriol. *185*, 726–734.

Merrell, D.S., Thompson, L.J., Kim, C.C., Mitchell, H., Tompkins, L.S., Lee, A., and Falkow, S. (2003). Growth phase-dependent response of *Helicobacter pylori* to iron starvation. Infect. Immun. *71*, 6510–6525.

Mogk, A., Homuth, G., Scholz, C., Kim, L., Schmid, F.X., and Schumann, W. (1997). The GroE chaperonin machine is a major modulator of the CIRCE heat shock regulon of *Bacillus subtilis*. EMBO J. *16*, 4579–4590.

Muhsen, K., and Cohen, D. (2008). *Helicobacter pylori* infection and iron stores, a systematic review and meta-analysis. Helicobacter *13*, 323–340.

Müller, S., Pflock, M., Schar, J., Kennard, S., and Beier, D. (2007). Regulation of expression of atypical orphan response regulators of *Helicobacter pylori*. Microbiol. Res. *162*, 1–14.

Narberhaus, F. (1999). Negative regulation of bacterial heat shock genes. Mol. Microbiol. *31*, 1–8.

Niehus, E., Gressmann, H., Ye, F., Schlapbach, R., Dehio, M., Dehio, C., Stack, A., Meyer, T.F., Suerbaum, S., and Josenhans, C. (2004). Genome-wide analysis of transcriptional hierarchy and feedback regulation in the flagellar system of *Helicobacter pylori*. Mol. Microbiol. *52*, 947–961.

Oh, J.D., Kling-Bäckhed, H., Giannakis, M., Xu, J., Fulton, R.S., Fulton, L.A., Cordum, H.S., Wang, C., Elliott, G., Edwards, J., Mardis, E.R., Engstrand, L.G., and Gordon, J.I. (2006). The complete genome sequence of a chronic atrophic gastritis *Helicobacter pylori* strain: evolution during disease progression. Proc. Natl. Acad. Sci. U S A. *103*, 9999–10004.

Olson, J.W., and Maier, R.J. (2002). Molecular hydrogen as an energy source for *Helicobacter pylori*. Science *298*, 1788–1790.

Parsonnet, J., Hansen, S., Rodriguez, L., Gelb, A.B., Warnke, R.A., Jellum, E., Orentreich, N., Vogelman, J.H., and Friedman, G.D. (1994). *Helicobacter pylori* infection and gastric lymphoma. N. Engl. J. Med. *330*, 1267–1271.

Pereira, L., and Hoover, T.R. (2005). Stable accumulation of sigma54 in *Helicobacter pylori* requiRes. the novel protein HP0958. J. Bacteriol. *187*, 4463–4469.

Pereira, L.E., Brahmachary, P., and Hoover, T.R. (2006). Characterization of *Helicobacter pylori* sigma54 promoter-binding activity. FEMS Microbiol. Lett. *259*, 20–26.

Pflock, M., Bathon, M., Schär, J., Müller, S., Mollenkopf, H., Meyer, T.F., and Beier, D. (2007a). The orphan response regulator HP1021 of *Helicobacter pylori* regulates transcription of a gene cluster presumably involved in acetone metabolism. J. Bacteriol. *189*, 2339–2349.

Pflock,. M., Dietz, P., Schar, J., and Beier, D. (2004). Genetic evidence for histidine kinase HP165 being an acid sensor of *Helicobacter pylori*. FEMS Microbiol. Lett. *234*, 51–61.

Pflock, M., Finsterer, N., Joseph, B., Mollenkopf, H., Meyer, T.F., and Beier, D. (2006a). Characterization of the ArsRS regulon of *Helicobacter pylori*, involved in acid adaptation. J. Bacteriol. *188*, 3449–3462.

Pflock, M., Kennard, S., Delany, I., Scarlato, V., and Beier, D. (2005). Acid-induced activation of the urease promoters is mediated directly by the ArsRS two-component system of *Helicobacter pylori*. Infect. Immun. *73*, 6437–6445.

Pflock, M., Kennard, S., Finsterer, N., and Beier, D. (2006b). Acid-responsive gene regulation in the human pathogen *Helicobacter pylori*. J. Biotechnol *126*, 52–60.

Pflock, M., Muller, S., and Beier, D. (2007b). The CrdRS (HP1365-HP1364). two-component system is not involved in ph-responsive gene regulation in the *Helicobacter pylori* Strains 26695 and G27. Curr. Microbiol. *54*, 320–324.

Reischl, S., Wiegert, T., and Schumann, W. (2002). Isolation and analysis of mutant alleles of the *Bacillus subtilis* HrcA repressor with reduced dependency on GroE function. J. Biol. Chem. *6*, 32659–32667.

Roncarati, D., Danielli, A., Spohn, G., Delany, I., and Scarlato, V. (2007a). Transcriptional regulation of stress response and motility functions in *Helicobacter pylori* is mediated by HspR and HrcA. J. Bacteriol. *189*, 7234–7243.

Roncarati, D., Spohn, G., Tango, N., Danielli, A., Delany, I., and Scarlato, V. (2007b). Expression, purification and characterization of the membrane-associated HrcA repressor protein of *Helicobacter pylori*. Protein Expr. Purif. *51*, 267–275.

Rust, M., Borchert, S., Niehus, E., Kuehne, S.A., Gripp, E., Bajceta, A., McMurry, J.L., Suerbaum, S., Hughes, K.T., and Josenhans, C. (2009). The *Helicobacter pylori* anti-sigma factor FlgM is predominantly cytoplasmic and cooperates with the flagellar body protein FlhA. J. Bacteriol. *191*, 4824–4834.

Sachs, G., Weeks, D.L., Melchers, K., and Scott, D.R. (2003). The gastric biology of *Helicobacter pylori*. Annu. Rev. Physiol. *65*, 349–369.

Salgado, H., Martínez-Flores, I., López-Fuentes, A., García-Sotelo, J.S., Porrón-Sotelo, L., Solano, H., Muñiz-Rascado, L., and Collado-Vides, J. (2012). extracting regulatory networks of *Escherichia coli* from RegulonDB. Methods Mol. Biol. *804*, 179–195.

Scarlato, V., Delany, I., Spohn, G., and Beier, D. (2001). Regulation of transcription in *Helicobacter pylori*, simple systems or complex circuits? Int. J. Med. Microbiol. *291*, 107–117.

Schauer, K., Gouget, B., Carriere, M., Labigne, A., and de Reuse, H. (2007). Novel nickel transport mechanism across the bacterial outer membrane energized by the TonB/ExbB/ExbD machinery. Mol. Microbiol. *63*, 1054–1068.

Schreiber, S., Konradt, M., Groll, C., Scheid, P., Hanauer, G., Werling, H.O., Josenhans, C., and Suerbaum, S. (2004). The spatial orientation of *Helicobacter pylori* in the gastric mucus. Proc. Natl. Acad. Sci. U.S.A. *6*, 5024–5029.

Schulz, A., and Schumann, W. (1996). hrcA, the first gene of the *Bacillus subtilis* dnaK operon encodes a negative regulator of class I heat shock genes. J. Bacteriol. *178*, 1088–1093.

Scott, D.R., Marcus, E.A., Weeks, D.L., and Sachs, G. (2002). Mechanisms of acid resistance due to the urease system of *Helicobacter pylori*. Gastroenterology *123*, 187–195.

Scott, D.R., Marcus, E.A., Wen, Y., Oh, J., and Sachs, G. (2007). Gene expression in vivo shows that *Helicobacter pylori* colonizes an acidic niche on the gastric surface. Proc. Natl. Acad. Sci. U S A *104*, 7235–7240.

Servant, P., and Mazodier, P. (2001). Negative regulation of the heat shock response in *Streptomyces*. Arch. Microbiol. *176*, 237–242.

Seshasayee, A.S., Bertone, P., Fraser, G.M., and Luscombe, N.M. (2006). Transcriptional regulatory networks in bacteriA., from input signals to output responses. Curr. Opin. Microbiol. *9*, 511–519.

Sharma, C.M., Hoffmann, S., Darfeuille, F., Reignier, J., Findeiss, S., Sittka, A., Chabas, S., Reiche, K., and Hackermüller, J. (2010). The primary transcriptome of the major human pathogen *Helicobacter pylori*. Nature *464*, 250–255.

Shen-Orr, S.S., Milo, R., Mangan, S., and Alon, U. (2002). Network motifs in the transcriptional regulation network of *Escherichia coli*. Nat. Genet. *31*, 64–68.

Spohn, G., Danielli, A., Roncarati, D., Delany, I., Rappuoli, R., and Scarlato, V. (2004). Dual control of *Helicobacter pylori* heat shock gene transcription by HspR and HrcA. J. Bacteriol. *186*, 2956–2965.

Spohn, G., Delany, I., Rappuoli, R., and Scarlato, V. (2002). Characterization of the HspR-mediated stress response in *Helicobacter pylori*. J. Bacteriol. *184*, 2925–2930.

Spohn, G., and Scarlato, V. (1999a). Motility of *Helicobacter pylori* is coordinately regulated by the transcriptional activator FlgR, an NtrC homolog. J. Bacteriol. *181*, 593–599.

Spohn, G., and Scarlato, V. (1999b). The autoregulatory HspR repressor protein governs chaperone gene transcription in *Helicobacter pylori*. Mol. Microbiol. *34*, 663–674.

Stoof, J., Kuipers, E.J., and van Vliet, A.H. (2010). Characterization of NikR-responsive promoters of urease and metal transport genes of *Helicobacter mustelae*. Biometals *23*, 145–159.

Tomb, J.F., White, O., Kerlavage, A.R., Clayton, R.A., Sutton, G.G., Fleischmann, R.D., Ketchum, K.A., Klenk, H.P., Gill, S., Dougherty, B.A., *et al.* (1997). The complete genome sequence of the gastric pathogen *Helicobacter pylori*. Nature *388*, 539–547.

Tsuda, M., Karita, M., Morshed, M.G., Okita, K., and Nakazawa, T. (1994). A urease-negative mutant of *Helicobacter pylori* constructed by allelic exchange mutagenesis lacks the ability to colonize the nude mouse stomach. Infect. Immun. *62*, 3586–3589.

van Vliet, A.H., Ernst, F.D., and Kusters, J.G. (2004a). NikR-mediated regulation of *Helicobacter pylori* acid adaptation. Trends Microbiol. *12*, 489–494.

van Vliet, A.H., Kuipers, E.J., Stoof, J., Poppelaars, S.W., and Kusters, J.G. (2004b). Acid-responsive gene induction of ammonia-producing enzymes in *Helicobacter pylori* is mediated via a metal-responsive repressor cascade. Infect. Immun. *72*, 766–773.

van Vliet, A.H., Poppelaars, S.W., Davies, B.J., Stoof, J., Bereswill, S., Kist, M., Penn, C.W., Kuipers, E.J., and Kusters, J.G. (2002a). NikR mediates nickel-responsive transcriptional induction of urease expression in *Helicobacter pylori*. Infect. Immun. *70*, 2846–2852.

van Vliet, A.H., Stoof, J., Poppelaars, S.W., Bereswill, S., Homuth, G., Kist, M., Kuipers, E.J., and Kusters, J.G. (2003). Differential regulation of amidase- and formamidase-mediated ammonia production by the *Helicobacter pylori* fur repressor. J. Biol. Chem. *278*, 9052–9057.

van Vliet, A.H., Stoof, J., Vlasblom, R., Wainwright, S.A., Hughes, N.J., Kelly, D.J., Bereswill, S., Bijlsma, J.J., Hoogenboezem, T., Vandenbroucke-Grauls, C.M., Kist, M., Kuipers, E.J., and Kusters, J.G. (2002b). The role of the Ferric Uptake Regulator (Fur). in regulation of *Helicobacter pylori* iron uptake. Helicobacter *7*, 237–244.

Waidner, B., Melchers, K., Stahler, F.N., Kist, M., and Bereswill, S. (2005). The *Helicobacter pylori* CrdRS two-component regulation system (HP1364/HP1365). is required for copper-mediated induction of the copper resistance determinant CrdA. J. Bacteriol. *187*, 4683–4688.

Wang, G., Alamuri, P., and Maier, R.J. (2006). The diverse antioxidant systems of *Helicobacter pylori*. Mol. Microbiol. *61*, 847–860.

Wen, Y., Feng, J., Scott, D.R., Marcus, E.A., and Sachs, G. (2009). The pH-responsive regulon of HP0244 (FlgS), the cytoplasmic histidine kinase of *Helicobacter pylori*. J. Bacteriol. *191*, 449–460.

Wen, Y., Feng, J., Scott, D.R., Marcus, E.A., and Sachs, G. (2006). Involvement of the HP0165-HP0166 two-component system in expression of some acidic-pH-upregulated genes of *Helicobacter pylori*. J. Bacteriol. *188*, 1750–1761.

Wen, Y., Feng, J., Scott, D.R., Marcus, E.A., and Sachs, G. (2011). A cis-encoded antisense small RNA regulated by the HP0165-HP0166 two-component system controls expression of *ureB* in *Helicobacter pylori*. J. Bacteriol. *193*, 40–51.

Wolfram, L., Haas, E., and Bauerfeind. P., (2006). Nickel represses the synthesis of the nickel permease NixA of *Helicobacter pylori*. J. Bacteriol. *188*, 1245–1250.

Xiao, B., Li, W., Guo, G., Li, B., Liu, Z., Jia, K., Guo, Y., Mao, X., and Zou, Q. (2009a). Identification of small noncoding RNAs in *Helicobacter pylori* by a bioinformatics-based approach. Curr. Microbiol. *58*, 258–263.

Xiao, B., Li, W., Guo, G., Li, B.S., Liu, Z., Tang, B., Mao, X.H., and Zou, Q.M. (2009b). Screening and identification of natural antisense transcripts in *Helicobacter pylori* by a novel approach based on RNase I protection assay. Mol. Biol. Rep. *36*, 1853–1858.

Ye, F., Brauer, T., Niehus, E., Drlica, K., Josenhans, C., and Suerbaum, S. (2007). Flagellar and global gene regulation in *Helicobacter pylori* modulated by changes in DNA supercoiling. Int J. Med. Microbiol. *297*, 65–81.

Yu, H., Luscombe, N.M., Qian, J., and Gerstein, M. (2003). Genomic analysis of gene expression relationships in transcriptional regulatory networks. Trends Genet. *19*, 422–427.

Zambelli, B., Danielli, A., Romagnoli, S., Neyroz, P., Ciurli, S., and Scarlato, V. (2008). High-affinity Ni^{2+} binding selectively promotes binding of *Helicobacter pylori* NikR to its target urease promoter. J. Mol. Biol. *383*, 1129–1143.

Zaslaver, A., Mayo, A.E., Rosenberg, R., Bashkin, P., Sberro, H., Tsalyuk, M., Surette, M.G., and Alon, U. (2004). Just-in-time transcription program in metabolic pathways. Nat. Genet. *36*, 486–491.

The Transcriptional Regulatory Network of *Mycobacterium tuberculosis*

12

Gábor Balázsi, Oleg A. Igoshin and Maria Laura Gennaro

Abstract

Approximately one-third of the world's human population is infected with *Mycobacterium tuberculosis*. Most infected individuals have latent infection: they are symptom-free and carry mostly dormant bacteria that survived the immune response. When the immune response weakens, dormant bacteria can reactivate and cause a life-threatening disease. If the mechanisms of dormancy were better understood, disease reactivation and spreading could be prevented. Here we review our recent work on two mycobacterial survival strategies related to dormancy: environmental sensing followed by stress response and stochastically delayed switching into dormancy. We discuss the use of large-scale regulatory networks to infer how hypoxia affects the mycobacterial transcriptome at the genomic scale. We also show how sigma factor sequestration by the corresponding anti-sigma factor may generate a non-linear response, which results in bistability when combined with positive feedback.

Introduction

Tuberculosis (TB), which has plagued mankind for millennia, still causes almost 10 million new cases of disease and approximately 1.7 million deaths per year worldwide (http://www.who.int/tb/publications/global_report/2010/en/), mostly in developing countries. The vast majority of TB deaths (98%) occur among young adults; moreover, TB is the single leading cause of death among women of reproductive age (http://www.iuatld.org). Thus, this disease severely disrupts the social structure and economic potential of many emerging countries, with grave implications on global economy and security. TB has also resurfaced in developed countries owing to increased global travel, HIV comorbidity, and high-risk behaviours.

Current control strategies are insufficient to control TB effectively. National and international TB control programmes, while successful in many regions, have been unable to prevent the global incidence of TB from growing at an annual rate of 1%. The magnitude of the challenge was recognized in 2006, when a new Global Plan (2006–2015) was launched with a renewed focus on obtaining innovative diagnostics, drugs, and vaccines. According to the Global Plan, the target of eradicating TB by 2050 will become realistic only with strengthened research efforts that lead to more effective control tools (http://www.stoptb.org/globalplan). A prominent area of TB research is to understand the environment-dependent regulation of the life cycle of the causative agent, *Mycobacterium tuberculosis*.

M. tuberculosis can switch between replicative (growth) and non-replicative (dormancy) states in response to cues generated by host immunity. When infection has progressed enough for the host to produce adaptive immune responses, tubercle bacilli stop or drastically slow growth. The result is asymptomatic (latent) infection, a condition that affects one-third of the world's population (http://www.who.int/tb/). When host immunity falters, tubercle bacilli can resume growth and cause disease. The diseased individual sheds tubercle bacilli and spreads the infection to new hosts by coughing and sneezing. Thus, the ability of the microorganism to switch between growth and dormancy is key to its virulence as it determines bacterial persistence in the individual host and in the population. Additionally, dormant bacilli are phenotypically less susceptible to drugs than actively growing cells, requiring prolonged use of multiple antibiotics to cure TB. These burdensome therapeutic regimens create logistical problems that ultimately lead to genotypic acquisition of multidrug resistance, which has generated a global health emergency (http://www.who.int/tb/). Understanding and blocking the 'path to and from dormancy' of the tubercle *Bacillus* is critical to effectively fight TB.

In principle, switching between growth and dormancy can be due to (i) a stress-sensing strategy, in which bacterial adaptation is induced by environmental stress condition(s) or (ii) a bet-hedging strategy in which a small proportion of the bacterial population stochastically switches to the dormant state in anticipation of stress (Kussell and Leibler, 2005). It is likely that *M. tuberculosis* utilizes some combination of the two strategies to execute a complex regulatory programme involving transcriptional regulation and post-translational protein modification networks. Central to these networks is the arsenal of transcription factors and 13 sigma factors that regulate transcript synthesis as well as sensory kinases that modulate protein activity. *M. tuberculosis* exhibits the highest alternative sigma factor/genome size ratio among obligate pathogens (Rodrigue *et al.*, 2006), which is a tell-tale sign of the bacterium having a complex developmental programme to respond to stress (Fig. 12.1A).

Below, we briefly review two main classes of regulators described in *M. tuberculosis* and summarize our recent efforts to understand two key survival strategies in this pathogen during hypoxia: environmental sensing coupled with transcriptional response and stochastically delayed switching into dormancy.

Environmental sensing and the transcriptomic response

Biology of environmental sensing and response

Two modes of transcriptional regulation have been extensively studied in the context of environmental sensing and response in *M. tuberculosis*: two-component systems and sigma factors.

Two-component systems (TCS) are protein pairs that regulate responses to various stimuli and are common throughout the prokaryotic kingdom. One protein of the pair is a sensor kinase that responds to environmental cues by modulating the phosphorylation state of the second protein, which typically is a transcriptional response regulator (Mitrophanov and Groisman, 2008b; Stock *et al.*, 2000). The sensor kinase can exhibit catalytic activities such as (1) trans-autophosphorylation of a histidine residue via hydrolysis of ATP or other phospho-donors; and (2) phospho-transfer to a specific aspartate

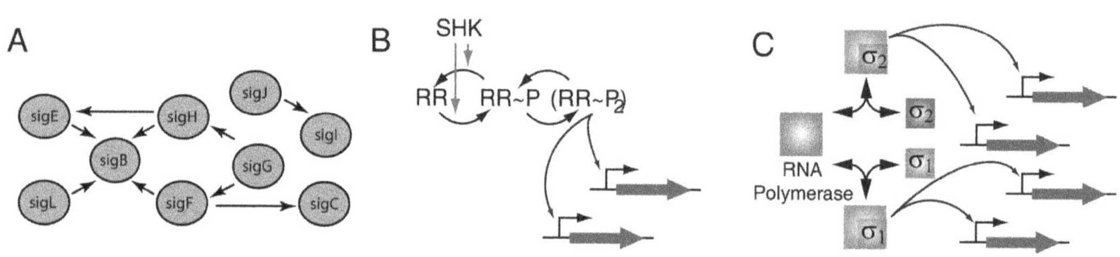

Figure 12.1 Master regulators and regulatory modes of *M. tuberculosis*. (A) The sigma factor network consists of several sigma factors that control each other through a set of complex regulatory interactions. The arrows indicate direct/indirect regulatory links. Adapted from (Sachdeva *et al.*, 2010). (B) Two-component system with sensor histidine kinase (SHK) and response regulator in unphosphorylated (RR), phosphorylated (RR~P) and dimer (RR~P_2) forms. (C) Alternative sigma factors associate with RNA polymerase to control gene expression in various conditions. A colour version of this figure is located in the plate section at the back of the book.

residue in the cognate regulator. Moreover, when non-phosphorylated, many sensor kinases display phosphatase activity towards their cognate response regulators (Fig. 12.1B).

The *M. tuberculosis* genome encodes 16 TCSs. Among the most investigated systems is dosRST (Saini *et al.*, 2004; Sherman *et al.*, 2001), which utilizes one sensor (dosT) to respond to hypoxia and another (dosS) to redox changes (Kumar *et al.*, 2007). In the mouse lung, the dosR regulon responds to conditions associated with expression of T helper cell (Th1)-mediated immunity (Shi *et al.*, 2003). The effect of dosRST inactivation on *M. tuberculosis* survival *in vivo* is not fully resolved [for example, (Converse *et al.*, 2009; Rustad *et al.*, 2008)], which raises the puzzling question of how a TCS responding to several stress signals may not be required for bacterial adaptation during infection. Another well-studied TCS is mprAB, which responds to cell-surface stress (Pang *et al.*, 2007). Polyphosphate, a molecule involved in bacterial stress responses (reviewed in Manganelli, 2007), donates phosphates for MprB, which then modulates the transition between MprA (inactive) and MprA~P (active) via its phosphotransfer and phosphatase activities (Sureka *et al.*, 2007; Zahrt *et al.*, 2003). MprA~P induces sigE, the gene encoding the alternative sigma factor E (Dona *et al.*, 2008; Sureka *et al.*, 2007). Another TCS that has received much attention is phoPR, which regulates genes involved in lipid metabolism (Ryndak *et al.*, 2008). The nature of the stress that activates phoPR is unknown, but it is clear that inactivation of phoPR leads to *M. tuberculosis* growth defects during mouse lung infection. Little is known about the role of another TCS, prrAB, which is expressed inside macrophages (Graham and Clark-Curtiss, 1999) during infection, since inactivation of prrAB causes only a transient growth defect in macrophages infected *in vitro* (Ewann *et al.*, 2002). Even more tenuous is the link between *M. tuberculosis* virulence and senX3regX3, since attenuation of a senX3regX3 deletion mutant in mice remains controversial (Ewann *et al.*, 2002; Parish *et al.*, 2003). The remaining TCSs remain an open area of investigation.

Critical to the bacterial stress response is RNA polymerase (RNAP) holoenzyme reprogramming by changing the sigma factor, a subunit of RNA polymerase that determines promoter specificity (Gruber and Gross, 2003). By associating itself with alternative sigma factors, the reprogrammed RNAP recognizes and transcribes sets of genes that are silent under unstressed conditions, thereby causing the expression of bacterial functions critical for survival under stress (Fig. 12.1C). *M. tuberculosis* encodes one 'housekeeping' sigma factor, sigA, and twelve alternative sigma factors, encoded by sigB through sigM that influence each other's expression, forming a 'sigma factor network' (Fig. 12.1A) (Rodrigue *et al.*, 2006; Sachdeva *et al.*, 2010). The regulons controlled by sigB, sigC, sigE, and sigF have been characterized by mutant transcriptomics (Rodrigue *et al.*, 2006; Sachdeva *et al.*, 2010). Mutants inactivated in sigC, sigD, sigE, sigF, sigH, or sigL exhibit attenuation of virulence in mice (according to a variety of phenotypic read-outs) and/or poor growth in cultured human and/or murine monocytes or macrophages (Calamita *et al.*, 2005; Chen *et al.*, 2000; Dainese *et al.*, 2006; Kaushal *et al.*, 2002; Sun *et al.*, 2004).

The network of alternative sigma factors is densely interconnected (a partial network is shown in Fig. 12.1A) [reviewed in (Sachdeva *et al.*, 2010)], indicating a possible integration of various stress signals, and coordinated sigma factor production under particular conditions. Examples of coordinated sigma factor expression in *M. tuberculosis* are sigD-sigE and sigD-sigG-sigH (Sachdeva *et al.*, 2010). Much attention has been devoted to the sigB–sigE–sigH network, in which both σ^E and σ^H control expression of sigB (Manganelli *et al.*, 2002; Manganelli *et al.*, 2001), while σ^H also controls sigE expression under some stress conditions (Manganelli *et al.*, 2002; Raman *et al.*, 2001). It has been proposed that sigB is also controlled by two additional sigma factors, σ^L and σ^F (Rodrigue *et al.*, 2006; Sachdeva *et al.*, 2010), leading to the proposition that σ^B is a hub node of signal integration.

As currently characterized, the connectivity in the sigma-factor transcriptional network is insufficient to describe the phenotypes of some deletion mutants. For example, even though sigB is regulated by multiple sigma factors, a sigB mutant of *M. tuberculosis* shows no defect in growth and persistence in mice (Fontan *et al.*, 2009). In contrast, a mutant in sigE, one of the sigma factors upstream of sigB, is attenuated for growth and/or virulence in mice (Ando *et al.*, 2003; Manganelli *et al.*, 2004). This example highlights the importance of mapping alternate routes of regulation by assembling and inferring increasingly larger transcriptional regulatory networks.

Network reconstruction

Regulatory network construction and assembly via the traditional approaches of promoter mapping, chromatin immunoprecipitation, and gel-shift assays take years to complete. The emergence of high-throughput transcriptomic profiling techniques has made it possible to accelerate the computational assembly and inference of such networks on a genomic scale (Veiga *et al.*, 2010). Network reconstruction from genome-scale data are based on a variety of network inference methods, including ordinary differential equations (ODEs), Bayesian statistics or information theory (Bonneau, 2008). Computational network reconstruction methods have improved to a degree that they can now infer large transcriptional regulatory networks with relatively high confidence based on transcriptome data (Faith *et al.*, 2007). To perform well, network inference algorithms require large amounts of microarray data collected in a wide variety of conditions; thus, they have been mainly tested on model organisms for which extensive datasets exist. Moreover, network inference tools require a 'gold standard' network with validated interactions as a training dataset or to assess their own performance. Owing to the recent assembly of such a network (Balázsi *et al.*, 2008) (see below) and the accelerated pace of microarray data collection for *M. tuberculosis*, the latest and best-performing algorithms can be used to infer the transcriptional regulatory network of this pathogen.

Assembly of a high-confidence TR network

A genome-scale *M. tuberculosis* transcriptional regulatory (TR) network (Fig. 12.2) has been assembled using three main sources: (i) 381 gene regulatory interactions were extracted from the literature, including 222 from an earlier database (Jacques *et al.*, 2005); (ii) an additional 223 *M. tuberculosis* regulator–target gene pairs were included because of their orthology to gene pairs with confirmed transcriptional-regulatory relationships in *Escherichia coli* (Babu, 2006); and (iii) the full list of *M. tuberculosis* operons (Roback *et al.*, 2007). The latter was utilized with the assumption that transcription factor binding to the promoter region affects the expression of all genes within an operon. The resulting TR network has 783 nodes (Fig. 12.2) corresponding to *M. tuberculosis* genes and their protein products, and 937 links corresponding to 45 transcription factors directly regulating the expression of their target genes. This TR network comprises ~20% of the *M. tuberculosis* genome (for reference, the *E. coli* gene regulatory network, which included 1364 genes, comprised ~35% of the *E. coli* genome) (Salgado *et al.*, 2006). Thus, the *M. tuberculosis* TR network (Balázsi *et al.*, 2008) had a size comparable to that of the well-studied model organism *E. coli*.

Network response to hypoxia

We used pre-defined modules to examine the response of the *M. tuberculosis* TR network during hypoxia. We defined these modules (origons) as the sets of genes regulated directly or indirectly by each transcription factor (Balázsi *et al.*, 2005). We refer to these origons by the name of the transcription factor at which they originate. Based on these pre-defined modules, we have recently developed a new method, NetReSFun (Network Response to Step Functions), which takes the TR network and time-course data as inputs and identifies a list of significantly affected origons for each time point as output. Compared to a similar approach used previously to analyse *E. coli* microarray data in the context of its genome-scale regulatory network (Balázsi *et al.*, 2005), NetReSFun reliably detects the time when a major expression change occurs in a group of genes. Using this tool and published time-course microarray data GSE8786 (Voskuil *et al.*, 2004), we identified 16 *M. tuberculosis* origons that were significantly affected during hypoxia-induced growth arrest (Balázsi *et al.*, 2008). We classified significantly responding origons as 'early', 'intermediate' or 'late' based on the peak in their responsiveness $Z_I(\tau)$ over the time course (Fig. 12.3A). For example, dosR, Rv0494,

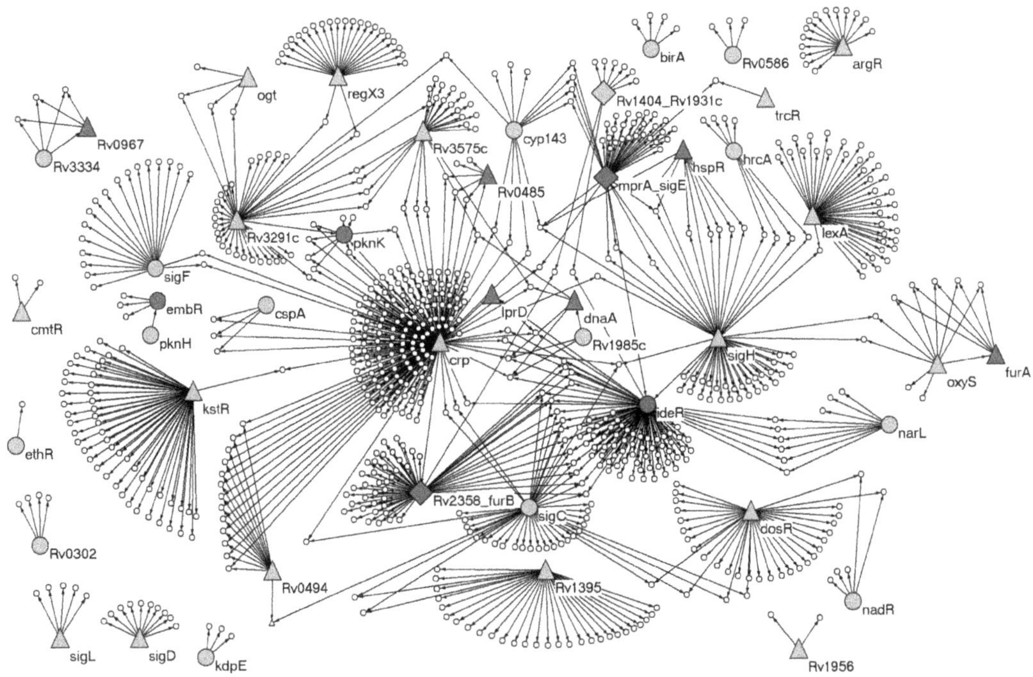

Figure 12.2 The *M. tuberculosis* gene regulatory network. The network was assembled from publicly available sources. Input nodes (TFs with no known transcriptional regulators) are shown in blue, while transit nodes (TFs with known transcriptional regulators) are shown in green. The white nodes represent output nodes (genes encoding proteins with no TF activity). Triangles = nodes that regulate their own expression. Diamonds = nodes that are part of two-gene feedback loops. Reproduced from (Balázsi *et al.*, 2008) according to a Creative Commons Attribution–Noncommercial–Share Alike 3.0 licence. A colour version of this figure is located in the plate section at the back of the book.

and sigD were early origons, because most of their constituent genes changed expression before day 6 of the hypoxic time course. Intermediate origons (including furB/zur, crp, sigH, kstR, and sigE-mprA) peaked at later time points, between days 8 and 14. Finally, late origons such as nadR, Rv1956, and hrcA were most responsive on or after day 20 (Fig. 12.3A). Interestingly, the dosR origon had a second prominent $Z_I(\tau)$ peak at day 80, corresponding to a gene expression change of sign opposite to that seen at the early time points (Balázsi *et al.*, 2008).

Responsive origons were also identified for: (i) time series data obtained in a different environmental condition; (ii) time series data collected at higher time resolution; and (iii) a computationally enlarged version of the current TR network. We briefly describe these results below.

First, we applied NetReSFun to time-course microarray data collected by Voskuil *et al.* (2004) at days 0, 6, 8, 14, 24, and 60 during the transition to stationary phase in aerated conditions. Eleven origons (nadR, hspR, Rv0494, sigE, sigC, furB, hrcA, ideR, dosR, sigD, and crp) responded significantly in both time courses (Fig. 12.3B). The most consistent early responder in both hypoxia and aerated conditions was the dosR origon, indicating that the response by this origon tends to immediately precede the arrest of bacterial growth. The origons sigD and Rv0494 appeared early, and they exhibited significant changes for a much longer time under these conditions than in hypoxia. This observation raises the possibility that these origons are condition-dependent initiators of growth arrest along with dosR. In contrast, the origons nadR, sigE, sigC, and furB reached maximal change after dosR in both time courses (Fig. 12.3C and D), suggesting that they may orchestrate the maintenance (rather than the initiation) of dormancy. Thus, NetReSFun can

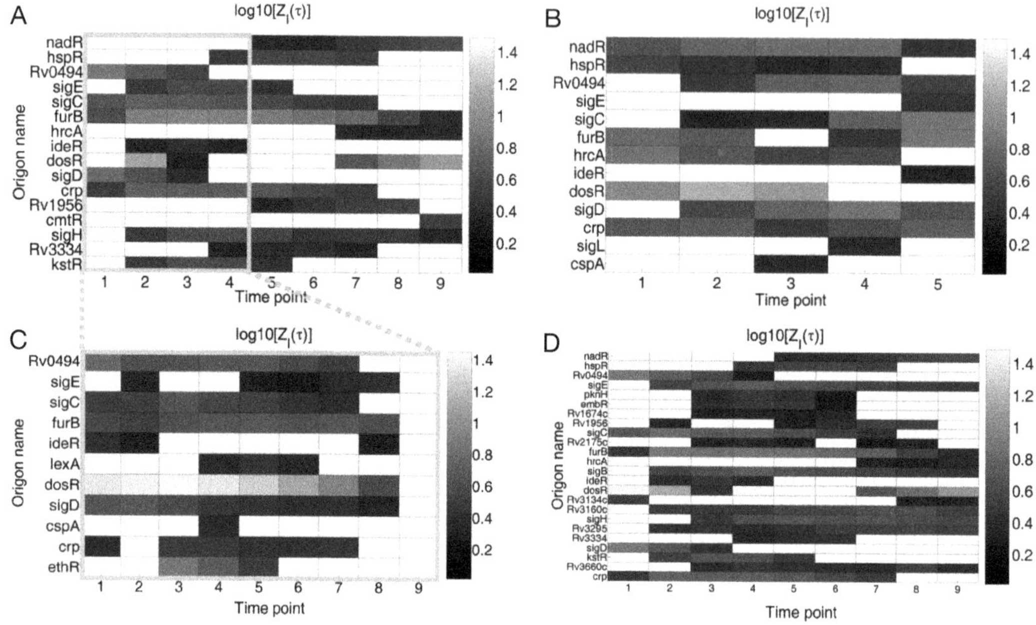

Figure 12.3 Network response to growth arrest. (A) Responsiveness $Z_l(t)$ of significantly responsive origons during growth arrest in hypoxia (time points correspond to 4, 6, 8, 10, 12, 14, 20, 30, 80 days) (B) Same as in (A), but during aerated growth (time points correspond to days 6, 8, 14, 24 and 60). Eleven origons (*nadR*, *hspR*, *Rv0494*, *sigE*, *sigC*, *furB*, *hrcA*, *ideR*, *dosR*, *sigD*, and *crp*) responded significantly in both (A) and B). (C) Same as in (A), but at higher time resolution (time points correspond to 23, 47, 73, 96, 121, 143, 164, 191 and 213 hours in hypoxia). (D) Same as in (A), but using a vastly enlarged version of the TR network obtained with the algorithm CLR. (A) and (B) are reproduced from (Balázsi *et al.*, 2008) according to a Creative Commons Attribution–Noncommercial–Share Alike 3.0 licence. A colour version of this figure is located in the plate section at the back of the book.

identify differences and similarities of network response to various environmental stimuli.

Second, we have also generated a hypoxia time course with samples collected at higher time resolution than published data (Voskuil *et al.*, 2004). As shown in Fig. 12.3C, the network response observed at higher time resolution with NetReSFun was consistent with that obtained with low sampling frequency (8 of the 11 origons in Fig. 12.3C are identical to the ones in Fig. 12.3A, and the times of their responsiveness agreed with those observed with lower sampling frequency). However, higher sampling frequency revealed three additional origons responsive during hypoxia-induced growth arrest: lexA, cspA and ethR, suggesting that some network-level changes may occur at shorter time scales of 1 day or below. These results indicate that NetResSFun reliably identifies significant origons, and that collecting microarray data with a high time resolution is important.

Additionally, using the computational network inference tool CLR (context likelihood of relatedness) and a large microarray dataset (Boshoff *et al.*, 2004), the network was computationally enlarged using CLR (Faith *et al.*, 2007) to comprise 1410 regulatory interactions among 1,036 genes. We tested the response of this enlarged network to hypoxia (Voskuil *et al.*, 2004) using NetReSFun. As shown in Fig. 12.3D, the origons significantly responsive using the smaller network remained significant after network enlargement. However, eight additional transcription factors were responsive during hypoxia compared to Fig. 12.3A: pknH, embR, Rv1674c, Rv1773c, Rv3134c, Rv3160c, Rv3295, Rv3660c. Thus, network enlargement can lead to the discovery of novel transcriptional modules involved in environment-induced growth arrest.

In summary, these analyses demonstrate that NetReSFun can reliably identify network modules responsive during growth arrest. Enlarging the current TR network or collecting more densely sampled data can generate new insights into the biology of the transition to dormancy by helping identify additional regulators of such transition.

Networks beyond transcriptional regulation: post-translational modulation of transcriptional activity

Stress response networks involve multiple auxiliary factors that do not directly interact with DNA but can nevertheless modulate the activity of transcriptional regulators via posttranslational interactions. For example, regulation of sigma factor activity involves anti-sigma factors, antagonist proteins that sequester sigma factors and prevent their association with RNA-polymerase, and anti-anti-sigma factors, which regulate anti-sigma factor activity by a partner-switching mechanism (upon binding by the anti-anti-sigma factor, the anti-sigma factor releases the sigma factor, which becomes available for interaction with core RNA polymerase) (Helmann, 1999). Of the anti-sigma factors of *M. tuberculosis* identified to date, at least three (RseA, RshA and RslA) belong to the zinc-associated anti-sigma family and are encoded by genes located immediately downstream of the cognate sigma factor genes (reviewed in Rodrigue *et al.*, 2006). While anti-sigma factors are typically co-expressed with the cognate sigma factor genes (for example, *sigH* and *rshA*; Song *et al.*, 2003), *sigE* and *rseA*, albeit adjacent, are independently expressed (White *et al.*, 2010). As shown below, such network architecture is essential for the observed network dynamics.

The activities of some transcriptional regulators can also be modulated by covalent modification, as is the case with two-component systems, or by interaction with small molecules (see for example Won *et al.*, 2009). Non-transcriptional regulation of transcription-factor concentration via small RNAs (Brantl, 2009) or regulated cleavage/degradation (Cox, 2007) can further affect stress responses. While essential for understanding the intricacies of the resultant stress-response dynamics (Ray *et al.*, 2011), non-transcriptional interactions in the networks are much more difficult to infer from high-throughput data. Below we describe how to make predictions and identify a possibly important part of the persistence regulatory switch by combining molecular details identified for a small network module with mathematical modelling.

Feedback regulation in the mycobacterial stress response

Many bacterial operons are autoregulated; consequently, the concentration of a protein encoded by one of the operon genes affects the rate of operon transcription (Clarke and Sperandio, 2005; Rosenfeld *et al.*, 2002; Soncini *et al.*, 1995). Statistical analysis of the transcriptional regulation network in *E. coli* confirmed that feedback occurs much more frequently than expected if regulatory links were rewired randomly, indicating that autoregulation may have functional significance (Shen-Orr *et al.*, 2002). The dynamic consequences of transcriptional feedback determine the functional effectiveness of some simple metabolic systems, as shown by Savageau (Igoshin *et al.*, 2007; Savageau, 1974). Some of these design principles have been extended and experimentally demonstrated (Brandman and Meyer, 2008; Kollmann *et al.*, 2005; Pang *et al.*, 2007; Rosenfeld *et al.*, 2002; Thattai and van Oudenaarden, 2001; Yu *et al.*, 2008). This research has indicated that negative feedback is likely to result in faster transient responses, robustness against fluctuations, and dynamical stability (Becskei and Serrano, 2000; Nevozhay *et al.*, 2009; Ray and Igoshin, 2010; Rosenfeld *et al.*, 2002). On the other hand, positive feedback is likely to result in slower response times, large range of responses, increased noise, and possibly multiple steady states (Becskei *et al.*, 2001; Igoshin *et al.*, 2007, 2008; Ray and Igoshin, 2010; Tiwari *et al.*, 2010; Veening *et al.*, 2008).

Feedback loops are also widespread in the transcriptional regulatory network of *M. tuberculosis*, since 31 of the 45 transcription factors included in the network are autoregulated (Fig. 12.2). Moreover, several of the sigma factors in Fig. 12.1 directly or indirectly upregulate their own transcription, thereby forming positive feedback loops, which can underlie bistability (Tiwari *et al.*, 2010). Since sigma factors function as monomers, the corresponding positive feedback loops are not expected to be cooperative and therefore may not lead to a bistable network output (Angeli *et al.*, 2004; Igoshin *et al.*, 2007; Tiwari *et al.*, 2010), unless cooperativity is generated via the post-translational modulation of sigma-factor activity (Tiwari *et al.*, 2010). In many cases, sigma and anti-sigma factors are expressed from a single operon that is controlled by an autoregulated promoter. The result is the stoichiometric combination of positive and negative feedback (Beaucher *et al.*, 2002; Igoshin *et al.*, 2007). This architecture is wide-spread in the stress response of Gram-positive bacteria, where it is associated with the increased regulation capacity required in stress-response (Igoshin *et al.*, 2007). Autoregulation is also common in TCSs, many of which are expressed from a single operon that is autoregulated by the activated response-regulator (Mitrophanov and Groisman, 2008a; Ray and Igoshin, 2010). TCS autoregulation also results in combined positive and negative feedback loops given the bifunctional role of the sensory kinase in modulating response-regulator activity. The MprA/MprB system is an example of autoregulated TCS in *M. tuberculosis* that is discussed below.

MprA/MprB/σ^E network in mycobacteria

As mentioned above, the alternative sigma factor σ^E and MprA/MprB are central to the stress response in *M. tuberculosis*. These regulators are highly conserved across mycobacterial species (Manganelli *et al.*, 2004), and disruption in the corresponding genes is associated with reduced virulence in mice (Ando *et al.*, 2003; Manganelli *et al.*, 2004; Zahrt *et al.*, 2003). These genes appear to function in a network module that involves mutual transcriptional regulation between activated MprA and σ^E (Fig. 12.4A). Mutual regulation gives rise to positive feedback, which is active only under conditions of stress that mimic dormancy-inducing environmental changes (Manganelli *et al.*, 2004). Stress-dependent induction of this module affects downstream genes essential for mycobacterial virulence and latency, such as relA (Balázsi *et al.*, 2008; Manganelli *et al.*, 1999; Sureka *et al.*, 2008; Zahrt and Deretic, 2001). Since the MprA/MprB/σ^E network plays an important role in controlling the execution of the gene expression programme during infection and in the switch to dormancy, its dynamical properties are crucial for understanding these processes. To study these properties it is essential to consider the interplay of transcriptional feedback loops and post-translational interactions. This requires detailed mathematical modelling.

As mentioned above, the MprA/MprB TCS consists of a sensor histidine kinase (MprB) and a transcriptional response regulator (MprA). MprB is a bifunctional enzyme that catalyses both the phosphorylation of the MprA response regulator and the dephosphorylation of phosphorylated MprA (Zahrt *et al.*, 2003). Phosphorylation of MprA is a two-step process in which MprB is first autophosphorylated in the presence of polyP as a phosphate donor, and then MprB phosphorylates MprA via a phosphotransfer reaction (Zahrt *et al.*, 2003). Unphosphorylated MprB also acts as a phosphatase on MprA. MprA can also be phosphorylated by small phosphodonor compounds, such as acetyl phosphate, in the absence of the cognate sensor kinase (Zahrt *et al.*, 2003). Phosphorylated MprA is transcriptionally active and induces its regulon, which includes mprAB and sigE (He and Zahrt, 2005).

Experiments involving sigE mutant and wild-type strains of mycobacteria suggest that σ^E regulates the transcription of the mprAB operon under conditions of stress, possibly by binding to an upstream promoter (Manganelli *et al.*, 2001). As a result, this network module contains two positive feedback loops. The uncommon occurrence of such multiple feedback architectures in bacteria (Shen-Orr *et al.*, 2002) raises the question of their physiological role.

Mathematical models and experiments with other networks have associated positive feedback with increased sensitivity and response range (Ferrell, 2008), slowing of transient dynamics (Savageau, 1974), amplification of gene expression noise (Hasty *et al.*, 2000), and possible bistability (Angeli *et al.*, 2004; Becskei *et al.*, 2001; Ozbudak *et al.*, 2004). In the mycobacterial stress response network, transcription of the downstream relA gene showed a bimodal distribution in a growing population of *M. smegmatis* (Sureka

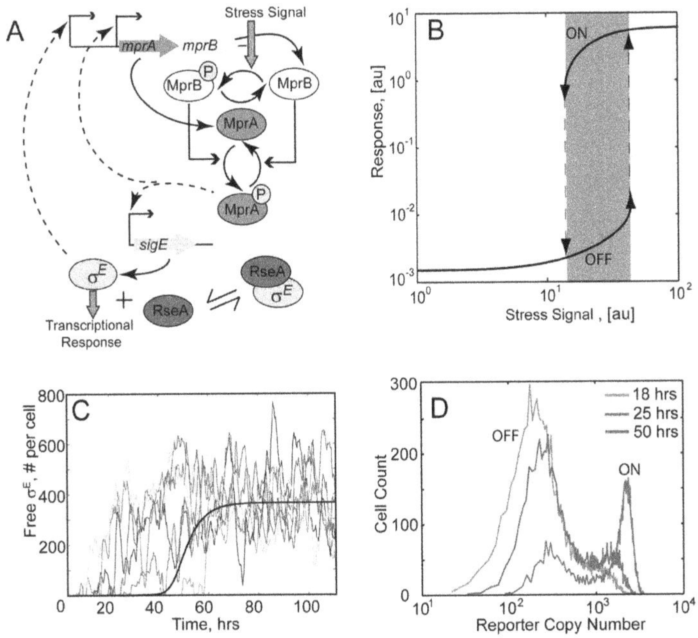

Figure 12.4 Architecture and dynamics of the mycobacterial stress-response circuit consisting of the MprA/MprB TCS and σE–RseA sigma/anti-sigma network. (A) The network involves posttranslational modulation of transcriptional activity of MprA via modulation of its phosphorylation state by MprB and modulation of σ^E–activity by RseA by sequestration. In addition, the circuit contains two feedback loops. Activated MprA~P upregulates the transcription of its own operon (direct loop) and activates production of σ^E which in its turn upregulates mprAB operon transcription form another promoter (indirect loop). Physiologically relevant signal and responses are indicated by grey arrows: the signal is modulation of MprB autophosphorylation (e.g. by polyp) whereas σ^E transcriptional activity of is a response. (B) Signal-response (bifurcation) diagram indicates that the network can display hysteretic bistability. Signal is modelled as autophosphorylation rate of MprB, the response is the concentration of free σ^E. For shaded signal range lines the cell can be in either of two stable steady states (OFF with low σ^E activity and ON with high σ^E activity). The fraction of cell population in either state will depend on the history initially uninduced or pre-induced population. At the boundaries of the bistable region small changes in signal lead to large jumps between steady states (indicated by arrows). (C) Noise in gene expression amplified by positive feedback leads to highly variable transitions between OFF and ON states. The black line denotes population-average (deterministic) response whereas coloured lines show stochastic trajectories of seven individual simulated cells. (D) As a result of slow and noisy switching kinetics the population will display bimodal distribution of σ^E-activity reporter such as relA. With time following the stress exposure the ON peak increases while OFF peak decreases. (B) to (D) are recreated from the data in Tiwari *et al.* (2009). A colour version of this figure is located in the plate section at the back of the book.

et al., 2008). With the help of an accompanying mathematical model, Sureka *et al* (2008) argued that bimodality is a consequence of bistability in the transcriptional response associated with the autoregulation of mprAB operon. This conclusion was based on the assumption that the phosphorylated form of MprB (MprB~P) possesses phosphatase activity towards MprA~P. However, experiments suggest that the non-phosphorylated (rather than the phosphorylated) form of MprB acts as a phosphatase in the above reaction (Zahrt *et al.*, 2003), as is the case with many other TCSs (Igoshin *et al.*, 2008; Shinar *et al.*, 2007). Indeed, when the corrected biochemical interaction and realistic induction ranges are considered (Tiwari *et al.*, 2010), the direct feedback due to the autoregulation of mprAB operon does not result in bistability. This is because the existence of positive feedback is necessary but not sufficient for network bistability (Angeli *et al.*, 2004). In addition to positive feedback, the signal propagation through the network must be cooperative. When the concentration of MprB, which possesses both phosphorylation and dephosphorylation activity towards MprA, is increased, the positive and negative effects counterbalance each other. As a result, the concentration of MprA~P will be independent of the operon expression level (Batchelor and Goulian, 2003; Shinar and Feinberg, 2010; Shinar *et al.*, 2007). Hence, the effect of feedback on system performance will be limited to the regime in which MprA phosphorylation is saturated and will be insufficient to produce network bistability under biochemically realistic parameter values (Tiwari *et al.*, 2010) (Fig. 12.4A).

The second feedback via σ^E does not help produce bistability either, because the two feedback loops act at different promoters and are therefore effectively coupled by an OR gate. That is, the increase in mprAB operon transcription is expected when either of the two feedback loops is activated. The effective cooperativity of the network was found to be a weighted average of the effective cooperativities of individual loops, and therefore it was smaller than the largest of the two (Tiwari *et al.*, 2010). The analysis also indicated that bistability in σ^E target can be associated with the post-translational regulation of σ^E by its anti-sigma factor RseA coupled with positive transcriptional feedback of σ^E to its own transcription via MprA/MprB TCS (Tiwari *et al.*, 2010). As a result, the network displays bistable and hysteretic response (concentration of free σ^E) as a function of MprA~P autophopshorylation rate (Fig. 12.4A).

Why is RseA so important for ultrasensitivity? In the network, the production of RseA is constitutive and is not regulated by σ^E or MprA~P, whereas the concentration of total σ^E present in the system increases with an increase in the concentration of active (unbound) σ^E due to indirect feedback via MprA/MprB. As long as the concentration of total σ^E is lower than that of RseA, most of it will be bound to RseA in an inactive complex, and very little active σ^E will be present in the system. When the synthesis rate of total σ^E exceeds that of RseA, the excess σ^E cannot be bound by RseA, thus resulting in a drastic increase in free σ^E (Fig. 12.4B). Hence, when the synthesis rate of total σ^E is equal to that of RseA, a small change in the synthesis rate of total σ^E can lead to a very large change in the concentrations of free σ^E. This 'ultrasensitive' mechanism is capable of generating very large network cooperativity when σ^E–RseA interaction strength is large.

What are the consequences of this bistability? When coupled with the noisy expression of network genes (Golding *et al.*, 2005), bistability manifests itself in the slow and noisy activation of downstream targets (such as relA) (Fig. 12.4C). As a result, their distribution in the population will be bimodal, with the first peak corresponding to cells that have not yet activated the sigE regulon and the second peak corresponding to activated cells (Fig. 12.4D). Even though the existence of such bimodal response and the resulting slow and noisy gene expression in *M. tuberculosis* are yet to be demonstrated, the physiological consequences would be quite interesting. For example, when subjected to a sudden stress signal, cells need to decide whether to execute the full-scale stress response immediately (e.g. switch to dormancy) or delay the response by waiting for conditions to improve. Either strategy can prove beneficial or detrimental depending on the duration and severity of the stress. Thus, by choosing a network architecture that amplifies noise and slows the average response time, the isogenic population of cells is able to hedge its bets against future uncertainty. We speculate that the MprAB-σ^E module is part of a larger dormancy switch network in which additional interlinked modules stochastically activated by their respective environmental signals ensure that the cell fate decision is made only in biologically desirable settings.

Conclusions

Advancing TB research requires quantitative approaches at a variety of different scales. One approach involves the analysis of the dynamics of small regulatory biomolecular circuits underlying critical cell fate decisions, such as latency, persistence, or stress response. A second approach involves further developing and utilizing a toolset for genome-scale (transcriptomic, proteomic) data analysis and interpretation. Both approaches (small-scale regulatory network dynamics and large-scale network utilization for data analysis) are needed, and they are complementary (Veiga *et al.*, 2010). Future work on TB dormancy should include investigating cell–cell communication and mapping interaction networks between bacterial cells and between bacteria and host cells.

Chapter highlights

- Understanding the environment-dependent, life cycle regulation of the pathogen, *Mycobacterium tuberculosis*, is key for the success of future anti-tuberculosis therapies.
- Two-component systems and alternative sigma factors are extensively studied in the context of environmental sensing and response.
- A genome-scale *M. tuberculosis* transcriptional regulatory network has been assembled from a variety of data sources and used to analyse network response to growth arrest due to hypoxia.
- Post-translational modulation of transcriptional activity is essential for mechanistic understanding of transcriptional network responses.
- Interplay of multiple feedback loops and post-translational regulation can result in bistability of the mycobacterial stress response network module containing the two-component system MprAB and alternative sigma factor σ^E.

Acknowledgements

We would like to thank Helena Boshoff for sharing her microarray data from (Boshoff *et al.*, 2004); Lanbo Shi for preparing the microarray data analysed in Fig. 12.3C, Abhinav Tiwari for useful comments and helping with Fig. 12.4; and Karl Drlica for comments on the manuscript. Support is acknowledged from NIH grants GM-096189, HL-106788, and AI-095924, and from the TB PAN-NET consortium funded by the European Commission's Seventh Framework Programme for Research (FP7-22368).

References

Ando, M., Yoshimatsu, T., Ko, C., Converse, P.J., and Bishai, W.R. (2003). Deletion of *Mycobacterium tuberculosis* sigma factor E results in delayed time to death with bacterial persistence in the lungs of aerosol-infected mice. Infect. Immun. *71*, 7170–7172.

Angeli, D., Ferrell, J.E., Jr., and Sontag, E.D. (2004). Detection of multistability, bifurcations, and hysteresis in a large class of biological positive-feedback systems. Proc. Natl. Acad. Sci. U.S.A. *101*, 1822–1827.

Babu, M., Teichmann, S., and Aravind, L. (2006). Evolutionary dynamics of prokaryotic transcriptional regulatory networks. J. Mol. Biol. *358*, 614–633.

Balázsi, G., Barabási, A.L., and Oltvai, Z.N. (2005). Topological units of environmental signal processing in the transcriptional regulatory network of *Escherichia coli*. Proc. Natl. Acad. Sci. U.S.A. *102*, 7841–7846.

Balázsi, G., Heath, A.P., Shi, L., and Gennaro, M.L. (2008). The temporal response of the *Mycobacterium tuberculosis* gene regulatory network during growth arrest. Mol. Syst. Biol. *4*, 225.

Batchelor, E., and Goulian, M. (2003). Robustness and the cycle of phosphorylation and dephosphorylation in a two-component regulatory system. Proc. Natl. Acad. Sci. U.S.A. *100*, 691–696.

Beaucher, J., Rodrigue, S., Jacques, P.E., Smith, I., Brzezinski, R., and Gaudreau, L. (2002). Novel *Mycobacterium tuberculosis* anti-sigma factor antagonists control sigmaF activity by distinct mechanisms. Mol. Microbiol. *45* 1527–1540.

Becskei, A., Seraphin, B., and Serrano, L. (2001). Positive feedback in eukaryotic gene networks: cell differentiation by graded to binary response conversion. EMBO J. *20*, 2528–2535.

Becskei, A., and Serrano, L. (2000). Engineering stability in gene networks by autoregulation. Nature *405*, 590–593.

Bonneau, R. (2008). Learning biological networks: from modules to dynamics. Nat. Chem. Biol. *4*, 658–664.

Boshoff, H.I., Myers, T.G., Copp, B.R., McNeil, M.R., Wilson, M.A., and Barry, C.E., 3rd (2004). The transcriptional responses of *Mycobacterium tuberculosis* to inhibitors of metabolism: novel insights into drug mechanisms of action. J. Biol. Chem. *279*, 40174–40184.

Brandman, O., and Meyer, T. (2008). Feedback loops shape cellular signals in space and time. Science *322*, 390–395.

Brantl, S. (2009). Bacterial chromosome-encoded small regulatory RNAs. Future Microbiol. *4*, 85–103.

Calamita, H., Ko, C., Tyagi, S., Yoshimatsu, T., Morrison, N.E., and Bishai, W.R. (2005). The *Mycobacterium tuberculosis* SigD sigma factor controls the expression of ribosome-associated gene products in stationary phase and is required for full virulence. Cell. Microbiol. *7*, 233–244.

Chen, P., Ruiz, R.E., Li, Q., Silver, R.F., and Bishai, W.R. (2000). Construction and characterization of a

Mycobacterium tuberculosis mutant lacking the alternate sigma factor gene, sigF. Infect. Immun. *68*, 5575–5580.

Clarke, M., and Sperandio, V. (2005). Transcriptional autoregulation by quorum sensing *Escherichia coli* regulators B and C (QseBC). in enterohaemorrhagic *E. coli* (EHEC). Mol. Microbiol. *58*, 441–455.

Converse, P.J., Karakousis, P.C., Klinkenberg, L.G., Kesavan, A.K., Ly, L.H., Allen, S.S., Grosset, J.H., Jain, S.K., Lamichhane, G., Manabe, Y.C., *et al.* (2009). Role of the dosR-dosS two-component regulatory system in *Mycobacterium tuberculosis* virulence in three animal models. Infect. Immun. *77*, 1230–1237.

Cox, M.M. (2007). Regulation of bacterial RecA protein function. Crit. Rev. Biochem. Mol. Biol. *42*, 41–63.

Dainese, E., Rodrigue, S., Delogu, G., Provvedi, R., Laflamme, L., Brzezinski, R., Fadda, G., Smith, I., Gaudreau, L., Palu, G., and Manganelli, R. (2006). Posttranslational regulation of *Mycobacterium tuberculosis* extracytoplasmic-function sigma factor sigma L and roles in virulence and in global regulation of gene expression. Infect. Immun. *74*, 2457–2461.

Dona, V., Rodrigue, S., Dainese, E., Palu, G., Gaudreau, L., Manganelli, R., and Provvedi, R. (2008). Evidence of complex transcriptional, translational, and posttranslational regulation of the extracytoplasmic function sigma factor sigmaE in *Mycobacterium tuberculosis*. J. Bacteriol. *190*, 5963–5971.

Ewann, F., Jackson, M., Pethe, K., Cooper, A., Mielcarek, N., Ensergueix, D., Gicquel, B., Locht, C., and Supply, P. (2002). Transient requirement of the PrrA-PrrB two-component system for early intracellular multiplication of *Mycobacterium tuberculosis*. Infect. Immun. *70*, 2256–2263.

Faith, J.J., Hayete, B., Thaden, J.T., Mogno, I., Wierzbowski, J., Cottarel, G., Kasif, S., Collins, J.J., and Gardner, T.S. (2007). Large-scale mapping and validation of *Escherichia coli* transcriptional regulation from a compendium of expression profiles. PLoS Biol. *5*, e8.

Ferrell, J.E., Jr. (2008). Feedback regulation of opposing enzymes generates robust, all-or-none bistable responses. Curr Biol *18*, R244–245.

Fontan, P.A., Voskuil, M.I., Gomez, M., Tan, D., Pardini, M., Manganelli, R., Fattorini, L., Schoolnik, G.K., and Smith, I. (2009). The *Mycobacterium tuberculosis* sigma factor sigmaB is required for full response to cell envelope stress and hypoxia *in vitro*, but it is dispensable for in vivo growth. J. Bacteriol. *191*, 5628–5633.

Golding, I., Paulsson, J., Zawilski, S.M., and Cox, E.C. (2005). Real-time kinetics of gene activity in individual bacteria. Cell *123*, 1025–1036.

Graham, J.E., and Clark-Curtiss, J.E. (1999). Identification of *Mycobacterium tuberculosis* RNAs synthesized in response to phagocytosis by human macrophages by selective capture of transcribed sequences (SCOTS). Proc. Natl. Acad. Sci. U.S.A. *96*, 11554–11559.

Gruber, T.M., and Gross, C.A. (2003). Multiple sigma subunits and the partitioning of bacterial transcription space. Annu. Rev. Microbiol. *57*, 441–466.

Hasty, J., Pradines, J., Dolnik, M., and Collins, J.J. (2000). Noise-based switches and amplifiers for gene expression. Proc. Natl. Acad. Sci. U.S.A. *97*, 2075–2080.

He, H., and Zahrt, T.C. (2005). Identification and characterization of a regulatory sequence recognized by *Mycobacterium tuberculosis* persistence regulator MprA. J. Bacteriol. *187*, 202–212.

Helmann, J.D. (1999). Anti-sigma factors. Curr. Opin. Microbiol. *2*, 135–141.

Igoshin, O.A., Alves, R., and Savageau, M.A. (2008). Hysteretic and graded responses in bacterial two-component signal transduction. Mol *68* Microbiol 1196–1215.

Igoshin, O.A., Brody, M.S., Price, C.W., and Savageau, M.A. (2007). Distinctive topologies of partner-switching signaling networks correlate with their physiological roles. J. Mol. Biol. *369*, 1333–1352.

Jacques, P.E., Gervais, A.L., Cantin, M., Lucier, J.F., Dallaire, G., Drouin, G., Gaudreau, L., Goulet, J., and Brzezinski, R. (2005). MtbRegList, a database dedicated to the analysis of transcriptional regulation in *Mycobacterium tuberculosis*. Bioinformatics *21*, 2563–2565.

Kaushal, D., Schroeder, B.G., Tyagi, S., Yoshimatsu, T., Scott, C., Ko, C., Carpenter, L., Mehrotra, J., Manabe, Y.C., Fleischmann, R.D., and Bishai, W.R. (2002). Reduced immunopathology and mortality despite tissue persistence in a *Mycobacterium tuberculosis* mutant lacking alternative sigma factor, SigH. Proc. Natl. Acad. Sci. U.S.A. *99*, 8330–8335.

Kollmann, M., Lovdok, L., Bartholome, K., Timmer, J., and Sourjik, V. (2005). Design principles of a bacterial signalling network. Nature *438*, 504–507.

Kumar, A., Toledo, J.C., Patel, R.P., Lancaster, J.R., Jr., and Steyn, A.J. (2007). *Mycobacterium tuberculosis* DosS is a redox sensor and DosT is a hypoxia sensor. Proc. Natl. Acad. Sci. U.S.A. *104*, 11568–11573.

Kussell, E., and Leibler, S. (2005). Phenotypic diversity, population growth, and information in fluctuating environments. Science *309*, 2075–2078.

Manganelli, R. (2007). Polyphosphate and stress response in mycobacteria. Mol. Microbiol. *65*, 258–260.

Manganelli, R., Dubnau, E., Tyagi, S., Kramer, F.R., and Smith, I. (1999). Differential expression of 10 sigma factor genes in *Mycobacterium tuberculosis*. Mol. Microbiol. *31*, 715–724.

Manganelli, R., Fattorini, L., Tan, D., Iona, E., Orefici, G., Altavilla, G., Cusatelli, P., and Smith, I. (2004). The extra cytoplasmic function sigma factor sigma(E). is essential for *Mycobacterium tuberculosis* virulence in mice. Infect. Immun. *72*, 3038–3041.

Manganelli, R., Voskuil, M.I., Schoolnik, G.K., Dubnau, E., Gomez, M., and Smith, I. (2002). Role of the extracytoplasmic-function sigma factor sigma(H). in *Mycobacterium tuberculosis* global gene expression. Mol. Microbiol. *45*, 365–374.

Manganelli, R., Voskuil, M.I., Schoolnik, G.K., and Smith, I. (2001). The *Mycobacterium tuberculosis* ECF sigma factor sigma E: role in global gene expression and survival in macrophages. Mol. Microbiol. *41*, 423–437.

Mitrophanov, A.Y., and Groisman, E.A. (2008a). Positive feedback in cellular control systems. Bioessays *30*, 542–555.

Mitrophanov, A.Y., and Groisman, E.A. (2008b). Signal integration in bacterial two-component regulatory systems. Genes Dev. *22*, 2601–2611.

Nevozhay, D., Adams, R.M., Murphy, K.F., Josic, K., and Balazsi, G. (2009). Negative autoregulation linearizes the dose–response and suppresses the heterogeneity of gene expression. Proc. Natl. Acad. Sci. U.S.A. *106*, 5123–5128.

Ozbudak, E.M., Thattai, M., Lim, H.N., Shraiman, B.I., and Van Oudenaarden, A. (2004). Multistability in the lactose utilization network of *Escherichia coli*. Nature *427*, 737–740.

Pang, X., Vu, P., Byrd, T.F., Ghanny, S., Soteropoulos, P., Mukamolova, G.V., Wu, S., Samten, B., and Howard, S.T. (2007). Evidence for complex interactions of stress-associated regulons in an mprAB deletion mutant of *Mycobacterium tuberculosis*. Microbiology *153*, 1229–1242.

Parish, T., Smith, D.A., Roberts, G., Betts, J., and Stoker, N.G. (2003). The senX3-regX3 two-component regulatory system of *Mycobacterium tuberculosis* is required for virulence. Microbiology *149*, 1423–1435.

Raman, S., Song, T., Puyang, X., Bardarov, S., Jacobs, W.R., Jr., and Husson, R.N. (2001). The alternative sigma factor SigH regulates major components of oxidative and heat stress responses in *Mycobacterium tuberculosis*. J. Bacteriol. *183*, 6119–6125.

Ray, J.C., and Igoshin, O.A. (2010). Adaptable functionality of transcriptional feedback in bacterial two-component systems. PLoS Comput Biol *6*, e1000676.

Ray, J.C., Tabor, J.J., and Igoshin, O.A. (2011). Non-transcriptional regulatory processes shape transcriptional network dynamics. Nat. Rev. Microbiol. *9*, 817–828.

Roback, P., Beard, J., Baumann, D., Gille, C., Henry, K., Krohn, S., Wiste, H., Voskuil, M.I., Rainville, C., and Rutherford, R. (2007). A predicted operon map for *Mycobacterium tuberculosis*. Nucleic Acids Res. *35*, 5085–5095.

Rodrigue, S., Provvedi, R., Jacques, P.E., Gaudreau, L., and Manganelli, R. (2006). The sigma factors of *Mycobacterium tuberculosis*. FEMS Microbiol. Rev. *30*, 926–941.

Rosenfeld, N., Elowitz, M.B., and Alon, U. (2002). Negative autoregulation speeds the response times of transcription networks. J. Mol. Biol. *323*, 785–793.

Rustad, T.R., Harrell, M.I., Liao, R., and Sherman, D.R. (2008). The enduring hypoxic response of *Mycobacterium tuberculosis*. PLoS ONE 3, e1502.

Ryndak, M., Wang, S., and Smith, I. (2008). PhoP, a key player in *Mycobacterium tuberculosis* virulence. Trends Microbiol. *16*, 528–534.

Sachdeva, P., Misra, R., Tyagi, A.K., and Singh, Y. (2010). The sigma factors of *Mycobacterium tuberculosis*: regulation of the regulators. Febs J *277*, 605–626.

Saini, D.K., Malhotra, V., Dey, D., Pant, N., Das, T.K., and Tyagi, J.S. (2004). DevR-DevS is a bona fide two-component system of *Mycobacterium tuberculosis* that is hypoxia-responsive in the absence of the DNA-binding domain of DevR. Microbiology *150*, 865–875.

Salgado, H., Santos-Zavaleta, A., Gama-Castro, S., Peralta-Gil, M., Penaloza-Spinola, M.I., Martinez-Antonio, A., Karp, P.D., and Collado-Vides, J. (2006). The comprehensive updated regulatory network of *Escherichia coli* K-12. BMC Bioinformatics 7, 5.

Savageau, M.A. (1974). Comparison of classical and autogenous systems of regulation in inducible operons. Nature 252, 546–549.

Shen-Orr, S.S., Milo, R., Mangan, S., and Alon, U. (2002). Network motifs in the transcriptional regulation network of *Escherichia coli*. Nat. Genet. *31*, 64–68.

Sherman, D.R., Voskuil, M., Schnappinger, D., Liao, R., Harrell, M.I., and Schoolnik, G.K. (2001). Regulation of the *Mycobacterium tuberculosis* hypoxic response gene encoding alpha -crystallin. Proc. Natl. Acad. Sci. U.S.A. *98*, 7534–7539.

Shi, L., Jung, Y.J., Tyagi, S., Gennaro, M.L., and North, R.J. (2003). Expression of Th1-mediated immunity in mouse lungs induces a *Mycobacterium tuberculosis* transcription pattern characteristic of nonreplicating persistence. Proc. Natl. Acad. Sci. U.S.A. *100*, 241–246.

Shinar, G., and Feinberg, M. (2010). Structural sources of robustness in biochemical reaction networks. Science *327*, 1389–1391.

Shinar, G., Milo, R., Martinez, M.R., and Alon, U. (2007). Input output robustness in simple bacterial signaling systems. Proc. Natl. Acad. Sci. U.S.A. *104*, 19931–19935.

Soncini, F.C., Véscovi, E.G., and Groisman, E.A. (1995). Transcriptional autoregulation of the Salmonella typhimurium phoPQ operon. 0021–9193 *177*, 4364–4371.

Song, T., Dove, S.L., Lee, K.H., and Husson, R.N. (2003). RshA, an anti-sigma factor that regulates the activity of the mycobacterial stress response sigma factor SigH. Mol. Microbiol. *50*, 949–959.

Stock, A.M., Robinson, V.L., and Goudreau, P.N. (2000). Two-component signal transduction. Annu Rev Biochem *69*, 183–215.

Sun, R., Converse, P.J., Ko, C., Tyagi, S., Morrison, N.E., and Bishai, W.R. (2004). *Mycobacterium tuberculosis* ECF sigma factor sigC is required for lethality in mice and for the conditional expression of a defined gene set. Mol. Microbiol. *52*, 25–38.

Sureka, K., Dey, S., Datta, P., Singh, A.K., Dasgupta, A., Rodrigue, S., Basu, J., and Kundu, M. (2007). Polyphosphate kinase is involved in stress-induced mprAB-sigE-rel signalling in mycobacteria. Mol. Microbiol. *65*, 261–276.

Sureka, K., Ghosh, B., Dasgupta, A., Basu, J., Kundu, M., and Bose, I. (2008). Positive feedback and noise activate the stringent response regulator rel in mycobacteria. PLoS One 3, e1771.

Thattai, M., and van Oudenaarden, A. (2001). Intrinsic noise in gene regulatory networks. Proc. Natl. Acad. Sci. U.S.A. *98*, 8614–8619.

Tiwari, A., Balazsi, G., Gennaro, M.L., and Igoshin, O.A. (2010). The interplay of multiple feedback loops with post-translational kinetics results in bistability of mycobacterial stress response. Phys Biol 7, 036005.

Veening, J.W., Igoshin, O.A., Eijlander, R.T., Nijland, R., Hamoen, L.W., and Kuipers, O.P. (2008). Transient heterogeneity in extracellular protease production by *Bacillus subtilis*. Mol Syst Biol 4, 184.

Veiga, D.F., Dutta, B., and Balazsi, G. (2010). Network inference and network response identification: moving genome-scale data to the next level of biological discovery. Mol Biosyst *6*, 469–480.

Voskuil, M.I., Visconti, K.C., and Schoolnik, G.K. (2004). *Mycobacterium tuberculosis* gene expression during adaptation to stationary phase and low-oxygen dormancy. Tuberculosis (Edinb). *84*, 218–227.

White, M.J., He, H., Penoske, R.M., Twining, S.S., and Zahrt, T.C. (2010). PepD participates in the mycobacterial stress response mediated through MprAB and SigE. J. Bacteriol. *192*, 1498–1510.

Won, H.S., Lee, Y.S., Lee, S.H., and Lee, B.J. (2009). Structural overview on the allosteric activation of cyclic AMP receptor protein. Biochim Biophys Acta *1794*, 1299–1308.

Yu, R., Pesce, G., Colman-Lerner, A., Lok, L., Pincus, D., Serra, E., Holl, M., Benjamin, K., Gordon, A., and Brent, R. (2008). Negative feedback that improves information transmission in yeast signalling. Nature *456*, 755–761.

Zahrt, T.C., and Deretic, V. (2001). *Mycobacterium tuberculosis* signal transduction system required for persistent infections. Proc. Natl. Acad. Sci. U.S.A. *98*, 12706–12711.

Zahrt, T.C., Wozniak, C., Jones, D., and Trevett, A. (2003). Functional analysis of the *Mycobacterium tuberculosis* MprAB two-component signal transduction system. Infect. Immun. *71*, 6962–6970.

Transcriptional Regulatory Network in *Pseudomonas aeruginosa* 13

Deepak Balasubramanian, Senthil Kumar Murugapiran, Eugenia Silva-Herzog, Lisa Schneper, Xing Yang, Gorakh Tatke, Giri Narasimhan and Kalai Mathee

Abstract

Pseudomonas aeruginosa is found in a wide range of habitats, primarily in soil and water and is the epitome of opportunistic human pathogens. A myriad of virulence factors produced by the bacterium ensure its success as a pathogen. *P. aeruginosa* has one of the largest genomes among eubacteria and transcriptional regulators constitute about 8% of the genome. Sequence analyses of the regulators belonging to different families shows clustering while network analysis shows extensive crosstalk, and reveals empirically identified and novel interactions between regulators. Gene expression in *P. aeruginosa* is an intricately interlinked process and is exemplified in the regulation of virulence factor expression. Major regulatory processes such as quorum sensing involving multiple regulators translate external signals perceived by the bacterium into gene expression/repression via regulatory cascades. Many global regulators have been identified that serve to link different virulence systems. Understanding the role of the as yet uncharacterized transcriptional regulatory proteins will provide important insights into the physiology of this important human pathogen and has potential therapeutic implications.

Introduction

Pseudomonas aeruginosa is an environmental saprophyte that can cause a number of opportunistic infections in humans including infections of the pulmonary and urinary tracts, and skin (Gregory *et al.*, 2007; Shigemura *et al.*, 2006; Takeyama *et al.*, 2002). It is notably known for its ability to cause acute (Estahbanati *et al.*, 2002; Mahar *et al.*, 2010; Marra *et al.*, 2006; Secher *et al.*, 2005) and chronic (Hassett *et al.*, 2010; Melendez *et al.*, 2010; Murray *et al.*, 2007) infections. *P. aeruginosa* is inherently resistant to antibiotics and can also acquire resistance genes by horizontal gene transfer (Kerr and Snelling, 2009). The ability of *P. aeruginosa* to infect animals, nematodes, plants and insects is owing to the expression of a wide range of virulence factors that are either cell-associated or secreted. Cell-associated factors are involved primarily in adherence and colonization (pili, LPS, flagella) and motility (type IV pili, flagella). In addition, some components have multiple functions. The LPS, for example induces the immune system and has been linked to dysregulation of the host response and sepsis (Cohen, 2002). In chronic infections, however, LPS modifications have been identified that make the bacteria less 'visible' to the immune system (Rocchetta *et al.*, 1999). In addition, *P. aeruginosa* secretes a plethora of extracellular virulence factors; a majority of them (LasA, LasB, protease IV, exotoxin A, lipases, phospholipase C) secreted via the Xcp Type II secretion system (Rosenau and Jaeger, 2000; Senf *et al.*, 2008). *P. aeruginosa* can also specifically deliver toxins directly into the host cell using the sophisticated Type III and Type VI secretion systems (T3SS and T6SS, respectively) (Filloux *et al.*, 2008; Hauser, 2009).

Pathogenesis of *P. aeruginosa* is thus due to the concerted action of multiple virulence mechanisms and is typically not due to the specific action of a single virulence factor. An intricate network of regulation involving multiple transcriptional regulators (transcriptional control), and small RNA-binding proteins and regulatory

RNA (post-transcriptional control) orchestrate the timed co-expression of multiple virulence mechanisms. At the heart of the *P. aeruginosa* pathogenesis network is quorum sensing (QS), a system of interconnected sensors and regulators that control expression of virulence and other bacterial traits in a cell-density dependent manner (Ng and Bassler, 2009). Key to the QS regulatory network are autoinducers (also known as quormones): small diffusible chemical signals that help bacteria sense self and non-self population densities and coordinate expression of virulence factors under high cell numbers thus increasing the chances of a successful infection (Ng and Bassler, 2009). To further enhance survival *P. aeruginosa*, like other bacteria, has evolved to combine signal recognition with transcriptional regulation for a quick and efficient response to changes in the environment. Critical to this process are the two-component systems (TCS) comprised of an inner membrane-bound sensor kinase that transduces the signal to a cytoplasmic response regulator, either directly or via a phosphorelay. In some cases, the two components are part of the same protein, the so-called hybrid sensor kinases that are known to be critical for *P. aeruginosa* pathogenesis (Goodman *et al.*, 2004; Laskowski *et al.*, 2004; Ventre *et al.*, 2006; Zolfaghar *et al.*, 2005). Auxiliary regulators connect histidine kinases with regulatory processes forming three-component regulatory signalling systems (Buelow and Raivio, 2010).

If genome sizes are any indication of complexity, *P. aeruginosa* is indeed a highly complex bacterium. With strains harbouring 6–7 MB chromosomes (Klockgether *et al.*, 2010; Mathee *et al.*, 2008), *P. aeruginosa* has one of the largest genomes in bacteria closely approaching that of lower eukaryotes such as *Saccharomyces*. In bacteria like *P. aeruginosa*, where the genome repertoire far exceeds many others, regulatory genes are acquired through horizontal gene transfer or evolution of paralogues that have new specificities for new sets of target genes. This is underscored by the comparative analyses of transcription factors from 175 prokaryotic genomes showing reduced conservation as compared to their target genes (Madan Babu *et al.*, 2006). However, such extensive analysis still precludes our ability to predict the new lineage-specific transcription factors and the new interactions that continue to evolve as the organisms fight for their survival in a myriad of environments. Thus it is not surprising that a single species, such as *P. aeruginosa* with its ability to thrive in diverse environments, has a mosaic genome consisting of a conserved core component interrupted in each strain by combinations of specific blocks of genes (Mathee *et al.*, 2008). These strain-specific segments of the genome are found in limited chromosomal locations, referred to as regions of genomic plasticity (RGP), and favours survival by enhancing the bacteria's metabolic and/or virulence capabilities (Mathee *et al.*, 2008).

Transcriptional regulation in *P. aeruginosa* is a complex and intricately interlinked process. Many 'global' regulators have been identified (Vfr, MvfR, AmpR, Rsm, LasR, RhlR) that have a profound effect on the gene expression pattern. Though many transcriptomic studies have looked at global scale expression of genes, understanding the role of transcription factors, structure–function and interaction analyses and the associated networks in *P. aeruginosa* still lags behind *Escherichia coli*. However, certain networks such as QS and alginate regulation are much better characterized in *P. aeruginosa* than in others. The key to understanding *P. aeruginosa* pathogenesis is to have a unified view of how the different regulators and regulatory networks are linked. This review will focus on the genome composition of the transcription factors in *P. aeruginosa* and attempt to glean meaningful interactions.

Families of transcriptional regulators in *P. aeruginosa*

Transcriptional regulators are grouped into numerous families based on their functional domains, particularly DNA and effector binding domains. Even though many regulators have been studied in *P. aeruginosa* and *in silico* analyses group them into different families (Fig. 13.1), experimental evidence assigning function for most of the 434 regulators is still lacking. Phylogenetic analysis shows that not all members fall into tight groups indicating that sequence homology is not sufficient to cluster all the members (Fig. 13.1).

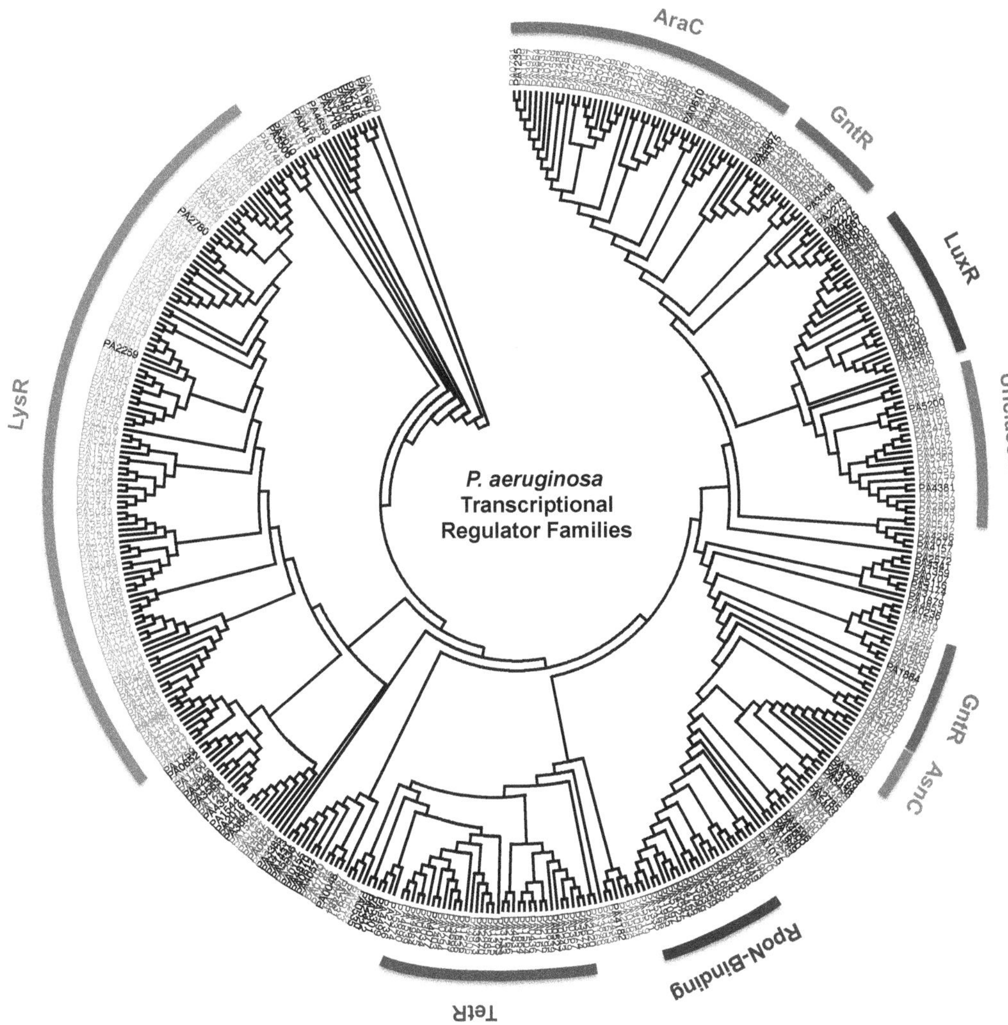

Figure 13.1 *P. aeruginosa* transcriptional regulatory protein families. A total of 434 amino acid sequences of transcription factors (Table A.1) were aligned using ClustalW (Version 2.1.10) (Larkin *et al.*, 2007). The first tree out of 31 most parsimonious trees (length = 69259) is shown. The consistency index is 0.155474 (0.147175), the retention index is 0.298198 (0.218198), and the composite index is 0.046362 (0.043887) for all the sites and parsimony-informative sites (in parentheses). The maximum parsimony tree was obtained using the Close-Neighbour-Interchange algorithm (Nei and Kumar, 2000) with search level 2 (Eck and Dayhoff, 1966; Nei and Kumar, 2000) in which the initial trees were obtained with the random addition of sequences (10 replicates). All alignment gaps were treated as missing data. There were a total of 1137 positions in the final dataset, out of which 743 were parsimony informative. Phylogenetic analyses were conducted in MEGA4 (Tamura *et al.*, 2007). The different families that cluster together are coloured. Members of families that do not form tight clusters (IclR, LacI/GalR, MerR, Cro-C1, CRP, ArsR, H-NS, OmpC, RpiR, and Dks/TraR) are coloured black. A colour version of this figure is located in the plate section at the back of the book.

Some families, like MerR and Cro-cI (labelled in black in Fig. 13.1), are scattered all over the tree. Others, like the GntR family, form multiple clusters, while some members of well-formed clusters fall in other clusters. For example, PA5261 of the LysR family is grouped with one of the GntR clusters (Fig. 13.1). Interestingly, the unclassified proteins, which do not have a known signature sequence, cluster together (Fig. 13.1). Experimental data such as structure analysis will help classify these proteins as well as the outliers into families.

This section deals with the major transcriptional regulator families in *P. aeruginosa* PAO1 and their prominent members. The characteristics of the major families are listed in Table 13.1.

LysR family

LysR-type transcriptional regulators (LTTRs) are the largest family of prokaryotic transcriptional regulators (Maddocks and Oyston, 2008; Schell, 1993). Many LTTRs are global activators/repressors of target gene expression. The targets are either divergently transcribed from them or are part of unlinked regulons at different locations on the chromosome (Hernandez-Lucas *et al.*, 2008; Heroven and Dersch, 2006). Divergent transcription from the target gene allows for autoregulation, though there are exceptions (Kong *et al.*, 2005; Maddocks and Oyston, 2008). This family is the namesake for the prototypic member LysR that activates *lysA* transcription involved in lysine biosynthesis in *E. coli* (Stragier *et al.*, 1983). LTTR members typically have a N-terminal winged

Table 13.1 Characteristics of the major families of transcriptional regulators in *P. aeruginosa*

Family	# in *P. aeruginosa*	# in *E. coli*	DNA binding motif	Other domains	Regulation	Oligomerization	DNA-binding without effector	Prototype
LysR	127	45	N-term wHTH	C-term effector-binding; a DNA/effector-binding	Activator/repressor	Dimer/tetramer	Yes	*E. coli* LysR
AraC/XylS	61	21	C-term HTH	N-term dimerization/sugar-binding	Activator	Dimer	Yes	*E. coli* AraC
TetR	39	11	N-term HTH	C-term effector-binding/oligomerization	Repressor	Dimer/octamer	Yes	Enterobacterial TetR
LuxR	32	17	C-term HTH	N-term regulatory domain	Activator	Multimer	Yes	*V. fisheri* LuxR
GntR	31	3	N-term HTH	C-term effector-binding/oligomerization	Repressor/activator	Homodimer	Yes	*B. subtilis* GntR
RpoN-binding	22	2	C-term HTH	Central RpoN-interacting; N-term effector-binding	Activator	Dimer/multimer	Yes	*E. coli* NtrC
Cro-C1	17	3	N-term HTH	C-term oligomerization	Repressor	Homodimer	Yes	Phage 434 (Cro); Phage lambda (C1)
IclR	9	8	N-term wHTH	C-term effector-binding/oligomerization	Repressor/activator	Dimer/tetramer	Yes	*E. coli* IclR
AsnC	9	3	N-term HTH	C-term effector-binding/oligomerization	Repressor/activator	Multimer	Yes	*E. coli* AsnC
LacI	7	15	N-term HTH	C-term effector-binding/oligomerization	Repressor	Dimer/tetramer	Yes	*E. coli* LacI
MerR	7	5	N-term HTH	C-term effector-binding	Activator/repressor	Dimer	Yes	*E. coli* MerR
ArsR	4	2	N-term HTH	C-term effector-binding/dimerization	Repressor	Homodimer	Yes	*E. coli* ArsR

helix–turn–helix (wHTH)-DNA binding domain (residues 1–65), C-terminus effector-binding domain (residues 100–173 and 196–206) and a domain that is needed for both functions (residues 227–253) (Schell, 1993). DNA binding, either as dimers or tetramers, occurs at the target promoters even in the absence of the effector signals. In most cases, co-inducer binding causes a change in footprint resulting in DNA bending and transcription activation (Maddocks and Oyston, 2008).

P. aeruginosa PAO1 has 127 LTTR members that despite controlling diverse functions (Table A.1) group together on amino acid sequence analysis (Fig. 13.1). Though the functions of many members remain unknown, the few that are known are regulators of critical processes including virulence in *P. aeruginosa*. MvfR (PA1003), for example regulates QS in a LasIR/RhlIR-independent manner by modulating PqsE (PA1000), the quinolone response signal protein (Cao *et al.*, 2001; Deziel *et al.*, 2005). The chromosomal ß-lactamase regulator AmpR (PA4109) is a global regulator of virulence in *P. aeruginosa*. In addition to regulating ß-lactam and non-ß-lactam resistance, AmpR also positively modulates expression of many acute virulence factors and negatively regulates chronic infection phenotypes (Balasubramanian *et al.*, 2011, 2012; Kong *et al.*, 2005). OxyR (PA5344) is an example of a regulator controlling expression of multiple related phenotypes. Under oxidative stress, OxyR enhances expression of antioxidant genes like *katB*, *ahpCF* and *ahpB* (Ochsner *et al.*, 2000). In addition, OxyR also plays a role in virulence (Lau *et al.*, 2005), cytotoxicity (Melstrom *et al.*, 2007), pyoverdine utilization, and pyocyanin and rhamnolipid production (Vinckx *et al.*, 2008, 2010). Turner *et al.* recently identified the bistable regulon of a LysR-type regulator BexR (PA2432) that regulates virulence factors like AprA (Turner *et al.*, 2009). Other characterized LysR-type members include an efflux pump regulator, MexT (PA2492) that has been recently shown to regulate other genes (Jin *et al.*, 2010), and MetR (PA3587), a regulator of genes involved in methionine synthesis (Yeung *et al.*, 2009). Except for PA0207 (RGP1), PA1223 (RGP11), PA2054 (RGP20), PA2056 (RGP20), and PA2220 (RGP53), all LysR members are part of the core genome (Table A.1).

IclR family

The isocitrate lyase regulator (IclR) family controls gene expression in prokaryotes in a signal-dependent manner (Molina-Henares *et al.*, 2006). IclR members are distributed in over 46 species of eubacteria and archea and typically regulate genes and operons involved in carbon metabolism or virulence. In the absence of specific substrates, IclR regulators typically repress specific catabolic genes, and in the presence of excess signalling molecules, de-repress expression (Krell *et al.*, 2006). Two domains in the protein bring about regulation: a winged HTH domain at the N-terminus that binds DNA, and an effector-binding domain at the C-terminus. IclR-like regulators form dimers or tetramers and recognize palindromic sequences on the DNA (Molina-Henares *et al.*, 2006). Some of the well-studied members of this family include the family namesake in *E. coli* IclR, a glyoxalate shunt repressor of the *aceBAK* operon involved in acetate utilization, the glycerol catabolism pathway repressor GlyR in *Streptomyces coelicolor* and KdgR, the pectin degradation pathway repressor in *Erwinia chrysanthemi*. A *Thermatoga maritima* IclR member TM0065, involved in xylulose metabolism was the first member of this family to be crystallized (Zhang *et al.*, 2002).

P. aeruginosa PAO1 has nine IclR homologues and all of them are yet to be characterized. A PA0155 (PcaR) homologue in *P. putida* has been demonstrated to regulate the *pca* operon involved in *p*-hydroxybenzoate degradation by binding both the –35 and -10 regions in the promoter (Romero-Steiner *et al.*, 1994). PA1015 is 42% similar to GlyR, a glycerol catabolic operon (*gylABX*) regulatory protein in *S. coelicolor* and is present immediately upstream of the *gylABX* promoter (Smith and Chater, 1988). In *P. aeruginosa* however, there is no *gylABX* operon in the immediate vicinity and the role of PA1015 remains unclear. All IclR members are part of the core genome with the exception of PA3508 (RGP34) (Table A.1).

LacI/GalR family

The LacI/GalR family members are metabolic regulators in prokaryotes that recognize sugar inducers. The prototypic members are the *E. coli* lactose repressor LacI and the galactose operon repressor GalR (Nguyen and Saier, 1995). They are typically localized in the cytoplasm and possess a N-terminal HTH motif of about 50–60 residues. The C-terminus contains regions that are involved in oligomerization and signal molecule recognition (Swint-Kruse and Matthews, 2009). LacI members function as either dimers or tetramers (Fukami-Kobayashi *et al.*, 2003). Well-studied proteins with a LacI-type HTH include the *Bacillus subtilis* CcpA and CcpB involved in repression of several operons (Chauvaux *et al.*, 1998), the *Haemophilus influenzae* galactose repressor GalR (Maskell *et al.*, 1992), the *E. coli* lactose operon repressor LacI (Wilson *et al.*, 2007), and the *Salmonella typhimurium* fructose repressor FruR (Vartak *et al.*, 1991).

Five of the seven LacI homologues in *P. aeruginosa*, all of which are part of the core genome, are involved in carbohydrate transport/metabolism (Table A.1). A fructose transport repressor (FruR) homologue is found in *P. aeruginosa* PAO1 (PA3563). However, others are involved in virulence. NfxB (PA4600), a regulator of the MexCD-OprJ efflux pump has also been shown to affect motility, probably through its effect on the motility regulator MorA (PA4601), and QS-dependent phenotypes by inappropriate MexCD-OprJ efflux expression (Stickland *et al.*, 2010). Interestingly, there is another LacI-like regulator in the same locus, *PA4596*, upstream of the MexCD-OprJ operon that is 74% similar to NfxB. Whether this is also involved in regulating MexCD-OprJ remains to be determined.

PtxS (PA2259) is an example of a transcriptional regulator controlling two very different pathways in *P. aeruginosa*. It is a repressor of the exotoxin A regulator PtxR (PA2258) (Colmer and Hamood, 1999) not by direct binding at P_{ptxR} from which it is divergently transcribed but through an unknown mechanism (Colmer-Hamood *et al.*, 2006). PtxS is also part of a five-gene *kgu* operon (*PA2259-PA2263*) involved in 2-ketoglutonate utilization, which it regulates by binding to two different operator sequences (Swanson *et al.*, 2000).

GntR family

The prototypic member of this family is *Bacillus subtilis* GntR, a repressor of the gluconate operon (Buck and Guest, 1989; Haydon and Guest, 1991). Many of the members of this family have been shown experimentally to be autoregulatory, enabling the prediction of operator sites and the discovery of *cis/trans* relationships (Rigali *et al.*, 2004). This family is further subdivided into four major (FadR, HutC, PlmA, MocR) and two minor (YtrA, AraR) subfamilies, which regulate various biological processes and important bacterial metabolic pathways (Rigali *et al.*, 2002). DNA binding by GntR members is induce independent. For example, GntR binds to the promoter of the operon in the absence of gluconate but not in the presence (Peekhaus and Conway, 1998). This family of proteins has a wHTH domain about 60–70 residues at the N-terminus whereas the C-terminus contains a subfamily-specific effector-binding domain and/or an oligomerization domain (Haydon and Guest, 1991). The GntR family of regulators binds as homodimers to two-fold symmetrical DNA elements (van Aalten *et al.*, 2001; Xu *et al.*, 2001).

P. aeruginosa PAO1 genome harbours 31 regulators belonging to the GntR family, of which many are grouped together by sequence analysis (Fig. 13.1) and only two are part of RGPs (PA2032 and PA2100 in RGP20 and RGP21, respectively). Based on sequence analysis, six of the transcriptional regulators are presumed to be involved in glycolate (PA5356), lactate (PA4769), gluconate (PA2320), histidine (PA5105), phosphate (PA3381) and *N*-acetyl-D-glucosamine (PA3757) catabolism. Of the 31, only the role of PA5499 (Np20) has been experimentally elucidated. Expression of *np20* is inducible *in vitro* by respiratory mucus, *in vivo* during infection of mice and was shown to be important in bacterial virulence in a neutropenic mouse infection model (Wang *et al.*, 1996). However, no structure–function studies are available.

LuxR family

The LuxR-type transcription regulator superfamily derives its name from the *Vibrio fisheri* LuxR protein, which alters transcription in response to QS to produce bioluminescence (Engebrecht and Silverman, 1984, 1987; Engebrecht *et al.*, 1983; Fuqua *et al.*, 1994). LuxR proteins contain two functional domains: a C-terminus HTH DNA binding domain and an N-terminal regulatory domain (Nasser and Reverchon, 2007). The DNA binding domain consists of a four-helix bundle. The LuxR proteins bind promoters as multimers (Ducros *et al.*, 2001). Based upon the N-terminus, the LuxR family may be further subdivided into at least four groups: the FixJ group (Birck *et al.*, 2002; Da Re *et al.*, 1994; David *et al.*, 1988; Kahn and Ditta, 1991), the LuxR group (reviewed in Nasser and Reverchon, 2007)), the large ATP-binding regulators of the LuxR superfamily (LAL) group (Raibaud and Richet, 1987; Richet and Raibaud, 1989), and the autonomous effector domain LuxR-type regulators (Ducros *et al.*, 1998; Ducros *et al.*, 2001). Many proteins that have been classified as LuxR superfamily members do not belong to any of these groups. Moreover, new groups are still being identified (de Bruijn and Raaijmakers, 2009).

P. aeruginosa PAO1 harbours 32 members of the LuxR superfamily (Suppl. Table 13.1) that group together in neighbour-joining trees (Fig. 13.1). Half of these appear to belong to the FixJ group. Of these, six have been demonstrated to function as TCS response regulators and ten are predicted to be response regulators based upon PFAM analysis (Table A.1). A LuxR member, NarL (PA3879), acting in concert with the sensor kinase NarX (PA3878), represses expression of proteins involved in the arginine fermentation pathway for ATP generation. In addition, NarL also activates proteins in the nitrate reductase operon (NarH, PA3874), the porin OprE (PA0291) and the class II ribonucleotide reductase components, NrdJa (PA5497) and NrdJb (PA5496, Benkert *et al.*, 2008). RocA1 (PA3948), and its cognate sensor kinase RocS1 (PA3946), were identified as regulators of fimbril cup genes (*cupB* and *cupC*, Kulasekara *et al.*, 2005). Other studied members of this family include BfiR (PA4196, (Petrova and Sauer, 2009, 2010)), ErdR (PA3604, (Mern *et al.*, 2010)), GacA (PA2586, Reimmann *et al.*, 1997), and the QS regulators LasR (PA1430), RhlR (PA3477), QscR (PA1898), and VqsR (PA2591) (see section on QS below) (Brint and Ohman, 1995; Chugani *et al.*, 2001; Gambello and Iglewski, 1991; Juhas *et al.*, 2004). All LuxR superfamily members are encoded by the *P. aeruginosa* core genome (Table A.1).

RpoN-binding family

This family of transcriptional regulators activates the expression of genes from promoters recognized by core RNA polymerase associated with the alternative sigma-54 factor (RpoN, PA4462). RpoN enhancer binding proteins share a conserved domain of about 230 residues involved in the ATP-dependent interaction with RpoN (Morett and Segovia, 1993). The RpoN-binding domain, which has ATPase activity, contains an atypical ATP-binding motif A (P-loop), and a motif B (Austin and Dixon, 1992). These proteins are part of the ATPases associated with diverse cellular activities clan: proteins that perform chaperone-like functions and assist in the assembly, operation, or disassembly of protein complexes (Sigrist *et al.*, 2010).

In *P. aeruginosa*, there are 22 RpoN-binding proteins, encoded by the core genome, which are involved in the regulation of metabolism, production of the exopolysaccharide alginate, virulence as well as biofilm formation (Table A.1). Nine of the 22 RpoN interacting proteins belong to signal transduction TCS and contain an N-terminal domain that can be phosphorylated by a sensor-kinase protein. Almost all of these proteins possess a HTH DNA binding domain in their C-terminus. At least six are NtrC-like RpoN enhancer binding proteins (PilR (PA4547), CbrB (PA4726), NtrC (PA5125), AlgB (PA5483), MifR (PA5511) and FleQ (PA1097)). NtrC transcriptional activators, first described in the regulation of nitrogen metabolism of enteric bacteria, possess three primary domains: a RpoN-interacting conserved central domain, a C-terminal domain which contains the HTH DNA binding motif and a homologous receiver N-terminal domain (Magasanik, 1993).

The transcriptional regulator CbrB is part of the TCS CbrAB (PA4725-PA4726), involved in regulating the carbon–nitrogen balance in *P. aeruginosa*. Phosphorylated CbrB activates the expression of the *aot–argR* operon (arginine catabolism) and the *hiu* operon (histidine catabolism) (Nishijyo *et al.*, 2001). Two biofilm regulators, MifR (PA5511), part of a signal transduction network in microcolony formation (Petrova and Sauer, 2009), and RocR of the RocSAR (SadARS) three-component signalling system involved in biofilm maturation (Kuchma et al., 2005), have been defined. RocSAR consists of the histidine kinase RocS1 and two response regulators, RocA1 (PA3948) and RocR (PA3947). The RocSAR system controls bacterial biofilm formation and virulence gene expression by regulating the transcription of various genes, including the *cup* fimbrial-gene clusters and Type III secretion system genes (Kuchma *et al.*, 2005). AlgB (PA5483), also a part of TCS, activates alginate production in mucoid *P. aeruginosa*. Alginate has been implicated in the pathogenesis of *P. aeruginosa* including resistance to phagocytosis and adherence mechanisms (Hoiby *et al.*, 2010).

MerR family

The classical MerR proteins are dimeric transcriptional regulators of Gram-negative mercury resistance operons present in transposons Tn*21* and Tn*50* (Brown *et al.*, 2003). Like many of the regulators already discussed, these regulators consist of two domains, a 70-amino acid N-terminal DNA-binding domain and a C-terminal effector-binding domain. The carboxy-terminus contains specific effector binding regions to respond to environmental stimuli such as heavy metals, oxidative stress, antibiotics or metal ions (Brown *et al.*, 2003; Heldwein and Brennan, 2001). The MerR proteins differ from many other classes of regulators in that they function as both an activator and repressor of the same gene. They achieve this by binding to the same *cis* element between −10 and −35 promoter regions containing a critical 19–bp region essential for MerR-dependent activation (Brown *et al.*, 2003). MerR acts as a weak repressor in the absence of ligand (Hg(II)) and as an activator when bound to Hg(II) (Helmann *et al.*, 1989).

P. aeruginosa contains seven MerR regulators, two of which, SoxR (PA2273) and CueR (PA4778) are characterized. Like its *E. coli* homologue, the *P. aeruginosa* SoxR contains a [2Fe–2S] centre (Kobayashi and Tagawa, 2004). In *E. coli*, SoxR regulates genes involved in the oxidative stress response to superoxide anions or nitric oxide by activating an AraC-type regulator SoxS (Amabile-Cuevas and Demple, 1991; Wu and Weiss, 1991), which is not conserved in *P. aeruginosa*. SoxR, however, binds weakly and activates transcription of the promoter from its divergently transcribed neighbour, *PA2274*, encoding a hypothetical protein (Kobayashi and Tagawa, 2004). CueR controls a copper resistance regulon consisting of five operons and 11 genes (Thaden *et al.*, 2010), and is under regulation of LasR (PA1430), a LuxR member (Thaden *et al.*, 2010). PA3689 is similar to CadR, a MerR family member from *P. putida* involved in cadmium resistance (Lee *et al.*, 2001). All MerR proteins are encoded as part of the *P. aeruginosa* core genome (Suppl. Table 13.1).

TetR family

The TetR family is named after the enterobacterial TetR repressor protein that confers resistance to tetracycline (Aramaki *et al.*, 1995) and subsequently a family profile was developed (Ramos *et al.*, 2005). TetR members are typically repressors that contain a 60-residue N-terminal HTH DNA binding domain and a C-terminus effector-binding/oligomerization domain that bind DNA either as dimers or octamers (Engohang-Ndong *et al.*, 2004; Hinrichs *et al.*, 1994; Orth *et al.*, 2000). TetR, a well-studied example, is a repressor of the enterobacterial tetracycline exporter TetA, and binds to the operator sites in P_{tetA}. Binding of the tetracycline–Mg complex to DNA-bound TetR leads to depression of *tetA* and confer tetracycline resistance (Grkovic *et al.*, 2002). TetR members primarily control multidrug efflux pumps, biosynthesis of antibiotics, catabolic pathways and virulence in both Gram-positive and Gram-negative bacteria.

There are 39 TetR members in *P. aeruginosa* PAO1, all of which are encoded by the core genome (Table A.1). The expression of the MexAB-OprM efflux pump (PA0425–PA0427) that confers resistance to β-lactams, in addition to diverse substrates, is under the control of two TetR members NalC (PA3721) and NalD (PA3574), of which NalC is thought to act via the primary repressor MexR (PA0424) (Cao *et al.*, 2004; Masuda *et al.*, 2000; Morita *et al.*, 2006). PsrA (PA3006) has been demonstrated to control expression of the *P. aeruginosa* T3SS (Shen *et al.*, 2006), probably through its regulatory effect on RpoS (Hogardt *et al.*, 2004). A TetR member, CifR (PA2931) represses expression of a toxin, CFTR inhibitory factor (Cif, PA2934)

(MacEachran *et al.*, 2008). Cif has been demonstrated to reduce expression of the CFTR protein in cell lines and proposed to play a role in aiding initial colonization by *P. aeruginosa* (MacEachran *et al.*, 2007). CifR regulation of Cif thus aids in the later stages on infection when Cif function is no longer required (MacEachran *et al.*, 2008). Other characterized *P. aeruginosa* TetR members (AguR (PA0294), AtuR (PA2885), DesT (PA4890), BetI (PA5374)) are primarily metabolic repressors (Table A.1) (Forster-Fromme *et al.*, 2006; Lamark *et al.*, 1991; Nakada *et al.*, 2001; Zhu *et al.*, 2006).

AraC family

The *E. coli* L-arabinose operon activator, AraC exemplifies this group of transcriptional regulators with which members share a 99-residue C-terminus homology (Gallegos *et al.*, 1997; Greenblatt and Schleif, 1971; Schleif, 1969). AraC proteins harbour a C-terminus HTH motif, a N-terminus dimerization/sugar-binding domain and primarily activate expression of genes involved in sugar metabolism, stress response and virulence (Martin and Rosner, 2001; Ramos *et al.*, 2005). Most AraC members, together with cAMP-receptor protein (CRP), activate promoters, either as monomers or dimers (Johnson and Schleif, 2000; Martin *et al.*, 1999).

P. aeruginosa PAO1 has 61 AraC members (Table A.1). PchR (PA4227), an AraC member, is a regulator of the biosynthesis of an iron uptake protein, pyochelin, and acts both as a positive and negative regulator of *fptA* (*PA4221*) and *pchR* expression (Heinrichs and Poole, 1996). PchR binds to a 32 bp PchR box in the promoter of genes that it regulates (Michel *et al.*, 2005). ExsA (PA1713), a *Yersinia enterocolitica* VirF homologue, was initially identified as part of a trans-regulatory locus involved in the expression of the T3SS effector exoenzyme S (Frank and Iglewski, 1991; Yahr and Frank, 1994) and has since been shown to regulate expression of the T3SS genes (reviewed in Yahr and Wolfgang, 2006). MmsR (PA3571) regulates the *mmsAB* operon (*PA3569-PA3570*) that encodes enzymes involved in valine metabolism (Steele *et al.*, 1992). Other than PA1380 (RGP15) and PA2047 (RGP20), AraC protein-encoding genes belong to the core genome of *P. aeruginosa* (Table A.1).

Cro-C1 family

This superfamily is named after the Cro and C1 repressors of the lysogenic phages 434 and lambda, respectively, where they are part of a binary switch that regulates lytic/lysogenic growth of the phage by differential binding to the operator sites (Aggarwal *et al.*, 1988; Guarente *et al.*, 1980). They generally have a 50–60 residue HTH domain at the N-terminus of the protein while oligomerization occurs at the C-terminus. The lambda Cro repressor is a 66 amino acid protein and binds DNA as a homodimer (Anderson and Cygler, 1985).

P. aeruginosa has 17 regulators belonging to this family, all of which are encoded by the core genome (Table A.1). Pyocin production in *P. aeruginosa* is under the tight regulation of PrtR (PA0611). As part of the SOS response, RecA (PA3617) degrades PrtR, which inhibits transcription of the pyocin gene activator, PrtN (PA0610; Matsui *et al.*, 1993). PrtR also modulates activity of a Dks/TraR member PtrB (PA0612), a repressor of the T3SS (Wu and Jin, 2005), linking secretion and pyocin synthesis in response to DNA damage. Two other hypothetical Cro/C1 members, PA0906 and PA4077 are 66% similar to PrtR, and PA4077 is co-transcribed with a hypothetical inner membrane protein PA4076 (Winsor *et al.*, 2009). Whether these are also involved in pyocin regulation is unclear. PA1359, a hypothetical regulator, bears 49% similarity to *E. coli* HipB, a known regulator of persister formation by modulation of HipA activity (Black *et al.*, 19991, 1994; Korch and Hill, 2006). Interestingly, PA1359 is part of a two-gene operon (Winsor *et al.*, 2009), similar to the *hipBA* operon in *E. coli*. None of the other members of this family have been characterized in PAO1.

AsnC family

This family of transcriptional regulators is typified by asparagine synthase C of *E. coli* and is part of the feast/famine regulatory proteins (FFRP) that includes Lrp and Asn (Yokoyama *et al.*, 2006). Members of this family are distributed in both archea and eubacteria. They typically contain a HTH DNA-binding domain at the N-terminus and a dimerization/effector-binding domain at the C-terminus, which also plays a role in assembly of the protein. FFRP members bind DNA as dimers, tetramers, octamers or hexamers (Brinkman *et al.*, 2003; Leonard *et al.*, 2001) and regulate genes in response to diverse signals. Identification of FFRPs is often complicated by the fact that the amino acid homologies between members are often only about 30% (Yokoyama *et al.*, 2006).

P. aeruginosa has nine AsnC members (Table A.1), all of which except PA2028 (RGP20) are encoded by the core genome. PA0513, an as yet uncharacterized regulator, is part of an 11-gene *nir* operon that is involved in the denitrification pathway (Kawasaki *et al.*, 1997). PA0515 is also part of the same operon and is listed as a probable AsnC-type transcriptional regulator in the *Pseudomonas* database (Winsor *et al.*, 2009) but has been excluded from our list (Table A.1) because it lacked a DNA binding motif in our analysis. BkdR (PA2246) is divergently transcribed from the *bkd* operon that is involved in valine, leucine, isoleucine degradation and has been characterized in *P. putida* (Madhusudhan *et al.*, 1993). The *P. aeruginosa* homologue of Lrp (PA5308), which regulates the global leucine response in *E. coli*, regulates the *dadA* (*PA5304*) and *dadX* (*PA5302*) genes that are required for L-alanine utilization (Chou *et al.*, 2008).

ArsR family

The *E. coli* ArsR that represses the *ars* operon in response to arsenic is the prototype for this family of transcriptional repressors (Busenlehner *et al.*, 2003). This family of proteins binds the promoter region of genes that they regulate (primarily metal transport and detoxification) and dissociate upon effector binding due to allosteric changes, allowing transcription by derepression (Osman and Cavet, 2010). The metal-binding site is at the C-terminus of the protein whereas the DNA-binding domain is at the N-terminus. ArsR members typically function as homodimers (Eicken *et al.*, 2003).

There are four members in *P. aeruginosa* encoded by the core genome (Table A.1). *arsR* (*PA2277*) is part of a four-gene cluster that has been demonstrated to confer resistance to arsenic and antimony when expressed in a heterologous *E. coli ars* mutant (Cai *et al.*, 1998). PA0547 is an uncharacterized ArsR-type regulator that is part of a two-gene operon with the S-adenosylmethionine synthetase gene, *metK* (*PA0546*). The other two ArsR members in *P. aeruginosa* are also uncharacterized (Table A.1).

Other families

Other *P. aeruginosa* regulatory families include the Dks/TraR family [PtrB (PA0612) and DksA (PA4723)], H-NS silencing proteins [MvaT (PA4315) and MvaU (PA2667)], CRP family [Vfr (PA0652) and Anr (PA1544)], OmpC family (AmgR; PA5200), OmpR-PhoP class (PA4381) and the RpiR family (PA5506).

Unclassified regulators

In addition to the families listed above, 37 regulators do not contain the signature motif for any family but still are significant contributors to the regulatory processes in *P. aeruginosa* (Table A.1). Of these, 27 have been characterized. CreB (PA0463), the regulator of the CreBC TCS operon, along with CreD (PA0465), encoded immediately downstream, are involved in colicin resistance in *E. coli* (Drury and Buxton, 1988). The CreBC TCS is involved in regulating β-lactam resistance in *Aeromonas* spp. (Avison *et al.*, 2004) while in *P. aeruginosa* it contributes to resistance only in a Δ*PBP4* background (Moya *et al.*, 2009). LexA (PA3007) is a repressor of RecA and is induced upon DNA damage, as part of the SOS response (Calero *et al.*, 1993). IrlR (PA4885) is part of a two-component system, a homologue of which has been shown to be involved in heavy metal homeostasis in *Alcaligenes eutrophus* (van der Lelie *et al.*, 1997). A few other members (KdpE (PA1637), DauR (PA3864), GltR (PA3192)) are metabolic regulators.

P. aeruginosa transcriptional regulatory network

Studying regulatory networks aid in elucidating interactions between the different transcriptional regulators and provide important insights into processes and pathways. To better understand regulatory networks and identify network motifs, analysis of network structure has proved useful. Two major network motifs that aid in bacterial response to environmental signals have been described: feed-forward motifs, and single-input and multiple-input motifs (Madan Babu *et al.*, 2006). Using published data on bacterial regulatory networks (Madan Babu *et al.*, 2006) and STRING analysis (Jensen *et al.*, 2009) of *P. aeruginosa* PAO1 regulators, a transcriptional regulatory network was obtained (Fig. 13.2a) consisting of 288 regulators and 593 target genes. As seen in regulatory networks in other bacteria (Madan Babu *et al.*, 2006), the *P. aeruginosa* network also consists of feed-forward motifs, and single- and multiple-input motifs (Fig. 13.2a). Vfr (PA0652), the *P. aeruginosa* homologue of *E. coli* cyclic AMP receptor protein and a regulator of virulence and QS, has the highest number of interacting partners, followed by the anaerobic growth and cyanide synthesis regulator Anr, an *E. coli* Fnr homologue (Fig. 13.2a). Interestingly, both these regulators belong to the diverse Crp family of transcriptional regulators that have a N-terminal nucleotide-binding domain similar to CRP, in addition to a C-terminus HTH (Korner *et al.*, 2003).

Conclusions about hierarchy among the regulators are difficult to draw from this network but we used the data to identify interactions between the regulators alone without their corresponding

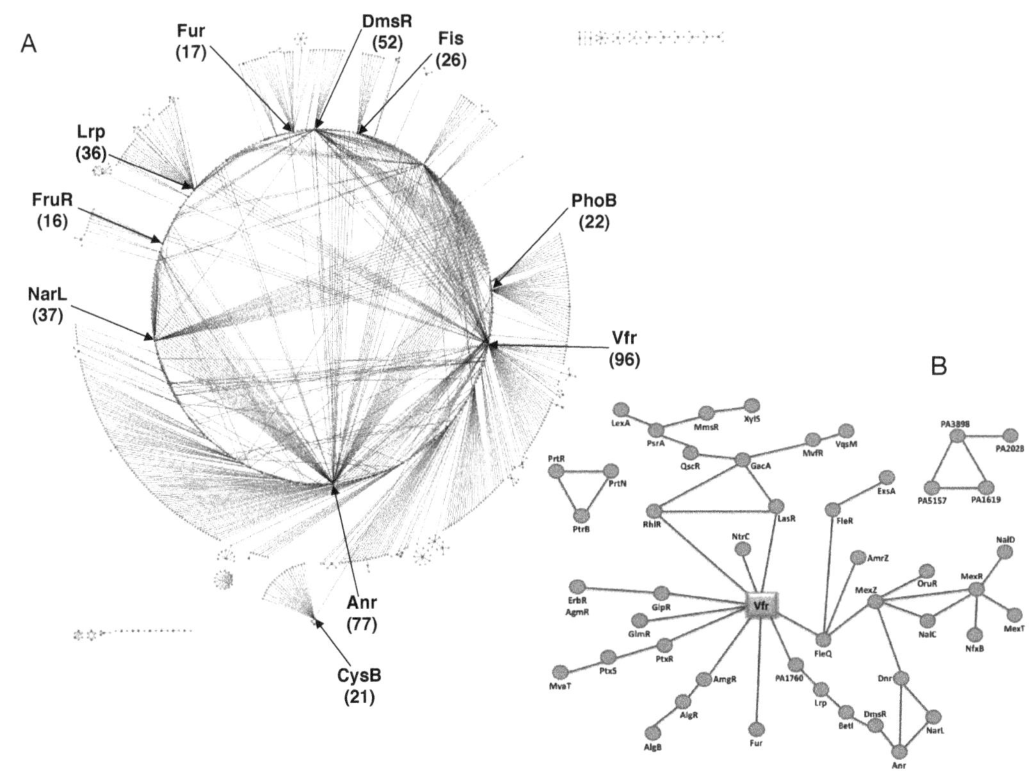

Figure 13.2 Transcriptional regulatory networks. (A) Transcriptional regulators and interacting partners. Previously inferred regulators of *P. aeruginosa* (Madan Babu *et al.*, 2006) and those obtained by text mining using STRING (Jensen *et al.*, 2009) analysis, were used to construct the transcriptional regulatory network using Cytoscape (Version 2.7.0) (Shannon *et al.*, 2003). The network consists of 288 regulators and 593 target genes resulting in 1095 interactions. The transcriptional regulators with the highest interactions are shown in the figure (see also Table A.2). (B) Interactions among transcriptional regulators. VisANT (Version 3.86) (Hu *et al.*, 2009) was used to visualize a total of 63 interactions between the regulators from the 1095 interactions of Fig. 13.2a. Vfr emerges as the key regulator. Other interactions such as the pyocin induction network and an uncharacterized network appear isolated from these interactions. A colour version of this figure is located in the plate section at the back of the book.

target genes. The Vfr regulatory hub stood out in this analysis (Fig. 13.2b). Vfr has been demonstrated experimentally to regulate multiple virulence mechanisms in *P. aeruginosa*. Network analysis shows that Vfr is linked to various other critical pathogenesis networks including alginate production, efflux systems, quorum sensing, iron uptake, flagella, T3SS, and metabolic regulons (Fig. 13.2b). While some of these links have already been established (Davinic *et al.*, 2009; Fuchs *et al.*, 2010; Jones *et al.*, 2010), it would be interesting to see whether the other interactions pan out. Our Cytoscape analysis also revealed the established interactions between PrtR, PrtN and PtrB to regulate pyocin production and T3SS in *P. aeruginosa* (Fig. 13.2b) (Matsui *et al.*, 1993; Wu and Jin, 2005). Moreover, PA5157, an uncharacterized regulator that bears 60% similarity to the *E. coli* MarR protein involved in regulating multiple antibiotic resistance and oxidative stress (Ariza *et al.*, 1994), interacts with three other hypothetical regulators PA1619, PA3898 and PA2028 (Fig. 13.2b). Whether these genes are part of a network contributing to the antibiotic resistance profile of *P. aeruginosa* is not known.

Virulence regulatory systems in *P. aeruginosa*

Network modelling, while providing us with a framework to study, has its limitations in requiring confirmation. The absolute proof of a regulatory interaction comes from empirical data, which serves to either confirm or negate findings from the bioinformatics analyses. We chose the QS system in *P. aeruginosa*, to highlight the similarities in complexity between modelled and established regulatory networks.

Quorum sensing (QS)

QS, a cell population density-dependent signalling system important in host–pathogen interaction, is involved in regulating virulence factor production, swarming motility, biofilm maturation, and efflux pump expression (reviewed in Ng and Bassler, 2009). Two types of QS signals are produced by *P. aeruginosa*, *N*-acylhomoserine lactones and 4-quinolones (Pearson *et al.*, 1994, 1995; Pesci *et al.*, 1999). Central to *N*-acylhomoserine lactone QS signalling are LasR (PA1430) and RhlR (PA3477), two LuxR-type regulators that respectively bind *N*-(3-oxododecanoylhomoserine lactone (3-oxo-C12-HSL) and *N*-butanoylhomoserine lactone (C4-HSL) (Pearson *et al.*, 1994, 1995). These HSLs are synthesized by the cognate synthetases, LasI and RhlI (Brint and Ohman, 1995; Ochsner and Reiser, 1995). Expression of *lasI* and *rhlI* is regulated by positive feedback loops mediated by 3-oxo-C12-HSL:LasR and C4-HSL:RhlR. LasR is higher up in the *P. aeruginosa* QS hierarchy and 3-oxo-C12-HSL:LasR activates *rhlR* and *rhlI* expression (Latifi *et al.*, 1996; Pesci *et al.*, 1997). The binding site for LasR and RhlR has been termed the *las-rhl* box (CT-N_{12}-AG) (Gilbert *et al.*, 2009; Schuster and Greenberg, 2007) and LasR binding sites have been identified in the promoters of many genes including known and putative regulators such as MvfR (PA1003), RsaL (PA1431), VqsR (PA2591), RhlR (PA3477), PvdS (PA2426), AmrZ (PA3385), PA2588, PA4778, and PA1760 (Gilbert *et al.*, 2009). In addition, three microarray studies identified a common set of 102 genes in the LasR/RhlR regulons (Hentzer *et al.*, 2003; Schuster and Greenberg, 2006; Schuster *et al.*, 2003; Wagner *et al.*, 2003, 2004).

Most QS-regulated genes are expressed during stationary phase and are dependent on the stationary phase sigma factor RpoS (PA3622; Schuster and Greenberg, 2007; Schuster *et al.*, 2004). Several additional regulators also play a role in QS (Fig. 13.3). Vfr binds and activates transcription of P_{lasR} (Albus *et al.*, 1997) and P_{rhlR} (Medina *et al.*, 2003), and represses the 4-hydroxy-2-quinolone-dependent QS pathway (Whitchurch *et al.*, 2005). Transcriptome analysis indicates that Vfr affects the expression of more than 100 genes in *P. aeruginosa* including T3SS genes (Wolfgang *et al.*, 2003). RsaL (PA1431), which is divergently transcribed from the *lasI* gene, encodes an autoregulating repressor that inhibits not only *lasI* transcription, but also genes involved in pyocyanin and cyanide synthesis (Rampioni *et al.*, 2006, 2007). Interestingly, RsaL repression is dominant over 3-oxo-C12-HSL:LasR activation of *lasI* to counteract the positive feedback loop (Rampioni *et al.*, 2006). RsaL binds as a dimer (Rampioni *et*

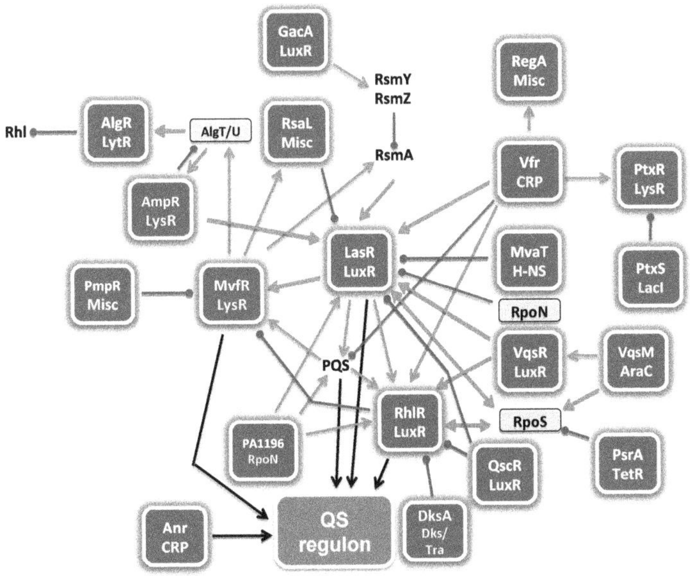

Figure 13.3 Association between QS regulators in *P. aeruginosa*. Interactions between the different transcriptional regulators (blue boxes) and sigma factors (yellow boxes) to control QS gene expression are depicted. The regulation is positive (green lines), negative (red lines), or both (black lines) and can be direct or indirect. PQS refers to the *Pseudomonas* quinolone signalling molecule. See text for details and references of individual interactions. A colour version of this figure is located in the plate section at the back of the book.

al., 2007) and regulates expression of 130 genes, including those involved in virulence, independent of its QS effect (Rampioni *et al.*, 2007). Another transcription factor that appears to co-regulate QS genes with LasR and RhlR is the anaerobic regulator Anr (PA1544). Together with LasR and RhlR, Anr directly activates the hydrogen cyanide biosynthetic *hcnABC* genes (*PA2193-PA2195*; Pessi and Haas, 2000). Bioinformatic analysis identified putative Anr consensus binding sites, the Fnr/Anr box, in about 25% of QS-regulated genes, suggesting that Anr may regulate these genes under oxygen limiting conditions (Schuster and Greenberg, 2006). Interestingly, studies suggest that the growth conditions affect the *las-rhl* hierarchy (Dekimpe and Deziel, 2009; Duan and Surette, 2007).

P. aeruginosa synthesizes over 50 4-quinolones, however, two, 2-heptyl-4-quinolone (HHQ) and 2-heptyl-3-hydroxy-4-quinolone (PQS) have been shown to act as QS signals (Deziel *et al.*, 2004; Diggle *et al.*, 2007; Pesci *et al.*, 1999). Synthesis of these 4-quinolones is dependent upon the *pqsABCDE* (*PA0996-PA1000*) and *phnAB* (*PA1001-PA1002*) operons (Deziel *et al.*, 2004). The *pqsE* gene is dispensable for 4-quinolone synthesis but is required for virulence (Diggle *et al.*, 2003; Gallagher *et al.*, 2002). A LTTR member, MvfR (also known as PqsR; PA1003), regulates the *pqs* operon (Xiao *et al.*, 2006). MvfR expression is positively regulated by LasR:3-oxo-C12-HSL, and negatively regulated by RhlR (Wade *et al.*, 2005) and the PqsR-mediated PQS regulator, PmpR (PA0964, Liang *et al.*, 2008). Transcriptome analysis revealed an extensive MvfR regulon that includes RsaL (PA1431), AlgT/U (PA0762), RsmA (PA0905), and LasR (PA1430) (Deziel *et al.*, 2005). Promoter binding of MvfR is enhanced by either HHQ or PQS (Xiao *et al.*, 2006). HHQ is converted to PQS by PqsH (PA2587) whose expression is LasR:3-oxo-C12-HSL-dependent (Deziel *et al.*, 2004; Gallagher *et al.*, 2002). PQS differs from HHQ in that it can sequester iron and induce an iron starvation response by activating genes involved in siderophore biosynthesis and iron scavenging (Bredenbruch *et al.*, 2006; Diggle *et al.*, 2007). Thus PQS affects gene expression in multiple ways: *mvfR*- and *pqsE*-dependent expression, *mvfR*-dependent and *pqsE*-independent expression, and through iron sequestration in a *mvfR*- and *pqsE*-independent manner (Diggle *et al.*, 2007).

QscR (PA1898) and VqsR (PA2591) are two additional LuxR-type QS regulators in *P. aeruginosa* (Fig. 13.3). QscR closely resembles the HSL-responsive LuxR homologues and has been proposed to control the timing of QS-regulated genes by repressing *lasI* expression (Chugani *et al.*, 2001). Transcriptome analysis identified 400 genes as being part of the QscR regulon (Lequette *et al.*, 2006). Additional experiments suggest that QscR can also act by mediating a response to 3-oxo-C10-HSL produced by other bacteria or form heterodimers with LasR or RhlR (Ledgham *et al.*, 2003). The *vqsR* promoter contains a *las-rhl* box and *in vivo* occupancy of LasR has been demonstrated (Gilbert *et al.*, 2009; Juhas *et al.*, 2004). Consistent with this finding, *vqsR* mutants failed to produce C4-HSL or 3-oxo-C12-HSL, showed reduced pyocyanin and protease production, and reduced virulence in a *C. elegans* model system due to a negative impact on expression of the QS regulon (Juhas *et al.*, 2004). VqsR functions downstream of VqsM (PA2227), an AraC type regulator (Dong *et al.*, 2005; Juhas *et al.*, 2004). Mutation in *vqsM* reduced expression of *rhlR, rsaL, vqsR, mvfR, pprB, rpoS, lasI* and *rhlI* (Dong *et al.*, 2005).

Other regulators have also been implicated in QS signalling. Mutations in the gene encoding the FIS regulator, *PA1196*, downregulate expression of *rhlI* and *rhlR* in addition to *pqsA*, which encodes the enzyme that modifies anthranilate for entry into the *Pseudomonas* quinolone signal (PQS) biosynthetic pathway (Liang *et al.*, 2009). Transposon insertional mutants of *PA5499* (*np20*) reduce *rhlI* expression as well as phenazine and cyanide production (Gallagher *et al.*, 2002). Mutations in a LTTR AmpR have also been shown to modulate QS activity, linking QS with β-lactam resistance in *P. aeruginosa* (Balasubramanian *et al.*, 2011, 2012; Kong *et al.*, 2005).

QS is also regulated post-transcriptionally through the action of the small regulatory RNAs, *rsmZ* (PA3621.1) and *rsmY* (PA0527.1; Brencic *et al.*, 2009; Kay *et al.*, 2006). *rsmYZ* antagonize the RNA-binding protein RsmA (PA0905), which negatively regulates QS genes. The GacSA TCS is activated at high cell densities and most of the effects of the Gac/Rsm pathway on QS can be attributed to increased C4-HSL levels (Heeb and Haas, 2001; Kay *et al.*, 2006). QteE (PA2593), though not a transcriptional regulator, controls the QS threshold by lowering LasR protein stability without affecting *lasR* transcription or translation, and by reducing RhlR levels (Siehnel *et al.*, 2010). The stringent response also affects the QS regulon. Under starvation conditions, RelA (PA0934) enhances production of guanosine tetraphosphate (ppGpp) (Greenway and England, 1999). Overexpression of RelA, also caused by decreased membrane fluidity (Baysse *et al.*, 2005), resulted in premature expression of *rpoS, lasR* and *rhlR* and premature synthesis of 3-oxo-C12-HSL and C4-HSL (van Delden *et al.*, 2001). Both expression and stringent response of expression of 3-oxo-C12-HSL is induced by a serine amino acid analogue (van Delden *et al.*, 2001).

Although there have been many advances, much remains unknown in our understanding of QS regulation in *P. aeruginosa*. Recent studies by Chugani and Greenberg showed that a *lasR, rhlR,* and *qscR* triple mutant retained the ability to regulate 37 genes, including those in the kynurenine pathway, in response to acylhomoserine lactones (Chugani and Greenberg, 2010). The kynurenine pathway serves as the source of anthranilate for the synthesis of the PQS (Chugani and Greenberg, 2010; Coleman *et al.*, 2008).

Conclusions and perspectives

Computational analyses and mathematical models have greatly advanced our understanding of regulatory networks (Alon, 2007; Madan Babu *et al.*, 2006; Martinez-Nunez *et al.*, 2010), some of which have been confirmed by empirical data. Our enhanced understanding of gene regulation in bacteria has come a long way from transcriptional regulators alone, as traditionally thought, to include post-transcriptional and post-translational control, including regulatory RNAs, RNA-binding proteins, sigma factors, DNA supercoiling, proteolysis and others. In *P. aeruginosa*, as with numerous other bacteria, a majority of the transcriptional regulators and open reading frames have yet to be characterized in terms of gene function and mode of regulation, which in itself undermines our efforts in predicting gene networks. Nevertheless, such models provide us with insight into potential interactions and give us a framework to design experiments to elucidate complex networks, and further reinforce empirical data. In addition, they also help identify paralogues and orthologues between genera. The wealth of data that we now possess from numerous *P. aeruginosa* whole genome transcriptome analyses (Balasubramanian and Mathee, 2009; Goodman and Lory, 2004), though critical in our understanding of the system and in identifying potential therapeutic targets, further exposes the gap in our knowledge in accurately determining links between signals recognized by the regulators and how they are transduced to their target genes. Analyses of regulator-mutant transcriptomes helps fill in the gap, but warrant further studies, especially in linking metabolic regulation with virulence. It is exciting to know that complementing data from traditional molecular biological experiments with data from high-throughput, whole genome analyses such as microarray, RNA deep sequencing, and network modelling will undoubtedly advance our understanding of how bacteria regulate gene expression, as individuals and as communities in response to environmental signals.

Chapter highlights

- About 8% of the *P. aeruginosa* genome encodes transcriptional regulators, highlighting the versatility of this bacterium in adapting to diverse habitats. However, a majority of the 434 transcriptional regulators remain uncharacterized.
- The transcriptional regulators are classified into 18 families on the basis of signature sequences including those involved in DNA and effector binding.
- Sequence analysis, data mining, and STRING and network analyses, supported by experimental evidence, provide insight into the complex regulatory circuitries involved in metabolic and virulence gene expression.
- The transcriptional regulatory network involves a complex interplay between multiple regulators belonging to different families, and is exemplified by the QS network controlling virulence factor expression.
- Network analysis reveals that the two regulators with the most nodes are Vfr and Anr.
- Identifying the roles of the uncharacterized regulators and other hypothetical proteins will elucidate the adaptive and pathogenic processes of this versatile pathogen that, ultimately, will help in designing better treatment strategies.

References

Aggarwal, A.K., Rodgers, D.W., Drottar, M., Ptashne, M., and Harrison, S.C. (1988). Recognition of a DNA operator by the repressor of phage 434: a view at high resolution. Science *242*, 899–907.

Albus, A.M., Pesci, E.C., Runyen-Janecky, L.J., West, S.E., and Iglewski, B.H. (1997). Vfr controls quorum sensing in *Pseudomonas aeruginosa*. J. Bacteriol. *179*, 3928–3935.

Alon, U. (2007). Network motifs: theory and experimental approaches. Nat. Rev. Genet. *8*, 450–461.

Amabile-Cuevas, C.F., and Demple, B. (1991). Molecular characterization of the soxRS genes of *Escherichia coli*: two genes control a superoxide stress regulon. Nucleic Acids Res. *19*, 4479–4484.

Anderson, W.F., and Cygler, M. (1985). Computer modeling studies of the structure of a repressor. Biosystems *18*, 3–14.

Aramaki, H., Yagi, N., and Suzuki, M. (1995). Residues important for the function of a multihelical DNA binding domain in the new transcription factor family of Cam and Tet repressors. Protein Eng. *8*, 1259–1266.

Ariza, R.R., Cohen, S.P., Bachhawat, N., Levy, S.B., and Demple, B. (1994). Repressor mutations in the marRAB operon that activate oxidative stress genes and multiple antibiotic resistance in *Escherichia coli*. J. Bacteriol. *176*, 143–148.

Austin, S., and Dixon, R. (1992). The prokaryotic enhancer binding protein NTRC has an ATPase activity which is phosphorylation and DNA dependent. EMBO J. *11*, 2219–2228.

Avison, M.B., Niumsup, P., Nurmahomed, K., Walsh, T.R., and Bennett, P.M. (2004). Role of the 'cre/blr-tag' DNA sequence in regulation of gene expression by the Aeromonas hydrophila beta-lactamase regulator, BlrA. J Antimicrob. Chemother. *53*, 197–202.

Balasubramanian, D., and Mathee, K. (2009). Comparative transcriptome analyses of *Pseudomonas aeruginosa*. Hum. Genomics *3*, 349–361.

Balasubramanian, D., Kong, K.F., Jayawardena, S.R., Leal, S.M., Sautter, R.T., and Mathee, K. (2011). Co-regulation of β-lactam resistance, alginate production and quorum sensing in *Pseudomonas aeruginosa*. J. Med. Microbiol. *60*, 147–156.

Balasubramanian, D., Schneper, L., Merighi, M., Smith, R., Narasimhan, G., Lory, S., and Mathee, K. (2012). The regulatory repertoire of *Pseudomonas aeruginosa* AmpC β-lactamase regulator AmpR includes virulence genes. PLoS ONE *7*, e34067.

Baysse, C., Cullinane, M., Denervaud, V., Burrowes, E., Dow, J.M., Morrissey, J.P., Tam, L., Trevors, J.T., and O'Gara, F. (2005). Modulation of quorum sensing in *Pseudomonas aeruginosa* through alteration of membrane properties. Microbiology *151*, 2529–2542.

Benkert, B., Quack, N., Schreiber, K., Jaensch, L., Jahn, D., and Schobert, M. (2008). Nitrate-responsive NarX-NarL represses arginine-mediated induction of the *Pseudomonas aeruginosa* arginine fermentation arcDABC operon. Microbiology *154*, 3053–3060.

Birck, C., Malfois, M., Svergun, D., and Samama, J. (2002). Insights into signal transduction revealed by the low resolution structure of the FixJ response regulator. J. Mol. Biol. *321*, 447–457.

Black, D.S., Irwin, B., and Moyed, H.S. (1994). Autoregulation of hip, an operon that affects lethality due to inhibition of peptidoglycan or DNA synthesis. J. Bacteriol. *176*, 4081–4091.

Black, D.S., Kelly, A.J., Mardis, M.J., and Moyed, H.S. (1991). Structure and organization of hip, an operon that affects lethality due to inhibition of peptidoglycan or DNA synthesis. J. Bacteriol. *173*, 5732–5739.

Bredenbruch, F., Geffers, R., Nimtz, M., Buer, J., and Haussler, S. (2006). The *Pseudomonas aeruginosa* quinolone signal (PQS) has an iron-chelating activity. Environ. Microbiol. *8*, 1318–1329.

Brencic, A., McFarland, K.A., McManus, H.R., Castang, S., Mogno, I., Dove, S.L., and Lory, S. (2009). The GacS/GacA signal transduction system of *Pseudomonas aeruginosa* acts exclusively through its control over the transcription of the RsmY and RsmZ regulatory small RNAs. Mol. Microbiol. *73*, 434–445.

Brinkman, A.B., Ettema, T.J., de Vos, W.M., and van der Oost, J. (2003). The Lrp family of transcriptional regulators. Mol. Microbiol. *48*, 287–294.

Brint, J., and Ohman, D. (1995). Synthesis of multiple exoproducts in *Pseudomonas aeruginosa* is under the control of RhlR-RhlI, another set of regulators in strain PAO1 with homology to the autoinducer-responsive LuxR-LuxI family. J. Bacteriol. *177*, 7155–7163.

Brown, N.L., Stoyanov, J.V., Kidd, S.P., and Hobman, J.L. (2003). The MerR family of transcriptional regulators. FEMS Microbiol. Rev. *27*, 145–163.

Buck, D., and Guest, J.R. (1989). Overexpression and site-directed mutagenesis of the succinyl-CoA synthetase of *Escherichia coli* and nucleotide sequence of a gene (g30). that is adjacent to the suc operon. Biochem J 260, 737–747.

Buelow, D.R., and Raivio, T.L. (2010). Three (and more). component regulatory systems – auxiliary regulators of bacterial histidine kinases. Mol. Microbiol. *75*, 547–566.

Busenlehner, L.S., Pennella, M.A., and Giedroc, D.P. (2003). The SmtB/ArsR family of metalloregulatory transcriptional repressors: Structural insights into prokaryotic metal resistance. FEMS Microbiol. Rev. *27*, 131–143.

Cai, J., Salmon, K., and DuBow, M.S. (1998). A chromosomal ars operon homologue of *Pseudomonas aeruginosa* confers increased resistance to arsenic and antimony in *Escherichia coli*. Microbiology *144*, 2705–2713.

Calero, S., Garriga, X., and Barbe, J. (1993). Analysis of the DNA damage-mediated induction of Pseudomonas putida and *Pseudomonas aeruginosa* lexA genes. FEMS Microbiol. Lett. *110*, 65–70.

Cao, H., Krishnan, G., Goumnerov, B., Tsongalis, J., Tompkins, R., and Rahme, L.G. (2001). A quorum sensing-associated virulence gene of *Pseudomonas aeruginosa* encodes a LysR-like transcription regulator with a unique self-regulatory mechanism. Proc. Natl. Acad. Sci. U.S.A. *98*, 14613–14618.

Cao, L., Srikumar, R., and Poole, K. (2004). MexAB-OprM hyperexpression in NalC-type multidrug-resistant *Pseudomonas aeruginosa*: identification and characterization of the nalC gene encoding a repressor of PA3720-PA3719. Mol 53 Microbiol 1423–1436.

Chauvaux, S., Paulsen, I.T., and Saier, M.H., Jr. (1998). CcpB, a novel transcription factor implicated in catabolite repression in *Bacillus subtilis*. J. Bacteriol. *180*, 491–497.

Chou, H.T., Kwon, D.H., Hegazy, M., and Lu, C.D. (2008). Transcriptome analysis of agmatine and putrescine catabolism in *Pseudomonas aeruginosa* PAO1. J. Bacteriol. *190*, 1966–1975.

Chugani, S., and Greenberg, E.P. (2010). LuxR homolog-independent gene regulation by acyl-homoserine lactones in *Pseudomonas aeruginosa*. Proc. Natl. Acad. Sci. U.S.A. *107*, 10673–10678.

Chugani, S.A., Whiteley, M., Lee, K.M., D'Argenio, D., Manoil, C., and Greenberg, E.P. (2001). QscR, a modulator of quorum-sensing signal synthesis and virulence in *Pseudomonas aeruginosa*. Proc. Natl. Acad. Sci. U.S.A. *98*, 2752–2757.

Cohen, J. (2002). The immunopathogenesis of sepsis. Nature *420*, 885–891.

Coleman, J.P., Hudson, L.L., McKnight, S.L., Farrow, J.M., 3rd, Calfee, M.W., Lindsey, C.A., and Pesci, E.C. (2008). *Pseudomonas aeruginosa* PqsA is an anthranilate-coenzyme A ligase. J. Bacteriol. *190*, 1247–1255.

Colmer, J.A., and Hamood, A.N. (1999). Expression of ptxR and its effect on toxA and regA expression during the growth cycle of *Pseudomonas aeruginosa* strain PAO1. Can J 45 Microbiol 1008–1016.

Colmer-Hamood, J.A., Aramaki, H., Gaines, J.M., and Hamood, A.N. (2006). Transcriptional analysis of the *Pseudomonas aeruginosa* toxA regulatory gene ptxR. Can J Microbiol 52, 343–356.

Da Re, S., Bertagnoli, S., Fourment, J., Reyrat, J.M., and Kahn, D. (1994). Intramolecular signal transduction within the FixJ transcriptional activator: *in vitro* evidence for the inhibitory effect of the phosphorylatable regulatory domain. Nucleic Acids Res. *22*, 1555–1561.

David, M., Daveran, M.L., Batut, J., Dedieu, A., Domergue, O., Ghai, J., Hertig, C., Boistard, P., and Kahn, D. (1988). Cascade regulation of nif gene expression in *Rhizobium meliloti*. Cell *54*, 671–683.

Davinic, M., Carty, N.L., Colmer-Hamood, J.A., San Francisco, M., and Hamood, A.N. (2009). Role of Vfr in regulating exotoxin A production by *Pseudomonas aeruginosa*. Microbiology *155*, 2265–2273.

de Bruijn, I., and Raaijmakers, J.M. (2009). Diversity and functional analysis of LuxR-type transcriptional regulators of cyclic lipopeptide biosynthesis in *Pseudomonas fluorescens*. Appl. Environ. Microbiol. *75*, 4753–4761.

Dekimpe, V., and Deziel, E. (2009). Revisiting the quorum-sensing hierarchy in *Pseudomonas aeruginosa*: the transcriptional regulator RhlR regulates LasR-specific factors. Microbiology *155*, 712–723.

Deziel, E., Lepine, F., Milot, S., He, J., Mindrinos, M.N., Tompkins, R.G., and Rahme, L.G. (2004). Analysis of *Pseudomonas aeruginosa* 4-hydroxy-2-alkylquinolines (HAQs). reveals a role for 4-hydroxy-2-heptylquinoline in cell-to-cell communication. Proc. Natl. Acad. Sci. U.S.A. *101*, 1339–1344.

Deziel, E., Gopalan, S., Tampakaki, A.P., Lepine, F., Padfield, K.E., Saucier, M., Xiao, G., and Rahme, L.G. (2005). The contribution of MvfR to *Pseudomonas aeruginosa* pathogenesis and quorum sensing circuitry regulation: multiple quorum sensing-regulated genes are modulated without affecting *lasRI, rhlRI* or the production of N-acyl-L-homoserine lactones. Mol. Microbiol. *55*, 998–1014.

Diggle, S.P., Winzer, K., Chhabra, S.R., Worrall, K.E., Camara, M., and Williams, P. (2003). The *Pseudomonas aeruginosa* quinolone signal molecule overcomes the cell density-dependency of the quorum sensing hierarchy, regulates rhl-dependent genes at the onset of stationary phase and can be produced in the absence of LasR. Mol. Microbiol. *50*, 29–43.

Diggle, S.P., Matthijs, S., Wright, V.J., Fletcher, M.P., Chhabra, S.R., Lamont, I.L., Kong, X., Hider, R.C., Cornelis, P., *et al.* (2007). The *Pseudomonas aeruginosa* 4-quinolone signal molecules HHQ and PQS play multifunctional roles in quorum sensing and iron entrapment. Chem. Biol. *14*, 87–96.

Dong, Y.H., Zhang, X.F., Soo, H.M., Greenberg, E.P., and Zhang, L.H. (2005). The two-component response regulator PprB modulates quorum-sensing signal production and global gene expression in *Pseudomonas aeruginosa*. Mol. Microbiol. *56*, 1287–1301.

Drury, L.S., and Buxton, R.S. (1988). Identification and sequencing of the *Escherichia coli* cet gene which codes for an inner membrane protein, mutation of which causes tolerance to colicin E2. Mol. Microbiol. *2*, 109–119.

Duan, K., and Surette, M.G. (2007). Environmental regulation of *Pseudomonas aeruginosa* PAO1 Las and Rhl quorum-sensing systems. J. Bacteriol. *189*, 4827–4836.

Ducros, V.M., Brannigan, J.A., Lewis, R.J., and Wilkinson, A.J. (1998). *Bacillus subtilis* regulatory protein GerE. Acta Crystallogr D Biol Crystallogr 54, 1453–1455.

Ducros, V.M., Lewis, R.J., Verma, C.S., Dodson, E.J., Leonard, G., Turkenburg, J.P., Murshudov, G.N., Wilkinson, A.J., and Brannigan, J.A. (2001). Crystal structure of GerE, the ultimate transcriptional regulator of spore formation in *Bacillus subtilis*. J. Mol. Biol. *306*, 759–771.

Eck, R.V., and Dayhoff, M.O. (1966). Atlas of protein sequence and structure. In National Biomedical Research Foundation. Silver Spring, Maryland.

Eicken, C., Pennella, M.A., Chen, X., Koshlap, K.M., VanZile, M.L., Sacchettini, J.C., and Giedroc, D.P. (2003). A metal-ligand-mediated intersubunit allosteric switch in related SmtB/ArsR zinc sensor proteins. J. Mol. Biol. *333*, 683–695.

Engebrecht, J., and Silverman, M. (1984). Identification of genes and gene products necessary for bacterial bioluminescence. Proc. Natl. Acad. Sci. U.S.A. *81*, 4154–4158.

Engebrecht, J., and Silverman, M. (1987). Nucleotide sequence of the regulatory locus controlling expression of bacterial genes for bioluminescence. Nucleic Acids Res. *15*, 10455–10467.

Engebrecht, J., Nealson, K., and Silverman, M. (1983). Bacterial bioluminescence: isolation and genetic analysis of functions from *Vibrio fischeri*. Cell *32*, 773–781.

Engohang-Ndong, J., Baillat, D., Aumercier, M., Bellefontaine, F., Besra, G.S., Locht, C., and Baulard, A.R. (2004). EthR, a repressor of the TetR/CamR family implicated in ethionamide resistance in mycobacteria, octamerizes cooperatively on its operator. Mol. Microbiol. *51*, 175–188.

Estahbanati, H.K., Kashani, P.P., and Ghanaatpisheh, F. (2002). Frequency of *Pseudomonas aeruginosa* serotypes in burn wound infections and their resistance to antibiotics. Burns *28*, 340–348.

Filloux, A., Hachani, A., and Bleves, S. (2008). The bacterial type VI secretion machine: yet another player for protein transport across membranes. Microbiology *154*, 1570–1583.

Forster-Fromme, K., Hoschle, B., Mack, C., Bott, M., Armbruster, W., and Jendrossek, D. (2006). Identification of genes and proteins necessary for catabolism of acyclic terpenes and leucine/isovalerate in *Pseudomonas aeruginosa*. Appl. Environ. Microbiol. *72*, 4819–4828.

Frank, D.W., and Iglewski, B.H. (1991). Cloning and sequence analysis of a trans-regulatory locus required for exoenzyme S synthesis in *Pseudomonas aeruginosa*. J. Bacteriol. *173*, 6460–6468.

Fuchs, E.L., Brutinel, E.D., Jones, A.K., Fulcher, N.B., Urbanowski, M.L., Yahr, T.L., and Wolfgang, M.C. (2010). The *Pseudomonas aeruginosa* Vfr regulator controls global virulence factor expression through cyclic AMP-dependent and -independent mechanisms. J. Bacteriol. *192*, 3553–3564.

Fukami-Kobayashi, K., Tateno, Y., and Nishikawa, K. (2003). Parallel evolution of ligand specificity between LacI/GalR family repressors and periplasmic sugar-binding proteins. Mol. Biol. Evol. *20*, 267–277.

Fuqua, W.C., Winans, S.C., and Greenberg, E.P. (1994). Quorum sensing in bacteria: the LuxR-LuxI family of cell density-responsive transcriptional regulators. J. Bacteriol. *176*, 269–275.

Gallagher, L.A., McKnight, S.L., Kuznetsova, M.S., Pesci, E.C., and Manoil, C. (2002). Functions required for extracellular quinolone signaling by *Pseudomonas aeruginosa*. J. Bacteriol. *184*, 6472–6480.

Gallegos, M.T., Schleif, R., Bairoch, A., Hofmann, K., and Ramos, J.L. (1997). Arac/XylS family of transcriptional regulators. Microbiol. Mol. Biol. Rev. *61*, 393–410.

Gambello, M.J., and Iglewski, B.H. (1991). Cloning and characterization of the *Pseudomonas aeruginosa* lasR gene, a transcriptional activator of elastase expression. J. Bacteriol. *173*, 3000–3009.

Gilbert, K.B., Kim, T.H., Gupta, R., Greenberg, E.P., and Schuster, M. (2009). Global position analysis of the *Pseudomonas aeruginosa* quorum-sensing transcription factor LasR. Mol. Microbiol. 73, 1072–1085.

Goodman, A.L., and Lory, S. (2004). Analysis of regulatory networks in *Pseudomonas aeruginosa* by genomewide transcriptional profiling. Curr. Opin. Microbiol. 7, 39–44.

Goodman, A.L., Kulasekara, B., Rietsch, A., Boyd, D., Smith, R.S., and Lory, S. (2004). A signaling network reciprocally regulates genes associated with acute infection and chronic persistence in *Pseudomonas aeruginosa*. Dev. Cell 7, 745–754.

Greenblatt, J., and Schleif, R. (1971). Arabinose C protein: regulation of the arabinose operon *in vitro*. Nat. New Biol. *233*, 166–170.

Greenway, D.L., and England, R.R. (1999). ppGpp accumulation in *Pseudomonas aeruginosa* and Pseudomonas fluorescens subjected to nutrient limitation and biocide exposure. Lett. Appl. Microbiol. *29*, 298–302.

Gregory, A.D., Hogue, L.A., Ferkol, T.W., and Link, D.C. (2007). Regulation of systemic and local neutrophil responses by G-CSF during pulmonary *Pseudomonas aeruginosa* infection. Blood *109*, 3235–3243.

Grkovic, S., Brown, M.H., and Skurray, R.A. (2002). Regulation of bacterial drug export systems. Microbiol. Mol. Biol. Rev. *66*, 671–701.

Guarente, L., Roberts, T.M., and Ptashne, M. (1980). A technique for expressing eukaryotic genes in bacteria. Science *209*, 1428–1430.

Hassett, D.J., Korfhagen, T.R., Irvin, R.T., Schurr, M.J., Sauer, K., Lau, G.W., Sutton, M.D., Yu, H., and Hoiby, N. (2010). *Pseudomonas aeruginosa* biofilm infections in cystic fibrosis: insights into pathogenic processes and treatment strategies. Expert Opin Ther. Targets *14*, 117–130.

Hauser, A.R. (2009). The Type III secretion system of *Pseudomonas aeruginosa*: infection by injection. Nat. Rev. Microbiol. 7, 654–665.

Haydon, D.J., and Guest, J.R. (1991). A new family of bacterial regulatory proteins. FEMS Microbiol. Lett. *63*, 291–295.

Heeb, S., and Haas, D. (2001). Regulatory roles of the GacS/GacA two-component system in plant-associated and other Gram-negative bacteria. Mol. Plant Microbe Interact. *14*, 1351–1363.

Heinrichs, D.E., and Poole, K. (1996). PchR, a regulator of ferripyochelin receptor gene (fptA). expression in *Pseudomonas aeruginosa*, functions both as an activator and as a repressor. J. Bacteriol. *178*, 2586–2592.

Heldwein, E.E., and Brennan, R.G. (2001). Crystal structure of the transcription activator BmrR bound to DNA and a drug. Nature *409*, 378–382.

Helmann, J.D., Wang, Y., Mahler, I., and Walsh, C.T. (1989). Homologous metalloregulatory proteins from both Gram-positive and Gram-negative bacteria control transcription of mercury resistance operons. J. Bacteriol. *171*, 222–229.

Hentzer, M., Wu, H., Andersen, J.B., Riedel, K., Rasmussen, T.B., Bagge, N., Kumar, N., Schembri, M.A., Song, Z., *et al.* (2003). Attenuation of *Pseudomonas aeruginosa* virulence by quorum sensing inhibitors. EMBO J. *22*, 3803–3815.

Hernandez-Lucas, I., Gallego-Hernandez, A.L., Encarnacion, S., Fernandez-Mora, M., Martinez-Batallar, A.G., Salgado, H., Oropeza, R., and Calva, E. (2008). The LysR-type transcriptional regulator LeuO controls expression of several genes in *Salmonella enterica* serovar Typhi. J. Bacteriol. *190*, 1658–1670.

Heroven, A.K., and Dersch, P. (2006). RovM, a novel LysR-type regulator of the virulence activator gene rovA, controls cell invasion, virulence and motility of *Yersinia pseudotuberculosis*. Mol. Microbiol. *62*, 1469–1483.

Hinrichs, W., Kisker, C., Duvel, M., Muller, A., Tovar, K., Hillen, W., and Saenger, W. (1994). Structure of the Tet repressor–tetracycline complex and regulation of antibiotic resistance. Science *264*, 418–420.

Hogardt, M., Roeder, M., Schreff, A.M., Eberl, L., and Heesemann, J. (2004). Expression of *Pseudomonas aeruginosa* exoS is controlled by quorum sensing and RpoS. Microbiology *150*, 843–851.

Hoiby, N., Bjarnsholt, T., Givskov, M., Molin, S., and Ciofu, O. (2010). Antibiotic resistance of bacterial biofilms. Int. J. Antimicrob. Agents *35*, 322–332.

Hu, Z., Hung, J.H., Wang, Y., Chang, Y.C., Huang, C.L., Huyck, M., and DeLisi, C. (2009). VisANT 3.5: multi-scale network visualization, analysis and inference based on the gene ontology. Nucleic Acids Res. *37*, W115–121.

Jensen, L.J., Kuhn, M., Stark, M., Chaffron, S., Creevey, C., Muller, J., Doerks, T., Julien, P., Roth, A., *et al.* (2009). STRING 8--a global view on proteins and their functional interactions in 630 organisms. Nucleic Acids Res. *37*, D412–416.

Jin, Y., Yang, H., Qiao, M., and Jin, S. (2010). MexT regulates Type III secretion system through MexS and PtrC in *Pseudomonas aeruginosa*. J Bacteriol *193*, 399–410.

Johnson, C.M., and Schleif, R.F. (2000). Cooperative action of the catabolite activator protein and AraC *in vitro* at the araFGH promoter. J. Bacteriol. *182*, 1995–2000.

Jones, A.K., Fulcher, N.B., Balzer, G.J., Urbanowski, M.L., Pritchett, C.L., Schurr, M.J., Yahr, T.L., and Wolfgang, M.C. (2010). Activation of the *Pseudomonas aeruginosa* AlgU regulon through mucA mutation inhibits cyclic AMP/Vfr signaling. J. Bacteriol. *192*, 5709–5717.

Juhas, M., Wiehlmann, L., Huber, B., Jordan, D., Lauber, J., Salunkhe, P., Limpert, A.S., von Gotz, F., Steinmetz, I., *et al.* (2004). Global regulation of quorum sensing and virulence by VqsR in *Pseudomonas aeruginosa*. Microbiology *150*, 831–841.

Kahn, D., and Ditta, G. (1991). Modular structure of FixJ: homology of the transcriptional activator domain with the −35 binding domain of sigma factors. Mol. Microbiol. *5*, 987–997.

Kawasaki, S., Arai, H., Kodama, T., and Igarashi, Y. (1997). Gene cluster for dissimilatory nitrite reductase (nir). from *Pseudomonas aeruginosa*: sequencing and identification of a locus for heme d1 biosynthesis. J. Bacteriol. *179*, 235–242.

Kay, E., Humair, B., Denervaud, V., Riedel, K., Spahr, S., Eberl, L., Valverde, C., and Haas, D. (2006). Two GacA-dependent small RNAs modulate the quorum-sensing response in *Pseudomonas aeruginosa*. J. Bacteriol. *188*, 6026–6033.

Kerr, K.G., and Snelling, A.M. (2009). *Pseudomonas aeruginosa*: a formidable and ever-present adversary. J Hosp. Infect. *73*, 338–344.

Klockgether, J., Munder, A., Neugebauer, J., Davenport, C.F., Stanke, F., Larbig, K.D., Heeb, S., Schock, U., Pohl, T.M., *et al.* (2010). Genome diversity of *Pseudomonas aeruginosa* PAO1 laboratory strains. J. Bacteriol. *192*, 1113–1121.

Kobayashi, K., and Tagawa, S. (2004). Activation of SoxR-dependent transcription in *Pseudomonas aeruginosa*. J. Biochem. *136*, 607–615.

Kong, K.F., Jayawardena, S.R., Indulkar, S.D., Del Puerto, A., Koh, C.L., Hoiby, N., and Mathee, K. (2005). *Pseudomonas aeruginosa* AmpR is a global transcriptional factor that regulates expression of AmpC and PoxB beta-lactamases, proteases, quorum sensing, and other virulence factors. Antimicrob Agents Chemother *49*, 4567–4575.

Korch, S.B., and Hill, T.M. (2006). Ectopic overexpression of wild-type and mutant hipA genes in *Escherichia coli*: effects on macromolecular synthesis and persister formation. J. Bacteriol. *188*, 3826–3836.

Korner, H., Sofia, H.J., and Zumft, W.G. (2003). Phylogeny of the bacterial superfamily of Crp-Fnr transcription regulators: exploiting the metabolic spectrum by controlling alternative gene programs. FEMS Microbiol. Rev. *27*, 559–592.

Krell, T., Molina-Henares, A.J., and Ramos, J.L. (2006). The IclR family of transcriptional activators and repressors can be defined by a single profile. Protein Sci. *15*, 1207–1213.

Kuchma, S.L., Connolly, J.P., and O'Toole, G.A. (2005). A three-component regulatory system regulates biofilm maturation and type III secretion in *Pseudomonas aeruginosa*. J. Bacteriol. *187*, 1441–1454.

Kulasekara, H.D., Ventre, I., Kulasekara, B.R., Lazdunski, A., Filloux, A., and Lory, S. (2005). A novel two-component system controls the expression of *Pseudomonas aeruginosa* fimbrial cup genes. Mol. Microbiol. *55*, 368–380.

Lamark, T., Kaasen, I., Eshoo, M.W., Falkenberg, P., McDougall, J., and Strom, A.R. (1991). DNA sequence and analysis of the *bet* genes encoding the osmoregulatory choline-glycine betaine pathway of *Escherichia coli*. Mol Microbiol. *5*, 1049–1064.

Larkin, M.A., Blackshields, G., Brown, N.P., Chenna, R., McGettigan, P.A., McWilliam, H., Valentin, F., Wallace, I.M., Wilm, A., *et al.* (2007). Clustal W and Clustal X version 2.0. Bioinformatics *23*, 2947–2948.

Laskowski, M.A., Osborn, E., and Kazmierczak, B.I. (2004). A novel sensor kinase-response regulator hybrid regulates Type III secretion and is required for virulence in *Pseudomonas aeruginosa*. Mol Microbiol *54*, 1090–1103.

Latifi, A., Foglino, M., Tanaka, K., Williams, P., and Lazdunski, A. (1996). A hierarchical quorum-sensing cascade in *Pseudomonas aeruginosa* links the transcriptional activators LasR and RhIR (VsmR). to expression of the stationary-phase sigma factor RpoS. Mol 21 Microbiol 1137–1146.

Lau, G.W., Britigan, B.E., and Hassett, D.J. (2005). *Pseudomonas aeruginosa* OxyR is required for full virulence in rodent and insect models of infection and for resistance to human neutrophils. Infect. Immun. *73*, 2550–2553.

Ledgham, F., Ventre, I., Soscia, C., Foglino, M., Sturgis, J.N., and Lazdunski, A. (2003). Interactions of the quorum sensing regulator QscR: interaction with itself and the other regulators of *Pseudomonas aeruginosa* LasR and RhlR. Mol. Microbiol. *48*, 3365–3370.

Lee, S.W., Glickmann, E., and Cooksey, D.A. (2001). Chromosomal locus for cadmium resistance in *Pseudomonas putida* consisting of a cadmium-transporting ATPase and a MerR family response regulator. Appl. Environ. Microbiol. *67*, 1437–1444.

Leonard, P.M., Smits, S.H., Sedelnikova, S.E., Brinkman, A.B., de Vos, W.M., van der Oost, J., Rice, D.W., and Rafferty, J.B. (2001). Crystal structure of the Lrp-like transcriptional regulator from the archaeon *Pyrococcus furiosus*. EMBO J. *20*, 990–997.

Lequette, Y., Lee, J.H., Ledgham, F., Lazdunski, A., and Greenberg, E.P. (2006). A distinct QscR regulon in the *Pseudomonas aeruginosa* quorum-sensing circuit. J. Bacteriol. *188*, 3365–3370.

Liang, H., Li, L., Dong, Z., Surette, M.G., and Duan, K. (2008). The YebC family protein PA0964 negatively regulates the *Pseudomonas aeruginosa* quinolone signal system and pyocyanin production. J. Bacteriol. *190*, 6217–6227.

Liang, H., Li, L., Kong, W., Shen, L., and Duan, K. (2009). Identification of a novel regulator of the quorum-sensing systems in *Pseudomonas aeruginosa*. FEMS Microbiol. Lett. *293*, 196–204.

MacEachran, D.P., Stanton, B.A., and O'Toole, G.A. (2008). Cif is negatively regulated by the TetR family repressor CifR. Infect. Immun. *76*, 3197–3206.

MacEachran, D.P., Ye, S., Bomberger, J.M., Hogan, D.A., Swiatecka-Urban, A., Stanton, B.A., and O'Toole, G.A. (2007). The *Pseudomonas aeruginosa* secreted protein PA2934 decreases apical membrane expression of the cystic fibrosis transmembrane conductance regulator. Infect. Immun. *75*, 3902–3912.

Madan Babu, M., Teichmann, S.A., and Aravind, L. (2006). Evolutionary dynamics of prokaryotic transcriptional regulatory networks. J. Mol. Biol. *358*, 614–633.

Maddocks, S.E., and Oyston, P.C. (2008). Structure and function of the LysR-type transcriptional regulator (LTTR). family proteins. Microbiology *154*, 3609–3623.

Madhusudhan, K.T., Lorenz, D., and Sokatch, J.R. (1993). The *bkdR* gene of *Pseudomonas putida* is required for expression of the bkd operon and encodes a protein related to Lrp of *Escherichia coli*. J. Bacteriol. *175*, 3934–3940.

Magasanik, B. (1993). The regulation of nitrogen utilization in enteric bacteria. J. Cell. Biochem. *51*, 34–40.

Mahar, P., Padiglione, A.A., Cleland, H., Paul, E., Hinrichs, M., and Wasiak, J. (2010). *Pseudomonas aeruginosa* bacteraemia in burns patients: risk factors and outcomes. Burns *36*, 1228–1233.

Marra, A.R., Bar, K., Bearman, G.M., Wenzel, R.P., and Edmond, M.B. (2006). Systemic inflammatory response syndrome in adult patients with nosocomial bloodstream infection due to *Pseudomonas aeruginosa*. J Infect *53*, 30–35.

Martin, R.G., and Rosner, J.L. (2001). The AraC transcriptional activators. Curr. Opin. Microbiol. *4*, 132–137.

Martin, R.G., Gillette, W.K., Rhee, S., and Rosner, J.L. (1999). Structural requirements for marbox function in transcriptional activation of mar/sox/rob regulon promoters in *Escherichia coli*: sequence, orientation and spatial relationship to the core promoter. Mol. Microbiol. *34*, 431–441.

Martinez-Nunez, M.A., Perez-Rueda, E., Gutierrez-Rios, R.M., and Merino, E. (2010). New insights into the regulatory networks of paralogous genes in bacteria. Microbiology *156*, 14–22.

Maskell, D.J., Szabo, M.J., Deadman, M.E., and Moxon, E.R. (1992). The gal locus from Haemophilus influenzae: cloning, sequencing and the use of gal mutants to study lipopolysaccharide. Mol Microbiol *6*, 3051–3063.

Masuda, N., Sakagawa, E., Ohya, S., Gotoh, N., Tsujimoto, H., and Nishino, T. (2000). Substrate specificities of MexAB-OprM, MexCD-OprJ, and MexXY-oprM efflux pumps in *Pseudomonas aeruginosa*. Antimicrob Agents Chemother *44*, 3322–3327.

Mathee, K., Narasimhan, G., Valdes, C., Qiu, X., Matewiṣh, J.M., Koehrsen, M., Rokas, A., Yandava, C.N., Engels, R., *et al.* (2008). Dynamics of *Pseudomonas aeruginosa* genome evolution. Proc. Natl. Acad. Sci. U.S.A. *105*, 3100–3105.

Matsui, H., Sano, Y., Ishihara, H., and Shinomiya, T. (1993). Regulation of pyocin genes in *Pseudomonas aeruginosa* by positive (prtN). and negative (prtR). regulatory genes. J. Bacteriol. *175*, 1257–1263.

Medina, G., Juarez, K., Diaz, R., and Soberon-Chavez, G. (2003). Transcriptional regulation of *Pseudomonas aeruginosa* rhlR, encoding a quorum-sensing regulatory protein. Microbiology *149*, 3073–3081.

Melendez, J.H., Frankel, Y.M., An, A.T., Williams, L., Price, L.B., Wang, N.Y., Lazarus, G.S., and Zenilman, J.M. (2010). Real-time PCR assays compared to culture-based approaches for identification of aerobic bacteria in chronic wounds. Clin. Microbiol. Infect. *16*, 1762–1769.

Melstrom, K.A., Jr., Kozlowski, R., Hassett, D.J., Suzuki, H., Bates, D.M., Gamelli, R.L., and Shankar, R. (2007). Cytotoxicity of Pseudomonas secreted exotoxins requires OxyR expression. J. Surg. Res. *143*, 50–57.

Mern, D.S., Ha, S.W., Khodaverdi, V., Gliese, N., and Gorisch, H. (2010). A complex regulatory network controls aerobic ethanol oxidation in *Pseudomonas aeruginosa*: indication of four levels of sensor kinases and response regulators. Microbiology *156*, 1505–1516.

Michel, L., Gonzalez, N., Jagdeep, S., Nguyen-Ngoc, T., and Reimmann, C. (2005). PchR-box recognition by the AraC-type regulator PchR of *Pseudomonas aeruginosa* requires the siderophore pyochelin as an effector. Mol. Microbiol. *58*, 495–509.

Molina-Henares, A.J., Krell, T., Eugenia Guazzaroni, M., Segura, A., and Ramos, J.L. (2006). Members of the IclR family of bacterial transcriptional regulators function as activators and/or repressors. FEMS Microbiol. Rev. *30*, 157–186.

Morett, E., and Segovia, L. (1993). The sigma 54 bacterial enhancer-binding protein family: mechanism of action and phylogenetic relationship of their functional domains. J. Bacteriol. *175*, 6067–6074.

Morita, Y., Cao, L., Gould, V.C., Avison, M.B., and Poole, K. (2006). *nalD* encodes a second repressor of the *mexAB-oprM* multidrug efflux operon of *Pseudomonas aeruginosa*. J. Bacteriol. *188*, 8649–8654.

Moya, B., Dotsch, A., Juan, C., Blazquez, J., Zamorano, L., Haussler, S., and Oliver, A. (2009). Beta-lactam resistance response triggered by inactivation of a nonessential penicillin-binding protein. PLoS Pathog. *5*, e1000353.

Murray, T.S., Egan, M., and Kazmierczak, B.I. (2007). *Pseudomonas aeruginosa* chronic colonization in cystic fibrosis patients. Curr. Opin. Pediatr. *19*, 83–88.

Nakada, Y., Jiang, Y., Nishijyo, T., Itoh, Y., and Lu, C.D. (2001). Molecular characterization and regulation of the *aguBA* operon, responsible for agmatine utilization in *Pseudomonas aeruginosa* PAO1. J. Bacteriol. *183*, 6517–6524.

Nasser, W., and Reverchon, S. (2007). New insights into the regulatory mechanisms of the LuxR family of quorum sensing regulators. Anal. Bioanal. Chem. *387*, 381–390.

Nei, M., and Kumar, S. (2000). Molecular evolution and phylogenetics. Oxford University Press, New York.

Ng, W.L., and Bassler, B.L. (2009). Bacterial quorum-sensing network architectures. Annu. Rev. Genet. *43*, 197–222.

Nguyen, C.C., and Saier, M.H., Jr. (1995). Phylogenetic, structural and functional analyses of the LacI-GalR family of bacterial transcription factors. FEBS Lett. *377*, 98–102.

Nishijyo, T., Haas, D., and Itoh, Y. (2001). The CbrA-CbrB two-component regulatory system controls the utilization of multiple carbon and nitrogen sources in *Pseudomonas aeruginosa*. Mol. Microbiol. *40*, 917–931.

Ochsner, U.A., and Reiser, J. (1995). Autoinducer-mediated regulation of rhamnolipid biosurfactant synthesis in *Pseudomonas aeruginosa*. Proc. Natl. Acad. Sci. U.S.A. *92*, 6424–6428.

Ochsner, U.A., Vasil, M.L., Alsabbagh, E., Parvatiyar, K., and Hassett, D.J. (2000). Role of the *Pseudomonas aeruginosa* oxyR-recG operon in oxidative stress defense and DNA repair: OxyR-dependent regulation of *katB-ankB*, *ahpB*, and *ahpC-ahpF*. J. Bacteriol. *182*, 4533–4544.

Orth, P., Schnappinger, D., Hillen, W., Saenger, W., and Hinrichs, W. (2000). Structural basis of gene regulation by the tetracycline inducible Tet repressor-operator system. Nat. Struct. Biol. *7*, 215–219.

Osman, D., and Cavet, J.S. (2010). Bacterial metal-sensing proteins exemplified by ArsR-SmtB family repressors. Nat. Prod. Rep. *27*, 668–680.

Pearson, J.P., Passador, L., Iglewski, B.H., and Greenberg, E.P. (1995). A second N-acylhomoserine lactone signal produced by *Pseudomonas aeruginosa*. Proc. Natl. Acad. Sci. U.S.A. *92*, 1490–1494.

Pearson, J.P., Gray, K.M., Passador, L., Tucker, K.D., Eberhard, A., Iglewski, B.H., and Greenberg, E.P. (1994). Structure of the autoinducer required for expression of *Pseudomonas aeruginosa* virulence genes. Proc. Natl. Acad. Sci. U.S.A. *91*, 197–201.

Peekhaus, N., and Conway, T. (1998). Positive and negative transcriptional regulation of the *Escherichia coli* gluconate regulon gene *gntT* by GntR and the cyclic AMP (cAMP)-cAMP receptor protein complex. J. Bacteriol. *180*, 1777–1785.

Pesci, E.C., Pearson, J.P., Seed, P.C., and Iglewski, B.H. (1997). Regulation of *las* and *rhl* quorum sensing in *Pseudomonas aeruginosa*. J. Bacteriol. *179*, 3127–3132.

Pesci, E.C., Milbank, J.B., Pearson, J.P., McKnight, S., Kende, A.S., Greenberg, E.P., and Iglewski, B.H. (1999). Quinolone signaling in the cell-to-cell communication system of *Pseudomonas aeruginosa*. Proc. Natl. Acad. Sci. U.S.A. *96*, 11229–11234.

Pessi, G., and Haas, D. (2000). Transcriptional control of the hydrogen cyanide biosynthetic genes *hcnABC* by the anaerobic regulator Anr and the quorum-sensing regulators LasR and RhlR in *Pseudomonas aeruginosa*. J. Bacteriol. *182*, 6940–6949.

Petrova, O.E., and Sauer, K. (2009). A novel signaling network essential for regulating *Pseudomonas aeruginosa* biofilm development. PLoS Pathog *5*, e1000668.

Petrova, O.E., and Sauer, K. (2010). The novel two-component regulatory system BfiSR regulates biofilm development directly through CafA by its control over the small RNA *rsmZ*. J Bacteriol. *192*, 5275–5288.

Raibaud, O., and Richet, E. (1987). Maltotriose is the inducer of the maltose regulon of *Escherichia coli*. J. Bacteriol. *169*, 3059–3061.

Ramos, J.L., Martinez-Bueno, M., Molina-Henares, A.J., Teran, W., Watanabe, K., Zhang, X., Gallegos, M.T., Brennan, R., and Tobes, R. (2005). The TetR family of transcriptional repressors. Microbiol. Mol. Biol. Rev. *69*, 326–356.

Rampioni, G., Bertani, I., Zennaro, E., Polticelli, F., Venturi, V., and Leoni, L. (2006). The quorum-sensing negative regulator RsaL of *Pseudomonas aeruginosa* binds to the *lasI* promoter. J. Bacteriol. *188*, 815–819.

Rampioni, G., Schuster, M., Greenberg, E.P., Bertani, I., Grasso, M., Venturi, V., Zennaro, E., and Leoni, L. (2007). RsaL provides quorum sensing homeostasis and functions as a global regulator of gene expression in *Pseudomonas aeruginosa*. Mol Microbiol *66*, 1557–1565.

Reimmann, C., Beyeler, M., Latifi, A., Winteler, H., Foglino, M., Lazdunski, A., and Haas, D. (1997). The global activator GacA of *Pseudomonas aeruginosa* PAO positively controls the production of the autoinducer N-butyryl-homoserine lactone and the formation of the virulence factors pyocyanin, cyanide, and lipase. Mol. Microbiol. *24*, 309–319.

Richet, E., and Raibaud, O. (1989). MalT, the regulatory protein of the *Escherichia coli* maltose system, is an ATP-dependent transcriptional activator. EMBO J *8*, 981–987.

Rigali, S., Derouaux, A., Giannotta, F., and Dusart, J. (2002). Subdivision of the helix–turn–helix GntR family of bacterial regulators in the FadR, HutC, MocR, and YtrA subfamilies. J. Biol. Chem. *277*, 12507–12515.

Rigali, S., Schlicht, M., Hoskisson, P., Nothaft, H., Merzbacher, M., Joris, B., and Titgemeyer, F. (2004). Extending the classification of bacterial transcription factors beyond the helix–turn–helix motif as an alternative approach to discover new cis/trans relationships. Nucleic Acids Res. *32*, 3418–3426.

Rocchetta, H.L., Burrows, L.L., and Lam, J.S. (1999). Genetics of O-antigen biosynthesis in *Pseudomonas aeruginosa*. Microbiol. Mol. Biol. Rev. *63*, 523–553.

Romero-Steiner, S., Parales, R.E., Harwood, C.S., and Houghton, J.E. (1994). Characterization of the *pcaR* regulatory gene from *Pseudomonas putida*, which is required for the complete degradation of p-hydroxybenzoate. J. Bacteriol. *176*, 5771–5779.

Rosenau, F., and Jaeger, K. (2000). Bacterial lipases from *Pseudomonas*: regulation of gene expression and mechanisms of secretion. Biochimie *82*, 1023–1032.

Schell, M.A. (1993). Molecular biology of the LysR family of transcriptional regulators. Annu. Rev. Microbiol. *47*, 597–626.

Schleif, R. (1969). An L-arabinose binding protein and arabinose permeation in *Escherichia coli*. J. Mol. Biol. *46*, 185–196.

Schuster, M., and Greenberg, E.P. (2006). A network of networks: quorum-sensing gene regulation in *Pseudomonas aeruginosa*. Int J Med Microbiol *296*, 73–81.

Schuster, M., and Greenberg, E.P. (2007). Early activation of quorum sensing in *Pseudomonas aeruginosa* reveals the architecture of a complex regulon. BMC Genomics *8*, 287.

Schuster, M., Lostroh, C.P., Ogi, T., and Greenberg, E.P. (2003). Identification, timing, and signal specificity of *Pseudomonas aeruginosa* quorum-controlled genes: a transcriptome analysis. J. Bacteriol. *185*, 2066–2079.

Schuster, M., Hawkins, A.C., Harwood, C.S., and Greenberg, E.P. (2004). The *Pseudomonas aeruginosa* RpoS regulon and its relationship to quorum sensing. Mol. Microbiol. *51*, 973–985.

Secher, I., Hermes, I., Pre, S., Carreau, F., and Bahuet, F. (2005). Surgical wound infections due to *Pseudomonas aeruginosa* in orthopedic surgery. Med Mal Infect *35*, 149–154.

Senf, F., Tommassen, J., and Koster, M. (2008). Polar secretion of proteins via the Xcp Type II secretion system in *Pseudomonas aeruginosa*. Microbiology *154*, 3025–3032.

Shannon, P., Markiel, A., Ozier, O., Baliga, N.S., Wang, J.T., Ramage, D., Amin, N., Schwikowski, B., and Ideker, T. (2003). Cytoscape: a software environment for integrated models of biomolecular interaction networks. Genome Res *13*, 2498–2504.

Shen, D.K., Filopon, D., Kuhn, L., Polack, B., and Toussaint, B. (2006). PsrA is a positive transcriptional regulator of the Type III secretion system in *Pseudomonas aeruginosa*. Infect. Immun. *74*, 1121–1129.

Shigemura, K., Arakawa, S., Sakai, Y., Kinoshita, S., Tanaka, K., and Fujisawa, M. (2006). Complicated urinary tract infection caused by *Pseudomonas aeruginosa* in a single institution (1999–2003). Int J Urol *13*, 538–542.

Siehnel, R., Traxler, B., An, D.D., Parsek, M.R., Schaefer, A.L., and Singh, P.K. (2010). A unique regulator controls the activation threshold of quorum-regulated genes in *Pseudomonas aeruginosa*. Proc. Natl. Acad. Sci. U.S.A. *107*, 7916–7921.

Sigrist, C.J., Cerutti, L., de Castro, E., Langendijk-Genevaux, P.S., Bulliard, V., Bairoch, A., and Hulo, N. (2010). PROSITE, a protein domain database for functional characterization and annotation. Nucleic Acids Res. *38*, D161–166.

Smith, C.P., and Chater, K.F. (1988). Structure and regulation of controlling sequences for the *Streptomyces coelicolor* glycerol operon. J. Mol. Biol. *204*, 569–580.

Steele, M.I., Lorenz, D., Hatter, K., Park, A., and Sokatch, J.R. (1992). Characterization of the *mmsAB* operon of *Pseudomonas aeruginosa* PAO encoding methylmalonate-semialdehyde dehydrogenase and 3-hydroxyisobutyrate dehydrogenase. J. Biol. Chem. *267*, 13585–13592.

Stickland, H.G., Davenport, P.W., Lilley, K.S., Griffin, J.L., and Welch, M. (2010). Mutation of *nfxB* causes global changes in the physiology and metabolism of *Pseudomonas aeruginosa*. J. Proteome Res. *9*, 2957–2967.

Stragier, P., Borne, F., Richaud, F., Richaud, C., and Patte, J.C. (1983). Regulatory pattern of the *Escherichia coli lysA* gene: expression of chromosomal *lysA–lacZ* fusions. J. Bacteriol. *156*, 1198–1203.

Swanson, B.L., Hager, P., Phibbs, P., Jr., Ochsner, U., Vasil, M.L., and Hamood, A.N. (2000). Characterization of the 2-ketogluconate utilization operon in *Pseudomonas aeruginosa* PAO1. Mol. Microbiol. *37*, 561–573.

Swint-Kruse, L., and Matthews, K.S. (2009). Allostery in the LacI/GalR family: variations on a theme. Curr. Opin. Microbiol. *12*, 129–137.

Takeyama, K., Kunishima, Y., Matsukawa, M., Takahashi, S., Hirose, T., Kobayashi, N., Kobayashi, I., and Tsukamoto, T. (2002). Multidrug-resistant *Pseudomonas aeruginosa* isolated from the urine of patients with urinary tract infection. J. Infect. Chemother. *8*, 59–63.

Tamura, K., Dudley, J., Nei, M., and Kumar, S. (2007). MEGA4: Molecular Evolutionary Genetics Analysis (MEGA). software version 4.0. Mol. Biol. Evol. *24*, 1596–1599.

Thaden, J.T., Lory, S., and Gardner, T.S. (2010). Quorum-sensing regulation of a copper toxicity system in *Pseudomonas aeruginosa*. J. Bacteriol. *192*, 2557–2568.

Turner, K.H., Vallet-Gely, I., and Dove, S.L. (2009). Epigenetic control of virulence gene expression in *Pseudomonas aeruginosa* by a LysR-type transcription regulator. PLoS Genet. *5*, e1000779.

van Aalten, D.M., DiRusso, C.C., and Knudsen, J. (2001). The structural basis of acyl coenzyme A-dependent regulation of the transcription factor FadR. EMBO J. *20*, 2041–2050.

van Delden, C., Comte, R., and Bally, A.M. (2001). Stringent response activates quorum sensing and modulates cell density-dependent gene expression in *Pseudomonas aeruginosa*. J. Bacteriol. *183*, 5376–5384.

van der Lelie, D., Schwuchow, T., Schwidetzky, U., Wuertz, S., Baeyens, W., Mergeay, M., and Nies, D.H. (1997). Two-component regulatory system involved in transcriptional control of heavy-metal homoeostasis in *Alcaligenes eutrophus*. Mol. Microbiol. *23*, 493–503.

Vartak, N.B., Reizer, J., Reizer, A., Gripp, J.T., Groisman, E.A., Wu, L.F., Tomich, J.M., and Saier, M.H., Jr. (1991). Sequence and evolution of the FruR protein of Salmonella typhimurium: a pleiotropic transcriptional regulatory protein possessing both activator and repressor functions which is homologous to the periplasmic ribose-binding protein. Res. Microbiol. *142*, 951–963.

Ventre, I., Goodman, A.L., Vallet-Gely, I., Vasseur, P., Soscia, C., Molin, S., Bleves, S., Lazdunski, A., Lory, S., *et al.* (2006). Multiple sensors control reciprocal expression of *Pseudomonas aeruginosa* regulatory RNA and virulence genes. Proc. Natl. Acad. Sci. U.S.A. *103*, 171–176.

Vinckx, T., Matthijs, S., and Cornelis, P. (2008). Loss of the oxidative stress regulator OxyR in *Pseudomonas aeruginosa* PAO1 impairs growth under iron-limited conditions. FEMS Microbiol. Lett. *288*, 258–265.

Vinckx, T., Wei, Q., Matthijs, S., and Cornelis, P. (2010). The *Pseudomonas aeruginosa* oxidative stress regulator OxyR influences production of pyocyanin and rhamnolipids: protective role of pyocyanin. Microbiology *156*, 678–686.

Wade, D.S., Calfee, M.W., Rocha, E.R., Ling, E.A., Engstrom, E., Coleman, J.P., and Pesci, E.C. (2005). Regulation of Pseudomonas quinolone signal synthesis in *Pseudomonas aeruginosa*. J. Bacteriol. *187*, 4372–4380.

Wagner, V.E., Gillis, R.J., and Iglewski, B.H. (2004). Transcriptome analysis of quorum-sensing regulation and virulence factor expression in *Pseudomonas aeruginosa*. Vaccine 22, S15–20.

Wagner, V.E., Bushnell, D., Passador, L., Brooks, A.I., and Iglewski, B.H. (2003). Microarray analysis of *Pseudomonas aeruginosa* quorum-sensing regulons: effects of growth phase and environment. J. Bacteriol. *185*, 2080–2095.

Wang, J., Lory, S., Ramphal, R., and Jin, S. (1996). Isolation and characterization of *Pseudomonas aeruginosa* genes inducible by respiratory mucus derived from cystic fibrosis patients. Mol. Microbiol. *22*, 1005–1012.

Whitchurch, C.B., Beatson, S.A., Comolli, J.C., Jakobsen, T., Sargent, J.L., Bertrand, J.J., West, J., Klausen, M., Waite, L.L., *et al.* (2005). *Pseudomonas aeruginosa fimL* regulates multiple virulence functions by intersecting with Vfr-modulated pathways. Mol. Microbiol. *55*, 1357–1378.

Wilson, C.J., Zhan, H., Swint-Kruse, L., and Matthews, K.S. (2007). The lactose repressor system: paradigms for regulation, allosteric behavior and protein folding. Cell Mol. Life Sci. *64*, 3–16.

Winsor, G.L., Van Rossum, T., Lo, R., Khaira, B., Whiteside, M.D., Hancock, R.E., and Brinkman, F.S. (2009). Pseudomonas Genome Database: facilitating user-friendly, comprehensive comparisons of microbial genomes. Nucleic Acids Res. *37*, D483–488.

Wolfgang, M.C., Lee, V.T., Gilmore, M.E., and Lory, S. (2003). Coordinate regulation of bacterial virulence genes by a novel adenylate cyclase-dependent signaling pathway. Dev. Cell *4*, 253–263.

Wu, J., and Weiss, B. (1991). Two divergently transcribed genes, soxR and soxS, control a superoxide response regulon of *Escherichia coli*. J. Bacteriol. *173*, 2864–2871.

Wu, W., and Jin, S. (2005). PtrB of *Pseudomonas aeruginosa* suppresses the type III secretion system under the stress of DNA damage. J. Bacteriol. *187*, 6058–6068.

Xiao, G., Deziel, E., He, J., Lepine, F., Lesic, B., Castonguay, M.H., Milot, S., Tampakaki, A.P., Stachel, S.E., *et al.* (2006). MvfR, a key *Pseudomonas aeruginosa* pathogenicity LTTR-class regulatory protein, has dual ligands. Mol. Microbiol. *62*, 1689–1699.

Xu, Y., Heath, R.J., Li, Z., Rock, C.O., and White, S.W. (2001). The FadR–DNA complex. Transcriptional control of fatty acid metabolism in *Escherichia coli*. J. Biol. Chem. *276*, 17373–17379.

Yahr, T.L., and Frank, D.W. (1994). Transcriptional organization of the trans-regulatory locus which controls exoenzyme S synthesis in *Pseudomonas aeruginosa*. J. Bacteriol. *176*, 3832–3838.

Yahr, T.L., and Wolfgang, M.C. (2006). Transcriptional regulation of the *Pseudomonas aeruginosa* Type III secretion system. Mol. Microbiol. *62*, 631–640.

Yeung, A.T., Torfs, E.C., Jamshidi, F., Bains, M., Wiegand, I., Hancock, R.E., and Overhage, J. (2009). Swarming of *Pseudomonas aeruginosa* is controlled by a broad spectrum of transcriptional regulators, including MetR. J. Bacteriol. *191*, 5592–5602.

Yokoyama, K., Ishijima, S.A., Clowney, L., Koike, H., Aramaki, H., Tanaka, C., Makino, K., and Suzuki, M. (2006). Feast/famine regulatory proteins (FFRPs): *Escherichia coli* Lrp, AsnC and related archaeal transcription factors. FEMS Microbiol. Rev. *30*, 89–108.

Zhang, R.G., Kim, Y., Skarina, T., Beasley, S., Laskowski, R., Arrowsmith, C., Edwards, A., Joachimiak, A., and Savchenko, A. (2002). Crystal structure of *Thermotoga maritima* 0065, a member of the IclR transcriptional factor family. J. Biol. Chem. *277*, 19183–19190.

Zhu, K., Choi, K.H., Schweizer, H.P., Rock, C.O., and Zhang, Y.M. (2006). Two aerobic pathways for the formation of unsaturated fatty acids in *Pseudomonas aeruginosa*. Mol. Microbiol. *60*, 260–273.

Zolfaghar, I., Angus, A.A., Kang, P.J., To, A., Evans, D.J., and Fleiszig, S.M. (2005). Mutation of *retS*, encoding a putative hybrid two-component regulatory protein in *Pseudomonas aeruginosa*, attenuates multiple virulence mechanisms. Microbes Infect. 7, 1305–1316.

Transcriptional Regulation Network in Cyanobacteria: a Comparative Genomic View

14

Xizeng Mao, Fenglou Mao, Zhengchang Su, Yi Li and Ying Xu

Abstract

Cyanobacteria are the oldest and most diverse organisms of autotrophic photosynthesis on Earth. Compared to the most studied bacteria such *as E. coli*, the overall knowledge about the transcriptional regulation system of cyanobacteria is still limited and fragmented. The availability of a large number of fully sequenced cyanobacterial genomes, along with genome-scale transcriptomic data, provided unprecedented opportunities for researchers to elucidate the transcriptional regulation system in this group of organisms in a systematic manner. In this chapter, we first provide a brief introduction to the basics of cyanobacteria, including their diverse living habits and phylogenetic classification, and then describe the basic components of their transcription regulation system. We then proceed to describe in details the regulatory machinery of three key pathways, namely photosynthesis, nitrogen assimilation and osmoregulation, along with their cross-talk networks. To demonstrate how comparative genomics can help to elucidate the complexity of transcriptional regulation systems under diverse living environments, we will use six cyanobacteria to demonstrate what information can be derived through genome comparisons and how. The chapter ends with a list of available Internet resources along with a list of challenging problems in full elucidation of cyanobacterial transcription regulation systems.

Introduction

Cyanobacteria, also known as blue-green algae or blue-green bacteria, are the oldest and most diverse organisms of autotrophic photosynthesis on Earth. They were responsible for the creation of the oxygen atmosphere we have now, which largely defines the biodiversity of living organisms on Earth. They are also the ancestors of chloroplasts in plants and eukaryotic algae through symbiosis (Raven and Allen, 2003).

Compared to other bacteria, cyanobacteria can conduct photosynthesis. Specifically they can convert carbon dioxide and water into carbohydrate and oxygen through photosynthesis (Binder, 1982), and some of them even can do photosynthesis without producing oxygen but using sulfur or sulfite as the final electron acceptors (Whitton and Potts, 2000). Some cyanobacteria can perform nitrogen fixation to ammonia, nitrites or nitrates, which can be absorbed by plants to synthesize proteins (Fay, 1992). Many cyanobacteria display circadian rhythms that were once thought to exist only in eukaryotic cells (Golden *et al.*, 1997; Takigawa-Imamura and Mochizuki, 2006). Another distinct characteristic of cyanobacteria is that they can live in almost every terrestrial and aquatic habitat, from oceans to fresh water and from bare rock to soil. These unique properties of cyanobacteria make them an ideal model system to study a variety of important biological problems, such as photosynthesis with or without oxygen production, nitrogen fixation and circadian rhythms.

The current understanding about transcription regulation in cyanobacteria is that they generally use more regulatory elements than other bacteria in response to the internal and environmental changes. Many of the cyanobacteria, such as *Synechocystis* PCC6803, use three groups of sigma factors: group 1, consisting of the σ^A factor, is responsible for the regulation of house-keeping genes; group 2, consisting of other σ^{70}-like factors, responds to environmental changes, such as heat shock, daylight change and nitrogen starvation; and group 3 factors are related to cell motility and growth (Imamura and Asayama, 2009). Large numbers of transcription factors and two-component systems have been identified for some cyanobacteria, but their cellular-level functions are largely unknown (Liu *et al.*, 2006; Wu *et al.*, 2007). Compared to the most studied bacteria like *E. coli*, the overall knowledge about transcriptional regulation systems of cyanobacteria is limited and fragmented.

The availability of a large number of fully sequenced cyanobacterial genomes, along with genome-scale transcriptomic data, provided unprecedented opportunities for bacteriologists to elucidate the transcriptional regulation system in this group of organisms in a systematic manner. Comparative genomics represents a key and powerful tool for carrying out such studies. As of now, 29 strains of cyanobacteria have been fully sequenced and 61 additional ones are being sequenced (NCBI Genome Project). Through identification of conserved gene clusters and *cis* regulatory elements across different cyanobacterial genomes, many key elements in the transcriptional regulation system have been identified, which can be further validated based on transcriptomic data and through other experimental techniques.

In this chapter, we first provide a brief introduction to the basics of cyanobacteria, including their diverse living habits and phylogenetic classification, and then describe the basic components in their transcription regulation system: RNA polymerase sigma factors, transcription factors and terminators, and two-component signalling systems. We will then proceed to describe in details the regulatory machinery of three key pathways, namely photosynthesis, nitrogen assimilation and osmoregulation, along with their associated cross-talk networks. To demonstrate how comparative genomics can help to elucidate the complexity of transcriptional regulation systems under diverse living environments, we will use six cyanobacteria to show what information can be derived through genome comparisons and how. The chapter ends with a list of available Internet resources along with a list of challenging problems in the study of cyanobacterial transcription regulation.

Diversity of cyanobacteria

Cyanobacteria live in almost every terrestrial and aquatic habitat on Earth, from oceans to fresh water and from bare rock to soil. They can occur as planktonic cells or form phototrophic biofilms in fresh water and marine environments. Some even live in the fur of sloths. For example, *Synechocystis* PCC6803 is a freshwater motile cyanobacterium capable of both phototrophic growth by oxygenic photosynthesis in sunlight and heterotrophic growth by glycolysis and oxidative phosphorylation in dark (Ikeuchi, 1996). Transitions between daylight and dark are effectively anticipated through a circadian clock. *Synechococcus* WH8102, a marine cyanobacterium, is abundant across all oceans and is the major primary producers on a global scale (Palenik *et al.*, 2003). *Thermosynechococcus elongatus* BP-1 is a rod-shaped cyanobacterium and has a distinct thermophilic character (Nakamura *et al.*, 2002). *Trichodesmium erythraeum* IMS101 is a genus of marine diazotrophic, non-heterocystous cyanobacteria and plays a major role in the tropical and subtropical oceans both as primary producers and suppliers of 'new' nitrogen through fixing atmospheric dinitrogen (Shi *et al.*, 2007).

Cyanobacteria are generally classified into five groups: group I (*Chroococcales*) comprises solitary and colonial cyanobacteria, such as *Synechococcus* and *Gloecapsa*; group II (*Pleurocapsales*) consists of unicellular to pseudo-filamentous, thallus-forming cyanobacteria, with cells capable of fission; group III (*Pleurocapsales*) comprises filamentous cyanobacteria not capable of cell differentiation; group IV (*Nostocales*) consists of filaments and is capable of cell differentiation; and group V (*Strigonematales*) represents the cyanobacteria capable of cell differentiation and has a more complex cellular organization compared to other cyanobacteria (Rippka *et al.*, 1979). This group of bacteria clearly covers a wide range of cellular complexities, ranging from well-defined unicellular organizations to quasi-multicellular properties, providing an ideal group of organisms for studying transcription regulation and organism evolution. Fig. 14.1 shows a phylogeny among 23 cyanobacteria with fully sequenced genomes and relevant classification information (Herrero *et al.*, 2008).

Components of transcription regulation

Cyanobacteria have all the basic elements of the transcription regulation machinery of other ancient bacteria, but their detailed regulation mechanisms tend to be more complex than those of other bacteria.

Transcription in cyanobacteria is performed by the RNA polymerase (RNAP) holoenzyme, comprising a core enzyme and sigma (σ) factor that provides promoter selectivity, as in *E. coli*. The core enzyme in cyanobacteria consists of five subunits ($\alpha_2\beta\beta'\gamma$), whose architecture is more similar to that of plants than that of eubacteria ($\alpha_2\beta\beta'$). Most σ factors in cyanobacteria are σ^{70} type or σ^{70}-like and classified into three groups based on their phylogenies; the only other σ factor family (σ^N) covers nitrogen-related transcription which *E. coli* does not have. Table 14.1 shows the details of these σ factors (Imamura and Asayama, 2009). We use *PCC6803* as an example to explain the three groups of σ factors. Group 1 comprises σ^A (an analogue of σ^{70}) for the expression of house-keeping genes and cell viability-related genes. Group 2, consisting of four σ factors, is responsive to environmental changes; specifically, σ^B responds to heat shock, darkness and nitrogen starvation; σ^C is responsible for cell viability and heat acclimation in the stationary phase of growth; σ^D is a general responder to changes in light; and σ^E regulates the circadian system. In group 3, σ^F plays a critical role in cell motility towards light; σ^G is involved in regulation of cell growth; and the functionalities of σ^H and σ^I are currently unclear (see Fig. 14.2). The global transcription regulation network consists of σ factors from all the three groups in *PCC6803*; and a disruption of one σ may alter the expression of other σ factor(s). A distinct feature of the transcription regulation machinery in cyanobacteria is that they tend to use two σ factors simultaneously instead of one as in *E. coli*.

Transcription factors (TFs) represent another key component of transcription regulation as in other bacteria. They serve to recruit RNA polymerase or other regulatory proteins, or to inhibit/repress transcription through binding to *cis* regulatory elements in the DNA sequence within the promoter regions. As of now, 1,88 putative TFs have been computationally predicted, with some of them having been experimentally validated, in the 21 fully sequenced cyanobacterial genomes (Wu *et al.*, 2007) (see http://bioinformatics.zj.cn for details). All these TFs are stored in cTFbase (Wu *et al.*, 2007), a database of transcription factors in cyanobacteria.

Comparative analyses revealed that the number of TFs encoded in different cyanobacterial genomes varies greatly, possibly because of the extraordinarily diverse environments in which they live (Wu *et al.*, 2007). For example, cyanobacteria living in fresh water or soil tend to have considerably more TFs than those living in marine water. While all TFs have the basic DNA

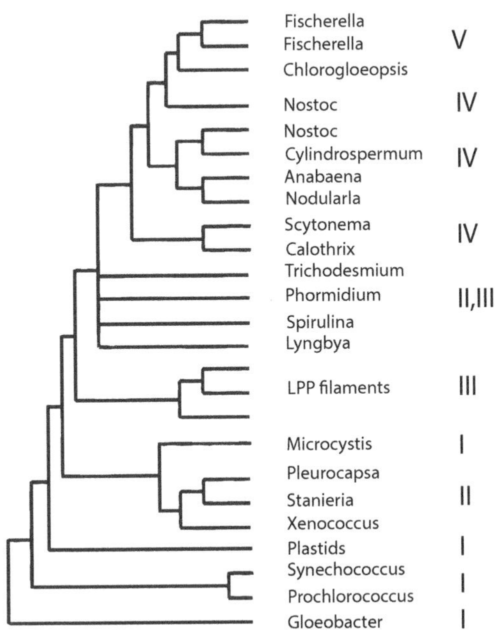

Figure 14.1 A phylogenetic tree of 23 cyanobacteria with fully sequenced genomes and relevant classification information (modified from Herrero *et al.*, 2008).

Table 14.1 A comparison of sigma factors across six cyanobacteria (modified from Imamura and Asayama, 2009)

Group	Symbol	PCC6803	NIES-843	PCC7002	PCC7942	BP-1	PCC7120	*E. coli*
1	σ^A	slr0653	MAE54470	SYNPCC7002_A2014	Synpcc7942_0649	tll0617	all5263	RpoD(BAE77118)
2	σ^B	sll0306	MAE14230	SYNPCC7002_A1202	Synpcc7942_1746	tll0831	all7615 alr3800 all7608 all7179	RpoS(BAE76818)
	σ^C	sll0184	MAE09350	SYNPCC7002_A0270	Synpcc7942_1849	tlr0499	all1692	
	σ^D	sll2012	MAE45250	SYNPCC7002_A1832	Synpcc7942_0672	tlr0264	alr3810	
	σ^E	sll1689	MAE52900	SYNPCC7002_A0364	–	–	alr4249	
	Others(M)	–	-	-	Synpcc7942_0569 Synpcc7942_1557	–	–	
3	σ^F	slr1564	MAE52920	SYNPCC7002_A1924	Synpcc7942_1510	tlr0738	all3853	RpoF(BAA15742)
	σ^G	slr1545	MAE02820	SYNPCC7002_A1970	Synpcc7942_1923	tlr2109	alr3280	RpoH(BAE77832)
	σ^H	sll0856	MAE49220	SYNPCC7002_A2111	–	tlr0317	-	RpoE(BAE76749)
	σ^I	sll0687	–	–	Synpcc7942_2004	–	all2193	FecI(BAE78284)
	σ^J	–	–	–	Synpcc7942_1784	–	alr0277	
RpoN		–	–	–	–	–	–	RpoN(BAE77246)

binding domains (DBD), some TFs have binding domains for a variety of substrates or ligands, and other TFs have their own enzymatic domains with different catalytic roles, suggesting the high level of sophistication in their transcription regulation machinery (Wu *et al.*, 2007). A comparative analysis revealed that 12 TF families appear across all the 21 cyanobacteria, namely BolA (cell cycle), DUF387 (chromosomal condensation and segregation), SfsA (sugar fermentation) and DnaA (chromosomal replication initiation), plus eight TF families relevant to response regulators of two-component signalling systems (OmpR, GerE), global nitrogen control (Crp), CO_2 fixation (LysR), metalloregulatory transcriptional repressor (arsR), iron homeostasis control (FUR), general metabolism (GntR) and DNA-binding protein (Bac_DNA_binding) (see Fig. 14.2) (Wu *et al.*, 2007).

Two types of terminators have been identified in prokaryotes: intrinsic transcription terminators containing a hairpin structure within its nascent transcript that disrupts the mRNA–RNAP complex and *Rho*-dependent transcription terminators that require the *Rho* factor to aid the disruption of mRNA–RNAP complex. While cyanobacteria use both terminator types, it has been reported that intrinsic terminators represent the dominating class of terminators in some cyanobacteria such as *Synechococcus* PCC7942 (Vijayan *et al.*, 2011).

Two-component systems represent the main class of signal transduction systems in prokaryotes. A typical such system, such as His-Asp, consists of a sensor kinase, with a sensor domain and a transmitter domain, and a response regulator having a receiver domain and regulation domain. A more complex form is His-Asp-His-Asp. The sensor kinase senses the relevant signals and activates the corresponding regulator(s) by auto-phosphorylating the His residue in its transmitter domain, thus transferring the phosphate cascade to the Asp residue in the response regulator (Liu *et al.*, 2006). Recent analyses show that cyanobacteria have the largest numbers of two-component systems compared to other bacteria with sequenced genomes. For example, *PCC6803* encodes 80 such genes (Liu *et al.*, 2006) and *PCC7120* has 195 (Ashby and Houmard, 2006), compared to 62 such genes in *E. coli* K12 (Mizuno, 1997). While these systems have been identified, only a few of them have been well studied even in the best-studied model organism PCC6803. Among the well-studied two-component systems are red and far-red light sensor (slr0473/slr0474), $Ni2^+$ or redox sensor (sll0798/sll0797), positive phototactic movement sensor (sll0043/sll0039/

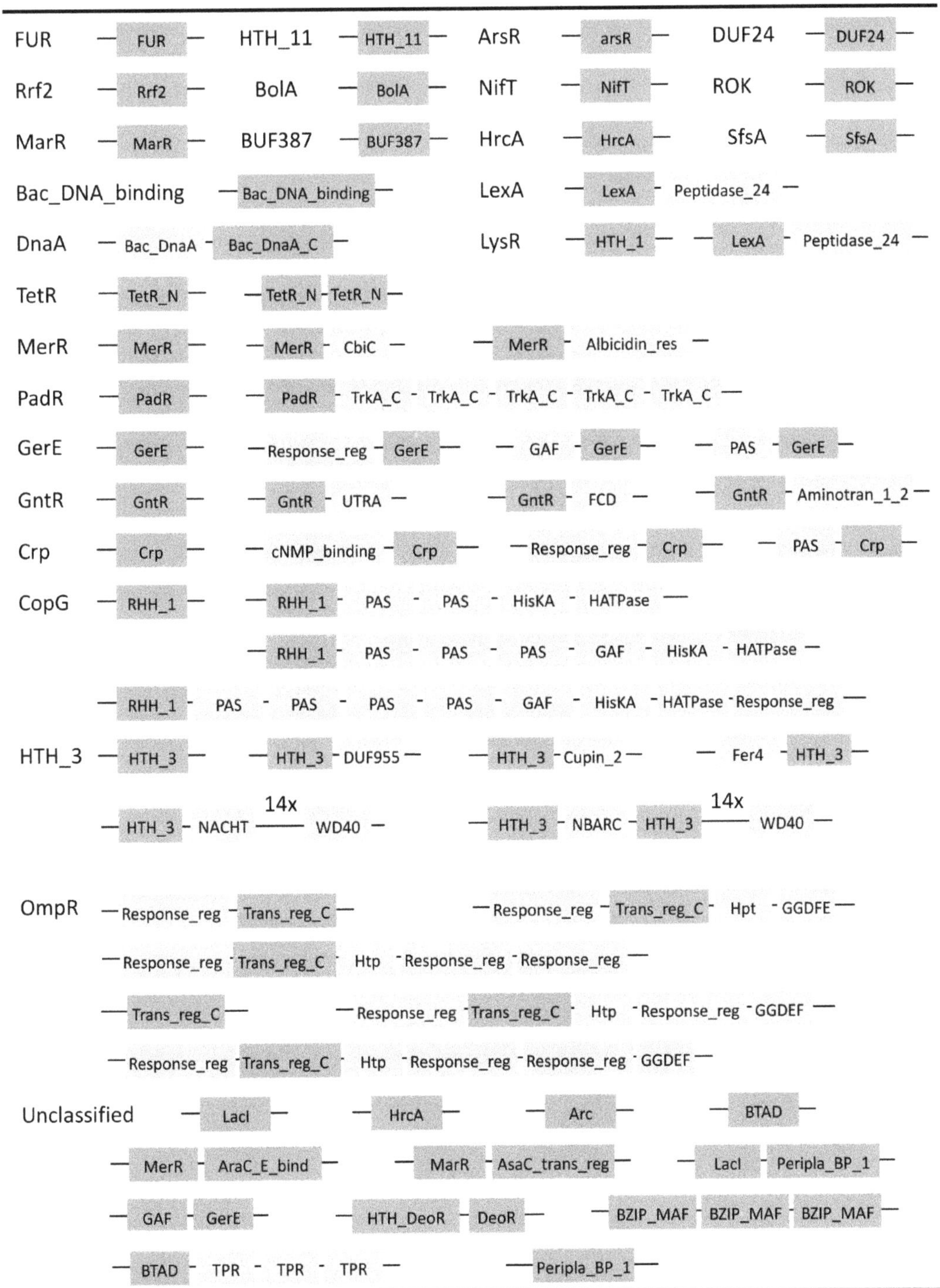

Figure 14.2 Domain architectures and classification of 12 TFs in 21 cyanobacterial genomes. Blue rectangles represent DBD domains, and yellow rectangles are the associated domains (modified from Wu *et al.*, 2007). A colour version of this figure is located in the plate section at the back of the book.

sll0038), Mn^{2+} sensor (slr0640/slr1837), phosphate sensor (sll0337/sll0081), low-temperature sensor (sll1905/sll0698/sll0038) and salt or osmotic stress sensor (sll0698/slr0115). The details about the known two-component systems in *PCC6803* are given in Table 14.2.

Elucidation of transcription regulation

We use three well-studied biological processes, photosynthesis, nitrogen assimilation and osmotic stress response, as examples to illustrate what can be possibly derived about the key elements in the related transcription regulation machinery through comparative genomics.

Information derivation through comparative omic analyses

To elucidate transcription regulation in a systematic manner, we have developed an effective procedure for deriving the key relevant elements, which involves the following steps: (i) collection of template networks for related model organisms having substantial experimental data and possibly known information about the target network through literature and/or database search, which is the only manual step in the procedure; (ii) prediction of operons, relevant transcription regulators and functional relatedness among genes in the target genome; and (iii) derivation of the operons regulated by the identified regulator(s) through identification of conserved *cis* regulatory motifs

Table 14.2 The well-studied two-component systems in *PCC6803* (modified from Liu *et al.*, 2006). See http://csbl.bmb.uga.edu/~xizeng/books/cyano/for details

Sensor kinase		Response regulator			
Synonym	Type	Synonym	Family	Physiological process	Reference(s)
slr1285	HK	slr1783	NarL	Salt, osmotic stress sensor	Shoumskaya *et al.* (2005)
slr1805 sll1229	HK HR	sll1708	NarL	Salt, osmotic stress sensor	Shoumskaya *et al.* (2005)
sll1590	HK	sll1592	NarL		
slr0311	HK	slr0312	NarL		
slr0533	HK	sll0649	OmpR	Salt, osmotic stress sensor	Shoumskaya *et al.* (2005)
slr0640	HK	slr1837	OmpR	Mn^{2+} sensor	Ogawa *et al.* (2002), Yamaguchi *et al.* (2002)
sll0337	HK	slr0081	OmpR	Phosphate sensor	Hirani *et al.* (2001)
sll0698	HK	slr0115	OmpR	Salt, osmotic stress sensor	Shoumskaya *et al.* (2005)
sll0798	HK	sll0797	OmpR	Ni^{2+} or redox sensor	Li and Sherman (2000), López-Maury *et al.* (2002)
sll0790	HK	sll0789	OmpR		
sll1296	HR	sll1292 sll1291	CheY PatA		
sll1905 sll0698	HR HK	sll0038	PatA	Low-temperature sensor	Suzuki *et al.* (2000)
slr0473	HK	slr0474	CheY	Red, far-red light sensor	Lamparter *et al.* (2001), Yeh *et al.* (1997)
sll0043	HR	sll0039 sll0038	CheY PatA	Positive phototactic movement sensor	Bhaya *et al.* (2001), Yoshihara *et al.* (2000)
slr1759	HR	slr1760	Others		
slr2098 slr2099 slr2104	HR HR HR	slr2100	Others		
sll1672	HR	sll1673	Others		
sll0698	HK	?		Drug sensor	Bartsevich and Shestakov (1995)
sll1124	HK	?		Blue light phototaxis	Wilde *et al.* (1997, 2002)

(and transcriptomic data if any); and (iv) mapping the template networks, if any, to the target genome through orthologous gene mapping, which is consistent with the predicted operons and gene associations (Mao *et al.*, 2010; Su *et al.*, 2006). This process can be largely automated as we have shown in our previous publications by the CINPER web server (Mao *et al.*, 2012).

Photosynthesis

The photosynthesis system in cyanobacteria is similar that those in plants (Herrero *et al.*, 2008). It has both photosystem I (PSI) and photosystem II (PSII). PSII split water into O_2, protons and electrons. The electrons will be transferred to ferredoxin through a chain of electron carriers in membrane and lumen, and used to reduce $NADP^+$ to make NADPH. A proton gradient will be generated in electron transfer, ATP synthase will utilize the gradient to make ATP. NADPH and ATP will be used for carbon fixation in the Calvin cycle. The photosynthesis system in cyanobacteria also has the following differences from those in plants:

1. Plants use chloroplast as the photosynthesis compartment, which cyanobacteria do not have. Photosynthesis in cyanobacteria is performed in thylakoid membrane (Vermaas, 2001). PSII, PSI and ATP synthase are located in the thylakoid membrane, and electron carriers are located in both the membrane and the thylakoid lumen while ferredoxin, thioredoxin and the Calvin cycle enzymes are all in cytoplasm;
2. Cyanobacteria utilize phycobilisome and a few additional genes with high sensitivity with light such as *pcb* as the antenna complex in addition to the basic light-harvesting complex *chl a* (Herrero *et al.*, 2008). Phycobilisome can absorb the light ignored by *chl a*, and can also perform chromatic acclimation (Gutu and Kehoe, 2012), which can absorb light with different colours by changing its pigment proteins. It can also move between PSII and PSI as an adaption between low and high light intensity (Joshua and Mullineaux, 2004; Mullineaux and Emlyn-Jones, 2005). These make phycobilisome flexible and efficient light-harvesting machinery, which is especially important for cyanobacteria living in deep water, where the light intensity is very low; and
3. RuBisCO, the enzyme catalysing the reaction of converting CO_2 to organic carbon, has a very low activity level in plants, which has even lower activity level in cyanobacteria. The possible reason is that in early atmosphere where CO_2 concentration was high while O_2 concentration is low, a high affinity RuBisCO was not necessary. Cyanobacteria developed a CO_2 concentrating mechanism (CCM), which can increase CO_2 concentration up to 1000 times in the close proximity of RuBisCO.

The current understanding is that photosynthesis and the closely related carbon fixation genes are regulated by light, O_2 and circadian cycles in cyanobacteria.

Redox (O_2) regulation

It has been long known that photosynthetic activity is affected by the O_2 level, but the mechanism was unclear until recently. The O_2 level can directly affect the redox level of the electron carrier ubiquinone (Wu and Bauer, 2008), which is an analogue of plastoquinone (PQ) and PQ is a mobile electron carrier between PSII and cytochrome b_6f, and is part of the photosynthesis electron transfer chain; in addition it is partially responsible for the formation of the proton gradient.

It was recently discovered that RegB/RegA is a global two-component system involved in transcription regulation of photosynthesis, carbon fixation, nitrogen fixation and electron transfer in *Rhodobacter capsulatus* and *Rhodobacter sphaeroides* (Elsen *et al.*, 2004; Wu and Bauer, 2008). A homologous two-component system has been identified in *Synechocystis* PCC6803 (Li and Sherman, 2000); and the speculation is that RegB/RegA is the global regulator in response to O_2 in cyanobacteria.

RegA is involved in regulation of *puc*, *puf* and *puh* operons, which encode a large portion of the light–harvesting complexes and reaction centre of photosynthesis (Elsen *et al.*, 2004; Wu and Bauer, 2008). RegA also affects tetrapyrrole synthesis, which is important for chlorophyll and haem synthesis. Chlorophyll binds to light-harvesting proteins and haem binds to electron carriers, both of which are important for photosynthesis. Putative binding sites of RegA have been found in some of these genes. RegA is also found to be able to bind to the promoter region of the two Cbb operons (Dubbs *et al.*, 2000; Dubbs and Tabita, 2003), which encode most genes involved in the Calvin cycle.

RegB, the sensor protein in the two-component system, senses the redox status of ubiquinone. It has been demonstrated that RegB activity in oxidized ubiquinone is 6-fold higher than in reduced ubiquinone (Swem *et al.*, 2006). RegB can be autophosphorylated, and then serve as a kinase by transferring the phosphate to RegA, which regulates a broad range of genes in many processes including photosynthesis and carbon fixation.

Light regulation

The expression levels of numerous genes, including those involved in photosynthesis, have been found to be affected by the level of light received by cyanobacteria cells while the detailed regulation mechanism remains not fully understood. It is now believed that the histidine kinase *Hik33* and transcriptional factor *RpaB* are the main two-component system responsible for gene regulation in response to change in light. It is known that *Hik33* can be autophosphorylated and can also phosphorylate *RpaB* (Muramatsu and Hihara, 2012).

The current understanding about light regulation is as follows. When the light intensity is low, the cyanobacterial cells will up-regulate the expression levels of genes involved in the light-harvesting complexes: phycobilisome and PSI (Mullineaux and Emlyn-Jones, 2005). RpaB activates the PSI genes through binding to and utilizing the high-light regulator 1 (HLR1) (Seino *et al.*, 2009). In response to high light intensity, the regulation machinery is more complicated (Muramatsu and Hihara, 2012). The cells need to first accelerate the PSII damage-repair process, decrease the light-harvesting rate, and then activate the photoprotection mechanisms as well as increase the energy consumption processes such as carbon and nitrogen fixation, as well as CO_2 and HCO_3^- transporters. The activated photoprotection genes include those involved in photorespiration and glycine decarboxylase complex (GDC), an essential enzyme in C_2 cycle. In addition, the orange carotenoid protein (OCP), for non-photochemical quenching, is also activated.

PSII consists of more than 20 domains, of which D1, D2 and cytochrome b_{559} form the reaction centre. Overexposure to light can damage the reaction centre domains such as the D1 domain. As a response, genes encoding D1 (psbA) will be considerably up regulated under high light conditions; so will genes encoding D2 (psbD). The response of the PSI genes is different as they will be down-regulated. To avoid over-absorption of the light energy, genes encoding the light harvesting complex (i.e. phycobilisome) will be down regulated; in addition, the phycobilisome complex will switch to be associated with PSI instead with PSII, with which phycobilisome binds under low light intensity. Many genes responsible for pigment (chlorophyll and phycocyanin) will also be down regulated.

Regulation responding to circadian rhythm

Cyanobacteria have a unique circadian mechanism different than all other organisms including plants. Three genes, *KaiA*, *KaiB* and *KaiC*, are involved in their circadian rhythm (Kageyama *et al.*, 2003). The mechanism of how the circadian rhythm affects photosynthesis has been unclear until recently. A recent study on the SasA/RpaA two-component system discovered a link between *KaiABC* and phycobilisome genes (Takai *et al.*, 2006). The phosphotransfer from SasA, a kinase known to interact with KaiC, to RpaA, known to be responsible for energy transfer from phycobilisome to both PSII and PSI, is significantly affected by the circadian state of the *Kai* protein complex, providing the first evidence that this two component system may be involved in transcription regulation of the photosynthesis system in response to light although further studies are required to derive the detailed mechanism. Using a comparative analysis of six cyanobacterial genomes and the aforementioned genomes of other species, we have derived various elements of the transcription regulation system of the photosynthesis system.

Table 14.3 shows PSII, PSI, phycobilisome, photosynthesis related electron carriers and circadian rhythm genes, and three two-component systems *RegB/RegA*, *SasA/RpaA*, and *Hik33/RpaB* in six cyanobacterial genomes. We can see most genes can be identified successfully.

Nitrogen assimilation

Nitrogen is an indispensable element for all forms of life, but it is limiting in many ecological niches such as the oliogotrophic oceanographic areas where cyanobacteria thrive (Capone, 2000). To overcome the issue, cyanobacteria have evolved several mechanisms to scavenge different forms of nitrogen-containing substances in the environments (Herrero *et al.*, 2001), including ammonium, nitrate/nitrite, urea, cyanate, amino acids and even dinitrogen through fixation (Herrero *et al.*, 2001). In general, ammonium is the preferred nitrogen source for cyanobacteria. When ammonium is replete in the environment, the expression of genes for the assimilation of other sources of nitrogen is suppressed, a phenomenon called *nitrogen control* (Herrero et al., 2001). When ammonium becomes limiting, however, genes involved in the uptake and metabolism of alternative sources of nitrogen will be induced. All forms of assimilated nitrogen are eventually converted to ammonium by relevant enzymes. Ammonium is then combined with 2-oxoglutarate produced by $NADP^+$-isocitrate dehydrogenase to form glutamine synthetase and glutamate synthase (GS-GOGAT) cycle (Muro-Pastor and Florencio, 1994), thereby ammonium is incorporated in the carbon backbones for other biosyntheses. Unlike in *E. coli*, cyanobacteria do not encode a 2-oxoglutarate dehydrogenase, and the only fate of 2-oxoglutarate is to be converted to glutamine (Herrero *et al.*, 2001). Hence, 2-oxoglutarate is widely believed to be an indicator of C/N balance in a cyanobacterial cell (Herrero *et al.*, 2001). Specifically, a higher level of 2-oxoglutarate signals a higher C/N ratio in a cell (Muro-Pastor *et al.*, 2001). It has been shown that most genes involved in nitrogen assimilation pathway in cyanobacteria are transcriptionally regulated by the global transcription factor NtcA that belongs to the cAMP receptor protein (CRP) family. Thus, it is different from the well-characterized NtrB-NtrC two-component system in *E. coli* and other enterics, where nitrogen assimilation genes are regulated by the transcription factor NtrC (Reitzer, 2003). It has been shown that 2-oxoglutarate is the effector molecule for NtcA activation (Flores and Herrero, 2005; Tanigawa *et al.*, 2002; Vazquez-Bermudez *et al.*, 2002, 2003). The activities of NtcA and some other nitrogen assimilation related proteins are also subject to modulations of the signal transduction protein P_{II} encoded by the *glnB* gene at transcriptional (Fadi Aldehni *et al.*, 2003) and/or posttranslational levels (Heinrich *et al.*, 2004). The activities of P_{II} are controlled by both its phosphorylation state at a seryl residue and the levels of 2-oxoglutrate and ATP in the cell (Lee *et al.*, 2000; Vazquez-Bermudez *et al.*, 2002, 2003). Albeit these studies, transcriptional regulatory networks of the nitrogen assimilation system and their relationship with other major pathways are far from being understood in cyanobacteria, in particular, in some newly sequenced cyanobacterial species.

The NtcA regulon

The current understanding of nitrogen assimilation pathways in cyanobacteria is mainly based on experimental investigations of a few relatively well-studied fresh water species. The experimental data about the marine strains are relatively scarce. However, the availability of genome sequences of an increasing number of marine strains and ecotypes have largely changed the understanding of nitrogen assimilation pathways in these organisms through a combination of computational and experimental investigations. Since NtcA is the key transcriptional regulator of the genes involved in the nitrogen assimilation process, computational reconstruction of nitrogen assimilation pathways can be started by predicting the possible members of the NtcA regulons in these genomes. This can be done by predicting putative NtcA binding sites in the upstream regions of predicted operons using phylogenetic foot-printing based algorithms (Blanchette and Tompa, 2002).

All the sequenced cyanobacterial genomes encode a NtcA gene (Scanlan *et al.*, 2009), even for the marine *Prochlorococcus* strains that have evolved a reduced genome, suggesting the importance of nitrogen regulation in their physiology.

Table 14.3 A comparison of the key components of photosynthesis network across six cyanobacteria, predicted by CINPER (Mao *et al.*, 2012). ++ indicates the system are fully predicted in the organism, and + indicates the system is partially predicted. See http://csbl.bmb.uga.edu/~xizeng/books/cyano/for details

Template	Models					
Gene/function	PCC 6803	NIES-843	WH 8102	PCC 7120	BP-1	IMS 101
Regulator						
Redox-responsive regulator RppAB	sll0797	MAE_28290	SYNW0947	all7584	tll1910	Tery_2305
	sll0798	MAE_49890	SYNW0948	all7583	tll1909	Tery_3210
Transcriptional regulator PsaB	slr1834	MAE_47560	SYNW2124	alr5154	tlr0731	Tery_4669
	slr1835	MAE_47570	SYNW2123	alr5314	tlr0732	Tery_4668
Ndhf3 operon transcriptional regulator	sll1594		SYNW2260	all0862	tlr2037	Tery_2009
Two-component response regulator RpaB/ Hik33	slr0947	MAE_22160	SYNW2246	all3822	tll2364	Tery_0954
	sll0698		SYNW0551	alr3511	tlr0437	Tery_4749
Transcriptional regulator NrdR	slr0780	MAE_43900	SYNW1984		tlr1861	Tery_4664
Rubiscooperon transcriptional regulator RbcR	sll1594	MAE_15110	SYNW2260	all0862	tlr2037	Tery_4333
LysR transcriptional regulator	sll0998	MAE_19370	SYNW2260	all3953	tlr1206	Tery_2009
Energy transfer regulator	slr0947		SYNW2246	all3822	tll2364	Tery_0954
Tetrapyrrole biosynthesis regulator HLIPs	ssr1789		SYNW0330	asl2354	tsr1755	Tery_1910
OCP regulator FRP	slr1964	MAE_18920	SYNW1369	all3148		
Ferric uptake regulator	slr1738	MAE_37080	SYNW0774	all1691	tll0048	Tery_1958
Transcriptional regulator PedR	ssl0564	MAE_54580	SYNW2268	asl2551	tsl1865	Tery_2735
PBS antenna regulator PmgA	sll1968	MAE_35690	SYNW1068	alr3655	tlr1461	Tery_2243
Antenna proteins	++	++	++	++	++	++
PSI						
Photosystem I P700 chlorophyll a apoprotein	++	++	++	++	++	++
Photosystem I subunit	++	++	++	++	++	++
Photosystem I reaction centre subunit	+	+	+	+	+	+
PSII						
Photosystem II protein	+	+	+	+	+	+
Cytochrome b_{559} subunit	++	++	++	++	++	++
Photosystem II reaction centre protein	++	++	++	++	++	++
Manganese-stabilizing polypeptide	++	++	++	++	++	++
Cytochrome c_{550}	++	++	++	++	++	++
Extrinsic protein precursor	++	++	++	++	++	++
Electron transfer chain						
F_0F_1 ATP synthase subunit	++	++	++	++	++	++
Apocytochrome *f*	++	++	++	++	++	++
Cytochrome b_6–*f* complex subunit	+	+	+	+	+	+
Ferredoxin	++	++	++	++	++	++
Ferredoxin-NADP oxidoreductase	++	++	++	++	++	++
Carbon fixation	+	+	+	+	+	+
Circadian rhythm						
Circadian clock protein KaiABC	slr0756	MAE_31730	SYNW0548	alr2884	tlr0481	Tery_3803
	slr0757	MAE_31740	SYNW0549	alr2885	tlr0482	Tery_3804
	slr0758	MAE_31750	SYNW0550	alr2886	tlr0483	Tery_3805
Two-component response regulator	slr0115	MAE_14910	SYNW2236	all0129	tlr2423	Tery_4937
Adaptive-response sensory kinase	sll0750	MAE_60820	SYNW0753	all3600	tlr0029	Tery_3802

NtcA sequences have similar phylogenetic relationships to the 16S rDNA, therefore they are likely vertically inherited from their last common ancestor. The DNA binding domains of NtcA are highly conserved in cyanobacteria (Su *et al.*, 2006), which may explain the high similarity of the NtcA binding sites with a palindromic structure $GTAN_8TAC$ across different species (Herrero *et al.*, 2001). This also forms the basis for prediction of NtcA binding sites by the phylogenetic foot-printing approach (Blanchette and Tompa, 2002). In addition to this binding site, the promoter regions of known NtcA-activated genes also contain a −10, *E. coli*-like box in the form of TAN_3T, with the NtcA binding site replacing the −35 box in the *E. coli*-type promoters (Herrero *et al.*, 2001). This information can be used to reduce the false predictions of NtcA binding sites by looking for a putative TAN_3T box in the neighbourhood of the binding site (Su *et al.*, 2006). To further reduce false predictions, similar motifs should also be considered in the regulatory regions of only the orthologous genes. We found this approach can reduce the false rates by 40 times on average compared to a conventional phylogenetic foot-printing procedure (Su *et al.*, 2006).

Using this method, high-scoring NtcA promoters (an NtcA binding site plus a downstream TAN_3T box) are found for at least a copy of the ammonium transporters of the *amt* family in all the cyanobacterial genomes analysed, suggesting that ammonium uptake is tightly under the NtcA regulation in these species. These NtcA promoters might be responsible for the Amt induction under low ammonia concentrations as it has been shown in *Synechococcus* PCC7942 (Vazquez-Bermudez *et al.*, 2002). Nitrate is the most common alternative source of nitrogen for cyanobacteria. However, it seems that not all cyanobacteria can utilize it since genomes of all sequenced marine *Prochlorococcus* strains do not encode any known genes responsible for nitrate/nitrite uptake and their subsequent reduction to ammonium (Scanlan *et al.*, 2009; Su *et al.*, 2006), which is consistent with the published experimental data (Moore *et al.*, 2002). This might be related to their ammonium-rich ecological niches in deep seawater (Rocap *et al.*, 2003; Ting *et al.*, 2002). However, it has been reported that some high-light adapted *Prochlorococcus* ecotypes are able to use nitrate/nitrite (Martiny *et al.*, 2009). On the other hand, two types of nitrate/nitrite transporters are encoded in the other species/strains, namely, the ABC transporters *nrtABCD* in freshwater strains such as *Nostoc* PCC7120, *Gloeobacter violaceus* PCC7421, *Synechocystis* PCC6803, *Synechococcus elongatus* PCC6301 and *Thermosynechococcus elongatus* BP-1, and the major facilitator type *nrtP* in marine *Synechococcus* ecotypes such as *Synechococcus* WH 8102. Moreover, fumarate/nitrite transporters are encoded in some low-light marine *Prochlorococcus* ecotypes such as *Prochlorococcus* MIT9313, MIT9303, NATL1A and NATL2A (Scanlan *et al.*, 2009; Su *et al.*, 2006). These transporters are often clustered with genes encoding the relevant metabolic enzymes, i.e. the nitrate reductase *narB* and nitrite reductase *nirA*. High scoring NtcA promoters are frequently found in the regulatory regions of these nitrate/nitrite transporter and enzyme genes, suggesting that they are under tight control of NtcA (Su *et al.*, 2006). It has been shown that *nirA*, *nrtABCD* and *narB* are also regulated by the LysR family protein NtcB in some species (Aichi *et al.*, 2001; Frias *et al.*, 2000; Frias *et al.*, 2003; Maeda *et al.*, 1998). Interestingly, NtcB is also regulated by NtcA in all the genomes that harbour the gene as binding sites, which are found in operons containing the gene (Su *et al.*, 2006) as suggested previously in *Nostoc* PCC7120 (Aichi *et al.*, 2001; Frias *et al.*, 2000; Maeda *et al.*, 1998).

Urea is another common form of nitrogen source for cyanobacteria that thrive in fresh water and even in oligotrophic oceans where the concentration of urea is at a level of 0.1–1 mol/l (Collier *et al.*, 1999). High affinity ABC urea transporters UrtABCDE are encoded in most fresh water species such as *Nostoc* PCC7120, *Synechocystis* PCC6803 and *Thermosynechococcus elongates* BP-1 and *Thermosynechococcus*, and both marine *Synechococcus* and *Prochlorococcus* strains such as *Prochlorococcus* MIT 9313, *Prochlorococcus* MED4 and *Synechococcus* WH 8102. Moreover, the three subunits of urease UreABC (urea amidohydrolase) and accessory proteins involved in urease assembly (UreDEFG) are encoded in these genomes harbouring *utr* genes. In *Prochlorococcus* MED4, *Prochlorococcus* MIT9313 and

Synechococcus WH 8102, the *ure* genes form two operons in the forms of *ureDABC* and *ureEFG* next to the *urt* genes. Though NtcA binding sites are found for the urea transporter genes, no NtcA binding sites are found for the *ure* genes except for a few cases such as *ureG* in *Synechocystis* PCC6803 and *ureEFG* in *Prochlorococcus* MED4, suggesting that *ure* genes are not subject to NtcA regulation at least in most of the genomes analysed. This is consistent with the finding that the transcription of urease in *Nostoc* PCC7120 is not subject to NtcA regulation; instead, it is constitutively expressed (Valladares *et al.*, 2002).

Cyanase (CynS) that catalyses the decomposition of cyanate (NCO^-) to CO_2 and ammonium is encoded in the genomes of some fresh water species (e.g. *Nostoc* PCC7120, *Synechococcus elongatus* PCC6301 and *Synechosystis* PCC6803), the majority of marine *Synechococcus* ecotypes and some *Prochlorococcus* ecotypes (*Prochlorococcus* MED4, NATL1A and NATL2A) (Scanlan *et al.*, 2009; Su *et al.*, 2006). Most of these genomes also encode an ABC cyanate transporter *cynABC/D* clustered together with the *synS* gene, which form an operon in some genomes such as *Synechocystis* PCC6801 and *Prochlorococcus* MED4. Thus, these genomes are likely to use cyanate as a nitrogen source as it has been demonstrated in *Synechococcus* PCC7942 (Harano *et al.*, 1997) and *Synechococcus elongatus* PCC6301 (Anderson *et al.*, 1990). High-scoring NtcA promoters are predicted for the *cynABC/D/S* operons across many *cyanobacterial* genomes, suggesting that they are under NtcA regulation as it has been demonstrated in *Synechococcus* PCC9742 (Harano *et al.*, 1997). Interestingly, it has been shown that the CynABC/D transporter can take up nitrite in *Synechococcus elongatus* PCC6301, suggesting that CynABC/D might play a role in nitrite assimilation by cyanobacteria that lack a high-affinity transporter for nitrite (Maeda and Omata, 2009) such as the marine *Prochlorococcus* ecotypes mentioned above.

Some cyanobacterial species are diazotrophic, and are responsible the major nitrogen fixation activity in ocean (Scanlan *et al.*, 2009). In the case of filamentous species such as *Nostoc* PCC7120, a specialized cell form called *a heterocyst* is developed for this purpose (Flores and Herrero, 2009). It has been shown that several nitrogen fixation (*nif*) genes and genes involved in the development of heterocyst such as *devBCA*, *xisA* and *hetC* are under the direct regulation of NtcA (Flores and Herrero, 2009). High-scoring NtcA promoters are found for other genes involved in nitrogen fixation (Su *et al.*, 2006). Although the positive regulator of the heterocyst development, HetR in PCC7120, is impaired in an NtcA mutant (Frias *et al.*, 1994), it is unlikely to be regulated directly by NtcA, since no NtcA prompter is found for this gene in our analyses. This suggests that a complex transcription regulatory network is under the control of NtcA, which mediates the indirect regulation effects of NtcA. High-scoring NtcA promoters are also found for *glnA* (glutamine synthetases gene) and *glnB* in most cyanobacterial genomes examined (Su *et al.*, 2006), suggesting that they are under the regulation of NtcA in these genomes as has already been shown many species (Herrero *et al.*, 2001). Lastly, high-scoring NtcA promoters are found for *ntcA* itself in mostly genome examined, suggesting that *ntcA* is autoregulated in these organisms.

A working model of the nitrogen assimilation network in *Synechococcus* WH8102

The model involves 133 genes predicted by computational methods and the 338 genes identified by the microarray experiments (Su *et al.*, 2006). In this model, the main regulator NtcA (SYNW0275) is activated by a high level of 2-oxoglutarate, indicative of a high C/N ratio, or relatively lower availability of nitrogen supply in the environment. 2-oxoglutarate functions as a signal from photosynthesis to the nitrogen assimilation process. The core of the nitrogen assimilation network consists of members of the NtcA regulon, which includes genes responsible for the uptake of various sources of nitrogen and their subsequent reduction and incorporation into the carbon skeleton, such as Nrtp (SYNW2462–2463), NirA (SYN2477), CynCBA (SYNW2486–2487), GlnB or P_{II} (SYNW0462), GlnA (SYNW1079), Amt1 (SYNW0253), UrtEDCBA (SYNW2438–2442), the glutamate transporter GltS (SYNW0882), the porin Som (SYNW2224), as well as the genes that are under NtcA control but also have a function in other biological pathways such as photosynthesis. These genes bridge the nitrogen assimilation process with various other cellular activities. The next layer of the network consists of proteins such as NarB (SYNW2464), UreGFE (SYNW2443–2445), UreDABC (SYNW2446–2449), CynS (SYNW2490), GOGAT (glutamate synthase), RbcL (SYNW1718) and Icd (SYNW0166), which are not members of the NtcA regulon but are functionally relevant to the network. Some of these genes are probably regulated by the regulators that are under NtcA regulation. The activity of NtcA is likely to be regulated by other proteins such as the hypothetical two-component response regulator SYNW2289, predicted to physically interact with NtcA, and hypothetical proteins SYNW0273 and SYNW0274 that are predicted to be located in the same operon of NtcA. Furthermore, Amt1 is likely to be regulated by P_{II} through a physical interaction, reminiscent of a recent finding that the *N*-acetyl-L-glutamate kinase is regulated by P_{II} through a physical interaction in *Synechococcus elongatus* (Heinrich *et al.*, 2004), and this has been experimentally proven in *E. coli* (Javelle and Merrick, 2005; Javelle *et al.*, 2004). The function of the nitrate reductase NarB might be regulated by the ferredoxin-NADP reductase PetH (SYNW0751) and SYNW2511. The sigma factors RpoD (SYNW2496) and SigA (SYNW1783), which were down- and up-regulated by ammonium, respectively (Su *et al.*, 2006), are probably involved in the transcription of some genes that are down- or up-regulated by ammonium. Of the 429 genes recruited into the network, 204 are hypothetical ones; their possible functions in nitrogen assimilation related processes warrant further experimental investigations.

Table 14.4 shows a comparative study (Mao *et al.*, 2012) about nitrogen assimilation genes among six cyanobacteria. Most ammonium, urea, nitrite and nitrate utilizing genes and transporters can be found in all the six organisms although our other studies have shown that some cyanobacteria living in ammonium rich environment might have lost the capability to utilize nitrite/nitrate. Nif genes are found in only two out of the six species, which can utilize gas phase nitrogen directly. Since gas phase nitrogen has a triple bond, which is hard to be broken thus hard to be fixed, this is probably the reason that nitrogen fixation system is rare.

Osmoregulation

Osmotic stress refers to stress induced by sudden changes of impermeable solute concentrations around a cell. Specifically the excessive solute imposes a pressure on the cell membrane and thus affects the equilibrium with the solution inside the cell. When this happens, water molecules will move in and out of the cells by diffusion through the cell membrane or aquaporin channels to regain the equilibrium. The induced water flow across the cell membrane may cause changes in the cell volume, specifically the cytoplasmic volume, and hence induce a number of cellular responses to maintain the homeostasis of the cell's water content. When the solute concentration inside a cell is higher than the environment around the cell, i.e. referred to as *hypo-osmotic stress*, water molecules will flow inwards, hence increasing the turgor pressure in cells. Alternatively when the solute concentration inside a cell is lower, i.e. under *hyperosmotic stress*, water will flow outwards, causing shock to the cell. Typically when a cell is under hyperosmotic stress, Na^+ and Cl^- quickly move into the cellular cytoplasm within seconds; the excessive toxic Na^+ will be actively exported by Na^+/H^+ antiporters and the non-toxic K^+ are also transported into the cell from the environment through its K^+ transporters to maintain the osmolarity needed by the cell in the first hour; and then transport some compatible osmolytes into the cell from the environment or synthesize them in the cell to replace the K^+ surplus within the following hours.

A working model of osmoregulation network in *Synecoccocus* WH8102

Mao *et al.* (2010) have built a model of the osmotic regulation network for *Synechococcus* WH8102 through comparative genomic analyses of five bacterial genomes and function association analysis. The working model consists of 114 genes, 15 of which have been validated based on microarray experiments. The model consists of all the key elements of the osmoregulation system, including transporters, signal sensors, transcriptional regulators and key enzymes. The details are as

Table 14.4 A comparison of the key components of the nitrogen assimilation network across the six cyanobacteria, predicted by CINPER (Mao *et al.*, 2012). See http://csbl.bmb.uga.edu/~xizeng/books/cyano/ for details.

Template			Model					
Synonym	Symbol	Function	PCC6803	NIES-843	WH8102	PCC7120	BP-1	IMS-101
alr3667	ureA	Urease subunit	slr1256	MAE_45220	SYNW2447	alr3667	tlr0981	Tery_0746
alr3668	ureB		sll0420	MAE_45230	SYNW2448	alr3668	tll0330	Tery_0747
alr3670	ureC		sll1750	MAE_61330	SYNW2449	alr3670	tlr0005	Tery_0752
alr0733	ureE	Urease accessory protein	slr1219	MAE_41100	SYNW2445	alr0733	tll1366	
alr4392	ntcA	Two-component system responsive for nitrogen	sll1423	MAE_01830	SYNW0275	alr4392	tll1650	Tery_2023
all0602	ntcB		slr0395	MAE_09380	SYNW2260	all0602	tll1359	Tery_4333
alr0608	nrtA	Nitrate transport protein	sll1450	MAE_14800	SYNW2487	alr0608	tlr1350	
alr0609	nrtB		sll1451	MAE_20000	SYNW2486	alr0609	tlr1351	Tery_1625
alr0610	nrtC		sll1452	MAE_20010	SYNW1417	alr0610	tlr1352	Tery_1308
alr0611	nrtD		sll1453	MAE_20020	SYNW2485	alr0611	tlr1354	Tery_1308
alr0607	nirA	Ferredoxin-nitrite reductase	slr0898	MAE_18410	SYNW2477	alr0607	tlr1349	Tery_1068
all1517	nifB	Nitrogen fixation protein				all1517		Tery_4133
all1456	nifU					all1456		Tery_4135
all1436	nifX					all1436		Tery_4140
all1455	nifH	Nitrogenase reductase	slr0749	MAE_16230	SYNW1725	all1455	tll2347	Tery_4136
asr1409	nifT	Nift protein				asr1409		Tery_4128
sll1454	narB	Nitrate reductase	sll1454	MAE_53960	SYNW2464	alr0612	tlr1355	Tery_1070
alr2817	hetC	Heterocyst differentiation protein	sll1180	MAE_35850	SYNW0193	alr2817		Tery_0225
alr2339	hetR					alr2339		Tery_1921
all1432	hesA	Hes protein	sll1536	MAE_32660	SYNW2053	all1432	tlr2403	Tery_4143
all1431	hesB		slr1565	MAE_52040	SYNW2214	all1431	tll0867	Tery_4144
all4968	gor	Glutathione reductase	slr1849	MAE_46260	SYNW1533	all4968	tll1608	Tery_4747
alr2328	glnA	Glutamate-ammonia ligase	slr1756	MAE_19270	SYNW1073	alr2328	tll1588	Tery_3834
all2319	glnB	Nitrogen regulatory protein P-II	ssl0707	MAE_57460	SYNW0462	all2319	tll0591	Tery_2842
alr3712	devA	Heterocyst specific ABC-transporter,	sll0484	MAE_21240	SYNW1087	alr3712	tll0643	Tery_4480
alr4682	amt1	Ammonium transporter protein 1	sll0895	MAE_12590	SYNW1898	alr4682	tlr0927	Tery_1792
alr1407	aksA	Trans-homoaconitate synthase	slr0186	MAE_54460	SYNW0730	alr1407	tll1397	Tery_2253
all1430	fdxH	Heterocyst ferredoxin	ssl0020	MAE_36530	SYNW0535	all1430	tsl1009	Tery_4145
alr0735	17228230	Urease accessory protein	sll0643	MAE_24230	SYNW2443	alr0735	tlr0057	Tery_0748
alr0734	17228229		slr1899	MAE_41820	SYNW2444	alr0734	tll1365	Tery_0751
alr3666	17231158	Urease accessory protein D	sll1639	MAE_04510	SYNW2446	alr3666	tlr0536	Tery_0749
sll1423	ntcA	Global nitrogen regulator	sll1423	MAE_01830	SYNW0275	alr4392	tll1650	Tery_2023
slr0395	ntcB	Transcriptional activator protein ntcb	slr0395	MAE_09380	SYNW2260	all0602	tll1359	Tery_4333
sll1450	nrtA	Nitrate transport protein	sll1450	MAE_14800	SYNW2487	alr0608	tlr1350	
sll1451	nrtB		sll1451	MAE_14790	SYNW2486	alr2878	tlr1351	Tery_1625
sll1452	nrtC		sll1452	MAE_14780	SYNW1417	alr0610	tlr1352	Tery_1308
sll1453	nrtD		sll1453	MAE_14770	SYNW2485	alr0611	tlr1354	Tery_1308
slr1756	glnA	Glutamate-ammonia ligase	slr1756	MAE_19270	SYNW1073	alr2328	tll1588	Tery_3834
slr0288	glnN		slr0288	MAE_09050				
ssl0707	glnB	Nitrogen regulatory protein P-II	ssl0707	MAE_57460	SYNW0462	all2319	tll0591	Tery_2842
alr0612	narB	Nitrate reductase	sll1454	MAE_53960	SYNW2464	alr0612	tlr1355	Tery_1070
alr1827	icd	Isocitrate dehydrogenase	slr1289	MAE_62530	SYNW0166	alr1827		Tery_0071
sll0108	amt1	Ammonium/methylammonium permease	sll0108	MAE_40010	SYNW0253	alr0991	tll1985	Tery_4477
alr1524	rbcL	Ribulose bisphosphate carboxylase	slr0009	MAE_47890	SYNW1718	alr1524	tll1506	Tery_4410
all4121	petH	Ferredoxin--NADP(+) reductase	slr1643	MAE_12570	SYNW0751	all4121	tlr1211	Tery_3658

follows: (i) *global regulator*: SYNW1621 (σ^{38}) is activated by the accumulated glutamate potassium, and regulates the expression of osmolyte transporter and synthetase genes, which bridges K^+ uptake and osmolyte accumulation processes, and coordinates osmoregulation with other biological processes; (ii) *two-component signal transduction systems*: SYNW0807–0808 (EnvZ and OmpR) regulate water across the outer membrane and responses by some osmoregulatory elements such as otsAB for synthesis of trehalose (Kaasen *et al.*, 1992) and proU/proP for transporting glycine betaine and proline (Gowrishankar, 1985); SYNW0551 and SYNW2246 may be responsible for sensing external osmotic stress and/or activating a number of genes relevant to osmoregulation; (iii) *ion transporters*: SYNW0157 functions as an Na^+/H^+ antiporter to export the excessive Na^+ out of the cell under osmotic stress; SYNW0663 and SYNW2168–2169 make the Ktr, a H^+-dependent K^+ transporter, to uptake K^+ from the extracellular environment under osmotic stress; (iv) *osmolyte transporters or synthetases*: SYNW0229 (BetT), SYNW1915–1917 (ProVWX) and SYNW2494 (ProP) are the up-take system for the major osmolyte, betaine when available in the environment, while SYNW1913–1914 (SdmT and GsmT) are responsible for synthesizing betaine from glycine when needed; SYNW2436 (GpgS) and SYNW2434 (GpgP) are used to synthesize glucosyl glycerate or mannosylglycerate from glucose; SYNW1281 (GgpS) and SYNW0860 (GgpP) are used to synthesize glucosylglycerol from glucose, while SYNW1283–1287 (GgtABCD) are used to uptake glucosylglycerol when available in the environment; SYNW2359 (SpeA) is for synthesizing the minor osmolyte arginine; and SYNW2520 (Sps) for synthesizing the minor osmolyte sucrose.

Comparison among osmoregulation networks across multiple cyanobacteria

Scanlan *et al.* (2009) compared the transporters and synthetases of the compatible osmolytes across 23 organisms in two genera of seawater cyanobacteria: *Prochlorococcus* and *Synechococcus* (see Table 14.1). Virtually all these organisms have orthologous genes of ggptP for synthesizing glucosyl-glycerol (GG), ggpPS for synthesizing GG-phosphate, and spsA for synthesizing sucrose, indicating that GG, GG-phosphate and sucrose are widely used in *Prochlorococcus* and *Synechococcus*. *Synechococcus* has considerably more osmolyte synthetases than *Prochlorococcus*, which is consistent with the fact that *Synechococcus* lives in more diverse and complex environments than *Prochlorococcus* (Scanlan *et al.*, 2009). In addition to the main metabolites, our analysis identified three additional osmolytes, glycine betaine, polyamine and putrescine (see Table 14.5). Specifically, glycine betaine is found only in WH8102 and IMS101. In addition, we also identified the Kdp potassium transport system, EnvZ/Omp two-component system and σ^S (σ^{38}) transcription regulator, which seem to be universal across all sequenced cyanobacteria.

Cross-talk networks

A surprising finding in our comparative genomic analysis is that high-scoring NtcA promoters (see 'Photosynthesis') are found in all genomes examined for many genes involved in the photosynthesis and carbon fixation processes from light harvesting in the antenna complex to electron transfers in the photosystems II and I, and from CO_2/HCO_3^- uptake and concentration to key reactions in the Calvin cycle (Su *et al.*, 2006). These results strongly suggest that these photosynthetic genes are likely to be regulated by NtcA. In fact, it has been shown that nitrogen assimilation and photosynthesis are highly co-orchestrated processes (Commichau *et al.*, 2006; Marsac *et al.*, 2001). On one hand, the uptake of nitrogen containing compounds is powered by ATP generated by photophosphorylation. Reduced ferredoxin from photosynthesis acts as an electron donor to nitrate and nitrite reductases, and the reducing power is necessary for the action of *GOGAT* (glutamate synthase). Furthermore, it has been shown that the expression of NtcA in *Synechocystis* PCC6803 is regulated by the redox state of the cell (Alfonso *et al.*, 2001). As NtcA is activated by the C/N balance indictor 2-oxoglutarate, the end product of the dark-reaction of photosynthesis (Muro-Pastor *et al.*, 1996), a higher level of 2-oxoglutarate

Table 14.5 A comparison of the key components of osmoregulation network across six cyanobacteria, predicted by CINPER (Mao *et al.*, 2012). See http://csbl.bmb.uga.edu/~xizeng/books/cyano/for details

Template			Model					
Synonym	Symbol	Function	PCC6803	NIES-843	WH 8102	PCC7120	BP-1	IMS 101
cg1016	betP	Glycine betaine transporter						
b0020	nhaR	DNA-binding transcriptional activator	sll0998	MAE_19370	SYNW2260	all3953	tlr1206	
b0311	betA	Betaine synthetases with proline as the substrate			SYNW1531			
b0312	betB		sll1561	MAE_52300	SYNW1956	alr3771	tlr0221	Tery_2599
b0695	kdpD	Two-component regulatory system activating Kdp	slr1731	MAE_53250	SYNW0753	all4242	tlr0029	Tery_3802
b0694	kdpE		sll0797	MAE_52640	SYNW2246	all4503	tlr0589	Tery_2902
b1126	potA	Polyamine transporter subunit	sll0240	MAE_12460	SYNW1544	alr1384	tll1492	Tery_2815
b1125	potB					all5042		Tery_2818
b1124	potC					alr1334		Tery_2817
b1123	potD		slr0401	MAE_10300		all5043	tll0716	Tery_2816
b0854	potF	Putrescine transporter	slr0401	MAE_10300		alr0299	tll0716	Tery_2804
b0855	potG	Subunit	sll1878	MAE_12460	SYNW1544	all5044	tll1492	Tery_2805
b0856	potH					all5042		Tery_2803
b0857	potI					alr1334		Tery_2802
b3405	ompR	Two-component regulatory system regulating ompc and ompf	slr0947	MAE_22160	SYNW0808	all3822	tll2364	Tery_0954
b3404	envZ		sll0798	MAE_42250	SYNW0807	all4502	tll1367	Tery_0544
b2677	proV	Glycine betaine transporter subunit			SYNW1915			Tery_4109
b2678	proW				SYNW1916			Tery_4110
b2679	proX							Tery_1459
b2741	rpoS	Sigma 38 factor	sll0184	MAE_54470	SYNW1621	alr4249	tll0831	Tery_4426
b2938	speA	Arginine decarboxylase	slr0662	MAE_46810	SYNW2359	all3401	tll1807	Tery_1142
PERMA_1578	225851111	Glucosylglycerate synthetase			SYNW2434			
PERMA_1582	225851114				SYNW2436			
slr1312	speA	Arginine decarboxylase	slr1312	MAE_46810	SYNW2359	all3401	tll1807	Tery_1142
slr1728	kdpA	Potassium-transporting ATPase subunit	slr1728	MAE_59800		all4246		
slr1729	kdpB		slr1729	MAE_59820		all4245		
slr1730	kdpC		slr1730	MAE_59870		all4243		
slr1731	kdpD		slr1731	MAE_53250		all4242		
slr1508	16330607	Hypothetical protein	slr1508	MAE_18670	SYNW0663	alr4178	tll0327	Tery_4615
slr1509	ntpJ	Na⁺-atpase subunit J	slr1509	MAE_18660	SYNW2168	all1802	tll0303	Tery_3883
sll0306	rpoD	Sigma 70 factor	sll0306	MAE_14230	SYNW0102	alr3800	tll0831	Tery_0784
sll0689	16331534	Na/H⁺ antiporter	sll0689	MAE_55560	SYNW0157	all1303	tlr0449	Tery_1631
sll0045	sps	Sucrose phosphate synthase	sll0045	MAE_32820	SYNW2520	all4985	tlr0582	Tery_0399
slr0529	16332018	Hypothetical protein	slr0529					
slr0746	stpA	Glucosylglycerolphosphate phosphatase	slr0746		SYNW0860			
slr0747	ggtA	ATP-binding subunit	slr0747		SYNW1285	alr4781	tlr1756	
sll1546	ppx	Exopolyphosphatase	sll1546	MAE_53740	SYNW1846	all3552	tll1680	Tery_0231

(high C/N ratio) might be an indicator of high photosynthesis activity. Therefore, 2-oxoglutarate might serve as a messenger from photosynthesis to nitrogen assimilation to speed up the latter, so that the activities of the two systems are synchronized. On the other hand, the intensity of photosynthesis depends on the availability of nitrogen in the environment (Morel and Price, 2003). It is also known that nitrogen-deprivation inhibits photosynthesis by inducing degradation of the photosynthetic apparatus, often referred to as *chlorosis* (Gorl *et al.*, 1998). Although chlorosis is a complex multistage global response (Gorl *et al.*, 1998), it has been recently shown that nitrogen starvation-induced chlorosis is an adaptation of a cell to long term survival (Gorl *et al.*, 1998; Sauer *et al.*, 1999), in which the photosynthesis is kept at a low level (Sauer *et al.*, 2001) to match the low availability of nitrogen. Interestingly, upon nitrogen becoming available, photosynthesis resumes rapidly (Gorl *et al.*, 1998; Sauer *et al.*, 2001). These facts strongly suggest that the photosynthesis process is able to sense the availability of nitrogen to the cell, and responds accordingly. It is also known that nitrogen starvation-induced chlorosis in *PCC7942* is strictly dependent on NtcA, and an NtcA mutant fails to reinitiate photosynthesis when nitrogen becomes available (Sauer *et al.*, 1999), suggesting that NtcA plays an important role in coordinating the activities of the nitrogen assimilation and photosynthesis processes. It is likely that the photosynthesis-related genes and probably some others that bear NtcA promoters might play a role in coordinating these two important processes. There are several lines of evidence that support this hypothesis. First, the expression of a few photosynthetic genes such as allophycocyanin *apc*, phycocyanin *cpc* and thioredoxin *trxM* are down-regulated, whereas porins *somA* and *somB* are up-regulated in the early stage of the chlorosis inducted by nitrogen starvation (Herrero *et al.*, 2001), while *apc/cpc* and *trxM* are maintained at a somewhat higher level in the late stage of chlorosis induced by nitrogen starvation (Sauer *et al.*, 2001). In good agreement with this, NtcA promoters are found for the orthologues of these genes in a number of cyanobacterial genomes analysed (Su *et al.*, 2006). The active expression of these photosynthesis-related genes during the later stage of chlorosis might be mediated by NtcA, since they are necessary for the re-initiation of photosynthesis upon the availability of nitrogen, while an *ntcA* mutant fails to reinitiate photosynthesis even when nitrogen is available. (Sauer *et al.*, 1999; Sauer *et al.*, 2001). Second, it has been shown recently that the large subunit of ribulose bisphosphate carboxylase RbcL in the Calvin cycle is up-regulated by nitrogen starvation (Bird and Wyman, 2003). Interestingly, the *rbcLS* operon is repressed under the same condition (Fadi Aldehni *et al.*, 2003; Ramasubramanian *et al.*, 1994). Regulation of the *rbcLS* operon by NtcA through an NtcA promoter has been experimentally verified (Ramasubramanian *et al.*, 1994), and strong NtcA promoters are found for *rbcL* in some species (Su *et al.*, 2006).

Although prediction of the NtcA regulon members is a powerful way to reconstruct nitrogen assimilation networks, it can potentially miss some members of the involved genes due either to the failure to predict their binding sites, or to genes losing their binding sites during the course of evolution. Additional genes can be recruited into the networks if a functional association can be established between a gene in the network and one that is not in yet using either computational or experimental methods (Su *et al.*, 2006; Wu *et al.*, 2005).

Internet resources

There are a number of useful data and information sources to related cyanobacteria, which may prove to be handy for cyanobacteria researchers.

1. The on-line database of cyanobacterial genera, CyanoDB.cz (http://www.cyanodb.cz/), provides the general description about all cyanobacteria such as their living conditions, shapes and phylogeny among other things. It also marks those species with complete genome sequences, but does not contain the genome sequence and gene information.
2. CyanoBase (http://genome.kazusa.or.jp/cyanobase) represents one of the most comprehensive databases for cyanobacteria, which includes information of genome sequences, gene annotation, protein–protein interaction and transcriptomic data, to name a few.
3. Cyanobacteria Gene Annotation Database CYORF (http://cyano.genome.ad.jp/) contains human-curated genome annotation information of all sequenced cyanobacteria.

4 CyanoClust (http://cyanoclust.c.u-tokyo.ac.jp/) is a database of homologous proteins in cyanobacteria and plastids. The protein clusters are generated using the Gclust software from ~39 cyanobacteria and ~10 other bacteria. The clusters are annotated semi-automatically.
5 CYANOSITE (http://www-cyanosite.bio.purdue.edu/index.html) is a website for general knowledge of cyanobacteria, including organism images, related literature and toxin cyanobacteria.
6 CyanoBIKE (http://cyanobike-community.csbc.vcu.edu/) is a general website for cyanobacteria research.

Summary

Cyanobacteria are the most ancient and diverse organisms of autotrophic photosynthesis on Earth. They can live in almost every terrestrial and aquatic habitat, and have diverse morphological forms ranging from unicellular form to filament quasi-multicellular form. Their diversity and relatively complexity in their cellular capabilities compared to other bacteria make the transcription regulation machinery of cyanobacteria considerably more complex than those of other bacteria, hence making them an ideal group of organisms for studying organism evolution and evolution of transcription regulation systems. The availability of a large number of fully sequenced cyanobacterial genomes makes it possible to carry out elucidation of transcription regulation in a systematic and semi-automated manner.

Transcriptions regulation in cyanobacteria consists of numerous elements such as RNA polymerase and sigma factors, transcription factors, transcription terminators and two-component systems. Cyanobacteria have more sub-types of sigma factors that allow transcription regulation in more accurate and sensitive level, compared to other bacteria. Similarly, transcription factors in cyanobacteria are more complex in terms of their protein domain architectures than those in other bacteria. A computational study revealed that intrinsic terminators may represent the dominating class of terminators in cyanobacteria as has been shown in PCC6803. In addition, cyanobacteria have the largest numbers of two-component systems and transcription factors compared to other bacteria, indicating their higher capability in handling more diverse and more complex habitat environments.

When genomic sequences (and transcriptomic data) are available, it is realistically possible to derive the general organization of the transcription regulation system for a specific biological process, which could be as complex as the photosynthesis system as we have repeatedly demonstrated in our previous work (Mao *et al.*, 2010; Su *et al.*, 2006). The key steps of this process include (i) collection of template networks for related model organisms, which have substantial experimental data and possibly known information about the target network; (ii) prediction of operons, transcription regulators and functional relatedness among genes in the target genome; (iii) derivation of the operons regulated by the identified regulator(s) through identification of conserved *cis* regulatory motifs; and (iv) mapping the template networks to the target genome through orthologous gene mapping that is consistent with the predicted operons and gene associations (Mao *et al.*, 2010; Su *et al.*, 2006). This process can be largely automated as we have shown in our previous publications by the CINPER web server (Mao *et al.*, 2012).

A number of new insights have been gained about transcription regulation of cyanobacteria through comparative genomic analysis of the photosynthesis, nitrogen assimilation and osmoregulation systems. They include (i) cyanobacteria have a flexible antenna complex phycobilisome for adaption under variety of light conditions, and high sensitive antenna proteins for survival in extreme low-light location such as deep ocean, and they also have carboxysome for increasing the CO_2 concentration at close proximity of RuBisCO; (ii) NtcA is the universal main regulator and activated by a high level of 2-oxoglutarate; (iii) NtcA also plays an important role in the coordination of photosynthesis and nitrogen assimilation under changing environmental conditions; (iv) σ^{38} functions as a global regulator under hyperosmotic stress, which bridges the K^+ uptake and osmolyte accumulation processes, and to coordinate osmoregulation with other biological processes; and (v) cyanobacteria in deep ocean use glycine betaine as their major osmolyte for possible larger osmotic stress changes.

Challenging issues

Integration of information derived from multiple sources

A large number of elements/functional relationships in in cyanobacteria's transcription regulation systems have been identified using computational techniques through comparative analyses of genomes, providing a skeleton for the target and previously unclear transcription regulation system in cyanobacteria. To fill the more details, it will require additional information from other sources of experimental data including transcriptomic and metabolomic data. Integration of information derived from multiple types of *omic* data for derivation of more detailed models for transcription regulation systems represents a highly challenging problem (Wu *et al.*, 2005). Further development is clearly needed aimed at building a capability for construction of high-resolution models for transcription regulation and biological systems in general through integrated data mining, hypothesis formulation and rationale experimental design in the most efficient manner.

From gene lists to wiring diagrams to real biological systems

Often the derived models using our current prediction capability do not have all the detailed relationships, particularly subtle relationships, among various identified elements/components. How to convert such coarse models to a well-defined network model or even better a chemistry reaction model? For bacteria, we anticipate that this could be done on the metabolic systems within the near future because the current techniques for identification of enzyme-encoding genes and the general knowledge about the relationships among substrates, products and enzymes are there for carry out realizing such capabilities although some further developments on the computational side are still needed. However to accomplish this for regulatory and signalling systems may represent a different level of challenge. Additional experimental data are clearly needed, such as DNA–protein interaction and protein–protein interaction data. In addition, to fully realize this goal, modelling capabilities for realistic simulation of kinetics of complex chemical reactions are also needed. We are clearly witnessing that these different areas start to merge all aimed to develop capabilities for driving realistic models of transcription regulation and other biological systems.

Chapter highlights

- We describe the basic components in cyanobacterial transcription regulation system: RNA polymerase sigma factors, transcription factors and terminators, and two-component signalling systems.
- We describe the major regulatory machinery of three key pathways, namely photosynthesis, nitrogen assimilation and osmoregulation, along with their associated cross-talk networks.
- Comparative genomics helps to elucidate the complexity of transcriptional regulation systems of cyanobacteria under diverse living environments.
- We provide a list of Internet resources along with a list of challenging problems in the study of cyanobacterial transcription regulation.

Acknowledgements

This work was supported in part by the National Science Foundation (DEB-0830024, EF0849615, CCF1048261, and DBI-0542119), and the DOE BioEnergy Science Center grant (DE-PS02-06ER64304), which is supported by the Office of Biological and Environmental Research in the Department of Energy Office of Science. Any options, findings, and conclusions or recommendations expressed in this material are those of authors and do not necessarily reflect the views of the sponsoring institutions.

References

Aichi, M., Takatani, N., and Omata, T. (2001). Role of NtcB in activation of nitrate assimilation genes in the cyanobacterium *Synechocystis* sp. strain PCC 6803, J. Bacteriol. *183*, 5840–5847.

Alfonso, M., Perewoska, I., and Kirilovsky, D. (2001). Redox control of ntcA gene expression in *Synechocystis* sp. PCC 6803. Nitrogen availability and electron transport regulate the levels of the NtcA protein. Plant Physiol. *125*, 969–981.

Anderson, P.M., Sung, Y.C., and Fuchs, J.A. (1990). The cyanase operon and cyanate metabolism. FEMS Microbiol. Rev. *7*, 247–252.

Ashby, M.K., and Houmard, J. (2006). Cyanobacterial two-component proteins: structure, diversity, distribution, and evolution. Microbiol. Mol. Biol. *reviews: 70*, 472–509.

Bartsevich, V.V., and Shestakov, S.V. (1995). The dspA gene product of the cyanobacterium Synechocystis sp. strain PCC 6803 influences sensitivity to chemically different growth inhibitors and has amino acid similarity to histidine protein kinases. Microbiology *141*, 2915–2920.

Bhaya, D., Takahashi, A., and Grossman, A.R. (2001). Light regulation of type IV pilus-dependent motility by chemosensor-like elements in *Synechocystis* PCC6803. Proc. Natl. Acad. Sci. U.S.A. *98*, 7540.

Binder, A. (1982). Respiration and photosynthesis in energy-transducing membranes of cyanobacteria. J. Bioenerget. Biomembr. *14*, 271–286.

Bird, C., and Wyman, M. (2003). Nitrate/nitrite assimilation system of the marine picoplanktonic cyanobacterium *Synechococcus* sp. strain WH 8103: effect of nitrogen source and availability on gene expression. Appl. Environ. Microbiol. *69*, 7009–7018.

Blanchette, M., and Tompa, M. (2002). Discovery of regulatory elements by a computational method for phylogenetic footprinting. Genome Res. *12*, 739–748.

Capone, D.G. (2000). The marine nitrogen cycle. In Kirchman, D. (ed.), Microbial Ecology of the Ocean. Wiley-Liss, New York, pp. 455–493.

Collier, J.L., Brahamsha, B., and Palenik, B. (1999). The marine cyanobacterium *Synechococcus* sp. WH7805 requires urease (urea amidohydrolase, EC 3.5.1.5). to utilize urea as a nitrogen source: molecular-genetic and biochemical analysis of the enzyme, Microbiology *145*, 447–459.

Commichau, F.M., Forchhammer, K., and Stulke, J. (2006). Regulatory links between carbon and nitrogen metabolism. Curr. Opin. Microbiol. 9, 167–172.

Dubbs, J.M., Bird, T.H., Bauer, C.E., and Tabita, F.R. (2000). Interaction of CbbR and RegA* transcription regulators with the *Rhodobacter sphaeroides* cbbIPromoter-operator region. J. Biol. Chem. *275*, 19224–19230.

Dubbs, J.M., and Tabita, F.R. (2003). Interactions of the cbbII promoter-operator region with CbbR and RegA (PrrA). regulators indicate distinct mechanisms to control expression of the two cbb operons of *Rhodobacter sphaeroides*. J. Biol. Chem. *278*, 16443–16450.

Elsen, S., Swem, L.R., Swem, D.L., and Bauer, C.E. (2004). RegB/RegA, a highly conserved redox-responding global two-component regulatory system. Microbiol. Mol. Biol. Rev.s *68*, 263–279.

Fadi Aldehni, M., Sauer, J., Spielhaupter, C., Schmid, R., and Forchhammer, K. (2003). Signal transduction protein P(II). is required for NtcA-regulated gene expression during nitrogen deprivation in the cyanobacterium *Synechococcus elongatus* strain PCC 7942. J. Bacteriol., *185*, 2582–2591.

Fay, P. (1992). Oxygen relations of nitrogen fixation in cyanobacteria. Microbiol. Rev.*56*, 340–373.

Flores, E., and Herrero, A. (2005). Nitrogen assimilation and nitrogen control in cyanobacteria. Biochem. Soc. Trans. *33*, 164–167.

Flores, E., and Herrero, A. (2009). Compartmentalized function through cell differentiation in filamentous cyanobacteria. Nat. Rev. Microbiol. *8*, 39–50.

Forchhammer, K. (2004). Global carbon/nitrogen control by PII signal transduction in cyanobacteria: from signals to targets. FEMS Microbiol. Rev. *28*, 319–333.

Frias, J.E., Flores, E., and Herrero, A. (1994). Requirement of the regulatory protein NtcA for the expression of nitrogen assimilation and heterocyst development genes in the cyanobacterium *Anabaena* sp. PCC 7120. Mol. Microbiol. *14*, 823–832.

Frias, J.E., Flores, E., and Herrero, A. (2000). Activation of the *Anabaena* nir operon promoter requires both NtcA (CAP family). and NtcB (LysR family). transcription factors. Mol. Microbiol. *38*, 613–625.

Frias, J.E., Herrero, A., and Flores, E. (2003). Open reading frame all0601 from *Anabaena* sp. strain PCC 7120 represents a novel gene, cnaT, required for expression of the nitrate assimilation nir operon. J. Bacteriol. *185*, 5037–5044.

Golden, S.S., Ishiura, M., Johnson, C.H., and Kondo, T. (1997). Cyanobacterial circadian rhythms. Annu. Rev. Plant Physiol. Plant Mol. Biol. *48*, 327–354.

Gorl, M., Sauer, J., Baier, T., and Forchhammer, K. (1998). Nitrogen-starvation-induced chlorosis in *Synechococcus* PCC 7942: adaptation to long-term survival. Microbiology *144*, 2449–2458.

Gowrishankar, J. (1985). Identification of osmoresponsive genes in *Escherichia coli*: evidence for participation of potassium and proline transport systems in osmoregulation. J. Bacteriol., *164*, 434–445.

Gutu, A., and Kehoe, D.M. (2012). Emerging perspectives on the mechanisms, regulation, and distribution of light color acclimation in cyanobacteria. Mol. Plant 5, 1–13.

Harano, Y., Suzuki, I., Maeda, S., Koneko, T., Tabata, S., and Omata, T. (1997). Identification and nitrogen regulation of the cyanase gene from the cyanobacteria *Synechocystis* sp. strain PCC 6803 and *Synechococcus* sp. strain PCC 7942. J. Bacteriol., *179*, 5744–5750.

Heinrich, A., Maheswaran, M., Ruppert, U., and Forchhammer, K. (2004). The *Synechococcus elongatus* P signal transduction protein controls arginine synthesis by complex formation with N-acetyl-L-glutamate kinase. Mol. Microbiol. 52, 1303–1314.

Herrero, A., Flores, E., and Flores, F.G. (2008). The Cyanobacteria: Molecular Biology, Genomics, and Evolution. Caister Academic Press.

Herrero, A., Muro-Pastor, A.M., and Flores, E. (2001). Nitrogen control in cyanobacteria, J. Bacteriol., *183*, 411–425.

Hirani, T.A., Suzuki, I., Murata, N., Hayashi, H., and Eaton-Rye, J.J. (2001). Characterization of a two-component signal transduction system involved in the induction of alkaline phosphatase under phosphate-limiting conditions in Synechocystis sp. PCC 6803. Plant Mol. Biol. *45*, 133–144.

Ikeuchi, M. (1996). [Complete genome sequence of a cyanobacterium *Synechocystis* sp. PCC 6803, the oxygenic photosynthetic prokaryote], *Tanpakushitsu kakusan koso*. Protein, Nucleic Acid, Enzyme *41*, 2579–2583.

Imamura, S., and Asayama, M. (2009). Sigma factors for cyanobacterial transcription. Gene Reg. Systems Biol. *3*, 65–87.

Javelle, A., and Merrick, M. (2005). Complex formation between AmtB and GlnK: an ancestral role in prokaryotic nitrogen control. Biochem. Soc. Trans. *33*, 170–172.

Javelle, A., Severi, E., Thornton, J., and Merrick, M. (2004). Ammonium sensing in *Escherichia coli*. Role of the ammonium transporter AmtB and AmtB–GlnK complex formation. J. Biol. Chem. *279*, 8530–8538.

Joshua, S., and Mullineaux, C.W. (2004). Phycobilisome diffusion is required for light-state transitions in cyanobacteria. Plant Physiol.*135*, 2112–2119.

Kaasen, I., Falkenberg, P., Styrvold, O.B., and Strom, A.R. (1992). Molecular cloning and physical mapping of the otsBA genes, which encode the osmoregulatory trehalose pathway of *Escherichia coli*: evidence that transcription is activated by katF (AppR). J. Bacteriol. *174*, 889–898.

Kageyama, H., Kondo, T., and Iwasaki, H. (2003). Circadian formation of clock protein complexes by KaiA, KaiB, KaiC, and SasA in cyanobacteria. J. Biol. Chem.*278*, 2388–2395.

Lamparter, T., Esteban, B., and Hughes, J. (2001). Phytochrome Cph1 from the cyanobacterium *Synechocystis* PCC6803. Eur. J. Biochem. *268*, 4720–4730.

Lee, H.M., Flores, E., Forchhammer, K., Herrero, A., and Tandeau De Marsac, N. (2000). Phosphorylation of the signal transducer PII protein and an additional effector are required for the PII-mediated regulation of nitrate and nitrite uptake in the cyanobacterium *Synechococcus* sp. PCC 7942. Eur. J. Biochem. *267*, 591–600.

Li, H., and Sherman, L.A. (2000). A redox-responsive regulator of photosynthesis gene expression in the cyanobacterium *Synechocystis* sp. strain PCC 6803. J. Bacteriol. *182*, 4268–4277.

Li, H., and Sherman, L.A. (2000). A redox-responsive regulator of photosynthesis gene expression in the cyanobacterium *Synechocystis* sp. Strain PCC 6803. J Bacteriol. *182*, 4268–4277.

Liu, X., Huang, W., and Wu, Q. (2006). Two-component signal transduction systems in the cyanobacterium. Tsinghua Science and Technology, *11*, 379–390.

Lópe-Maury, L., García-Domínguez, M., Florencio, F.J., and Reyes, J.C. (2002). A t-component signal transduction system involved in nickel sensing in the cyanobacterium *Synechocystis* sp. PCC 6803, Mol. Microbiol., *43*, 247–256.

Maeda, S., Kawaguchi, Y., Ohe, T.A., and Omata, T. (1998). cis-acting sequences required for NtcB-dependent, nitrite-responsive positive regulation of the nitrate assimilation operon in the cyanobacterium *Synechococcus* sp. strain PCC 7942. J. Bacteriol. *180*, 4080–4088.

Maeda, S., and Omata, T. (2009). Nitrite transport activity of the ABC-type cyanate transporter of the cyanobacterium *Synechococcus elongatus*. J. Bacteriol. *191*, 3265–3272.

Mao, X., Chen, X., Zhang, Y., Pangle, S., and Xu, Y. (2012). CINPER: an interactive web system for pathway prediction for prokaryotes. Submitted.

Mao, X., Olman, V., Stuart, R., Paulsen, I.T., Palenik, B., and Xu, Y. (2010). Computational prediction of the osmoregulation network in *Synechococcus* sp. WH8102. BMC Genomics *11*, 291.

Marsac, N.T.d., Lee, H.M., Hisbergues, M., Castets, A.M., and Bédu, S. (2001). Control of nitrogen and carbon metabolism in cyanobacteria. J. Appl. Phycol. *13*, 287–292.

Martiny, A.C., Kathuria, S., and Berube, P.M. (2009). Widespread metabolic potential for nitrite and nitrate assimilation among *Prochlorococcus* ecotypes, Proc. Natl. Acad. Sci. U.S.A. *106*, 10787–10792.

Mizuno, T. (1997). Compilation of all genes encoding two-component phosphotransfer signal transducers in the genome of *Escherichia coli*. DNA Res. *4*, 161–168.

Moore, L.R., Post, A.F., Rocap, G., and Chisholm, S.W. (2002). Utilization of different nitrogen sources by the marine cyanobacteria *Prochlorococcus* and *Synechococcus*. Limnol. Oceanogr. *47*, 989–996.

Morel, F.M., and Price, N.M. (2003). The biogeochemical cycles of trace metals in the oceans. Science *300*, 944–947.

Mullineaux, C.W., and Emlyn-Jones, D. (2005). State transitions: an example of acclimation to low-light stress. J. Exp. Bot. *56*, 389–393.

Muramatsu, M., and Hihara, Y. (2012). Acclimation to high-light conditions in cyanobacteria: from gene expression to physiological responses. J. Plant Res.*125*, 11–39.

Muro-Pastor, M.I., and Florencio, F.J. (1994). NADP(+)-isocitrate dehydrogenase from the cyanobacterium *Anabaena* sp. strain PCC 7120: purification and characterization of the enzyme and cloning, sequencing, and disruption of the icd gene. J. Bacteriol.,*176*, 2718–2726.

Muro-Pastor, M.I., Reyes, J.C., and Florencio, F.J. (1996). The $NADP^+$-isocitrate dehydrogenase gene (icd). is nitrogen regulated in cyanobacteria. J. Bacteriol. *178*, 4070–4076.

Muro-Pastor, M.I., Reyes, J.C., and Florencio, F.J. (2001). Cyanobacteria perceive nitrogen status by sensing intracellular 2-oxoglutarate levels. J. Biol. Chem. *276*, 38320–38328.

Nakamura, Y., Kaneko, T., Sato, S., Ikeuchi, M., Katoh, H., Sasamoto, S., Watanabe, A., Iriguchi, M., Kawashima, K., Kimura, T., *et al.* (2002). Complete genome structure of the thermophilic cyanobacterium *Thermosynechococcus elongatus* BP-1 (supplement). DNA Res. *9*, 135–148.

Ogawa, T., Bao, D.H., Katoh, H., Shibata, M., Pakrasi, H.B., and Bhattacharyya-Pakrasi, M. (2002). A two-component signal transduction pathway regulates manganese homeostasis in Synechocystis 6803, a photosynthetic organism, J. Biol. Chem. *277*, 28981–28986.

Palenik, B., Brahamsha, B., Larimer, F.W., Land, M., Hauser, L., Chain, P., Lamerdin, J., Regala, W., Allen, E.E., McCarren, J., *et al.* (2003). The genome of a motile marine *Synechococcus*. Nature *424*, 1037–1042.

Ramasubramanian, T.S., Wei, T.F., and Golden, J.W. (1994). Two *Anabaena* sp. strain PCC 7120 DNA-binding factors interact with vegetative cell- and heterocyst-specific genes. J. Bacteriol. *176*, 1214–1223.

Raven, J.A., and Allen, J.F. (2003). Genomics and chloroplast evolution: what did cyanobacteria do for plants? Genome Biol. *4*, 209.

Reitzer, L. (2003). Nitrogen assimilation and global regulation in *Escherichia coli*. Annu. Rev. Microbiol. *57*, 155–176.

Rippka, R., Deruelles, J., Waterbury, J.B., Herdman, M., and Stanier, R.Y. (1979). Generic assignments, strain histories and properties of pure cultures of cyanobacteria. Microbiology *111*, 1–61.

Rocap, G., Larimer, F.W., Lamerdin, J., Malfatti, S., Chain, P., Ahlgren, N.A., Arellano, A., Coleman, M., Hauser, L., Hess, W.R., *et al.* (2003). Genome divergence in two *Prochlorococcus* ecotypes reflects oceanic niche differentiation. Nature *424*, 1042–1047.

Sauer, J., Gorl, M., and Forchhammer, K. (1999). Nitrogen starvation in *Synechococcus* PCC 7942: involvement of glutamine synthetase and NtcA in phycobiliprotein degradation and survival. Arch. Microbiol. *172*, 247–255.

Sauer, J., Schreiber, U., Schmid, R., Volker, U., and Forchhammer, K. (2001). Nitrogen starvation-induced chlorosis in *Synechococcus* PCC 7942. Low-level photosynthesis as a mechanism of long-term survival, Plant Physiol. *126*, 233–243.

Scanlan, D.J., Ostrowski, M., Mazard, S., Dufresne, A., Garczarek, L., Hess, W.R., Post, A.F., Hagemann, M., Paulsen, I., and Partenski, F. (2009). Ecological genomics of marine picocyanobacteria. Microbiol. Mol.Biol.Rev. *73*, 249–299.

Seino, Y., Takahashi, T., and Hihara, Y. (2009). The response regulator RpaB binds to the upstream element of photosystem I genes to work for positive regulation under low-light conditions in *Synechocystis* sp. Strain PCC 6803. J. Bacteriol., *191*, 1581–1586.

Shi, T., Sun, Y., and Falkowski, P.G. (2007). Effects of iron limitation on the expression of metabolic genes in the marine cyanobacterium *Trichodesmium erythraeum* IMS101. Env. Microbiol. *9*, 2945–2956.

Shoumskaya, M.A., Paithoonrangsarid, K., Kanesaki, Y., Los, D.A., Zinchenko, V.V., Tanticharoen, M., Suzuki, I., and Murata, N. (2005). Identical Hik-Rre systems are involved in perception and transduction of salt signals and hyperosmotic signals but regulate the expression of individual genes to different extents in *Synechocystis*. J. Biol. Chem. *280*, 21531–21538.

Su, Z., Dam, P., Mao, F., Chen, X., Olman, V., Jiang, T., Palenik, B., and Xu, Y. (2006). Computational inference and experimental validation of nitrogen assimilation regulatory networks in cyanobacterium *Synechococcus* sp. WH8102. Nucleic Acids Res. *34*, 1050–1065.

Su, Z., Olman, V., Mao, F., and Xu, Y. (2006). Comparative genomics analysis of NtcA regulons in cyanobacteria: regulation of nitrogen assimilation and its coupling to photosynthesis. Nucleic Acids Res. *33*, 5156–5171.

Suzuki, I., Los, D.A., Kanesaki, Y., Mikomi, K., and Murata, N. (2000). The pathway for perception and transduction of low-temperature signals in *Synechocystis*. Science's STKE *19*, 1327.

Swem, L.R., Gong, X., Yu, C.A., and Bauer, C.E. (2006). Identification of a ubiquinone-binding site that affects autophosphorylation of the sensor kinase RegB. J. Biol. Chem. *281*, 6768–6775.

Takai, N., Nakajima, M., Oyama, T., Kito, R., Sugita, C., Sugita, M., Kondo, T., and Iwasaki, H. (2006). A KaiC-associating SasA-RpaA two-component regulatory system as a major circadian timing mediator in cyanobacteria. Proc. Natl. Acad. Sci. U.S.A. *103*, 12109–12114.

Takigawa-Imamura, H., and Mochizuki, A. (2006). Transcriptional autoregulation by phosphorylated and non-phosphorylated KaiC in cyanobacterial circadian rhythms. J. Theoret. Biol. *241*, 178–192.

Tanigawa, R., Shirokane, M., Maeda Si, S., Omata, T., Tanaka, K., and Takahashi, M. (2002). Transcriptional activation of NtcA-dependent promoters of *Synechococcus* sp. PCC 7942 by 2-oxoglutarate *in vitro*. Proc. Natl. Acad. Sci. U.S.A. *99*, 4251–4255.

Ting, C.S., Rocap, G., King, J., and Chisholm, S.W. (2002). Cyanobacterial photosynthesis in the oceans: the origins and significance of divergent light-harvesting strategies. Trends Microbiol. *10*, 134–142.

Valladares, A., Montesinos, M.L., Herrero, A., and Flores, E. (2002). An ABC-type, high-affinity urea permease identified in cyanobacteria. Mol. Microbiol. *43*, 703–715.

Vazquez-Bermudez, M.F., Herrero, A., and Flores, E. (2002). 2-Oxoglutarate increases the binding affinity of the NtcA (nitrogen control). transcription factor for the Synechococcus glnA promoter. FEBS Lett. *512*, 71–74.

Vazquez-Bermudez, M.F., Herrero, A., and Flores, E. (2003). Carbon supply and 2-oxoglutarate effects on expression of nitrate reductase and nitrogen-regulated genes in Synechococcus sp. strain PCC 7942. FEMS Microbiol Lett. *221*, 155–159.

Vazquez-Bermudez, M.F., Paz-Yepes, J., Herrero, A., and Flores, E. (2002). The NtcA-activated amt1 gene encodes a permease required for uptake of low concentrations of ammonium in the cyanobacterium Synechococcus sp. PCC 7942. Microbiology *148*, 861–869.

Vermaas, W. (2001). Photosynthesis and Respiration in Cyanobacteria. John Wiley and Sons, Inc., eLS.

Vijayan, V., Jain, I.H., and O'Shea, E.K. (2011). A high resolution map of a cyanobacterial transcriptome, *Genome biology*, *12*, R47–R47.

Whitton, B.A., and Potts, M. (2000). The Ecology of Cyanobacteria: Their Diversity in Time and Space. Kluwer Academic.

Wilde, A., Churin, Y., Schubert, H., and Börner, T. (1997). Disruption of a Synechocystis sp. PCC 6803 gene with partial similarity to phytochrome genes alters growth under changing light qualities. *FEBS letters*, *406*, 89.

Wilde, A., Fiedler, B., and Börner, T. (2002). The cyanobacterial phytochrome Cph2 inhibits phototaxis towards blue light. Mol. Microbiol. *44*, 981–988.

Wu, H., Su, Z., Mao, F., Olman, V., and Xu, Y. (2005). Prediction of functional modules based on comparative genome analysis and gene ontology application. Nucl. Acids Res. *33*, 2822–2837.

Wu, J., and Bauer, C.E. (2008). RegB/RegA, a global redox-responding two-component system. Adv. Exp. Med. Biol. *631*, 131–148.

Wu, J., Zhao, F., Wang, S., Deng, G., Wang, J., Bai, J., Lu, J., Qu, J., and Bao, Q. (2007). cTFbase: a database for comparative genomics of transcription factors in cyanobacteria. BMC Genomics *8*, 104.

Yamaguchi, K., Suzuki, I., Yamamoto, H., Lyukevich, A., Bodrova, I., Los, D.A., Piven, I., Zinchenko, V., Kanehisa, M., and Murata, N. (2002). A two-component Mn2+-sensing system negatively regulates expression of the mntCAB operon in *Synechocystis*. Plant Cell Online *14*, 2901–2913.

Yeh, K.C., Wu, S.H., Murphy, J.T., and Lagarias, J.C. (1997). A cyanobacterial phytochrome two-component light sensory system. Science *277*, 1505–1508.

Yoshihara, S., Suzuki, F., Fujita, H., Geng, X.X., and Ikeuchi, M. (2000). Novel putative photoreceptor and regulatory genes required for the positive phototactic movement of the unicellular motile cyanobacterium *Synechocystis* sp. PCC 6803. Plant Cell Physiol. *41*, 1299–1304.

Appendix

The following tables can also be accessed online at http://biorg.cis.fiu.edu/Deepak2012/Book_Suppl_Tables.xlsx

Table A.1 Family-wise list of transcription factors found in *P. aeruginosa* PAO1. The table shows the genomic location of the protein (core/RGP) (Mathee *et al.* 2008) and the different PFAM domains (Winzor *et al.* 2009, http://pfam.sanger.ac.uk/and http://expasy.org/prosite/). ORF (PA) numbers, protein name, homologue and amino acid information are from the *Pseudomonas* database (Winzor *et al.* 2009)

PA no.	Protein name	No. of amino acids	Subfamily	Nearest homologue	Core/ RGP	Domain 1	Domain 1 description	Domain 2	Domain 2 description	Other domains	Domain description
AraC											
PA0306					Core						
PA0163		265		52% similar to putative regulatory protein YeaM (*Escherichia coli*)	Core	PF00165	HTH_AraC				
PA0248	PobR	288		87% similar to putative regulator PobR (*Pseudomonas aeruginosa*)	Core	PF00165	HTH_AraC				
PA0416	ChpD	264		40% similar to regulatory protein mmsR which is a member of the AraC/XylS family of transcriptional regulators (*Pseudomonas aeruginosa*)	Core	PF02311	AraC_binding	PF00165	HTH_AraC		
PA0564		303		57% similarity to hypothetical transcriptional regulator HI1052 from (*Haemophilus influenzae*)	Core	PF00165	HTH_AraC				
PA0748		306		55% similar to regulatory protein mmsR (*Pseudomonas aeruginosa*)	Core	PF00165	HTH_AraC				
PA0780	PruR	250		54% similar to lumQ gene product (*Photobacterium leiognathi*)	Core	PF00165	HTH_AraC				
PA0791		262		50% similarity to hypothetical yeaM gene product of (*Escherichia coli*), a putative regulatory protein.	Core	PF00165	HTH_AraC				
PA0831	OruR	339		100% identical to OruR, transcriptional regulator required for growth on ornithine as the sole carbon source (*Pseudomonas aeruginosa*)	Core	PF00165	HTH_AraC				
PA0864		262		50% similar to putative regulatory protein YeaM (*Escherichia coli*)	Core	PF00165	HTH_AraC				
PA0893	ArgR	329		0	Core	PF00165	HTH_AraC				
PA1109		259		52% similar to putative regulatory protein YeaM (*Escherichia coli*)	Core	PF00165	HTH_AraC				
PA1182		333		47% similar to transcriptional regulator OruR (*Pseudomonas aeruginosa*)	Core	PF00165	HTH_AraC				
PA1229	YeaM	268		67% similar to putative regulatory protein YeaM (*Escherichia coli*)	Core	PF00165	HTH_AraC				
PA1235		260		51% similar to putative regulatory protein YeaM (*Escherichia coli*)	Core	PF00165	HTH_AraC				
PA1261		224		62% similar to a region of probable transcriptional regulator LumQ (*Photobacterium leiognathi*)	Core	PF00165	HTH_AraC				
PA1380		321		60% similar to putative regulatory protein YqhC (*Escherichia coli*)	RGP 15	PF00165	HTH_AraC				

PA1599		260	56% similar to putative regulatory protein YeaM (*Escherichia coli*)	Core	PF00165	HTH_AraC				
PA1619		274	45% similar to putative transcriptional regulator ((*Streptomyces coelicolor*)	Core	PF00165	HTH_AraC				
PA1713	ExsA	278	71% similar to thermoregulatory transcription factor LcrF (*Yersinia pestis*); 72% similar to transcriptional activator VirF (*Yersinia enterocolitica*); 99% similar to exoenzyme S synthesis regulatory protein ExsA (*Pseudomonas aeruginosa*)	Core	PF00165	HTH_AraC				
PA1850		334	57% similar to AraC-like protein (*Azorhizobium caulinodans*)	Core	PF00165	HTH_AraC				
PA2047		329	51% similar to putative regulatory proteinYqhC (*Escherichia coli*)	RGP 20	PF00165	HTH_AraC				
PA2096		344	51% similar to hypothetical transcriptional regulator CY21B4.12 (*Mycobacterium tuberculosis*)	Core	PF00165	HTH_AraC				
PA2227	VqsM	325	47% similar to regulator OruR (*Pseudomonas aeruginosa*)	Core	PF00165	HTH_AraC				
PA2276		296	62% similar (over C-terminal 3/4ths of ORF) to yqhC gene product of (*Escherichia coli*), a putative transcriptional regulator.	Core	PF00165	HTH_AraC				
PA2281		284	64% similar to the C-terminal end of regulatory protein PchR *Synechocystis* sp.	Core	PF00165	HTH_AraC				
PA2332		301	46% similar to putative regulator HI1052 (*Haemophilus influenzae* Rd)	Core	PF00165	HTH_AraC				
PA2337	MtlR	301	84% similar to putative activator protein MtlR (*Pseudomonas fluorescens*)	Core	PF00165	HTH_AraC				
PA2488		254	47% similar to putative regulatory protein YeaM (*Escherichia coli*)	Core	PF00165	HTH_AraC				
PA2489		271	46% similar to putative transcriptional regulator (*Streptomyces coelicolor*)	Core	PF00165	HTH_AraC				
PA2511		333	50% similar to putative regulatory protein OxoS (*Pseudomonas putida*)	Core	PF00165	HTH_AraC				
PA2519	XylS	318	70% similar to positive regulatory protein XylS (TOL plasmid)	Core	PF00165	HTH_AraC				
PA2556		361	42% similar to regulator OruR (*Pseudomonas aeruginosa*)	Core	PF00165	HTH_AraC				
PA2588		342	48% similar to regulator OruR (*Pseudomonas aeruginosa*)	Core	PF00165	HTH_AraC				
PA2696		297	52% similar to regulatory protein GdhBR (*Pantoea citrea*)	Core	PF00165	HTH_AraC				
PA2704		339	54% similar to transcriptional regulator OruR (*Pseudomonas aeruginosa*)	Core	PF00165	HTH_AraC				
PA2917		278	47% similar to putative transcriptional regulator (*Streptomyces coelicolor*)	Core	PF00165	HTH_AraC				
PA3027		343	49% similar to transcriptional regulator OruR (*Pseudomonas aeruginosa*)	Core	PF00165	HTH_AraC	PF08281	Sigma70_r4_2	PF09312	SurA_N
PA3094		345	80% similar to putative virulence regulator (*Pseudomonas alcaligenes*)	Core	PF00165	HTH_AraC				

PA no.	Protein name	No. of amino acids	Subfamily	Nearest homologue	Core/ RGP	Domain 1	Domain 1 description	Domain 2	Domain 2 description	Other domains	Domain description
PA3215		337		52% similar to transcriptional regulator PhbR (*Pseudomonas* sp. 61–3)	Core	PF00165	HTH_AraC				
PA3220		254		48% similar to probable transcriptional regulator LumQ (*Photobacterium leiognathi*)	Core	PF00165	HTH_AraC	PF07883	Cupin_2		
PA3269		284		49% similar to putative transcriptional regulator (*Streptomyces coelicolor*)	Core	PF00165	HTH_AraC	PF02311	AraC_binding		
PA3423		247		52% similar to a region of putative araC-type regulatory protein (*Escherichia coli*)	Core	PF00165	HTH_AraC	PF07366	SnoaL		
PA3571	MmsR	307			Core	PF00165	HTH_AraC	PF00165	HTH_AraC	PF07883; PF03079	Cupin_2; ARD
PA3782		317		49% similar to a region of ArgR regulatory protein (*Pseudomonas aeruginosa*)	Core	PF00165	HTH_AraC	PF01965	DJ-1_PfpI		
PA3830		270		51% similar to putative regulatory protein YeaM (*Escherichia coli*)	Core	PF00165	HTH_AraC	PF07883	Cupin_2	PF02311	AraC_binding
PA3898		293		54% similar to regulatory protein GdhBR (*Pantoea citrea*)	Core	PF00165	HTH_AraC				
PA3927		262		53% similar to putative regulatory protein YeaM (*Escherichia coli*)	Core	PF00165	HTH_AraC	PF02311	AraC_binding	PF07883	Cupin_2
PA4070		316		45% similar to the regulatory protein for 2-phenylethylamine catabolism FeaR (*Escherichia coli*)	Core	PF00165	HTH_AraC				
PA4094		319		42% similar to nitrilase regulator (*Rhodococcus rhodochrous*)	Core	PF00165	HTH_AraC				
PA4120		303		63% similar to the HpaA gene product of *Escherichia coli*	Core	PF00165	HTH_AraC	PF02311	AraC_binding	PF07883	Cupin_2
PA4184		340		48% similar to ArgR regulatory protein (*Pseudomonas aeruginosa*)	Core	PF00165	HTH_AraC	PF01965	DJ-1_PfpI		
PA4227	PchR	296		100% similar to regulatory protein PchR (*Pseudomonas aeruginosa*)	Core	PF00165	HTH_AraC				
PA4288		267		51% similar to putative regulatory protein YeaM (*Escherichia coli*)	Core	PF00165	HTH_AraC	PF07883	Cupin_2	PF02311	AraC_binding
PA4436		299		45% similar to ArgR regulatory protein (*Pseudomonas aeruginosa*)	Core	PF00165	HTH_AraC				
PA4787		333		42% similar to regulator OruR (*Pseudomonas aeruginosa*)	Core	PF00165	HTH_AraC				
PA5032		357		53% similar to regulator OruR (*Pseudomonas aeruginosa*)	Core	PF00165	HTH_AraC				
PA5324		356		46% similar to hypothetical transcriptional regulator (*Mycobacterium tuberculosis*)	Core	PF00165	HTH_AraC	PF09312	SurA_N		
PA5342		266		46% similar to putative transcriptional regulator (*Streptomyces coelicolor*)	Core	PF00165	HTH_AraC	PF02311	AraC_binding	PF07883	Cupin_2
PA5380	GbdR	367		59% similar to ArgR regulatory protein (*Pseudomonas aeruginosa*)	Core	PF00165	HTH_AraC	PF01965	DJ-1_PfpI		
PA5389	CdhR	336		58% similar to ArgR regulatory protein (*Pseudomonas aeruginosa*)	Core	PF00165	HTH_AraC	PF01965	DJ-1_PfpI		

ArsR								
PA0279	YdfF	232	59% similar to hypothetical protein YdfF ((*Bacillus subtilis*))	Core	PS00846	Bacterial regulatory proteins, arsR family signature		
PA0547		333	49% similar to a region of unknown protein (*Streptomyces fradiae* Transposon Tn4556)	Core	PS00846	Bacterial regulatory proteins, arsR family signature	PF01022	HTH_5, Bacterial regulatory protein, arsR family
PA2277	ArsR	116	82% identical to arsR gene product of (*P. aeruginosa*); 73% similar (over three-quarters of ORF length) to *arsR* gene product of (*Escherichia coli*)	Core				
PA4354		100	76% similar to a region of hypothetical protein (*Agrobacterium radiobacter*)	Core	PS00846	Bacterial regulatory proteins, arsR family signature (weak)		
AsnC								
PA0513	NirG	147	77% similar to NirG protein (*Pseudomonas stutzeri*)	Core	PF01037	AsnC family		
PA2028		157	62% similar to AzlB protein (*Bacillus subtilis*)	RGP20	PF01037	AsnC family		
PA2082		158	57% similar to leucine-responsive regulatory protein Lrp (*Escherichia coli*)	Core	PF01037	AsnC family		
PA2246	BkdR	153	64% similar to BkdR protein (*P. putida*)	Core	PF01037	AsnC family		
PA2577		145	55% similar to hypothetical protein Rv3291c (*Mycobacterium tuberculosis*)	Core	PF01037	AsnC family		
PA3965		169	63% similar to Leucine-responsive regulatory protein Lrp (*Escherichia coli*)	Core	PF01037	AsnC family	PF01047	MarR family
PA4508		157	70% similar to transcription regulatory protein PdhR (*Ralstonia eutropha*)	Core	PF01037	AsnC family		
PA4784		155	57% similar to putative transcriptional regulator ((*Bacillus subtilis*))	Core	PF01037	AsnC family		
PA5308	Lrp, DadR	162	79% similar to global response regulator Lrp (*Escherichia coli*)	Core	PF01037	AsnC family		
Cro/cl								
PA0048		127	55% similar to a region of TrbA (plasmid RK2, (*Escherichia coli*))	Core				
PA0225		179	51% similar to putative epoxidase (*Methanobacterium thermoautotrophicum*)	Core				
PA0535		184	48% similar to putative aldehyde dehydrogenase (*Azotobacter vinelandii*)	Core				
PA0611	PrtR	256	0	Core				
PA0906		68	70% similar to hypothetical repressor protein AF1793 (*Archaeoglobus fulgidus*)	Core				
PA1359		103	0	Core				

PA no.	Protein name	No. of amino acids	Subfamily	Nearest homologue	Core/ RGP	Domain 1	Domain 1 description	Domain 2	Domain 2 description	Other domains	Domain description
PA1879		238		60% similar to regulator PhnR (*Salmonella typhimurium*)	Core						
PA1884		72		73% similar to hypothetical protein YozG (*Bacillus subtilis*)	Core						
PA2312		114			Core						
PA2780		193		54% similar to hypothetical protein (*Bacillus* sp.)	Core						
PA2785		183		51% similar to putative transcriptional regulator protein (*Bacillus* sp.)	Core						
PA3260		218		Small portion 64% similar to *Pvu*II restriction endonuclease regulator *pvu*IIC of (*Proteus vulgaris*); 49% similar to hypothetical *hipB* protein of (*Escherichia coli*)	Core						
PA4077		237		66% similar to pyocin synthesis negative regulatory protein PrtR (*Pseudomonas aeruginosa*)	Core						
PA4499		187		44% similar to hypothetical protein YcjC (*Escherichia coli*)	Core						
PA4987		199		53% similar to aldehyde dehydrogenase (*Azotobacter vinelandii*); 48% similar to hypothetical ycjC gene product of (*Escherichia coli*)	Core						
PA5301	YcjC	182		62% similar to hypothetical protein YcjC (*Escherichia coli*)	Core						
PA5403		68		77% similar to putative transcriptional regulator YozG (*Bacillus subtilis*)	Core						
DksA/TraR											
PA0612	PtrB	66		59% similar to hypothetical protein YbiI (over N-terminal 75% of ORF) (*Escherichia coli*)	RGP 3	PF01258	zf-dskA_traR, Prokaryotic dksA/traR C4-type zinc finger				
PA4723	DksA	148		86% similar to dosage-dependent dnaK suppressor protein DksA (*Escherichia coli*)	Core	PF01258	zf-dskA_traR, Prokaryotic dksA/traR C4-type zinc finger				
PA4870	YbiI	88		84% similar to hypothetical protein YbiI (*Escherichia coli*)	Core	PF01258	zf-dskA_traR, Prokaryotic dksA/traR C4-type zinc finger				
PA5536		134		59% similar to dnaK suppressor protein DksA (*Escherichia coli*)	Core	PF01258	zf-dskA_traR, Prokaryotic dksA/traR C4-type zinc finger				
RpoN-binding											
PA0873	PhhR	519			Core	PF00158	Sigma-54 interaction domain	PF00989	PAS, PAS fold		
PA1097	FleQ	490		66% similar to sigma 54 transcriptional activator (*Vibrio cholerae*)	Core	PF02954	HTH_8, Bacterial regulatory protein, Fis family	PF06490	FleQ, Flagellar regulatory protein FleQ	PF00158	Sigma-54 interaction domain

PA1099	FleR	473	68% similar to two-component response regulator (*Vibrio cholerae*)	Core	PF00158	Sigma-54 interaction domain	PF00072	Response regulator receiver domain	PF02954	HTH_8, Bacterial regulatory protein, Fis family
PA1196		466	53% similar to arginine catabolism positive regulator RocR (*Bacillus subtilis*)	Core	PF07728	AAA_5, ATPase family associated with various cellular activities (AAA)	PF08448	PAS_4, PAS fold	PF02954; PF00158	HTH_8, Bacterial regulatory protein, Fis family; Sigma-54 interaction domain
PA1335		425	67% similar to C4-dicarboxylate transport transcription regulator DctD (*Rhizobium meliloti*)	Core	PF02954	HTH_8, Fis family	PF00158	Sigma54_activat,	PF00072	Response regulator
PA1663		503	55% similar to response regulator of hydrogenase 3 activity HydG (*Escherichia coli*)	Core	PF00158	Sigma-54 interaction domain	PF02954	HTH_8, Bacterial regulatory protein, Fis family		
PA1945		442	57% similar to hydG gene product of (*Escherichia coli*)	Core	PF02954	HTH_8, Bacterial regulatory protein, Fis family	PF00158	Sigma-54 interaction domain		
PA2005		481	58% similar to Acetoacetate metabolism regulatory protein AtoC (*Escherichia coli*); 53% similar to nifA gene product of (*Azotobacter vinelandii*)	Core	PF00158	Sigma-54 interaction domain	PF02954	HTH_8, Bacterial regulatory protein, Fis family	PF08448	PAS_4, PAS fold
PA2354		376	58% similar to transcriptional protein FtrC (*Caulobacter crescentus*)	Core	PF00158	Sigma-54 interaction domain	PF07728	AAA_5, ATPase family associated with various cellular activities (AAA)		
PA2359		361	60% similar to a region of transcriptional protein FtrC (*Caulobacter crescentus*)	Core	PF07728	AAA_5, ATPase family associated with various cellular activities (AAA)	PF00158	Sigma-5 4 interaction domain		
PA2449		511	58% similar phenylalanine hydroxylase gene cluster transcription activator PhhR (*Pseudomonas aeruginosa*)	Core	PF00158	Sigma-54 interaction domain				
PA2665		517	63% similar to putative 2-component transcriptional regulator YgaA (*Escherichia coli*)	Core	PF00158	Sigma-54 interaction domain	PF01590	GAF, GAF domain		
PA3932		324	55% similar to regulator for prp operon PrpR (*Escherichia coli*)	Core	PF00158	Sigma-54 interaction domain				
PA4021		643	57% similar to transcription factor AcoR (*Alcaligenes eutrophus*)	Core	PF00158	Sigma-54 interaction domain	PF07728	AAA_5, ATPase family associated with various cellular activities (AAA)	PF02954	HTH_8, Bacterial regulatory protein, Fis family
PA4147	AcoR	625	59% similar to the acoR gene product of (*Alcaligenes eutrophus*)	Core	PF00158	Sigma-54 interaction domain	PF02954	HTH_8, Bacterial regulatory protein, Fis family		
PA4493	RoxR	186	67% similar to putative photosynthetic response regulator PrrA (*Rhodobacter sphaeroides*); 66% similar to RegA (*Rhodovulum sulfidophilum*)	Core	PF02954	HTH_8, Bacterial regulatory protein, Fis family	PF00072	Response regulator receiver domain		
PA4547	PilR	445		Core	PF02954	HTH_8, Bacterial regulatory protein, Fis family	PF00072	Response regulator receiver domain	PF00158	Sigma-54 interaction domain

PA no.	Protein name	No. of amino acids	Subfamily	Nearest homologue	Core/ RGP	Domain 1	Domain 1 description	Domain 2	Domain 2 description	Other domains	Domain description
PA4581	RtcR	531		79% similar to regulator RtcR (*Escherichia coli*)	Core	PF00158	Sigma-54 interaction domain	PF06956	RtcR, Regulator of RNA terminal phosphate cyclase		
PA4726	CbrB	478		99% similar to regulator CbrB (*Azotobacter vinelandii* DJ; 90% similar to NifA (*Alcaligenes faecalis*)	Core	PF07728	AAA_5, ATPase family associated with various cellular activities (AAA)	PF00158	Sigma-54 interaction domain	PF00072	Response regulator receiver domain
PA4853	Fis	107		80% similar to Fis protein (*Escherichia coli*)	Core	PF02954	HTH_8, Bacterial regulatory protein, Fis family				
PA5125	NtrC	476		79% similar to *S .typhimurium* NtrC	Core	PF02954	HTH_8, Bacterial regulatory protein, Fis family	PF07728	AAA_5, ATPase family associated with various cellular activities (AAA)	PF00158; PF00072	Sigma-54 interaction domain; Response regulator receiver domain
PA5166		462		67% similar to C4-dicarboxylate transport system regulatory protein DctD (*Rhizobium meliloti*)	Core	PF00072	Response regulator receiver domain	PF00158	Sigma-54 interaction domain		
PA5483	AlgB	449		100% similar to AlgB (*Azotobacter vinelandi*); 97% response regulator *Desulfomicrobium baculatum* DSM 4028)	Core	PF02954	HTH_8, Bacterial regulatory protein, Fis family	PF00158	Sigma-54 interaction domain	PF00072	Response regulator receiver domain
PA5511	MifR	447		69% similar to DctD gene product (*R. meliloti*)	Core	PF00158	Sigma-54 interaction domain	PF00072	Response regulator receiver domain		
GntR											
PA0120		228	FadR	45% similar to putative transcriptional regulator LldR (*Escherichia coli*); Part of two gene operon, upstream ORF is probable dicarboxylate transporter	Core	PF07729	FCD, FCD domain.				
PA0121		244	FadR	44% similar to hypothetical transcriptional regulator (*Streptomyces ambofaciens*)	Core	PF07729	FCD, FCD domain.				
PA0253		158	MarR	55% similar to MarR regulator (*Salmonella typhimurium*)	Core	PF01047	MarR family				
PA0275		228	GntR	48% similar to hypothetical protein Rv3676 (*Mycobacterium tuberculosis*)	Core	PF00027	cNMP_binding, Cyclic nucleotide-binding domain.				
PA0424	MexR	147	MarR	48% similar to HosA protein (*Escherichia coli*)	Core	PF01047	MarR family				
PA0487		252		54% similar to molybdate uptake regulatory protein ModE (*Escherichia coli*)	Core	PF03459	TOBE domain to bind ligands	PF00126	HTH_1	PF03459	TOBE
PA0797		240	FadR	48% similar YdhC protein (*Bacillus subtilis*); three-gene operon, first gene, the other two genes are involved in citrate synthase activity	Core	PF07729	FCD, FCD domain.				
PA0942		186	MarR	51% similar to a region of regulatory protein PecS (*Erwinia chrysanthemi*)	Core	PF01047	MarR family				

PA1142		238	HutC	65% similar to PhnR, probable repressor protein in phnXWRSTUV locus for uptake and metabolism of 2-aminoethylphosphonate (*Salmonella typhimurium*)	Core	PF07702	UTRA, UTRA domain.	
PA1269		222	FadR	44% similar to putative transcriptional regulator (*Streptomyces coelicolor*)	Core	PF07729	FCD, FCD domain.	
PA1285		149	MarR	56% similar to hypothetical protein Rv1049(*Mycobacterium tuberculosis*)	Core	PF01047	MarR family	
PA1374		168	GntR superfamily	51% similar to hypothetical protein Rv3095 (*Mycobacterium tuberculosis*)	Core	PF01638	HxlR-like HTH	Weak MarR
PA1467		323		52% similar to putative repressor (*Streptomyces peucetius*)	Core			
PA1490		258		59% similar to repressor GlpR (*Pseudomonas aeruginosa*)	Core			
PA1520		259	FadR	48% similar to putative transcriptional regulator (*Streptomyces coelicolor*)	Core	PF07729	FCD, FCD domain.	
PA1526		219	GntR superfamily	46% similar to hypothetical protein (*Escherichia coli*); 1st gene in a 2-gene operon (2nd gene – Hypo)	Core		No discernible second domain	
PA1603		140	MarR	50% similar to HosA protein (*Escherichia coli*)	Core	PF01047	MarR family	
PA1627		215	FadR	70% similar to unknown ORF (*Pseudomonas marginalis* pv. *alfalfae*); first gene in a two-gene operon; the other gene is probable major facilitator superfamily (MFS) transporter	Core	PF07729	FCD, FCD domain.	
PA1653		158	MarR	61% similar to PetP protein *Rhodobacter capsulatus*	Core	PF01047	MarR family	
PA2032	YjiR	474	MocR	72% similar to putative regulator YjiR (*Escherichia coli*)	RGP 20	PF00155	Aminotransferase class I and II.	
PA2100		477	MocR	49% similar to hypothetical transcriptional regulator (*Rhodobacter sphaeroides*); Human homologue	RGP 21	PF00155	Aminotransferase class I and II.	
PA2299		249	HutC	53% similar to putative transcriptional regulator YvoA (*Bacillus subtilis*); third gene in a three-gene operon; The other two code for probable ferridoxin and oxidoreductase	Core	PF07702	UTRA, UTRA domain.	
PA2320	GntR, GnuR	343		56% similarity to GntR (*Escherichia coli*)	Core	PF00532	Peripla_BP_1, Periplasmic binding proteins and sugar binding domain of the LacI family.	
PA2692		174	Rrf2 subfamily	55% similar to hypothetical protein (*Bacillus halodurans*); Homologous to NsrR (*Pectobacterium carotovorum* subsp. carotovorum PC1)	Core	PF02082	Rrf2, Transcriptional regulator.	
PA2802		239	HutC	56% similar to probable repressor protein PhnR (*Salmonella typhimurium*)	Core	PF07702	UTRA, UTRA domain.	
PA2825	OspR	163	MarR	69% similar to hypothetical protein (*Acinetobacter* sp. ADP1)	Core	PF01047	MarR family	
PA2849	OhrR	151	MarR	74% similar to hypothetical protein (*Acinetobacter* sp. ADP1)	Core	PF01047	MarR family	
PA2897		479	MocR	59% similar to putative regulator YjiR (*Escherichia coli*); Human Homologue;	Core	PF00155	Aminotransferase class I and II.	
PA3067		147	MarR	50% similar to hypothetical protein Rv0880 (*Mycobacterium tuberculosis*)		PF01047	MarR family	

PA no.	Protein name	No. of amino acids	Subfamily	Nearest homologue	Core/ RGP	Domain 1	Domain 1 description	Domain 2	Domain 2 description	Other domains	Domain description
PA3249		298	HutC	60% similar to regulator PhnR (*Salmonella typhimurium*)	Core	PF07702	UTRA, UTRA domain.				
PA3341		144	MarR	62% similar to transcriptional regulator SlyA (*Salmonella typhimurium*)	Core	PF01047	MarR family				
PA3381	PhnF	240	HutC	58% similar to PhnF protein (*Escherichia coli*); 10th gene in a 13-gene operon	Core	PF07702	UTRA, UTRA domain.				
PA3458		157	MarR	48% similar to transcriptional regulator YkoM ((*Bacillus subtilis*))	Core	PF01047	MarR family				
PA3583	GlpR	251			Core						
PA3757		247	HutC	55% similar to putative transcriptional regulator YvoA ((*Bacillus subtilis*)); First gene in a 5-gene operon involved in N-acetylglucosamine catabolism	Core	PF07702	UTRA, UTRA domain.				
PA3815	IscR	163	Rrf2 subfamily	84% similar to unknown hypothetical protein (*A. vinelandii*)	Core	PF02082	Rrf2, Transcriptional regulator				
PA4132		471	MocR	64% similar to putative transcriptional regulator (*Escherichia coli*)	Core	PF00155	Aminotransferase class I and II.				
PA4135		140	MarR	73% similar to homoprotocatechuate degradation operon regulator HpcR (*Escherichia coli*)	Core	PF01047	MarR family				
PA4165		496	MocR	53% similar to putative regulatory protein MocR (*Rhizobium leguminosarum* bv. *viciae*); human homologue	Core	PF00155	Aminotransferase class I and II.				
PA4169		142	Rrf2 subfamily	57% similar to hypothetical protein YwnA (*Bacillus subtilis*)	Core	PF02082	Rrf2, Transcriptional regulator				
PA4185		242	FadR	47% similar to putative transcriptional regulator (*Streptomyces coelicolor*)	Core	PF07729	FCD, FCD domain.				
PA4764	Fur	134	GntR superfamily	77% similar to Regulatory protein Fur (*Escherichia coli*)	Core	PF01475	FUR, Ferric uptake regulator family				
PA4769	LldR	257	FadR	66% similar to transcriptional regulator for pyruvate dehydrogenase complex PdhR (*Escherichia coli*)	Core	PF07729	FCD, FCD domain.				
PA4906	VanR	237	FadR	66% similar to hypothetical protein AF009672 (*Acinetobacter* sp. ADP1); part of a two-gene operon, upstream gene is a probable dehydrogenase	Core	PF07729	FCD, FCD domain.				
PA5105	HutC	250	HutC	94% similar to hutC gene product of *Pseudomonas putida*; first gene in a two-gene operon (second gene – Hypo)	Core	PF07702	UTRA, UTRA domain.				
PA5157	MarR	156	MarR	60% similar to MarR protein (*Escherichia coli*)	Core	PF01047	MarR family				
PA5283		458	MocR	48% similar to putative transcriptional regulator (*Bacillus subtilis*) (Human homologue)	Core	PF00155	Aminotransferase class I and II.				
PA5356	GlcC	251	FadR	78% similar to GlcC protein (*Escherichia coli*); first gene in a three-gene operon; second Hyp and third 4-hydroxybenzoate-octaprenyl transferase	Core	PF07729	FCD, FCD domain.				

PA5431		491	MocR	52% similar to hypothetical regulator MocR (*Rhizobium meliloti*); first gene in a two-gene operon (second gene – Hypo)	Core	PF00155	Aminotransferase class I and II.		
PA5499	Np20	167		54% similar to zinc-uptake regulator Zur (*Escherichia coli*); PMID: 8971720; first gene in a three-gene operon; second and third involved in Zn transport	Core	PF01475	FUR, Ferric uptake regulator family.		
PA5525		246	FadR	49% similar to *nta* operon transcriptional regulator (*Escherichia coli*); second gene in a part of two-gene operon; downstream gene is a probable short-chain dehydrogenase	Core	PF07729	FCD, FCD domain.		
PA5550	GlmR	257		60% similar to regulator for glucitol operon, SrlR (*Escherichia coli*)	Core				
PA0268		473	MocR	55% similar to putative regulatory protein MocR (*Rhizobium leguminosarum* bv. *viciae*): Has a human homologue	Core	PF00155	Aminotransferase class I and II.		
H-NS (silencing)									
PA2667	MvaU	117		64% similar to P16 subunit of heteromeric transcriptional activator MvaT (*Pseudomonas mevalonii*)	Core				
PA4315	MvaT	124		82% similar to heteromeric transcriptional activator MvaT P16 subunit (*Pseudomonas mevalonii*)	Core	PB003306	Pfam-B_3306	PF00816.14	Histone_HNS
IclR									
PA0155	PcaR	279		88% similar to regulatory protein PcaR (*Pseudomonas putida*)	other				
PA0236		259		49% similar to repressor protein IclR (*Escherichia coli*)	Core				
PA1015		266		42% similar to glycerol operon regulatory protein GylR (*Streptomyces coelicolor*)	Core				
PA1630		288		58% similar to hypothetical protein (*Bordetella pertussis*)	Core				
PA2010		267		49% similar to putative transcriptional regulatory protein OhbR (*Pseudomonas aeruginosa*)	Core				
PA3174		242		0	Core				
PA3508		277		57% similar to hypothetical protein (*Acinetobacter* sp. ADP1)	RGP34				
PA4157		262		51% similar to putative transcriptional regulatory protein OhbR (*Pseudomonas aeruginosa*)	Core				
PA4341		256		45% similar to transcriptional regulator for mhp operon (*Escherichia coli*)	Core				
LacI/GalR									
PA1949	RbsR	337		62% similar to regulator for rbs operon RbsR (*Escherichia coli*)	Core				
PA2259	PtxS	340		0	Core				
PA3184	HexR	285		76% similar to HexR protein (*Escherichia coli*)	Core	PF01418	RpiR family	PF01380	SIS (Sugar ISomerase) domain

PA no.	Protein name	No. of amino acids	Subfamily	Nearest homologue	Core/ RGP	Domain 1	Domain 1 description	Domain 2	Domain 2 description	Other domains	Domain description
PA3563	FruR	329		62% similar to transcriptional repressor of fru operon FruR (*Escherichia coli*)	Core						
PA4596		172		74% similar to nfxB gene product of (*Pseudomonas aeruginosa*)	Core						
PA4600	NfxB	187		95% similar to nfxB (excluding frameshift) (*P. aeruginosa*)	Core						
PA5438		293		63% similar to putative regulator HexR (*Pseudomonas aeruginosa*)	Core	PF01418	RpiR family	PF01380	SIS (Sugar ISomerase) domain		
LuxR											
PA0034		207		75% similar to BvgA-positive transcription regulator, putative (*Bordetella pertussis*)	Core	PF00072	Response regulator	PF00196	GerE		
PA0533		496		44% similar to a region of putative sensory transduction histidine kinase (*Methanobacterium thermoautotrophicum*)	Core	PF00196	GerE	PF00989	PAS	PF08448; PB001360; PB002546	PAS_4; Pfam-B_1360; Pfam-B_2546
PA0601		210		66% similar to VsrD protein (*Burkholderia solanacearum*)	Core	PF00072	Response regulator	PF08266	GerE	PB001041	Pfam-B_1041
PA1136		243		42% similar to transcriptional activator LasR (*Pseudomonas aeruginosa*)	Core	PF03472	Autoinducer binding	PF00196	GerE		
PA1347		230		65% similar to a region of putative regulator FimZ (*Salmonella typhimurium*)	Core	PF00196	GerE	PF08448	PAS_4		
PA1397		210		56% similar to nitrate/nitrite response regulator NarP (sensor NarQ) (*Escherichia coli*)	Core	PF00072	Response regulator	PF00196	GerE		
PA1430	LasR	239		100% identical to transcription activator LasR (*Pseudomonas aeruginosa*)	Core	PF03472	Autoinducer binding	PF00196	GerE		
PA1484		268		50% similar to DMSO reductase regulatory protein DorX (*Rhodobacter sphaeroides*)	Core	PF00196	GerE	PB000146	Pfam-B_146		
PA1759		901		44% similar to regulator AcoK (*Klebsiella pneumoniae*)	Core	PF00196	GerE	PB008287	Pfam-B_8287		
PA1760		907		44% similar to regulator AcoK (*Klebsiella pneumoniae*)	Core	PF00196	GerE	PB001429	Pfam-B_1429	PB008287	Pfam-B_8287
PA1898	QscR, PhzR	237		29% identity, 46% similarity to LasR (*Pseudomonas aeruginosa*);27% identity, 45% similarity to LuxR (*Vibrio fischeri*); 33% identity, 49% similarity to SolR (*Ralstonia solanacearum*); 49% similar to positive regulatory protein PhzR (*Pseudomonas aureofaciens*); 53% similar, 32% identity to regulatory protein RhlR (*Pseudomonas aeruginosa*)	Core	PF03472	Autoinducer binding	PF00196	GerE		
PA1978	ErbR, AgmR	221		100% identical to *agmR* gene product of (*P. aeruginosa*)	Core	PF00072	Response regulator	PF00196	GerE		
PA1980	EraR, ExaE	225		57% similar to uhpA gene product of (*Salmonella typhimurium*)	Core	PF00072	Response regulator	PF00196	GerE		
PA2376		213		54% similar to nitrate/nitrite regulatory protein NarL (*Pseudomonas aeruginosa*)	Core	PF00072	Response regulator	PF00196	GerE		
PA2586	GacA	214		99% similar to response regulator GacA (*Pseudomonas aeruginosa*)	Core	PF00072	Response regulator	PF00196	GerE	PB000645	Pfam-B_645

PA2591	VqsR	268	46% similar to DMSO reductase regulatory protein DorX (*Rhodobacter sphaeroides*)	Core	PF00196	GerE	PB000146	Pfam-B_146		
PA2899		212	53% similar to biosynthesis positive regulator RcsB (*Escherichia coli*)	Core	PF00072	Response regulator	PF00196	GerE		
PA3045		207	77% similar to putative transcription regulator BvgA (*Bordetella pertussis*)	Core	PF00072	Response regulator	PF00196	GerE		
PA3420		827		Core	PF00196	GerE	PB010363	Pfam-B_10363		
PA3477	RhlR	241		Core	PF03472	Autoinducer binding	PF00196	GerE		
PA3599		261	47% similar to dorX gene product of (*Rhodobacter sphaeroides*)	Core	PF00196	GerE	PB000146	Pfam-B_146	PB000645	Pfam-B_645
PA3604	ErdR	217	61% similar to glycerol metabolism activator (AGMR protein) (*P. aeruginosa*)	Core	PF00072	Response regulator	PF00196	GerE		
PA3714		213	62% similar to RcsB protein (*Erwinia amylovora*)	Core	PF00072	Response regulator	PF00196	GerE		
PA3771		325	54% similar to a region of putative regulatory protein (*Streptomyces coelicolor*)	Core	PF00196	GerE	PB005070	Pfam-B_5070		
PA3879	NarL	219	74% similar to *E. coli* NarL protein.	Core	PF00072	Response regulator	PF00196	GerE		
PA3921		906	42% similar to regulator AcoK (*Klebsiella pneumoniae*)	Core	PF00196	GerE	PB001429	Pfam-B_1429	PB001429; PB004550; PB009525; PB010363	Pfam-B_1429; Pfam-B_4550; Pfam-B_9525; Pfam-B_10363
PA3948	RocA1	209	74% similar to *bvgA* gene product of (*Bordetella pertussis*)	Core	PF00072	Response regulator	PF00196	GerE		
PA4074		222	54% similar to regulator protein MoaR (*Klebsiella aerogenes*)	Core	PF03472	Autoinducer binding	PF00196	GerE	PB000146	Pfam-B_146
PA4080	RcsB	214	63% similar to rcsB gene product of (*Escherichia coli*)	Core	PF00072	Response regulator	PF00196	GerE		
PA4196	bfiR	214	62% similar to transcriptional regulatory protein FixJ (*Azorhizobium caulinodans*)	Core	PF00196	GerE, Bacterial regulatory proteins, luxR family	PF08281	Sigma70_r4_2, Sigma-70, region 4	PF00072	Response regulator receiver domain
PA4296	PprB	275	54% similar to NtrC-type response regulator (*Eubacterium acidaminophilum*); 43% similar to regulatory components of sensory transduction system *Synechocystis* sp.	Core	PF00072	Response regulator	PF00196	GerE, Bacterial regulatory proteins, LuxR family		
PA4806		227	46% similar to DMSO reductase regulatory protein DorX (*Rhodobacter sphaeroides*)	Core	PF00196	GerE				
LysR										
PA0032		304	49% similar to glycine cleavage system transcriptional activator protein GcvA (*Escherichia coli*)	Core	PF00126	HTH_1	PF03466	LysR_substrate		
PA0037	TrpI	295	97% similarity to TrpI protein (*Pseudomonas aeruginosa*)	Core	PF00126	HTH_1	PF03466	LysR_substrate		
PA0056		306	55% similar to regulator GstR (*Rhizobium leguminosarum*)	Core	PF00126	HTH_1	PF03466	LysR_substrate		
PA0123		302	51% similar to transcriptional activator PtxR (*Pseudomonas aeruginosa*)	Core	PF00126	HTH_1	PF03466	LysR_substrate		
PA0133		305	47% similar to transcriptional regulator OxyR (*Mycobacterium xenopi*)	Core	PF00126	HTH_1	PF03466	LysR_substrate		

PA no.	Protein name	No. of amino acids	Subfamily	Nearest homologue	Core/ RGP	Domain 1	Domain 1 description	Domain 2	Domain 2 description	Other domains	Domain description
PA0152	PcaQ	275		56% similar to regulatory protein PcaQ (*Agrobacterium tumefaciens*)	Core	PF00126	HTH_1	PF03466	LysR_substrate		
PA0159		312		52% similar to regulator GstR (*Bradyrhizobium japonicum*)	Core	PF00126	HTH_1	PF03466	LysR_substrate		
PA0181		310		48% similar to putative transcriptional regulator YeaT (*Escherichia coli*)	Core	PF00126	HTH_1	PF03466	LysR_substrate		
PA0191		305		59% similar to transcriptional control factor SdsB *Pseudomonas* sp.	Core	PF00126	HTH_1	PF03466	LysR_substrate		
PA0207		297		51% similar to glycine cleavage activator protein GcvA (*Escherichia coli*)	RGP 1	PF00126	HTH_1	PF03466	LysR_substrate		
PA0217	MdcR	309		65% similar to malonate decarboxylase operon regulator MdcR (*Klebsiella pneumoniae*)	Core	PF00126	HTH_1	PF03466	LysR_substrate		
PA0218		306		49% similar to putative transcriptional regulator YvbU ((*Bacillus subtilis*))	Core	PF00126	HTH_1	PF03466	LysR_substrate		
PA0233		313		61% similar to putative transcriptional regulator (*Rhizobium leguminosarum* bv. *trifolii*)	Core	PF00126	HTH_1	PF03466	LysR_substrate		
PA0272		310		44% similar to MexT protein (*Pseudomonas aeruginosa*)	Core	PF00126	HTH_1	PF03466	LysR_substrate		
PA0289	GpuR	320		39% similar to LysR-type regulatory protein (*Comamonas* sp. JS765)	Core	PF00126	HTH_1	PF03466	LysR_substrate		
PA0448		302		55% similar to GcvA gene product of *E. coli*	Core	PF00126	HTH_1	PF03466	LysR_substrate		
PA0477		308		47% similar to DNA-binding protein NahR *Pseudomonas putida*; 48% similar to MexT protein (*Pseudomonas aeruginosa*)	Core	PF00126	HTH_1	PF03466	LysR_substrate		
PA0479		316		46% similar to putative transcriptional regulator YhaJ (*Escherichia coli*)	Core	PF00126	HTH_1	PF03466	LysR_substrate		
PA0491		308		46% similar to oxidative stress transcriptional regulator OxyR (*Xanthomonas campestris*)	Core	PF00126	HTH_1	PF03466	LysR_substrate		
PA0528		292		49% similar to positive regulator CynR (*Escherichia coli*)	Core	PF00126	HTH_1	PF03466	LysR_substrate		
PA0701		328		43% similar to iron-regulated virulence regulator (*Vibrio cholerae*)	Core	PF00126	HTH_1	PF03466	LysR_substrate		
PA0708		284		53% similar to putative transcriptional regulator YneJ (*Escherichia coli*)	Core	PF00126	HTH_1	PF03466	LysR_substrate		
PA0739		304		75% similar to SdsB (regulator of sdsA) *Pseudomonas* sp.	Core	PF00126	HTH_1	PF03466	LysR_substrate		
PA0784		309		50% similar to glycine cleavage system transcriptional activator GcvA (*Haemophilus influenzae* Rd)	Core	PF00126	HTH_1	PF03466	LysR_substrate		
PA0815		317		46% similar to putative transcriptional regulator YjiE (*Escherichia coli*)	Core	PF00126	HTH_1	PF03466	LysR_substrate		
PA0816		294		60% similar to Brg1 (*Bordetella pertussis*), a transcriptional regulator that is controlled by the oxygen-responsive transcriptional regulator Btr under anaerobic conditions	Core	PF00126	HTH_1	PF03466	LysR_substrate		

PA0876		314	54% similar to putative transcriptional regulator *Synechocystis* sp.	Core	PF00126	HTH_1	PF03466	LysR_substrate
PA0877		298	60% similar to hypothetical protein (*Acinetobacter lwoffii* K24)	Core	PF00126	HTH_1	PF03466	LysR_substrate
PA1003	MvfR, PqsR	332	99% identical to previously identified hypothetical protein (*Pseudomonas aeruginosa*)	Core	PF00126	HTH_1	PF03466	LysR_substrate
PA1067		304	54% similar to regulator GstR (*Bradyrhizobium japonicum*)	Core	PF00126	HTH_1	PF03466	LysR_substrate
PA1128		303	60% similar to YcaN, hypothetical transcriptional regulator (*Escherichia coli*)	Core	PF00126	HTH_1	PF03466	LysR_substrate
PA1138		291	49% similar to transcription activator (*Azospirillum brasilense*)	Core	PF00126	HTH_1	PF03466	LysR_substrate
PA1141		300	51% similar to probable transcriptional regulator LrhA (*Escherichia coli*)	Core	PF00126	HTH_1	PF03466	LysR_substrate
PA1145		300	40% similar to positive transcriptional control factor SdsB *Pseudomonas* sp.	Core	PF00126	HTH_1	PF03466	LysR_substrate
PA1184		296	59% similar to glycine cleavage activator protein GcvA (*Escherichia coli*)	Core	PF00126	HTH_1	PF03466	LysR_substrate
PA1201		304	60% similar to putative transcriptional regulator *Synechocystis* sp.	Core	PF00126	HTH_1	PF03466	LysR_substrate
PA1223		297	53% similar to regulator SyrM1 (*Rhizobium* sp. NGR234)	RGP 11	PF00126	HTH_1	PF03466	LysR_substrate
PA1264		288	47% similar to regulator HexA (*Erwinia carotovora* subsp. *carotovora*)	Core	PF00126	HTH_1	PF03466	LysR_substrate
PA1309		286	52% similar to putative transcriptional regulator YwfK ((*Bacillus subtilis*))	Core	PF00126	HTH_1	PF03466	LysR_substrate
PA1312		303	57% similar to PtxR (*Pseudomonas aeruginosa*)	Core	PF00126	HTH_1	PF03466	LysR_substrate
PA1328		302	55% similar to a region of putative transcriptional regulator YcaN (*Escherichia coli*)	Core	PF00126	HTH_1	PF03466	LysR_substrate
PA1399		306	49% similar to putative regulatory protein (*Pseudomonas aeruginosa*)	Core	PF00126	HTH_1	PF03466	LysR_substrate
PA1413	YneJ	290	69% similar to putative transcriptional regulator YneJ (*Escherichia coli*)	Core	PF00126	HTH_1	PF03466	LysR_substrate
PA1422	GbuR	297	49% similar to regulatory protein cfxR (*Alcaligenes eutrophus*)	Core	PF00126	HTH_1	PF03466	LysR_substrate
PA1570		293	54% similar to transcriptional regulator protein RbcR (*Chromatium vinosum*)	Core	PF00126	HTH_1	PF03466	LysR_substrate
PA1738		304	62% similar to putative transcriptional regulator YcaN (*Escherichia coli*)	Core	PF00126	HTH_1	PF03466	LysR_substrate
PA1754	CysB	324		Core	PF00126	HTH_1	PF03466	LysR_substrate
PA1826		301	50% similar to regulator of catechol degradation (CatM) (*Acinetobacter calcoaceticus*)		PF00126	HTH_1	PF03466	LysR_substrate

PA no.	Protein name	No. of amino acids	Subfamily	Nearest homologue	Core/ RGP	Domain 1	Domain 1 description	Domain 2	Domain 2 description	Other domains	Domain description
PA1853		287		80% similar to ORF286, hypothetical protein in the 30-kb denitrification gene cluster (*Pseudomonas stutzeri*)	Core	PF00126	HTH_1	PF03466	LysR_substrate		
PA1859		295		69% similar to putative transcriptional regulator YafC (*Escherichia coli*)	Core	PF00126	HTH_1	PF03466	LysR_substrate		
PA1915		515			Core	PF07905	PucR				
PA1961		311		50% similar to putative transcriptional regulator YafC (*Escherichia coli*)	Core	PF00126	HTH_1	PF03466	LysR_substrate		
PA1998	DhcR	306		52% similar to ybhD, yfeR, and yfhT hypothetical transcriptional regulatory proteins (*Escherichia coli*)	Core	PF00126	HTH_1	PF03466	LysR_substrate		
PA2054	CynR	295		66% similar to cynR gene product of (*Escherichia coli*)	RGP 20	PF00126	HTH_1	PF03466	LysR_substrate		
PA2056		300		54% similar to regulator GstR (*Bradyrhizobium japonicum*)	RGP 20	PF00126	HTH_1	PF03466	LysR_substrate		
PA2076		300		44% similar to putative LysR-type transcriptional regulator *Pseudomonas putida*	Core	PF00126	HTH_1	PF03466	LysR_substrate		
PA2115		317		47% similar to putative transcriptional regulator YrdL (*Bacillus subtilis*)	Core	PF00126	HTH_1	PF03466	LysR_substrate		
PA2121		301		50% similar to transcriptional activator nahR of (*Pseudomonas putida*-plasmid)	Core	PF00126	HTH_1	PF03466	LysR_substrate		
PA2123		310		55% similar to putative transcriptional regulator YhcS (*Haemophilus influenzae* Rd)	Core	PF00126	HTH_1	PF03466	LysR_substrate		
PA2206		315		54% similar to galactose binding protein regulator (*Azospirillum brasilense*)	Core	PF00126	HTH_1	PF03466	LysR_substrate		
PA2220	OprR	306		Identical to putative transcriptional regulator OprR (*Pseudomonas aeruginosa*)	RGP 23	PF00126	HTH_1	PF03466	LysR_substrate		
PA2258	PtxR	312			Core	PF00126	HTH_1	PF03466	LysR_substrate		
PA2267		298		45% similar to putative transcriptional activator MauR (*Paracoccus denitrificans*)	Core	PF00126	HTH_1	PF03466	LysR_substrate		
PA2316		297		55% similar to hypothetical *yafC* gene product of (*Escherichia coli*)	Core	PF00126	HTH_1	PF03466	LysR_substrate		
PA2334		303		56% similar to positive transcriptional control factor SdsB *Pseudomonas* sp.	RGP 53	PF00126	HTH_1	PF03466	LysR_substrate		
PA2383		306		50% similar to positive regulator of gcv operon GcvA (*Escherichia coli*)	Core	PF00126	HTH_1	PF03466	LysR_substrate		
PA2417		312		53% similar to transcriptional regulator PtxR (*Pseudomonas aeruginosa*)	Core	PF00126	HTH_1	PF03466	LysR_substrate		
PA2432	BexR	304		55% similar to putative transcriptional regulator *Synechocystis* sp.	Core	PF00126	HTH_1	PF03466	LysR_substrate		
PA2447		307		51% similar to hypothetical protein (*Haemophilus influenzae*)	Core	PF00126	HTH_1	PF03466	LysR_substrate		

PA2469		304	52% similar to galactose binding protein regulator (*Azospirillum brasilense*)	Core	PF00126	HTH_1	PF03466	LysR_substrate
PA2492	MexT	347	100% similar to a region of MexT protein (*Pseudomonas aeruginosa*)	Core	PF00126	HTH_1	PF03466	LysR_substrate
PA2497		292	52% similar to transcriptional regulator HexA (*Erwinia carotovora* subsp. *atroseptica*)	Core	PF00126	HTH_1	PF03466	LysR_substrate
PA2510	CatR	290	81% similar to catR regulatory protein *Pseudomonas putida*	Core	PF00126	HTH_1	PF03466	LysR_substrate
PA2534		303	58% similar to putative transcriptional regulator YhjC (*Escherichia coli*)	Core	PF00126	HTH_1	PF03466	LysR_substrate
PA2547		305	56% similar to regulator GstR (*Bradyrhizobium japonicum*)	Core	PF00126	HTH_1	PF03466	LysR_substrate
PA2551		310	54% similar to transcriptional activator NahR (*Pseudomonas putida* plasmid NAH7)	Core	PF00126	HTH_1	PF03466	LysR_substrate
PA2601		298	46% similar to activator, hydrogen peroxide-inducible genes, OxyR (*Escherichia coli*)	Core	PF00126	HTH_1	PF03466	LysR_substrate
PA2681		309	53% similar to regulator PtxR (*Pseudomonas aeruginosa*)	Core	PF00126	HTH_1	PF03466	LysR_substrate
PA2758		295	46% similar to DNA binding protein HpkR (*Pseudomonas syringae* pv. syringae)	Core	PF00126	HTH_1	PF03466	LysR_substrate
PA2834		319	47% similar to putative regulatory protein (*Pseudomonas aeruginosa*)	Core	PF00126	HTH_1	PF03466	LysR_substrate
PA2838		323	53% similar to regulator GstR (*Bradyrhizobium japonicum*)	Core	PF00126	HTH_1	PF03466	LysR_substrate
PA2846		284	56% similar to 2,2-dialkylglycine decarboxylase repressor protein (*Burkholderia cepacia*)	Core	PF00126	HTH_1	PF03466	LysR_substrate
PA2848		339	49% similar to regulator GstR (*Bradyrhizobium japonicum*)	Core	PF00126	HTH_1	PF03466	LysR_substrate
PA2877		297	62% similar to putative transcriptional regulator YhjC (*Escherichia coli*)	Core	PF00126	HTH_1	PF03466	LysR_substrate
PA2879	HpkR	296	74% similar to putative DNA binding protein HpkR (*Pseudomonas syringae* pv. *syringae*)	Core	PF00126	HTH_1	PF03466	LysR_substrate
PA2921		329	50% similar to regulator TsaR (*Comamonas testosteroni*)	Core	PF00126	HTH_1	PF03466	LysR_substrate
PA2930		308	52% similar to putative regulatory protein TrpI (*Pseudomonas syringae*)	Core	PF00126	HTH_1	PF03466	LysR_substrate
PA3122	YibL	297	93% similar to hypothetical gene product YibL (*Azotobacter vinelandii*)	Core	PF00126	HTH_1	PF03466	LysR_substrate
PA3124		303	59% similar to hypothetical gene product YafC (*Escherichia coli*)	Core	PF00126	HTH_1	PF03466	LysR_substrate
PA3135		306	55% similar to putative transcriptional regulator *Synechocystis* sp.	Core	PF00126	HTH_1	PF03466	LysR_substrate
PA3225		309	48% similar to putative transcriptional regulator Yeey (*Escherichia coli*)	Core	PF00126	HTH_1	PF03466	LysR_substrate
PA3321		304	52% similar to regulatory protein MexT (*Pseudomonas aeruginosa*)	Core	PF00126	HTH_1	PF03466	LysR_substrate

PA no.	Protein name	No. of amino acids	Subfamily	Nearest homologue	Core/ RGP	Domain 1	Domain 1 description	Domain 2	Domain 2 description	Other domains	Domain description
PA3398		308		59% similar to putative transcriptional regulator YeiE (*Escherichia coli*)	Core	PF00126	HTH_1	PF03466	LysR_substrate		
PA3433	Ywbl	297		59% similar to putative transcriptional regulator Ywbl ((*Bacillus subtilis*))	Core	PF00126	HTH_1	PF03466	LysR_substrate		
PA3565		306		50% similar to GstR (*Rhizobium leguminosarum*)	Core	PF00126	HTH_1	PF03466	LysR_substrate		
PA3587	MetR	306		63% similar to regulator MetR (*Escherichia coli*)	Core	PF00126	HTH_1	PF03466	LysR_substrate		
PA3594		295		55% similar to putative transcriptional regulator YnfL (*Escherichia coli*)	Core	PF00126	HTH_1	PF03466	LysR_substrate		
PA3630		302		53–56% similar to numerous transcriptional regulators	Core	PF00126	HTH_1	PF03466	LysR_substrate		
PA3711		301		43% similar to putative transcriptional regulator YgiP (*Escherichia coli*)	Core	PF00126	HTH_1	PF03466	LysR_substrate		
PA3776		302		52% similar to regulator BudR (*Klebsiella terrigena*)	Core	PF00126	HTH_1	PF03466	LysR_substrate		
PA3778		311		46% similar to regulator CbbRI (*Rhodobacter capsulatus*)	Core	PF00126	HTH_1	PF03466	LysR_substrate		
PA3845		298		52% similar to positive regulator GcvA (*Escherichia coli*)	Core	PF00126	HTH_1	PF03466	LysR_substrate		
PA3895		317		55% similar to putative transcriptional regulator *Synechocystis* sp.	Core	PF00126	HTH_1	PF03466	LysR_substrate		
PA3995		297		51% similar to the HexA gene product of (*Erwinia carotovora* subsp. atroseptica)	Core	PF00126	HTH_1	PF03466	LysR_substrate		
PA4109	AmpR	296			Core	PF00126	HTH_1	PF03466	LysR_substrate		
PA4145		296		48% similar to transcriptional activator PtxR (*Pseudomonas aeruginosa*)	Core	PF00126	HTH_1	PF03466	LysR_substrate		
PA4174		307		55% similar to putative transcriptional regulator YhjC (*Escherichia coli*)	Core	PF00126	HTH_1	PF03466	LysR_substrate		
PA4203		284		61% similar to putative transcriptional regulator YneJ (*Escherichia coli*)	Core	PF00126	HTH_1	PF03466	LysR_substrate		
PA4363	IciA	300		56% similar to inhibitor of chromosome initiation IciA (*Escherichia coli*)	Core	PF00126	HTH_1	PF03466	LysR_substrate		
PA4902		298		51% similar to putative regulatory protein (*Pseudomonas stutzeri*)	Core	PF00126	HTH_1	PF03466	LysR_substrate		
PA4914		312		53% similar to positive regulator GcvA (*Escherichia coli*)	Core	PF00126	HTH_1	PF03466	LysR_substrate		
PA4989		294		46% similar to DNA binding protein HpkR (*Pseudomonas syringae* pv. *syringae*)	Core	PF00126	HTH_1	PF03466	LysR_substrate		
PA5029	YnfL	301		72% similar to putative transcriptional regulator YnfL (*Escherichia coli*)	Core	PF00126	HTH_1	PF03466	LysR_substrate		
PA5085		318		47% similar to regulatory protein LysR (*Escherichia coli*)	Core	PF00126	HTH_1	PF03466	LysR_substrate		
PA5179		293		47% similar to putative LysR-type transcriptional regulator (*Mycobacterium leprae*)	Core	PF00126	HTH_1	PF03466	LysR_substrate		

PA5189		302	51% similar to putative regulatory protein (*Pseudomonas aeruginosa*)	Core	PF00126	HTH_1	PF03466	LysR_substrate		
PA5218		304	57% similar to hypothetical ynfL gene product of (*Escherichia coli*)	Core	PF00126	HTH_1	PF03466	LysR_substrate		
PA5293		305	53% similar to positive regulator GcvA (*Escherichia coli*)	Core	PF00126	HTH_1	PF03466	LysR_substrate		
PA5344	OxyR	310	60% similar to oxyR gene product of (*Escherichia coli*)	Core	PF00126	HTH_1	PF03466	LysR_substrate		
PA5382	YeiE	297	64% similarity to putative transcriptional regulator YeiE (*Escherichia coli*)	Core	PF00126	HTH_1	PF03466	LysR_substrate		
PA5428		302	56% similar to putative transcriptional regulator YjiE (*Escherichia coli*)	Core	PF00126	HTH_1	PF03466	LysR_substrate		
PA5437		311	62% similar to regulator CbbRI (*Rhodobacter capsulatus*)	Core	PF00126	HTH_1	PF03466	LysR_substrate		
PA5261	AlgR	248	88% similar to encystment and alginate biosynthesis regulatory protein (*Azotobacter vinelandii*)	Core	PF00072	Response regulator receiver domain	PF04397	LytTR, LytT DNA-binding domain		
MerR										
PA2016	LiuR, GnyR	134	85% identical to transcriptional regulator, putative (*P. putida* KT2440); 59% similar to putative transcriptional regulator YbbI (*Escherichia coli*)	Core	PF00376	MerR	PF09278	MerR-DNA-bind		
PA2273	SoxR	156	Identical to unpublished SoxR of *P. aeruginosa* (gi|2495412|sp|Q51506|); 77% similar to SoxR gene product of (*Escherichia coli*)	Core	PB000200	Pfam-B_200	PF00376	MerR	PB016009; PF09278	Pfam-B_16009; MerR-DNA-binding
PA2718		162	52% similar to putative transcriptional regulator YbbI (*Escherichia coli*)	Core	PF00376	MerR	PF09278	MerR-DNA-bind		
PA3689	YhdM	156	66% similar to putative regulatory protein YhdM (*Haemophilus influenzae*)	Core	PF00376	MerR	PF09278	MerR-DNA-bind		
PA4659		299	47% similar to putative transcriptional regulator YcgE (*Escherichia coli*)	Core	PB000200	Pfam-B_200	PF00376	MerR		
PA4778	CueR, YbbI	132	65% similar to putative transcriptional regulator YbbI (*Escherichia coli*)	Core	PF00376	MerR	PF05103	DivIVA	PF09278	MerR-DNA-binding
PA4878		270	47% similar to a region of multidrug efflux transporter regulator TipA (*Streptomyces lividans*)	Core	PF00376	MerR	PF06445	AraC-binding domain		
PA5116		141	49% similar to activator SoxR (*Escherichia coli*)	Core	PF00376	MerR	PF09278	MerR-DNA-bind		
OmpC family										
PA5200	AmgR	247	OmpR similar *E. coli*	Core	PF00486	Transcriptional regulatory protein, C terminal	PF00072	Response regulator receiver domain		
OmpR-PhoP class										
PA4032		238	52% similar to hypothetical protein Rv1033c (response regulator homolog TrcR) (*Mycobacterium tuberculosis*); 51% similar to phosphate regulatory protein PhoB (*Bradyrhizobium japonicum*)		PF0072	Response regulator receiver domain	PF00486	Transcriptional regulatory protein, C terminal		

PA no.	Protein name	No. of amino acids	Subfamily	Nearest homologue	Core/ RGP	Domain 1	Domain 1 description	Domain 2	Domain 2 description	Other domains	Domain description
PA4381		227		91% similar to ColR (*Pseudomonas fluorescens*)	Core	PF00072	Response regulator receiver domain	PF00486	Transcriptional regulatory protein, C terminal		
RpiR											
PA5506		285		44% similar to RpiR protein (*Escherichia coli*)	Core	PF01418	RpiR family	PF01380	SIS (Sugar ISomerase) domain		
TetR											
PA0167		221		60% similar to putative regulator YcdC (*Escherichia coli*)	Core	PF08362	TetR_C_3, YcdC-like protein, C-terminal region				
PA0243		222		44% similar to putative transcriptional regulator (*Methanobacterium thermoautotrophicum*)	Core	PF00440	TetR_N, Bacterial regulatory proteins, tetR family				
PA0294	AguR	221		49% similar to putative transcriptional regulator (*Streptomyces coelicolor*)	Core	PF00440	TetR_N, Bacterial regulatory proteins, tetR family				
PA0367		215		53% similar to hypothetical protein *Pseudomonas putida*	Core						
PA0436		206		54% similar to putative tet operon regulator YcdC (*Escherichia coli*)	Core	PF08362	TetR_C_3, YcdC-like protein, C-terminal region				
PA0475		189		47% similar to hypothetical protein Rv0196 (*Mycobacterium tuberculosis*)	Core						
PA0828		227		50% similar to a region of putative transcriptional regulator (*Aquifex aeolicus*)	Core						
PA0839		196		48% similar to hypothetical protein (*Escherichia coli*)	Core						
PA1226		202		45% similar to putative transcriptional regulator (*Streptomyces coelicolor*)	Core						
PA1241		186		46% similar to hypothetical protein YxaF (*Bacillus subtilis*)	Core						
PA1283		186		48% similar to hypothetical protein *Synechocystis* sp.	Core						
PA1290		197		54% similar to unidentified reading frame L (ORFL TetC protein) in transposon Tn10 (*Escherichia coli*)	Core						
PA1315		204		49% similar to MtrR protein (*Neisseria gonorrhoeae*)	Core						
PA1403		210		43% similar to a region of hypothetical protein Rv1963c (*Mycobacterium tuberculosis*)	Core						
PA1504		216		57% similar to putative tet operon regulator YcdC (*Escherichia coli*)	Core	PF08362	TetR_C_3, YcdC-like protein, C-terminal region				
PA1539		282		53% similar to a region of hypothetical protein YijC (*Escherichia coli*)		PF00440					

PA1836		193	45% similar to putative repressor (*Bacillus megaterium*)	Core		
PA1864		216	54% similar to putative tet operon regulator YcdC (*Escherichia coli*)	Core	PF08362	TetR_C_3, YcdC-like protein, C-terminal region
PA2020		210	72% similar to transcriptional regulator AmrR (*Burkholderia pseudomallei*)	Core	PF08361	TetR_C_2, MAATS-type transcriptional repressor, C-terminal region
PA2196		194	63% similar to hypothetical protein (*Escherichia coli*)	Core		
PA2270		196	52% similar to hypothetical protein *Synechocystis* sp.	Core		
PA2484		204	65% similar to SocA3 protein *Myxococcus xanthus*	Core		
PA2766		193		Core		
PA2885	AtuR	198	50% similar to hypothetical protein Rv3557c (*Mycobacterium tuberculosis*)	Core		
PA2931	CifR	196	63% similar to socA3 protein *Myxococcus xanthus*	Core		
PA2957		212	44% similar to putative transcriptional regulator (*Streptomyces coelicolor*)	Core		
PA3006	PsrA	233	90% similar to PsrA *Pseudomonas putida*	Core		
PA3034		185	54% similar to putative transcriptional regulator *Streptomyces lividans*	Core		
PA3133		180		Core		
PA3574	NalD	212	46% similar to acrR gene product, a probable transcriptional regulator from (*Escherichia coli*)	Core	PF08361	TetR_C_2, MAATS-type transcriptional repressor, C-terminal region
PA3678		212	47% similar to putative transcriptional regulator (*Streptomyces coelicolor*)	Core		
PA3699		237	65% similar to a region of hypothetical protein Rv3066 (*Mycobacterium tuberculosis*)	Core		
PA3721	NalC	213	43% similar to putative transcriptional regulator (*Streptomyces coelicolor*)	Core		
PA3973		215	45% similar to hypothetical protein Rv3557c (*Mycobacterium tuberculosis*)	Core		
PA4831		186	54% similar to a region of acrAB operon repressor AcrR (*Escherichia coli*)	Core		
PA4890	DesT	209	52% similar to hypothetical protein YijC (*Escherichia coli*)	Core		
PA4984		215	46% similar to regulatory protein BetI (*Escherichia coli*)	Core		
PA5059		205	86% similar to PhaD (*Pseudomonas* sp.); 99% similar to hypothetical protein 3 (phaC2 3′ region)	Core		
PA5374	BetI	197	71% similar to *E. coli betI* gene product	Core		

PA no.	Protein name	No. of amino acids	Subfamily	Nearest homologue	Core/ RGP	Domain 1	Domain 1 description	Domain 2	Domain 2 description	Other domains	Domain description
CRP											
PA0652	Vfr	214			Core	PF00325	Crp	PF00027.22	cNMP_binding		
PA1544	Anr	244		100% identical to transcriptional activator Anr (*Pseudomonas aeruginosa*)	Core	PF00325	Crp	PF00027.22	cNMP_binding		
Unclassified											
PA0116		157		65% similar to hypothetical protein Rv3095 (*Mycobacterium tuberculosis*)	Core	PF01638.10	HxlR				
PA0463	CreB	229		68% similar to *E. coli* catabolic regulation response regulator *creB* gene product; 71% similar to *Aeromonas jandaei* response regulator protein	Core	PF00486	Transcriptional regulatory protein, C terminal	PF00072	Response regulator receiver domain		
PA0527	Dnr	227		100% identical to Dnr protein (*Pseudomonas aeruginosa*)	Core	PF00027	cNMP_binding				
PA0610	PrtN	104			Core	PF11112.1	PyocinActivator				
PA0707	ToxR, RegA	259		99% identity to ToxR (exotoxin A regulatory protein) (*P. aeruginosa*)	Core	PF07720	TPR_3, tetratricopeptide repeat				
PA0756		223		72% similarity to trans-activating transcriptional regulatory protein tctD (*Salmonella typhimurium*)	Core	PF00486		PF00072	Response regulator receiver domain		
PA0929		239		53% similar to Response-regulator BaeR protein (*Escherichia coli*); 64% similar to PfeR protein (*Pseudomonas aeruginosa*)	Core	PF00486	Transcriptional regulatory protein, C terminal	PF00072	Response regulator receiver domain		
PA1157		236		63% similar to transcriptional regulatory protein RstA (RstB sensor) (*Escherichia coli*); 64% similar to response regulator OmpR (sensor, EnvZ)	Core	PF00072	Response regulator receiver domain	PF00486	Transcriptional regulatory protein, C terminal		
PA1179	PhoP	225		71% similar to regulatory protein PhoP (*Salmonella typhimurium*); 62% similar to trans-activating protein tctD (*Salmonella typhimurium*)	Core	PF00486	Transcriptional regulatory protein, C terminal	PF00072	Response regulator receiver domain		
PA1431	RsaL	80			??						
PA1437		229		81% similar to transcriptional activator protein CzcR (*Ralstonia eutropha*); 74% similar to transcriptional activator protein CopR (*Pseudomonas syringae*)	Core	PF00486	Transcriptional regulatory protein, C terminal	PF00072	Response regulator receiver domain		
PA1607		146		60% similar to hypothetical protein Rv3095(*Mycobacterium tuberculosis*)	Core	PF01638.10	HxlR				
PA1637	KdpE		230	69% similar to *kdpE* gene product of (*E. coli*)		PF00486	Transcriptional regulatory protein, C terminal	PF00072	Response regulator receiver domain		
PA1704	PcrR	144		64% similar to LcrR (*Yersinia enterocolitica*); 99% similar to PcrR (*Pseudomonas aeruginosa*); 64% similar to low calcium response locus protein LcrR (*Yersinia pestis*)	Core	PF09621.3	LcrR				

PA1799	ParR	235	64% similar to RstA (*Escherichia coli*)	Core	PF00072	Response regulator receiver domain	PF00486	Transcriptional regulatory protein, C terminal
PA2479		226	64% similar to putative 2-component transcriptional regulator YgiX (*Escherichia coli*)	Core	PF00486	Transcriptional regulatory protein, C terminal	PF00072	Response regulator receiver domain
PA2523		224	75% similar to regulator protein CzcR (*Ralstonia eutropha*)	Core	PF00486	Transcriptional regulatory protein, C terminal	PF00072	Response regulator receiver domain
PA2572		447	57% similar to a region of putative DNA-binding protein HupR1 *Rhodobacter capsulatus*	Core	PF00072	Response regulator receiver domain		
PA2657		223	62% similar to putative 2-component transcriptional regulator YgiX (*Escherichia coli*)	Core	PF00072	Response regulator receiver domain	PF00486	Transcriptional regulatory protein, C terminal
PA2686	PfeR	305	PfeR activator of the ferric enterobactin receptor (PfeA)	Core	PF00486	Transcriptional regulatory protein, C terminal	PF00072	Response regulator receiver domain
PA2713		159	61% similar to hypothetical protein Rv3095 (*Mycobacterium tuberculosis*)	Core	PF01638.10	HxlR		
PA2736		117		RGP 28	PF01047.15	MarR		
PA2809	CopR	226	82% similar to transcriptional activator CopR *Pseudomonas* syringae	Core	PF00072	Response regulator receiver domain	PF00486	Transcriptional regulatory protein, C terminal
PA2881								
PA3007	LexA	204	77% similar to LexA protein (*Escherichia coli*)	Core	PF01726	LexA_DNA_bind	PF00717.16	Peptidase_S24
PA3077		223	68% similar to ColR (*Pseudomonas fluorescens*)	Core	PF00486	Transcriptional regulatory protein, C terminal	PF00072	Response regulator receiver domain
PA3192	GltR	242	99% similar to GltR (*Pseudomonas aeruginosa*)	Core	PF00486	Transcriptional regulatory protein, C terminal	PF00072	Response regulator receiver domain
PA3204		225	65% similar to *E. coli cpxR* gene product	Core	PF00486	Transcriptional regulatory protein, C terminal	PF00072	Response regulator receiver domain
PA3385	AmrZ	108		Core	PF03869	Arc		
PA3864	DauR	210	44% similar to hypothetical protein ((*Streptomyces coelicolor*) A3(2))	Core	PF08348.4	PAS_6		
PA4057	NrdR, YbaD	154	82% similar to hypothetical ybaD gene product of (*Escherichia coli*)	Core	PF03477.9	ATP-cone		
PA4101	BfmR	246	63% similar to glucose uptake regulatory gene glrR (*Pseudomonas aeruginosa*)	RGP 44	PF00072	Response regulator receiver domain	PF00486	Transcriptional regulatory protein, C terminal
PA4182		212	51% similar to hypothetical protein YKL070w (*Saccharomyces cerevisiae*)	Core	PF04299.5	FMN_bind_2		

PA no.	Protein name	No. of amino acids	Subfamily	Nearest homologue	Core/ RGP	Domain 1	Domain 1 description	Domain 2	Domain 2 description	Other domains	Domain description
PA4776	PmrA	221		70% similar to putative 2-component transcriptional regulator YgiX (*Escherichia coli*)	Core	PF00486	Transcriptional regulatory protein, C terminal	PF00072	Response regulator receiver domain		
PA4885	IrlR	229		79% similar to IrlR gene product of *Burkholderia pseudomallei;* 81% similar to *czcR* (*Ralstonia eutropha*); 78% similar to transcriptional activator protein CopR (*Pseudomonas syringae*)	Core	PF00072	Response regulator receiver domain	PF00486	Transcriptional regulatory protein, C terminal		
PA4983		244		63% similar to DmsR (*Rhodobacter sphaeroides*)	Core	PF00072	Response regulator receiver domain	PF00486	Transcriptional regulatory protein, C terminal		
PA5360	PhoB	229		96% similar to *phoB* gene product of (*P. aeruginosa*)	Core	PF00072	Response regulator receiver domain	PF00486	Transcriptional regulatory protein, C terminal		

Table A.2 Transcription factor interacting partners. Genes that are differentially regulated by each transcriptional regulator were identified using the *Pseudomonas* database (Winsor *et al.*, 2009), published data (Madan Babu *et al.*, 2006) and data mining using STRING 8.3 (http://string.embl.de/). The data presented in this table was used as the input data for Cytoscape (www.cytoscape.org) to generate Fig. 13.2

ORF no.	Gene name	No. of regulated genes	Genes regulated
PA0652	*vfr*	96	PA0066, PA0165, PA0224, PA0265, PA0347, PA0376, PA0551, PA0552, PA0589, PA0653, PA0689, PA0783, PA0836, PA1092, PA1097, PA1183, PA1288, PA1337, PA1384, PA1430, PA1562, PA1580, PA1581, PA1582, PA1583, PA1584, PA1585, PA1586, PA1587, PA1588, PA1589, PA1629
PA1544	*anr*	77	PA0836, PA1172, PA1173, PA1174, PA1175, PA1176, PA1317, PA1318, PA1319, PA1320, PA1321, PA1337, PA1475, PA1476, PA1477, PA1479, PA1480, PA1481, PA1482, PA1483, PA1546, PA1561, PA1562, PA1581, PA1582, PA1583, PA1584, PA1585, PA1586, PA1588, PA1589, PA1780
PA2738	*himA*	57	PA0066, PA0353, PA0482, PA0836, PA0888, PA0962, PA0993, PA0994, PA1326, PA1339, PA1376, PA1580, PA1585, PA1586, PA1588, PA1589, PA1629, PA1631, PA1818, PA2634, PA2637, PA2638, PA2639, PA2640, PA2641, PA2642, PA2643, PA2644, PA2645, PA2646, PA2647, PA2648
PA4983		52	PA0265, PA0482, PA0599, PA0854, PA1183, PA1317, PA1318, PA1319, PA1320, PA1321, PA1376, PA1562, PA1580, PA1581, PA1582, PA1583, PA1584, PA1585, PA1586, PA1587, PA1588, PA1589, PA1787, PA2623, PA2634, PA2637, PA2638, PA2639, PA2640, PA2641, PA2642, PA2643
PA3879	*narL*	37	PA1172, PA1173, PA1174, PA1175, PA1176, PA1475, PA1476, PA1477, PA1479, PA1480, PA1481, PA1482, PA1483, PA1780, PA1781, PA2611, PA2637, PA2638, PA2639, PA2640, PA2641, PA2642, PA2643, PA2644, PA2645, PA2646, PA2647, PA2648, PA2649, PA3724, PA3872, PA3873
PA5308	*lrp*	36	PA0059, PA0316, PA0353, PA0993, PA0994, PA1070, PA1071, PA1072, PA1073, PA1074, PA1217, PA1326, PA1760, PA2443, PA3118, PA3120, PA3121, PA3167, PA3700, PA3972, PA4097, PA4503, PA4504, PA4505, PA4506, PA4695, PA4696, PA5013, PA5035, PA5036, PA5213, PA5214
PA4853	*fis*	26	PA1155, PA1156, PA1581, PA1582, PA1583, PA1584, PA1585, PA1586, PA1587, PA1588, PA1589, PA1787, PA1804, PA3769, PA3770, PA4022, PA4343, PA4538, PA4740, PA4741, PA4742, PA4743, PA4744, PA4745, PA4746, PA4852
PA5360	*phoB*	22	PA0688, PA3127, PA3296, PA3372, PA3373, PA3374, PA3375, PA3376, PA3377, PA3378, PA3379, PA3380, PA3381, PA3382, PA3383, PA3384, PA3981, PA4874, PA5361, PA5365, PA5366, PA5367
PA1754	*cysB*	21	PA0280, PA0281, PA0282, PA0932, PA1393, PA1493, PA1756, PA1838, PA2356, PA2357, PA2709, PA3444, PA3540, PA3935, PA3936, PA3937, PA3938, PA4130, PA4442, PA4443, PA4513
PA4764	*fur*	17	PA1165, PA1618, PA1902, PA1922, PA2466, PA2688, PA3899, PA3900, PA4158, PA4159, PA4160, PA4161, PA4228, PA4231, PA4468, PA5521, PA5531
PA3563	*fruR*	16	PA0551, PA0552, PA1376, PA1498, PA1770, PA1780, PA1781, PA2611, PA2623, PA2634, PA3181, PA3194, PA3560, PA3561, PA3562, PA5192
PA0893	*argR*	14	PA0895, PA0896, PA0897, PA0898, PA0899, PA0901, PA2042, PA3068, PA3537, PA3934, PA4976, PA4977, PA5152, PA5170
PA3007	*lexA*	14	PA0576, PA0577, PA0579, PA1886, PA2585, PA3008, PA3138, PA3617, PA4232, PA4234, PA4660, PA4763, PA4821, PA5443
PA0253		12	PA0254, PA0272, PA1928, PA1929, PA2002, PA2077, PA2113, PA2965, PA3703, PA3805, PA5237, PA5489
PA2510	*catR*	12	PA2513, PA2518, PA2740, PA3279, PA3280, PA3656, PA4408, PA4417, PA4418, PA4759, PA5045, PA5259
PA3898		12	PA0305, PA0854, PA1384, PA1619, PA2028, PA2854, PA3183, PA3764, PA4378, PA4468, PA4997, PA5157
PA0424	*mexR*	11	PA0425, PA0426, PA2020, PA2492, PA2495, PA3168, PA3574, PA3720, PA3721, PA4600, PA4964
PA1097	*fleQ*	11	PA1098, PA1099, PA1100, PA1443, PA1452, PA1453, PA1454, PA1455, PA2020, PA3385, PA4462
PA1179	*phoP*	11	PA1180, PA1343, PA1559, PA1560, PA1979, PA3552, PA3559, PA4773, PA4777, PA5199, PA5361
PA0272		10	PA1928, PA1929, PA2002, PA2077, PA2113, PA2965, PA3703, PA3805, PA5237, PA5489
PA0762	*algU*	10	PA0376, PA0576, PA0766, PA3337, PA4572, PA5011, PA5012, PA5077, PA5078, PA5261
PA1430	*lasR*	9	PA1432, PA1871, PA2586, PA2587, PA3476, PA3477, PA3622, PA3724, PA3861
PA3204		9	PA0766, PA3644, PA3645, PA3646, PA3725, PA3737, PA4953, PA4954, PA5489
PA0527	*dnr*	8	PA1544, PA1546, PA1781, PA1920, PA2020, PA3391, PA3392, PA3879
PA0547		8	PA0960, PA0962, PA0963, PA2326, PA4943, PA4946, PA4967, PA5471
PA0873	*phhR*	8	PA0292, PA1750, PA3000, PA3139, PA3166, PA4462, PA5039, PA5434
PA3477	*rhlR*	8	PA3478, PA3479, PA3622, PA3724, PA3861, PA4407, PA4408, PA4409
PA5125	*ntrC*	8	PA0888, PA1339, PA5075, PA5119, PA5124, PA5287, PA5288, PA5484
PA0001	*dnaA*	7	PA0002, PA0003, PA0376, PA1155, PA1156, PA3769, PA3770
PA1003	*mvfR*	7	PA1430, PA1690, PA1725, PA2227, PA2587, PA3477, PA3861
PA1099	*fleR*	7	PA1100, PA1101, PA1430, PA1454, PA1713, PA1714, PA4462
PA2273	*soxR*	7	PA2740, PA3280, PA3656, PA3718, PA4417, PA4759, PA5259
PA1619		6	PA0305, PA0854, PA3183, PA4468, PA4615, PA5157
PA1998	*dhcR*	5	PA1999, PA2000, PA5384, PA5388, PA5389

ORF no.	Gene name	No. of regulated genes	Genes regulated
PA2016	*liuR*	5	PA2320, PA2886, PA2888, PA2891, PA2893
PA2020		5	PA2642, PA3391, PA3721, PA4671, PA5471
PA2259	*ptxS*	5	PA2260, PA2261, PA2262, PA2263, PA4315
PA3583	*glpR*	5	PA0347, PA3581, PA3582, PA3584, PA5235
PA4280	*birA*	5	PA0420, PA0500, PA0501, PA0503, PA0504
PA0487		4	PA1861, PA1863, PA4210, PA4811
PA0610	*prtN*	4	PA0611, PA0612, PA0651, PA4590
PA0780	*pruR*	4	PA0781, PA0782, PA0783, PA0982
PA1432	*lasI*	4	PA1898, PA2227, PA3477, PA4296
PA1760		4	PA2163, PA2339, PA2340, PA2341
PA1949	*rbsR*	4	PA1946, PA1947, PA1948, PA1950
PA1979	*eraS*	4	PA1980, PA2586, PA3948, PA5483
PA2551		4	PA1217, PA3118, PA3120, PA3121
PA2586	*gacA*	4	PA3476, PA3477, PA3622, PA4856
PA2665		4	PA1779, PA3697, PA4811, PA4893
PA3540	*algD*	4	PA5125, PA5200, PA5261, PA5483
PA5200	*amgR*	4	PA0045, PA1288, PA5261, PA5361
PA5261	*algR*	4	PA5262, PA5322, PA5483, PA5484
PA5356	*glcC*	4	PA0482, PA5352, PA5353, PA5355
PA5380	*gbdR*	4	PA5398, PA5399, PA5410, PA5411
PA0037	*trpI*	3	PA0609, PA3166, PA4590
PA0152	*pcaQ*	3	PA0153, PA0154, PA0247
PA0928	*gacS*	3	PA1898, PA2586, PA3006
PA0996	*pqsA*	3	PA1003, PA1430, PA3477
PA1184		3	PA5213, PA5214, PA5215
PA1637	*kdpE*	3	PA1633, PA1634, PA1635
PA1713	*exsA*	3	PA1714, PA1716, PA3841
PA1898	*qscR*	3	PA2586, PA3006, PA3476
PA2054	*cynR*	3	PA2052, PA2053, PA2500
PA2118	*ada*	3	PA1686, PA3306, PA3972
PA2258	*ptxR*	3	PA2259, PA2426, PA3479
PA2320	*gntR*	3	PA2321, PA3181, PA3194
PA2809	*copR*	3	PA2810, PA5199, PA5361
PA2885	*atuR*	3	PA2886, PA2888, PA2891
PA3006	*psrA*	3	PA3007, PA3571, PA3622
PA3184		3	PA3193, PA3194, PA5036
PA3476	*rhlI*	3	PA3477, PA4723, PA5261
PA3587	*metR*	3	PA1843, PA4602, PA5124
PA3899		3	PA2914, PA3900, PA3901
PA4080		3	PA2978, PA3816, PA5563
PA4109	*ampR*	3	PA4110, PA4393, PA4522
PA4341		3	PA0247, PA4127, PA4152
PA4363	*iciA*	3	PA0001, PA0002, PA0003
PA4462	*rpoN*	3	PA4547, PA5125, PA5483
PA4547	*pilR*	3	PA4549, PA4556, PA5124
PA4600	*nfxB*	3	PA4964, PA5263, PA5348
PA4769		3	PA1587, PA5015, PA5016
PA5085		3	PA0836, PA2683, PA5277
PA5344	*oxyR*	3	PA0139, PA0140, PA0962
PA5374	*betI*	3	PA3933, PA5372, PA5373
PA5550	*glmR*	3	PA2274, PA4457, PA5552

ORF no.	Gene name	No. of regulated genes	Genes regulated
PA0036	*trpB*	2	PA0037, PA3587
PA0155	*pcaR*	2	PA0159, PA4974
PA0159		2	PA1294, PA4974
PA0287	*gpuP*	2	PA0289, PA1422
PA0288	*gpuA*	2	PA0289, PA1422
PA0289	*gpuR*	2	PA1418, PA1421
PA0370		2	PA2273, PA2510
PA0409	*pilH*	2	PA0416, PA3587
PA0416	*chpD*	2	PA0417, PA1456
PA0425	*mexA*	2	PA2020, PA3721
PA0482	*glcB*	2	PA2273, PA2809
PA0506		2	PA3006, PA3571
PA0519	*nirS*	2	PA0527, PA1544
PA0520	*nirQ*	2	PA0527, PA1544
PA0576	*rpoD*	2	PA0652, PA1544
PA0611	*prtR*	2	PA0612, PA4590
PA0831	*oruR*	2	PA2020, PA4402
PA0888	*aotJ*	2	PA0893, PA4315
PA1085	*flgJ*	2	PA1097, PA1099
PA1588	*sucC*	2	PA2273, PA2510
PA1589	*sucD*	2	PA2273, PA2510
PA1710	*exsC*	2	PA1713, PA3006
PA1957		2	PA0761, PA4919
PA1978	*erbR*	2	PA1982, PA3583
PA2012	*liuD*	2	PA2016, PA2885
PA2014	*liuB*	2	PA2016, PA2885
PA2219	*opdE*	2	PA2273, PA2510
PA2227	*vqsM*	2	PA3476, PA4296
PA2272	*pbpC*	2	PA2273, PA2510
PA2492	*mexT*	2	PA2493, PA2495
PA2523		2	PA2809, PA2810
PA2570	*lecA*	2	PA3477, PA4315
PA2571		2	PA3587, PA5261
PA2686	*pfeR*	2	PA2687, PA2688
PA2810	*copS*	2	PA5200, PA5360
PA2951	*etfA*	2	PA3006, PA3571
PA2952	*etfB*	2	PA3006, PA3571
PA2953		2	PA3006, PA3571
PA3574	*nalD*	2	PA0425, PA0426
PA4227	*pchR*	2	PA4228, PA4229
PA4581	*rtcR*	2	PA4583, PA4585
PA4726	*cbrB*	2	PA5105, PA5124
PA5199	*amgS*	2	PA5200, PA5360
PA0017		1	PA2586
PA0024	*hemF*	1	PA0527
PA0041		1	PA0652
PA0044	*exoT*	1	PA1713
PA0176	*aer2*	1	PA1544
PA0186		1	PA1754
PA0240		1	PA3782
PA0254		1	PA0272

ORF no.	Gene name	No. of regulated genes	Genes regulated
PA0286	*desA*	1	PA4890
PA0292	*aguA*	1	PA0294
PA0293	*aguB*	1	PA0294
PA0296	*spuI*	1	PA4726
PA0297	*spuA*	1	PA4726
PA0298	*spuB*	1	PA4726
PA0299	*spuC*	1	PA4726
PA0328		1	PA0893
PA0408	*pilG*	1	PA0416
PA0410	*pilI*	1	PA0416
PA0411	*pilJ*	1	PA0416
PA0412	*pilK*	1	PA0416
PA0413	*chpA*	1	PA0416
PA0414	*chpB*	1	PA0416
PA0415	*chpC*	1	PA0416
PA0523	*norC*	1	PA0527
PA0572		1	PA0652
PA0705	*migA*	1	PA5499
PA0755	*opdH*	1	PA0756
PA0756		1	PA0757
PA0763	*mucA*	1	PA5483
PA0779		1	PA2665
PA0844	*plcH*	1	PA5380
PA0870	*phhC*	1	PA0873
PA0871	*phhB*	1	PA0873
PA0872	*phhA*	1	PA0873
PA0887	*acsA*	1	PA1980
PA0889	*aotQ*	1	PA0893
PA0890	*aotM*	1	PA0893
PA0891		1	PA0893
PA0892	*aotP*	1	PA0893
PA0905	*rsmA*	1	PA2586
PA0932	*cysM*	1	PA2354
PA0969	*tolQ*	1	PA2686
PA1000	*pqsE*	1	PA1003
PA1002	*phnB*	1	PA5499
PA1077	*flgB*	1	PA1097
PA1078	*flgC*	1	PA1097
PA1082	*flgG*	1	PA1097
PA1084	*flgI*	1	PA1097
PA1093		1	PA1097
PA1094	*fliD*	1	PA1097
PA1098	*fleS*	1	PA1099
PA1174	*napA*	1	PA3879
PA1178	*oprH*	1	PA1179
PA1196		1	PA3861
PA1418		1	PA1422
PA1421	*gbuA*	1	PA1422
PA1458		1	PA5261
PA1461	*motD*	1	PA2259
PA1543	*apt*	1	PA1544

ORF no.	Gene name	No. of regulated genes	Genes regulated
PA1546	*hemN*	1	PA3879
PA1557	*ccoN2*	1	PA5261
PA1609	*fabB*	1	PA4890
PA1610	*fabA*	1	PA4890
PA1690	*pscU*	1	PA4109
PA1712	*exsB*	1	PA1713
PA1714	*exsD*	1	PA2519
PA1725	*pscL*	1	PA4109
PA1796	*folD*	1	PA4600
PA1850		1	PA2063
PA1920	*nrdD*	1	PA3879
PA1977		1	PA1978
PA1980	*eraR*	1	PA1982
PA1999	*dhcA*	1	PA5389
PA2000	*dhcB*	1	PA5389
PA2011	*liuE*	1	PA2016
PA2013	*liuC*	1	PA2016
PA2015	*liuA*	1	PA2016
PA2018		1	PA2020
PA2193	*hcnA*	1	PA5261
PA2354		1	PA3444
PA2491		1	PA2492
PA2495	*oprN*	1	PA4600
PA2507	*catA*	1	PA2510
PA2508	*catC*	1	PA2510
PA2511		1	PA2512
PA2515	*xylL*	1	PA2519
PA2519	*xylS*	1	PA3571
PA2522	*czcC*	1	PA2809
PA2561		1	PA5360
PA2585	*uvrC*	1	PA2586
PA2587	*pqsH*	1	PA5499
PA2611	*cysG*	1	PA4723
PA2637	*nuoA*	1	PA2686
PA2664	*fhp*	1	PA2665
PA2696		1	PA1850
PA3082	*gbt*	1	PA5380
PA3094		1	PA5322
PA3168	*gyrA*	1	PA4600
PA3175		1	PA5105
PA3176	*gltS*	1	PA3184
PA3182	*pgl*	1	PA3184
PA3183	*zwf*	1	PA3184
PA3257	*prc*	1	PA5483
PA3266	*capB*	1	PA3168
PA3385	*amrZ*	1	PA5262
PA3391	*nosR*	1	PA3587
PA3479	*rhlA*	1	PA5261
PA3545	*algG*	1	PA5483
PA3547	*algL*	1	PA5483
PA3569	*mmsB*	1	PA3571

ORF no.	Gene name	No. of regulated genes	Genes regulated
PA3570	*mmsA*	1	PA3571
PA3581	*glpF*	1	PA3583
PA3594		1	PA3004
PA3689		1	PA3690
PA3710		1	PA3782
PA3720		1	PA3721
PA3721	*nalC*	1	PA4598
PA3724	*lasB*	1	PA4723
PA3782		1	PA5065
PA3815	*iscR*	1	PA4236
PA3875	*narG*	1	PA3879
PA3878	*narX*	1	PA3879
PA3946	*two-compo*	1	PA3948
PA3947	*rocR*	1	PA3948
PA4101	*bfmR*	1	PA4208
PA4222		1	PA4227
PA4223		1	PA4227
PA4224	*pchG*	1	PA4227
PA4225	*pchF*	1	PA4227
PA4226	*pchE*	1	PA4227
PA4293	*pprA*	1	PA4296
PA4295	*fppA*	1	PA4296
PA4296	*pprB*	1	PA4305
PA4446	*algW*	1	PA5483
PA4498		1	PA4499
PA4525	*pilA*	1	PA4547
PA4550	*fimU*	1	PA5261
PA4597	*oprJ*	1	PA4600
PA4598	*mexD*	1	PA4600
PA4599	*mexC*	1	PA4600
PA4723	*dksA*	1	PA5338
PA4725	*cbrA*	1	PA4726
PA4844		1	PA5360
PA4888	*desB*	1	PA4890
PA4889		1	PA4890
PA5092	*hutI*	1	PA5105
PA5098	*hutH*	1	PA5105
PA5100	*hutU*	1	PA5105
PA5124	*ntrB*	1	PA5125
PA5157		1	PA1619
PA5255	*algQ*	1	PA5261
PA5259	*hemD*	1	PA5261
PA5260	*hemC*	1	PA5261
PA5292	*pchP*	1	PA5380
PA5373	*betB*	1	PA5374
PA5384		1	PA5389
PA5388		1	PA5389
PA5436		1	PA5437
PA5483	*algB*	1	PA5484
PA5549	*glmS*	1	PA5550

Index

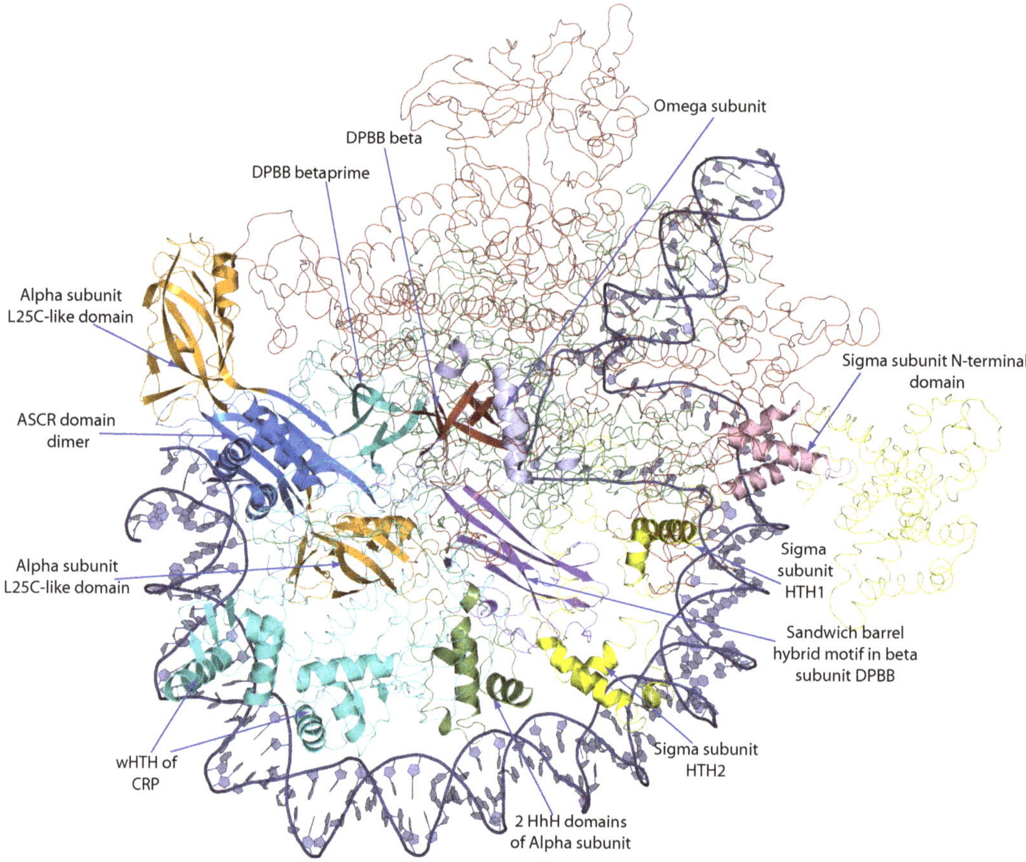

Figure 1.1 Structure of the bacterial transcription initiation complex. The cartoon representation was derived from an EM structure of the initiation complex (PDB: 3iyd) in association with DNA. The structure contains the α, β, β′, ω and σ^{70} of the bacterial DNA-dependent RNA polymerase and the wHTH domains of CRP (CAP) transcription factor. Only the key globular domains of these proteins are shown and labelled for better clarity. The remaining parts of the structure are shown as coils.

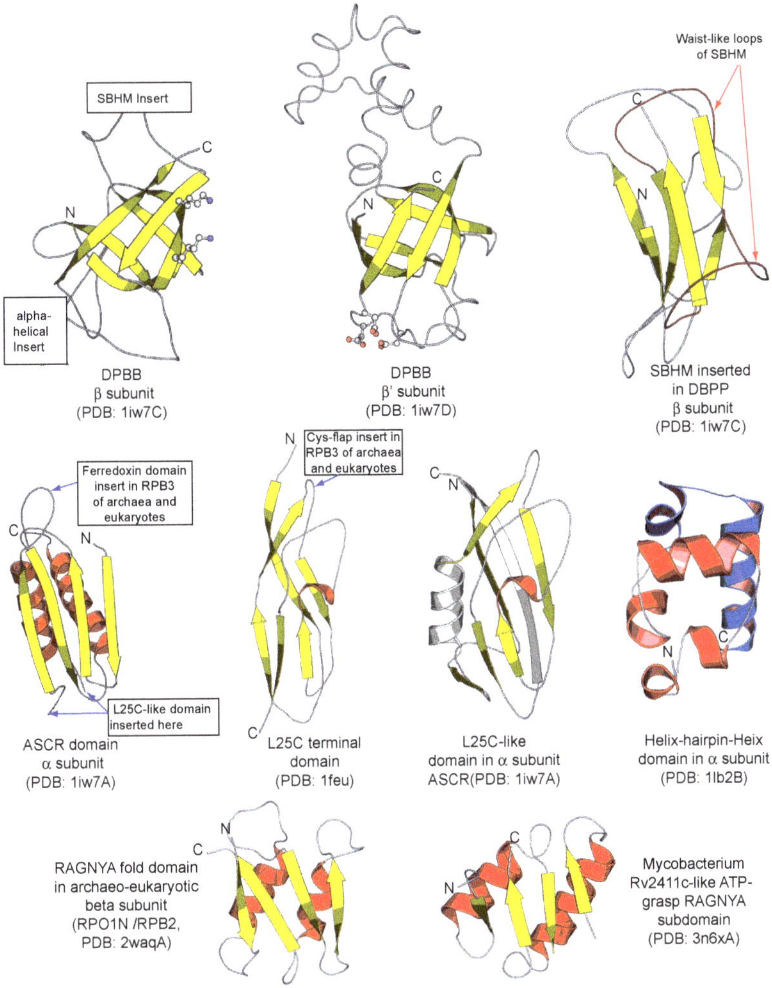

Figure 1.2 Structures of key conserved domains of the β, β′ and α subunits. Strands are coloured yellow, whereas helices are coloured red or blue. Only the core conserved regions of the domains are shown. Inserts in domains are mostly suppressed or excised as depicted. The C-terminal domain of the ribosomal L25 protein is depicted to illustrate its structural relationship with the conserved domain inserted into the ASCR domain of the α subunit (L25C-like domain). Also shown are the positions of the RPB3 (α) subunit-specific inserts in the archaeo-eukaryotes. Structural elements in the L25C-like domain of the α subunit that are not present in the ribosomal L25 protein are coloured grey.

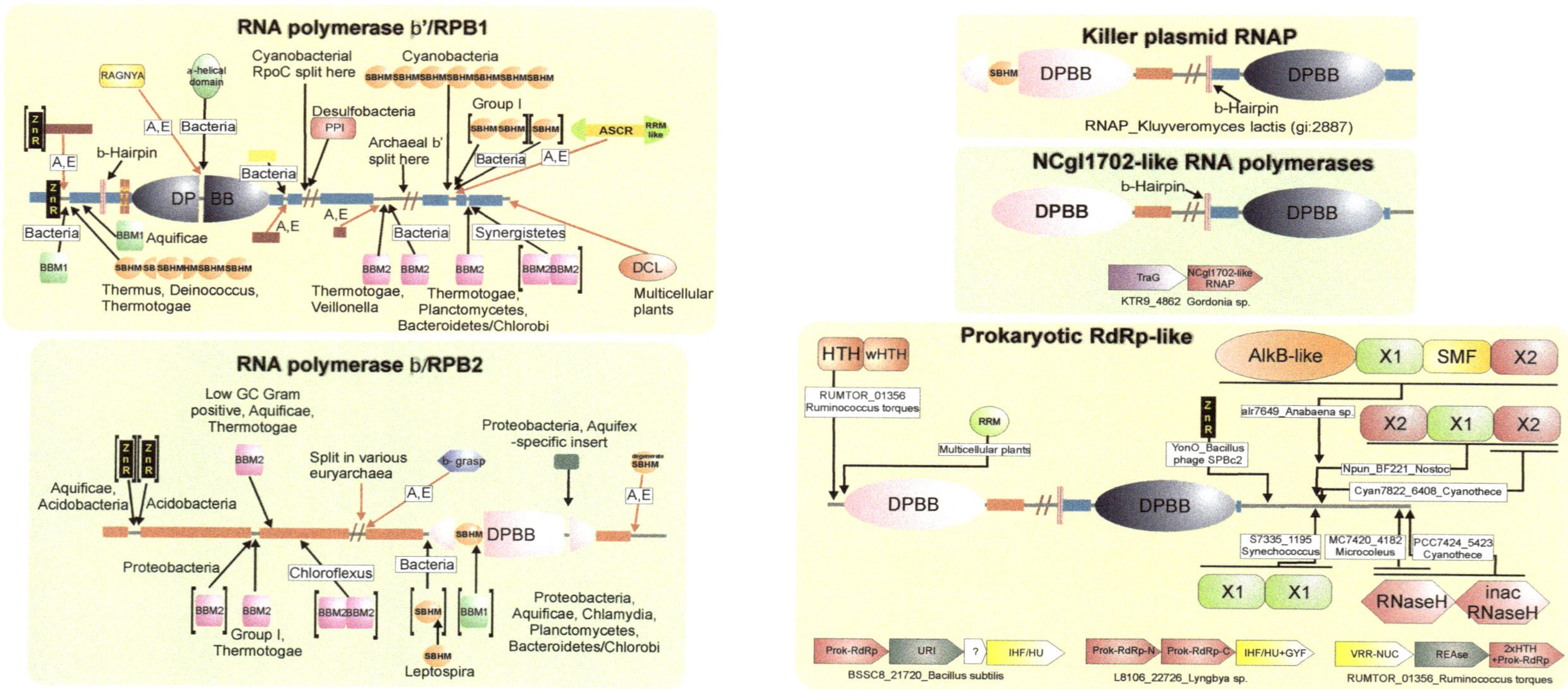

Figure 1.3 Domain architectures of the RNA polymerase β and β′ subunits, yeast killer plasmid RNA polymerase, NCgl1702-like RNA polymerases and the prokaryotic RdRP-like RNA polymerases. For the β and β′ subunits, the domain architecture reconstructed to the last universal common ancestor is shown in the centre and inserts in various lineages are shown around this core. Archaeo-eukaryotic domain inserts are indicated with a red arrow and bacterial inserts are marked with a black arrow. Lineages in which the inserts are observed are indicated near the arrows or architecture. Bacterial inserts, on occasions, differ within members of a closely related bacterial lineage. For a more detailed discussion of these variations, refer to (Lane and Darst, 2010a). A similar representation is used for the prokaryotic RdRP-like proteins, where lineage-specific inserts are marked with a representative gene and species name around a core conserved architecture. Genes in operons are shown as box-arrows with the arrow head pointing from the 5′ to the 3′ direction of the coding sequence. Operons are labelled with the gene name of the polymerase gene and species name. Refer to the supplement <AU: Is this referring to the supplementary material in Chapter 13?> for more detailed domain architectures and gene neighbourhoods. Standard abbreviations are used for domain and lineage names. The DCL domain is an RNA binding domain which is also found in a stand-alone form in bacteria and in several eukaryotic rRNA biogenesis proteins. Other abbreviations: A, E, archaea and eukaryotes; ASCR, alpha subunit core related; ATL, AT-hook like motifs; PPI, peptidyl prolyl isomerase; ZnR, zinc ribbon.

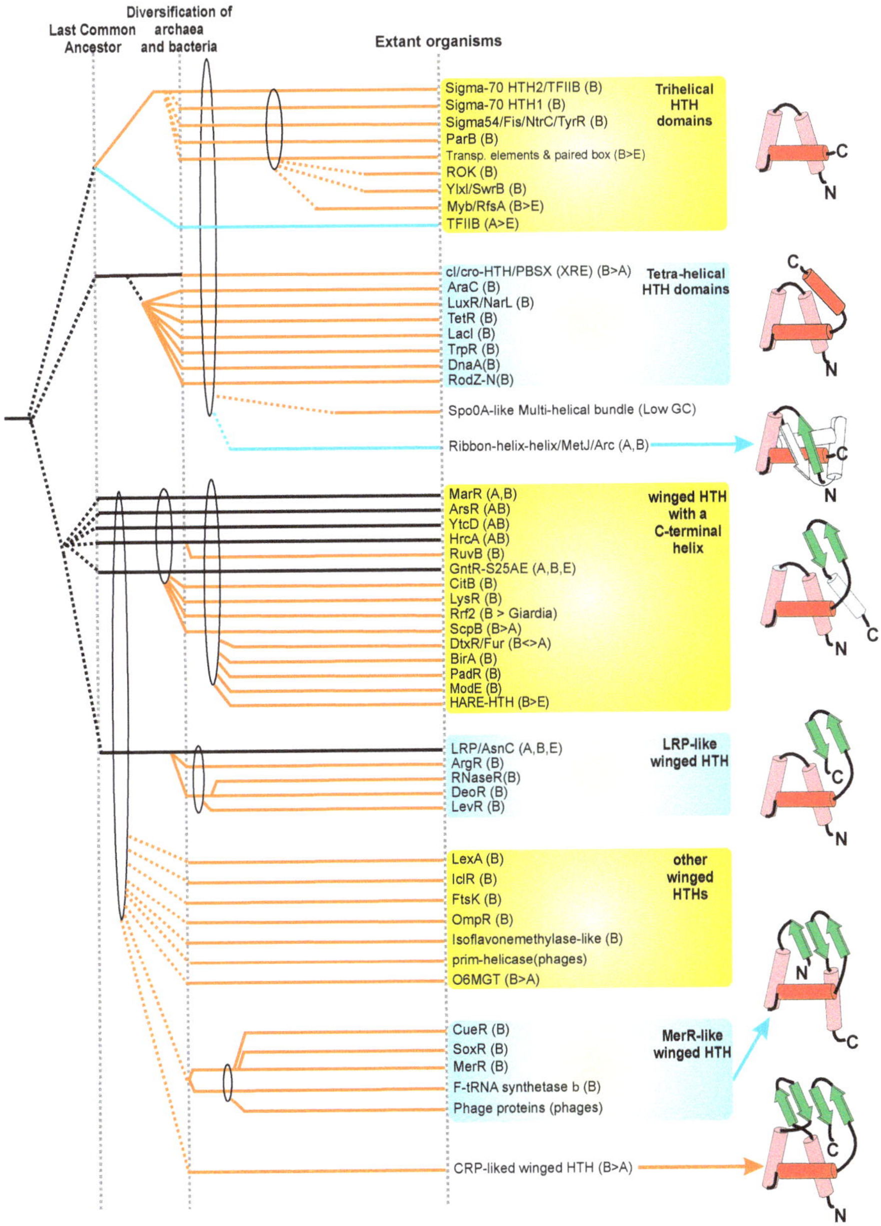

Figure 1.4 Higher order evolutionary relationships of bacterial specific transcription factors containing a HTH domain. The horizontal lines represent temporal epochs corresponding to major transitions in evolution of bacteria, namely the last universal common ancestor and the diversification of archaea and bacteria. Solid lines reflect the maximum depth of time to which a particular family can be traced. Broken lines indicate an uncertainty with respect to the exact point of origin of a lineage. The ellipses encompass groups of lineages from which a new lineage with relatively limited distribution could have potentially emerged. Lineages of archaeal origin are coloured blue, those of bacterial origin are coloured orange and those present in archaea and bacteria are coloured black. The phyletic distribution of the lineages are also shown in brackets. A, Archaea; B, bacteria; E, eukaryotes. The '>' reflects lateral transfer with the arrow head pointing to the potential direction of transfer. Also shown to the right are cartoon representations of the major structural types of HTH domains found in bacterial transcription factors. The TFIIB lineage of archaeo-eukaryotic HTHs is shown to illustrate its relationship with the sigma factor.

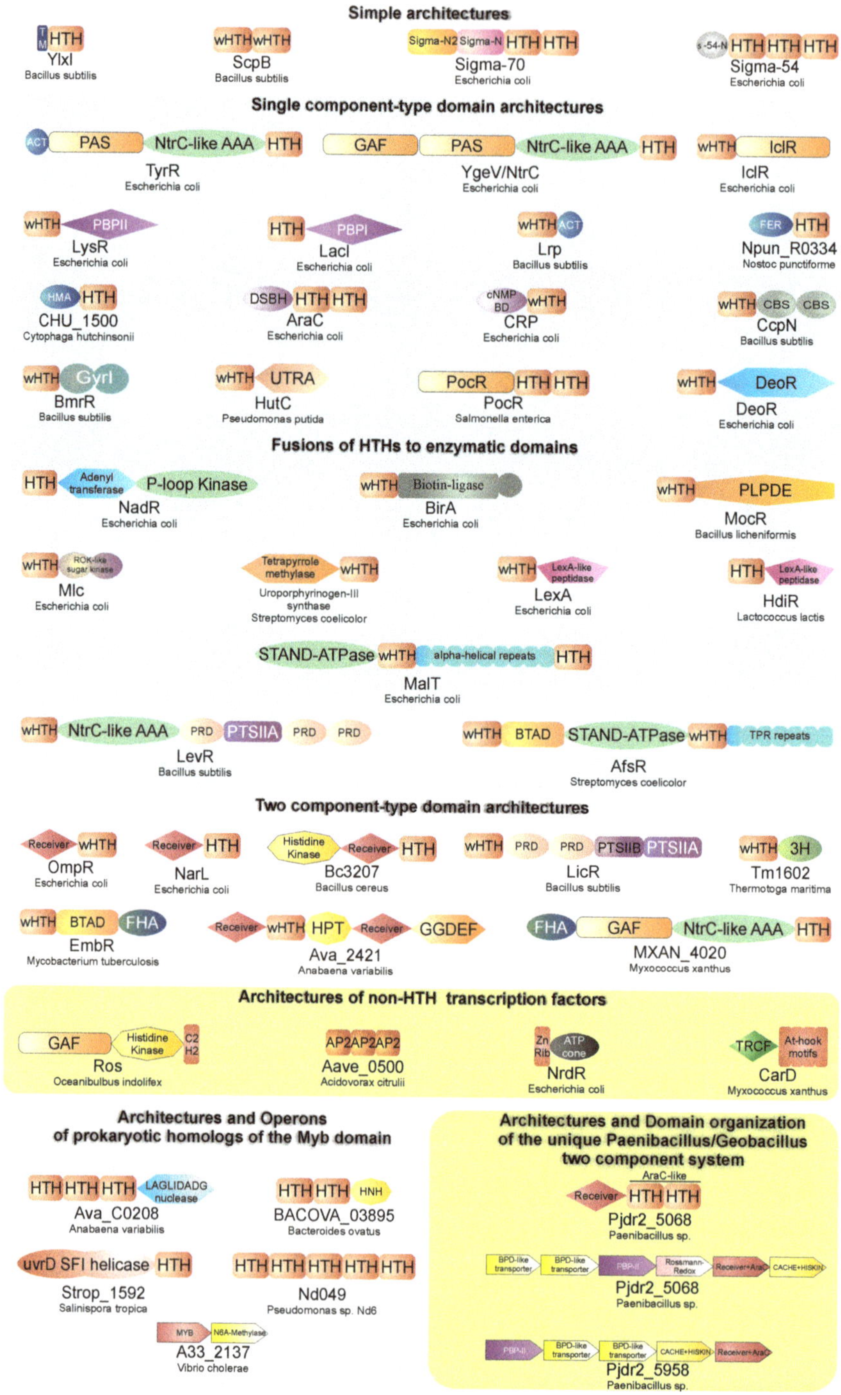

Figure 1.5 Examples of domain architectures of bacterial transcription factors described in the text. Proteins are labelled with their gene and species names. The domains are not drawn to scale. Standard nomenclatures were mostly used to depict the various domains. Additional abbreviations include: TM, transmembrane; σ-54N, globular domain found at the N-terminus of σ^{54}; Sigma-N2 and Sigma-N, conserved N-terminal domains found in σ^{70}; BTAD, conserved domain found in bacterial signalling proteins; ZnRib, zinc ribbon; FER, classical ferredoxin domain of the RRM fold.

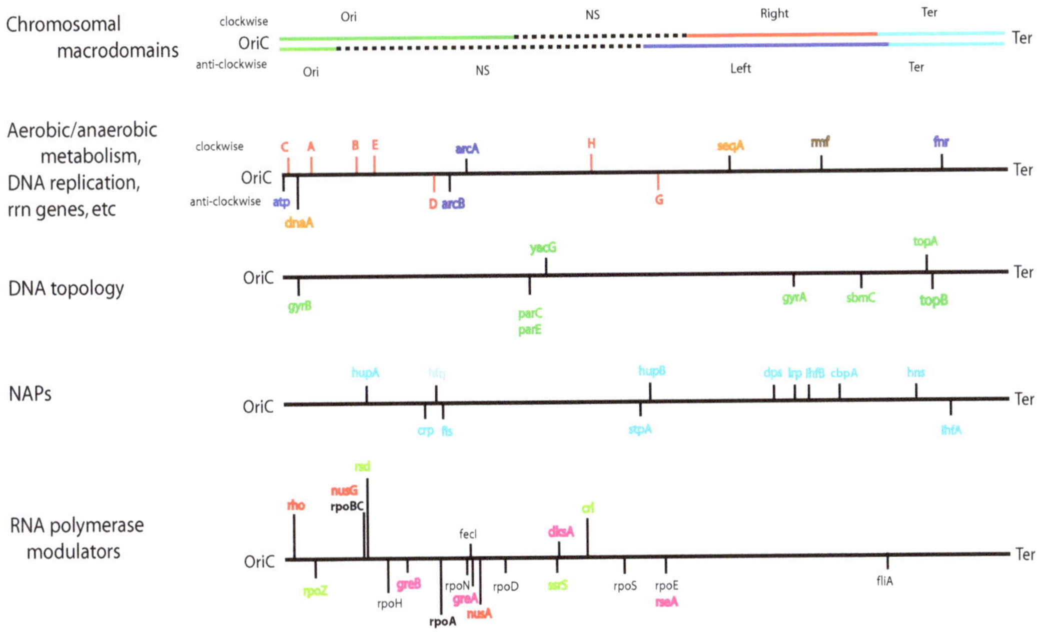

Figure 2.5 Spatial ordering of regulatory genes on the *E. coli* chromosome along the OriC–Ter axis. For full details see Sobetzko *et al.*, (2012). Top line: Correspondence of macrodomains defined by Valens *et al.*, (2004) to linear map. First bar: Selected genes involved in aerobic/anaerobic metabolism (dark blue), DNA replication (orange), *rrn* genes (red) and transition phase (brown). Genes on clockwise (right) replichore are above the bar and genes on anti-clockwise (left) replichore below the bar. Second bar: Selected genes involved in control of DNA topology. *gyrB*, a component of DNA gyrase, responsible for increasing negative superhelicity, maps close to the origin while gyrase inhibitor *sbmC*, and *topA* and *topB*, both responsible for relaxing DNA map either close to or within the Ter macrodomain. Third bar: Selected genes encoding NAPs. The NAP-encoding gene closest to OriC is *hupA*, encoding HUα. Its early expression relative to *hupB*, encoding HUβ , could buffer high negative superhelicity generated by DNA gyrase. Fourth bar: Selected genes involved in modulating RNA polymerase activity, including sigma factor utilization regulators, secondary channel binding proteins, termination/elongation factors and RNA polymerase subunits. Sigma factor utilization regulators (light green): *rpoZ*, mapping close to the origin, encodes the ω subunit of RNA polymerase, which confers a preference for utilization of σ^{70}.

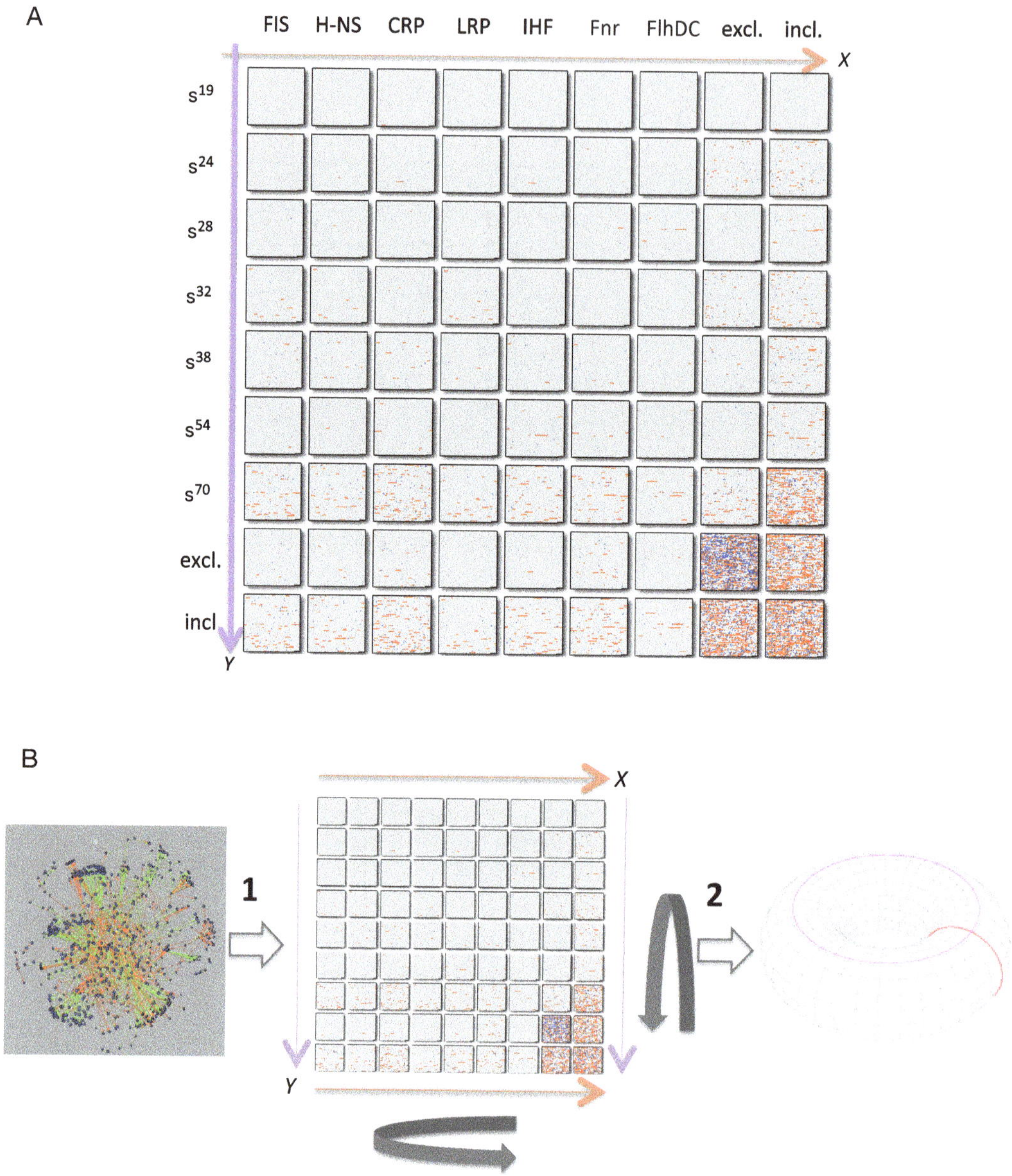

Figure 2.6 (A) Couplon matrix of σ factors and abundant NAPs. FlhDC is not a NAP but is incorporated in the matrix as a well-known regulator of a specific function – biosynthesis of flagella. Fnr is a global transcriptional regulator closely related to CRP. Each square or couplon holds all genes regulated by the two intersecting factors indicated on each axis. Genes organized in operons are shown in red colour, independent genes are in blue. The genes in each couplon are topologically ordered and arranged from left to right in rows for each couplon square. The 'excl' columns or rows represent the genes regulated by the column or row factor but by no listed factor intersecting with the column or row. The 'incl' columns or rows hold the genes that are regulated by the column or row factor without any additional restrictions. 'incl/excl' and 'excl/incl' couplons hold the sum of respective 'incl' row or column. 'incl/incl' holds all genes involved in creating this matrix e.g. all genes regulated by a sigma factor or a NAP. (B) The holistic model of topologically closed transcriptional regulation system. The RegulonDB dataset (Salgado *et al.*, 2006) is organized in the couplon matrix (1). Wrapping of the square scheme of couplon matrix (2) around both the vertical and horizontal axes (arrows) generates a torus of orthogonally operating regulon circuits (Muskhelishvili *et al.*, 2010). Note that this torus represents a logical construct, rather than mathematical object.

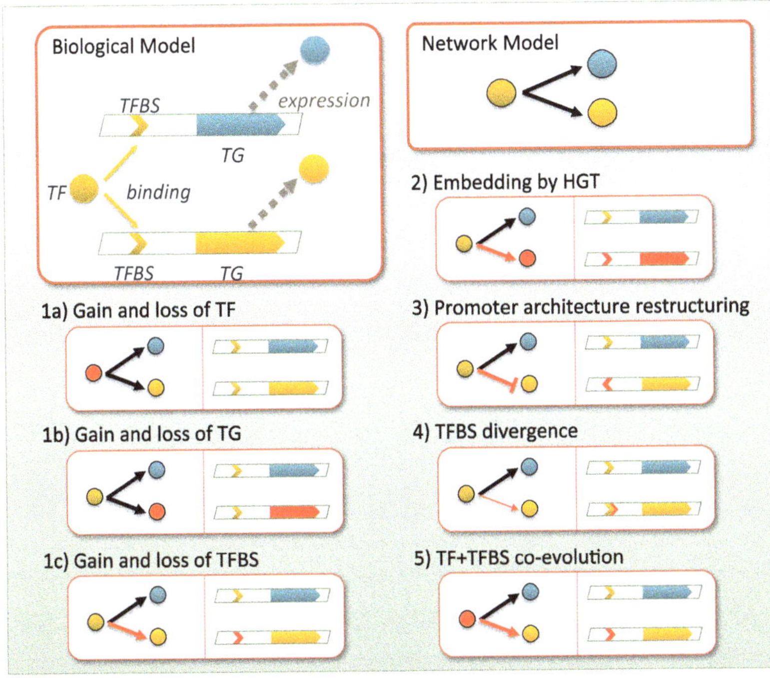

Figure 3.1 Evolutionary regulatory models in Bacteria. The biological model on the top left describes the regulation of two target genes (in blue and yellow) by the binding of a given TF (green sphere) on a specific TFBS (green chevron). Transcriptional regulatory networks use to represent a regulatory interaction by an edge directed from the TF to the TG (top right). We labelled from 1 to 5 the different models using red colour to highlight the regulatory change affecting one of the two genes being recently acquired in a bacterial species. We also represented how this corresponding novelty can affect the network model. Class 1: gain and loss of one of the regulatory actors such as (a) TF, (b) TG and (c) TFBS. Class 2: embedding of the new regulatory targets via HGT, i.e. the recently acquired TFBS and its new gene will evolve quickly to get regulated by a pre-existing TF. This situation is different to the class 1b where a new TG can be included within an operon with pre-existing TFBS. Class 3: Restructuring of the promoter architecture which implies changes in the orientation of the TFBS (as illustrated), but also the positioning of the TFBS, and to some extend their occurrence in the promoter region. The change in the orientation can have a consequence on the regulatory effect of a TF which can be seen on a transcriptional regulatory circuit showing activation and repression. Class 4: TFBS divergence such as mutation can lead to changes in the affinity of the TF or to gradually changes to an innovative TFBS. This can be observed in a weighted regulatory network where the weight on the edge indicates the strength of the regulation. Class 5: The TF and the TFBS can co-evolve together in general across different taxonomical groups.

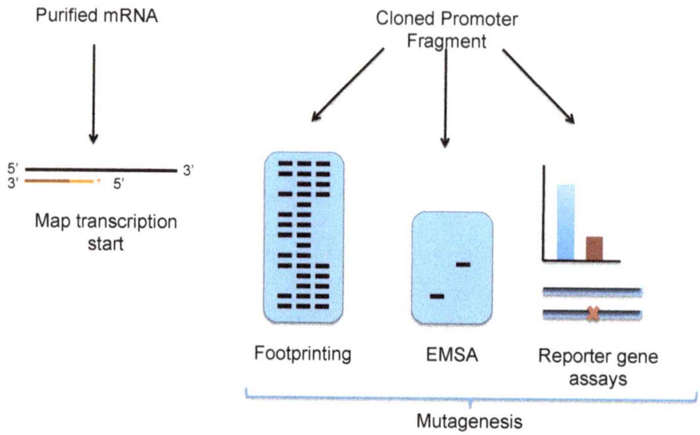

Figure 7.1 Bottom-up methodologies. Several approaches can be exploited to study the regulation of individual genes. Often the transcription start site is mapped by primer extension of purified mRNA. Once the promoter has been cloned it can be analysed using footprinting, electromobility shift assays (EMSA) and reporter gene assays. Further information can be obtained from mutational analysis.

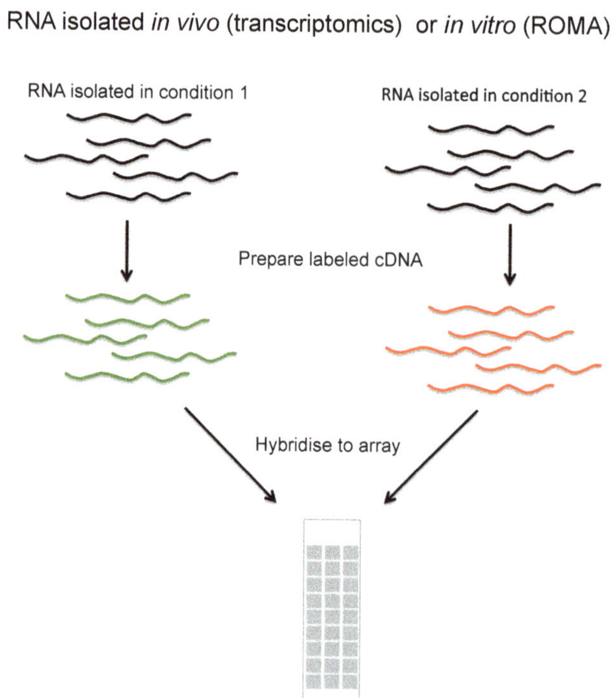

Figure 7.2 Transcriptomics and ROMA. mRNA levels can be determined on a genomic scale. RNA is purified either from cells (transcriptomics) or from an *in vitro* run-off transcription assays using genomic DNA as template (ROMA). Labelled cDNA is prepared and hybridized to a microarray to quantify different sequences, and hence different starting mRNAs. Two samples can be compared directly, for example, wild-type versus mutant or cells growing in different conditions.

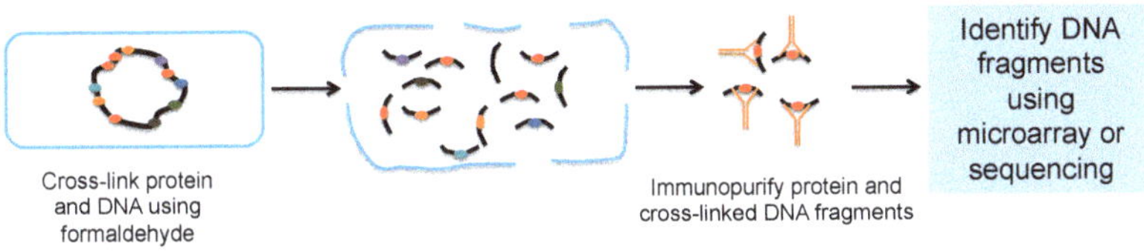

Figure 7.4 Chromatin immunoprecipitation (ChIP-ChIP and ChIP-Seq). Cells are grown and transcription factors cross-linked to their DNA binding sites. An antibody specific for the transcription factor of interest is used to purify the transcription factor and the cross-linked target DNA. The purified DNA target can be analysed using microarrays or by direct sequencing.

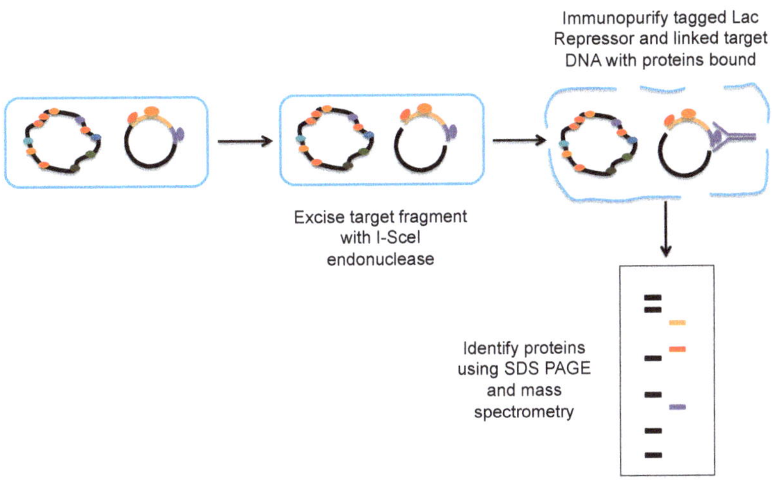

Figure 7.5 DNA sampling. The target DNA sequence is cloned into a vector adjacent to binding sites for the Lac Repressor and flanked by I-SceI recognition sites. During the assay, the target DNA is excised by I-SceI endonuclease from the vector, *in vivo*, in cells containing Lac repressor that is tagged with the FLAG epitope. The protein–DNA complex is then purified using antibodies specific for the tagged Lac Repressor. Following SDS PAGE, the proteins are identified by mass spectrometry.

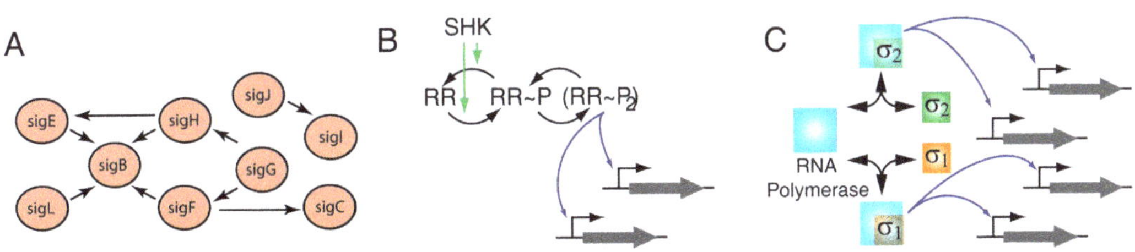

Figure 12.1 Master regulators and regulatory modes of *M. tuberculosis*. (A) The sigma factor network consists of one housekeeping sigma factor and 13 alternative sigma factors that control each other through a set of complex regulatory interactions. The arrows indicate direct/indirect regulatory links. Adapted from (Sachdeva *et al.*, 2010). (B) Two-component system with sensor histidine kinase (SHK) and response regulator in unphosphorylated (RR), phosphorylated (RR~P) and dimer (RR~P_2) forms. (C) Alternative sigma factors associate with RNA polymerase to control gene expression in various conditions.

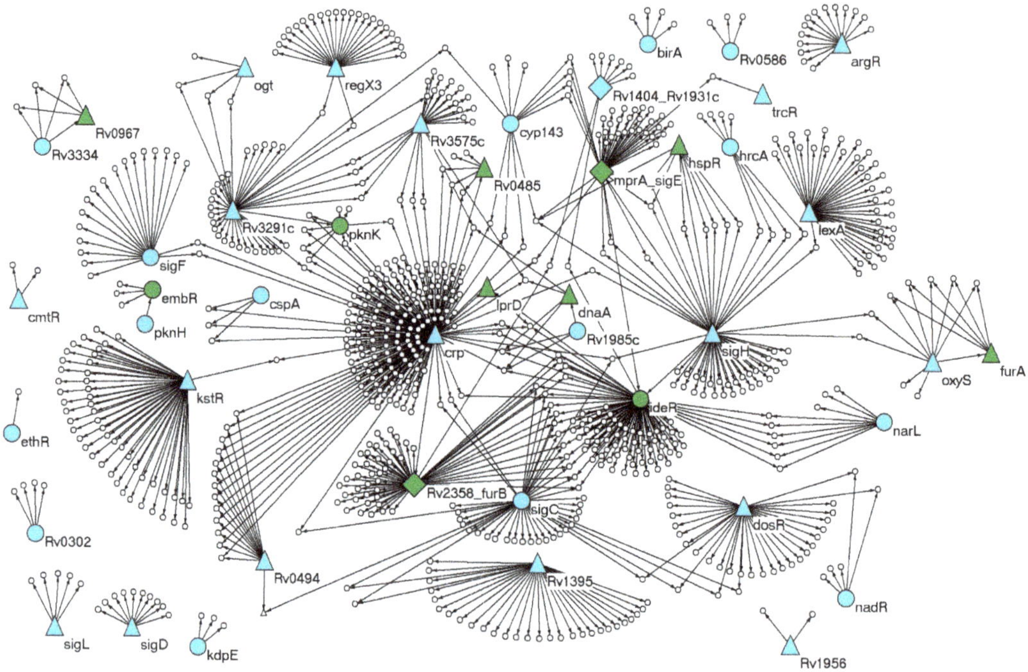

Figure 12.2 The current *M. tuberculosis* gene regulatory network. The network was assembled from publicly available sources. Input nodes (TFs with no known transcriptional regulators) are shown in blue, while transit nodes (TFs with known transcriptional regulators) are shown in green. The white nodes represent output nodes (genes encoding proteins with no TF activity). Triangles = nodes that autoregulate their own expression. Diamonds = nodes that are part of two-gene feedback loops. Reproduced from (Balázsi *et al.*, 2008) according to a Creative Commons Attribution–Noncommercial–Share Alike 3.0 licence.

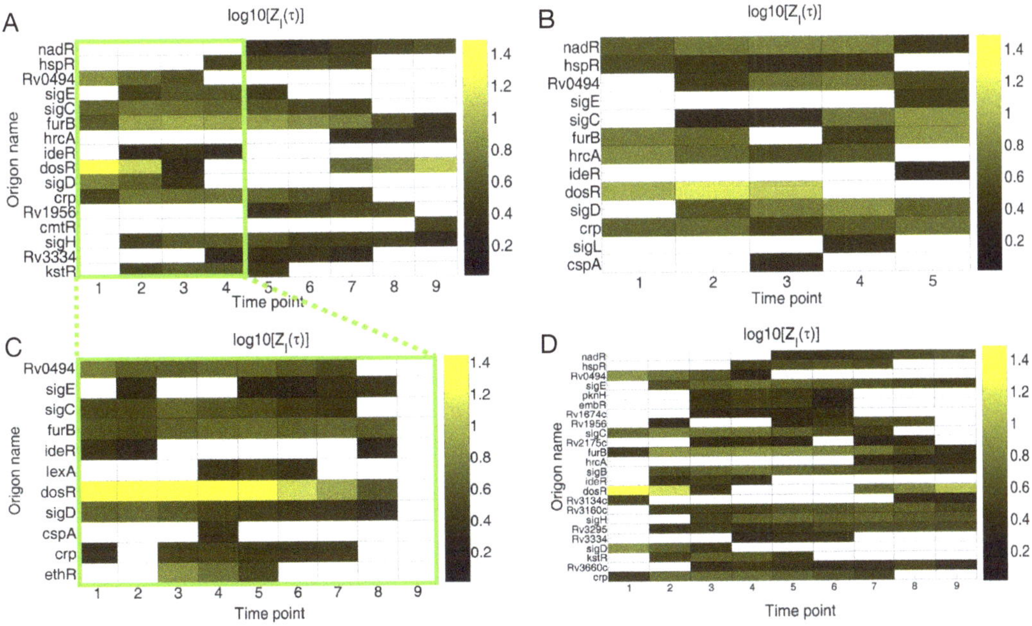

Figure 12.3 Network response to growth arrest. (A) Responsiveness $Z_I(t)$ of significantly responsive origons during growth arrest in hypoxia (time points correspond to 4, 6, 8, 10, 12, 14, 20, 30, 80 days) (B) Same as in (A), but during aerated growth (time points correspond to days 6, 8, 14, 24 and 60). Eleven origons (*nadR*, *hspR*, *Rv0494*, *sigE*, *sigC*, *furB*, *hrcA*, *ideR*, *dosR*, *sigD*, and *crp*) responded significantly in both (A) and B). (C) Same as in (A), but at higher time resolution (time points correspond to 23, 47, 73, 96, 121, 143, 164, 191 and 213 hours in hypoxia). (D) Same as in (A), but using a vastly enlarged version of the TR network using the algorithm CLR. (A) and (B) are reproduced from (Balázsi *et al.*, 2008) according to a Creative Commons Attribution–Noncommercial–Share Alike 3.0 licence.

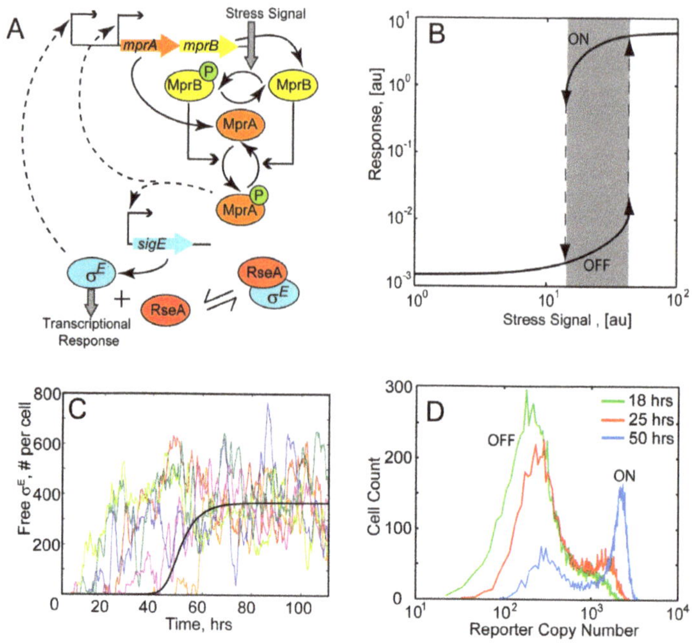

Figure 12.4 Architecture and dynamics of the mycobacterial stress-response circuit consisting of the MprA/MprB TCS and σE–RseA sigma/anti–sigma network. (A) The network involves posttranslational modulation of transcriptional activity of MprA via modulation of its phosphorylation state by MprB and modulation of σ^E–activity by RseA by sequestration. In addition, the circuit contains two feedback loops. Activated MprA~P upregulates the transcription of its own operon (direct loop) and activates production of σ^E which in its turn upregulates mprAB operon transcription form another promoter (indirect loop). Physiologically relevant signal and responses are indicated by grey arrows: the signal is modulation of MprB autophosphorylation (e.g. by polyp) whereas σ^E transcriptional activity of is a response. (B) Signal-response (bifurcation) diagram indicates that the network can display hysteretic bistability. Signal is modelled as autophosphorylation rate of MprB, the response is the concentration of free σ^E. For shaded signal range lines the cell can be in either of two stable steady states (OFF with low σ^E activity and ON with high σ^E activity). The fraction of cell population in either state will depend on the history initially uninduced or pre-induced population. At the boundaries of the bistable region small changes in signal lead to large jumps between steady states (indicated by arrows). (C) Noise in gene expression amplified by positive feedback leads to highly variable transitions between OFF and ON states. The black line denotes population-average (deterministic) response whereas coloured lines show stochastic trajectories of seven individual simulated cells. (D) As a result of slow and noisy switching kinetics the population will display bimodal distribution of σ^E-activity reporter such as relA. With time following the stress exposure the ON peak increases while OFF peak decreases. (B) to (D) are recreated from the data in Tiwari *et al.* (2009).

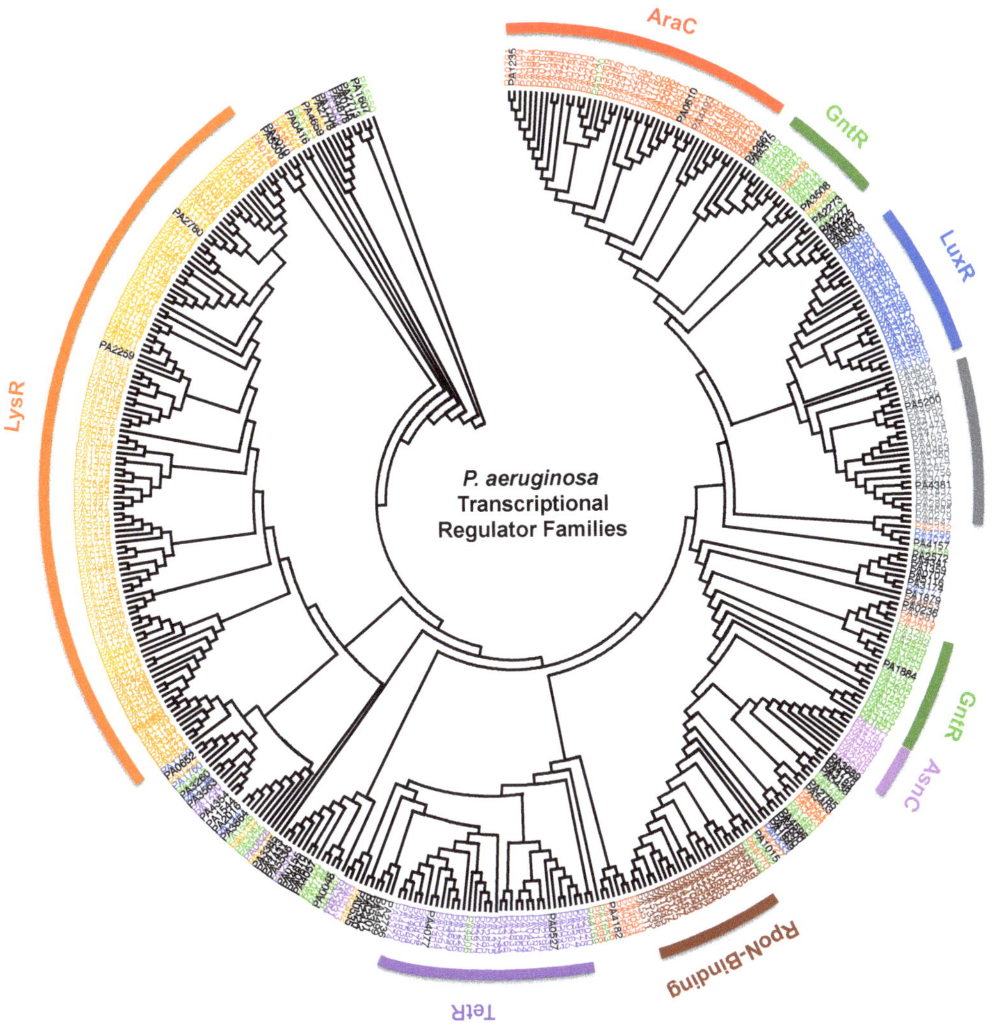

Figure 13.1 *P. aeruginosa* transcriptional regulatory protein families. A total of 434 amino acid sequences of transcription factors (Suppl. Table 13.1) were aligned using ClustalW (Version 2.1.10) (Larkin *et al.*, 2007). The first tree out of 31 most parsimonious trees (length = 69259) is shown. The consistency index is 0.155474 (0.147175), the retention index is 0.298198 (0.218198), and the composite index is 0.046362 (0.043887) for all the sites and parsimony-informative sites (in parentheses). The maximum parsimony tree was obtained using the Close-Neighbour-Interchange algorithm (Nei and Kumar, 2000) with search level 2 (Eck and Dayhoff, 1966; Nei and Kumar, 2000) in which the initial trees were obtained with the random addition of sequences (10 replicates). All alignment gaps were treated as missing data. There were a total of 1137 positions in the final dataset, out of which 743 were parsimony informative. Phylogenetic analyses were conducted in MEGA4 (Tamura *et al.*, 2007). The different families that cluster together are coloured. Members of families that do not form tight clusters (IclR, LacI/GalR, MerR, Cro-C1, CRP, ArsR, H-NS, OmpC, RpiR, and Dks/TraR) are coloured black.

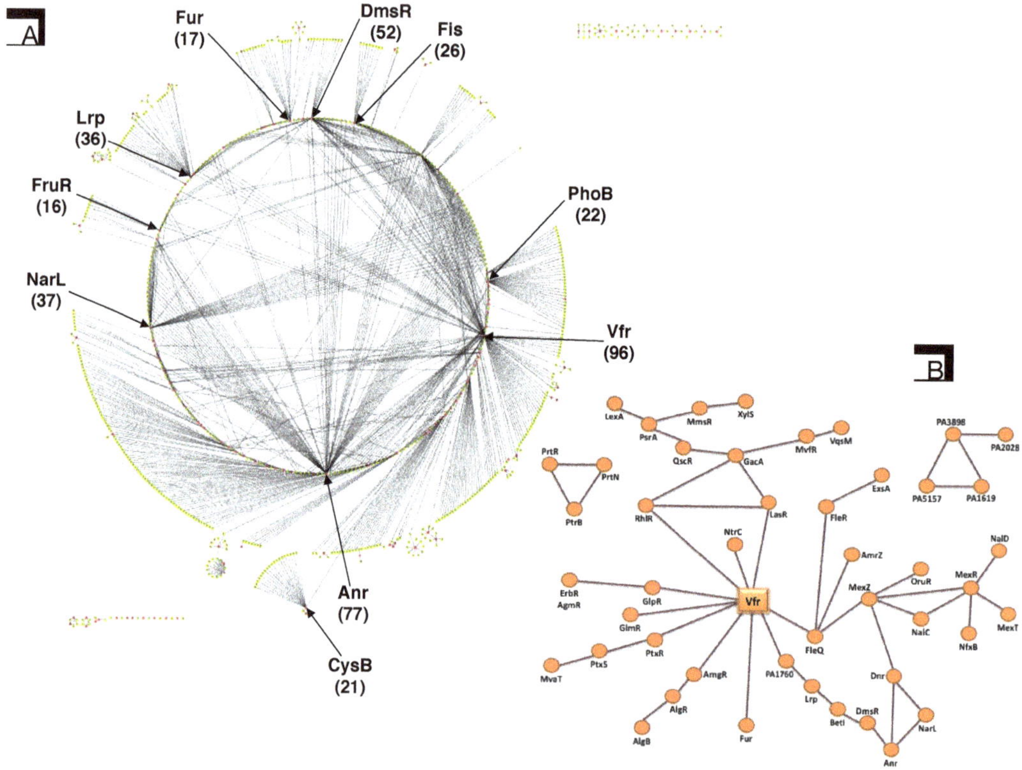

Figure 13.2 (a) Transcriptional regulators and interacting partners. Previously inferred regulators of *P. aeruginosa* (Madan Babu *et al.*, 2006) and those obtained by text mining using STRING (Jensen *et al.*, 2009) analysis, were used to construct the transcriptional regulatory network using Cytoscape (Version 2.7.0) (Shannon *et al.*, 2003). The network consists of 288 regulators and 593 target genes resulting in 1095 interactions. The transcriptional regulators with the highest interactions are shown in the figure (see also Supplementary Table 13.2). (b) Interactions among transcriptional regulators. VisANT (Version 3.86) (Hu *et al.*, 2009) was used to visualize a total of 63 interactions between the regulators from the 1095 interactions of Fig. 13.2a. Vfr emerges as the key regulator. Other interactions such as the pyocin induction network and an uncharacterized network appear isolated from these interactions.

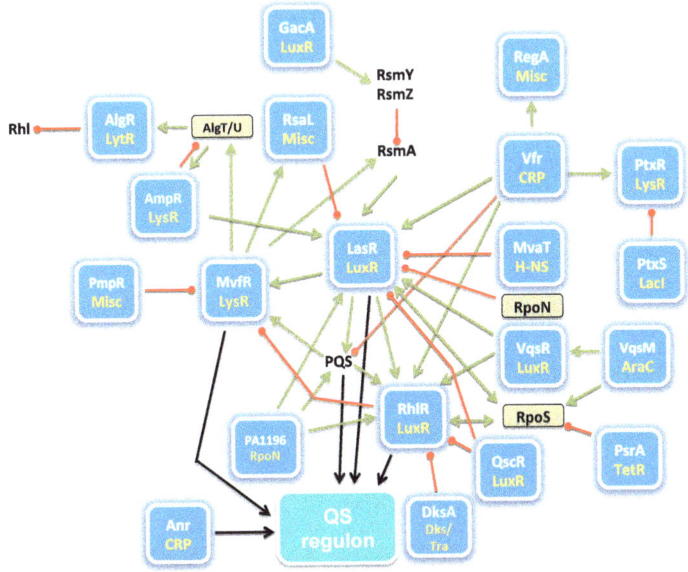

Figure 13.3 Association between QS regulators in *P. aeruginosa*. Interactions between the different transcriptional regulators (blue boxes) and sigma factors (yellow boxes) to control QS gene expression are depicted. The regulation is positive (green lines), negative (red lines), or both (black lines) and can be direct or indirect. PQS refers to the *Pseudomonas* quinolone signalling molecule. See text for details and references of individual interactions.

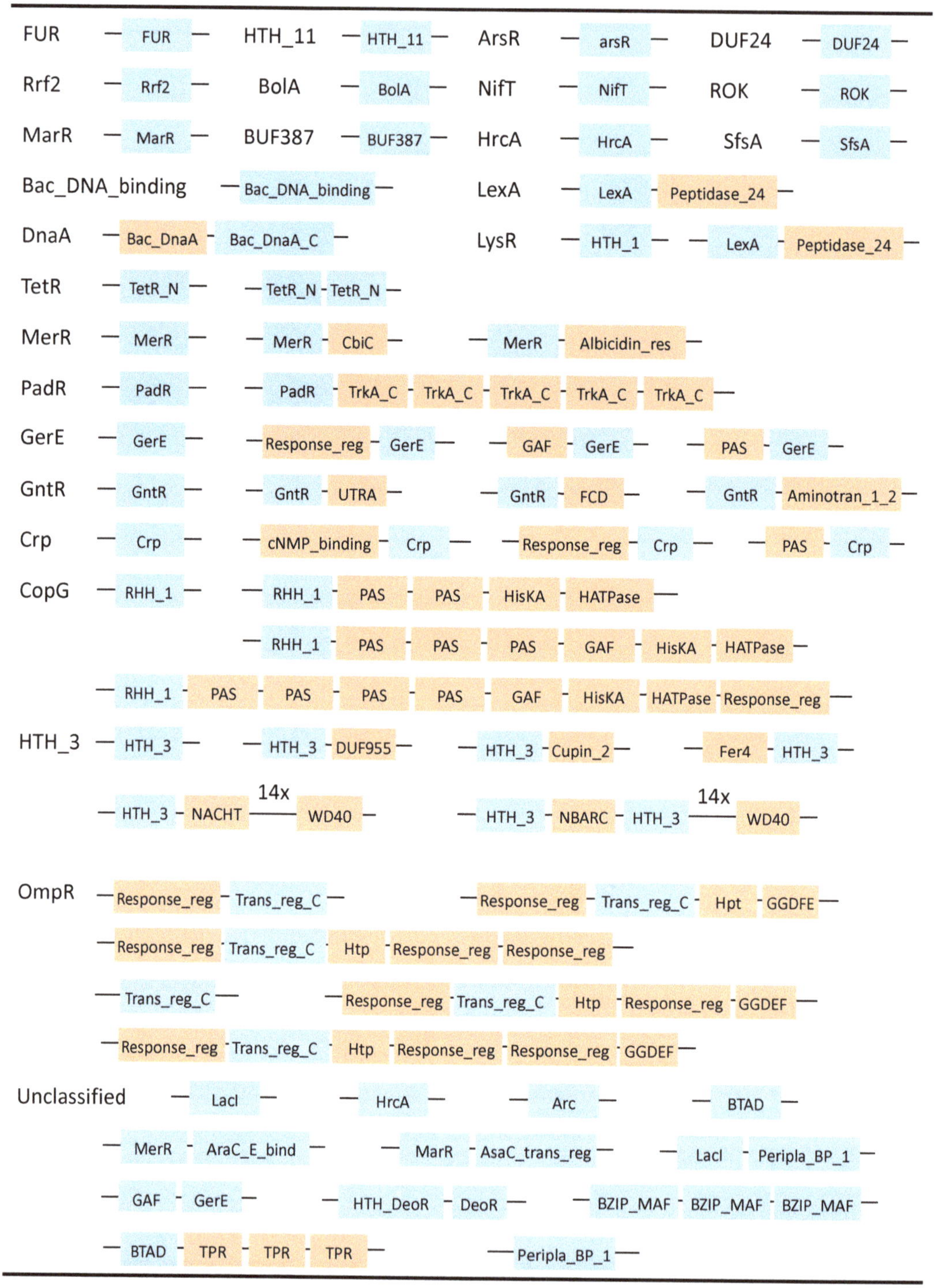

Figure 14.2 Domain architectures and classification of 12 TFs in 21 cyanobacterial genomes. Blue rectangles represent DBD domains, and yellow rectangles are the associated domains (modified from Wu *et al.*, 2007).